Catalyst Manufacture Recovery and Use 1972

Marshall Sittig

Thirty-Six Dollars

NOYES DATA CORPORATION
Noyes Building
Park Ridge, New Jersey 07656, U.S.A.

FOREWORD

The detailed, descriptive information in this book is based on U.S. patents since 1962 relating to the manufacture of catalysts, their recovery and use.

This book serves a double purpose in that it supplies detailed technical information and can be used as a guide to the U.S. patent literature in this field. By indicating all the information that is significant, and eliminating legalistic phraseology, this book presents an advanced, technically oriented review of manufacturing techniques for making, using and recovering catalysts on an industrial scale.

The U.S. patent literature is the largest and most comprehensive collection of technical information in the world. There is more practical, commercial, timely process information assembled here than is available from any other source. The technical information obtained from a patent is extremely reliable and comprehensive; sufficient information must be included to avoid rejection for "insufficient disclosure."

The patent literature covers a substantial amount of information not available in the journal literature. The patent literature is a prime source of basic, commercially useful information. This information is overlooked by those who rely primarily on the periodical literature. It is realized that there is a lag between a patent application on a new process development and the granting of a patent, but it is felt that this may roughly parallel or even anticipate the lag in putting that development into commercial practice.

Many of these patents are being utilized commercially. Whether used or not, they offer opportunities for technological transfer. Also, a major purpose of this book is to describe the number of technical possibilities available, which may open up profitable areas of research and development. One should have to go no further than this condensed information to establish a sound background before launching into research in this field.

The Table of Contents is organized in such a way as to serve as a subject index. Other indexes by company, inventor, and patent number help in providing easy access to the information contained in this book.

CONTENTS AND SUBJECT INDEX

INTRODUCTION

The factors that determine the effectiveness of a particular catalyst in a particular situation are still obscured despite the expenditure of enormous amounts of R & D dollars. Development of a new catalyst tends to be frustrating, time consuming, and expensive. Attempts to apply available catalysts to systems other than those in which they have been proved effective might well be as difficult and as expensive as the development of completely new catalysts.

There is general agreement that the key to catalyst activity lies in the catalyst structure, both microscopic and macroscopic. The activity of a given catalyst frequently depends on the precise method of preparation and activation, which may greatly influence the physical and chemical properties. The history of a solid material is important since the material may be exposed to chemical poisons, or may be rendered unusable as in the flaking caused by a highly exothermic reaction.

There is presently no generalized theoretical framework for systematic selection of catalysts. Hence, simple practical knowledge must dominate any process of choosing catalyst candidates for further testing. The existing knowledge of past catalyst types serves as a guide. Once the reaction type has been established, certain classes of catalysts can be chosen immediately on the basis of past successful applications to similar type reactions.

	Catalyst	
Reaction	Type	Examples
Oxidation	metal oxide	V_2O_5, WO_3, Ag_2O, Fe_2O_3, NiO
Hydrogenation	metals	Ni, Pt, Pd, Fe, Cu, Ag
Dehydration	metal oxides	Al_2O_3, SiO_2, MgO, ThO_2
Cracking	acidic	SiO_2-Al_2O_3 (molecular sieves)

Thus, in this review, an attempt will be made to cite numerous pertinent examples of current industrial operating practice as a guide to present applications and as a help to future selection of catalytic materials for the same or similar processes.

Many general practices are encountered over and over in the use of catalysts, whether in acrylonitrile manufacture, petroleum cracking or a variety of other processes.

> The physical form of catalysts must be prepared to suit the process. This may involve the preparation of pellets or spheres of particular composition, size, shape and porosity.

> The contaminants which decrease catalyst activity must be controlled when present in the feed stream and removed from the catalyst when accumulated or produced there in the course of the reaction. This frequently involves controlled combustion of carbonaceous deposits.

> The active ingredients in the catalytic body must be recovered in many cases, both because of the value of the material and also because of pollution which may be involved in their escape to the atmosphere.

Since these are problems common to many processes, a review of specific solutions should be of common interest in industry.

The general arrangement used in this book is to discuss classes of catalysts starting with the alkali-metal catalysts in Group Ia of the Periodic Table and going progressively through to the platinum catalysts in Groups VIII of the Periodic

Table. Then under each type of basic catalyst component, a variety of specific reactions are discussed with specific examples of catalyst manufacture, recovery or use for the process in question. Thus, a panoramic view of practical industrial examples of catalyst preparation, handling and utilization is presented which it is hoped, will be useful to the student and the practicing industrial technologist alike.

Of supplemental interest to the reader will be a "Survey of Commercially Available Catalyst and Sorbent Materials" by R.H. Cherry et al of Battelle Memorial Institute in Natl. Air Pollution Control Admin. Report APTD-0685 (July 17, 1969). Also of interest will be books dealing with catalytic applications in specific industrial fields such as "Catalytic Conversion of Automobile Exhaust" by J. McDermott, Noyes Data Corp., Park Ridge, New Jersey (July 1971).

ALKALI METAL-CONTAINING CATALYSTS

Metallic sodium with or without various supports has been proposed as a catalyst for a number of chemical processes. For example, metallic sodium has been proposed as a catalyst for the polymerization of monoolefins, diolefins (dienes) and skeletal isomerization of chemical compounds wherein a change in the physical conformation of a carbon chain is obtained.

For other information on the use of metallic sodium as a catalyst, the reader is referred to "Sodium, Its Manufacture, Properties and Uses", American Chemical Society Monograph 133, Reinhold Publishing Corp., New York, (1956) by Marshall Sittig.

For broader information on catalytic uses of the alkali metals in general, the reader is referred to "Handling and Uses of the Alkali Metals", Advances in Chemistry Series 19, American Chemical Society, Washington, D.C., (1957).

OLEFIN ISOMERIZATION

Standard Oil (Indiana)

A process developed by T.M. O'Grady and A.N. Wennerberg; U.S. Patent 3,260,679; July 12, 1966; assigned to Standard Oil Company (Indiana) pertains to a catalyst containing metallic sodium on a gamma-type alumina support wherein metallic sodium is promoted with at least a transition metal compound.

In research leading to this process there has been developed a nonpyrophoric active and selective supported promoted sodium catalyst which can be readily prepared and is safe to handle in chemical processing equipment. This catalyst comprises 1 to 25 weight percent metallic sodium and 0.1 to 10 weight percent transition metal compound, preferably a compound of iron on a metal oxide support, such as activated aluminas, silica, magnesia and the like as well as naturally occurring materials, such as bauxite, kieselguhr and boehmite among others. Desirably the supports contain mainly alumina converted or convertible to gamma-type aluminas.

One preferred method of catalyst preparation is as follows: A weighed amount of the calcined support is charged to a suitable 3-neck round-bottom flask provided with a stirrer, inert gas inlet and outlet tubes, and a thermocouple directly contacting the mixing materials. The flask is heated, while stirring the support, to the temperature at which the alkali metal is to be dispersed on the support. Temperatures above the melting point of the alkali metal are usually and preferably used so that it is dispersed as a molten liquid for better contacting. Other methods of dispersing the alkali metal, such as a physical admixture of it and the support, can be used, but they are not as effective. The support and liquid alkali metal are mixed at the preparation temperature for 5 minutes to 3 hours, usually 5 to 30 minutes. Predried promoters are added, and the mixing is continued at the preparation temperature for a specified time which is discussed later. The catalyst, cooled quickly to room temperature or below while mixing, is ready for use.

Preparation of optimum catalysts may be controlled colorimetrically. Where sodium and promoter are dispersed on alumina, a series of color changes occur. Initially, the color progresses from white to darkening shades of blue and sometimes, depending on preparation conditions and concentrations, to black. It then regresses to lighter shades of blue, to gray and finally to an off-white. During the latter color sequence a shade of blue is obtained at which catalyst activity is a maximum. These color changes undoubtedly relate to the reactions of the sodium and promoter on the support.

A process developed by W.F. Wolff; U.S. Patent 3,405,196; October 8, 1968; assigned to Standard Oil Co. (Indiana) involves a catalyst preparation method in a process for the isomerization of terminal olefins wherein a catalyst comprising an alkali metal dispersed upon a high-surface, substantially inert, solid support is contacted with an activating gas at a temperature between 5° and 50°C. in sufficient quantity to provide an oxygen-to-alkali metal ratio between 0.01 and

2.0 atoms of oxygen per atom of alkali metal. The activating gas is selected from a molecular oxygen-containing gas, nitrous oxide, mixtures thereof, and mixtures thereof with inert gases. The catalyst used in this process contains an alkali metal dispersed on a high-surface, substantially inert support. The alkali metal may be selected from a group comprising sodium, potassium, rubidium and cesium. Sodium is particularly desirable. The supporting material should possess a high surface area, large pores, and be only slightly acidic. The supporting material should be calcined to drive out the water. This calcination may be carried out at temperatures of from 150° to 650°C., 450° to 600°C. being particularly desirable, and a pressure of from 0.1 mm. of mercury to 1.0 atmosphere for 0.1 to 50 hours. The supporting material is desirable in a granular or powdered form. Activated alumina is a particularly desirable supporting material.

This alumina generally has a surface area ranging from 50 to 1,000 square meters per gram. The alumina used in the preparation of the catalyst has a surface area of 200 to 210 square meters per gram, a packed bulk density of 68 pounds per cubic feet and a particle size of 80 to 200 mesh. This granular alumina has an ignition loss of 6.8% at 1100°F. and is composed of 92.0% Al_2O_3, 0.8% Na_2O, approximately 0.1% SiO_2 and approximately 0.1% Fe_2O_3.

The supported alkali metal catalyst is prepared by contacting the high surface area supporting material with the alkali metal while the latter is in the molten state. This contacting is done in the presence of agitation under an inert atmosphere, such as argon, nitrogen or helium. This so-called inert atmosphere must be such that it will not react with the alkali metal to transform the alkali metal into a derivative that will not catalyze isomerization. The alkali metal may be added to the calcined support at a temperature of from 10° to 35°C. under an inert atmosphere. The resulting mixture is then heated to a temperature of from 150° to 500°C., 300° to 400°C. being particularly desirable. The mixture is continuously agitated at the elevated temperature until the alkali metal appears to be evenly distributed upon the surface of the supporting material. Even distribution usually will be attained within a period of from approximately 10 minutes to 2 hours.

If sodium is being distributed on activated alumina, a uniform layer of sodium is indicated by a blue-black color which occurs over the entire surface of the alumina. The catalyst may contain between 1 and 40% by weight of the alkali metal, depending upon the particular alkali metal and the particular supporting material employed. Generally, that amount of alkali metal which is sufficient to form a monomolecular layer on the support is desired. When sodium is to be on activated alumina, the resulting catalyst will usually contain from 2 to 15% by weight of sodium.

The improvement of this process comprises the pretreating of the supported alkali metal catalyst with an activating gas at a temperature within the range of 5° to 50°C., 20° to 35°C. being particularly desirable. Preferably a positive flow of the activating gas through the catalyst bed is maintained. Care should be taken during the pretreatment to avoid general or localized overheating of the catalyst. The activating gas is selected from the group consisting essentially of a molecular oxygen-containing gas, nitrous oxide (N_2O), mixtures thereof and mixtures thereof with inert gases. These inert gases are gases which will not chemically react with the sodium, e.g., nitrogen, argon, neon and helium.

It is desirable that the activating gas be used in an amount which is sufficient to provide a quantity of oxygen which will furnish an oxygen-to-alkali metal ratio within the range between 0.01 and 2.0 atoms of oxygen per atom of alkali metal. Of course, the atoms of oxygen would be the atoms of oxygen in the activating gas and the atoms of sodium are the atoms of sodium present in the catalyst that is being used to isomerize the terminal olefins. Preferably, the quantity of activating gas should be maintained at a level which will provide an oxygen-to-alkali metal ratio within the range between 0.1 and 0.4 atom of oxygen per atom of alkali metal. The rate of flow of the activating gas should be maintained at a level which will not result in the creation of excessive temperatures.

Figure 1 shows a suitable form of apparatus for the conduct of this process. The desired supported alkali metal catalyst, a sodium-on-alumina catalyst, is charged under an inert atmosphere to cylindrical vessel 1. Vessel 1 is insulated and contains coil 2 through which steam or a standard coolant can be circulated, depending upon whether heating or cooling of the contents of vessel 1 is desired. The inert atmosphere may be nitrogen. The catalyst exists in vessel 1 as a fluid bed 3. The particle size distribution of the catalyst is such that it will permit adequate fluidization so that efficient and uniform contacting of the catalyst particles with either a pretreating gas or the hydrocarbon being converted can be maintained.

At the lower end of vessel 1 is a support for the catalyst bed. This support is made up of a grid 4, a fine screen 5 and several layers 6 and 7 of different sized Alundum beads. The catalyst must be maintained under an inert atmosphere, which can be nitrogen. The inert gas is introduced into vessel 1 through lines 8 and 9 and drier 10. Such gas not only acts as an inert atmosphere, but also operates to keep the catalyst bed 3 fluidized. As the inert gas need not be at an elevated temperature at this time, heater 11 is not needed. Therefore, valves 12 and 13 remain closed while valves 14, 15 and 16 remain open. Drier 10 contains molecular sieves or some other suitable desiccant, and is used to remove moisture from the various gases which pass through it prior to their introduction into vessel 1.

Before the catalyst is used to convert hydrocarbons, it receives a pretreatment with an activating atmosphere. The activating atmosphere comprises an oxygen-containing gas. The flow of inert gas to vessel 1 is stopped by closing valve 16.

Valve 17 is opened and the oxygen-containing gas is passed through lines 18 and 9 and drier 10 into vessel 1. The design of the vessel and the selection of the flow rate of gas should be such as to maintain a suitable fluidized catalyst bed. The fluidized bed will permit the most efficient and uniform treatment of the catalyst.

The oxygen that is passed through the bed 3 should be held to an amount that is slightly less than that which is required to convert half of the available alkali metal to its oxide. Tests have indicated that excessive treatments in oxygen will not produce optimum catalyst activity, but it should be noted that even the excessive oxygen treatments do result in activity improvement. The exhaust gas will exit from vessel 1 through line 19. When the pretreatment has been completed, valve 17 is closed. This prevents the introduction of additional oxygen-containing gas into vessel 1. Valves 14, 15 and 16 are open while valves 12 and 13 remain closed. This permits inert gas to flow through lines 8 and 9 and drier 10 into vessel 1.

Since in this specific embodiment the hydrocarbon feed will be such that an elevated temperature of the catalyst bed 3 is not desired, heater 11 will not be used. The hydrocarbon feed will be a refinery B–B stream. If heavier hydrocarbons were used, the heater 11 would be needed. The catalyst bed should be cooled to a temperature of approximately 20° to 25°F. When the catalyst bed is at the desired temperature, the hydrocarbon feed stock is introduced into vessel 1 through drier 20, line 21, valve 22 and inlet 23. The inlet 23 is so designed as to introduce the hydrocarbon without impinging it upon the sides of vessel 1. In this embodiment, a B–B stream is used as a hydrocarbon feed. The products from the reaction are withdrawn from vessel 1 through line 19 into a separation zone and appropriate recovery equipment. Samples of product are removed periodically from line 19 through valve 24 and outlet 25 so that the conversion can be monitored. When the results of appropriate tests performed on these samples indicate that the conversion has been reduced as a result of catalyst deactivation, the run may be halted, vessel 1 opened, used catalyst withdrawn and new catalyst installed. On the other hand, if a second vessel is present in the installation and is hooked in parallel with vessel 1, and if this second vessel contains pretreated catalyst, the hydrocarbon conversion can be switched to this second vessel while the catalyst in vessel 1 is changed. The parallel vessels may be operated alternately to furnish continuous operation.

FIGURE 1:   APPARATUS FOR THE ISOMERIZATION OF TERMINAL OLEFINS USING ALKALI METAL CATALYST

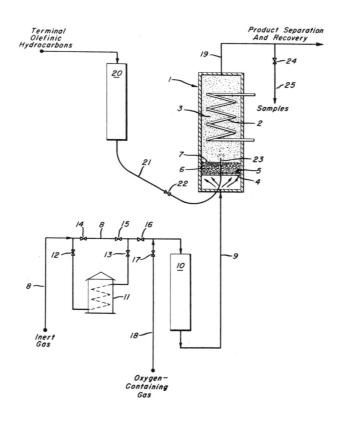

Source: W.F. Wolff; U.S. Patent 3,405,196; October 8, 1968

OLEFIN DIMERIZATION

Shell Oil

A process developed by C.W. Bittner and G. Holzman; U.S. Patent 3,075,027; January 22, 1963; assigned to Shell Oil Company involves the regeneration of unsupported alkali metal catalysts utilized in condensation reactions which comprises heating the catalyst in the absence of modifying gases such as free hydrogen but in the presence of a substantially inert normally liquid carrier (in other words, in an inert medium) at a temperature between 200° and 400°C. for a period of time between 15 minutes and 24 hours. It has been found that this heat treatment unexpectedly and greatly improves the selectivity of an alkali metal containing catalyst with moderate regeneration of total activity of the catalyst. When the total activity of the catalyst has decreased to an undesirable extent, at least part thereof may be regenerated under more severe conditions comprising vaporizing the alkali metal from the residue with which it is usually associated after its use in condensation reactions.

A preferred cyclic process therefore comprises utilizing an unsupported alkali metal catalyst in the condensation of organic molecules containing an active hydrogen atom and an organic molecule having a double bond (in some cases the same type of molecule) by contacting the two organic molecules at a temperature between 100° and 350°C. and at a pressure from 5 to 200 atmospheres with a catalytic amount of at least 1 alkali metal catalyst, separating the catalyst from the condensation product so formed, regenerating the separated catalyst by heating at a temperature between 200° and 400°C. for a period between 1/4 and 24 hours in the absence of hydrogen and in the presence of a liquid carrier and contacting the regenerated catalyst with further quantities of the condensable organic compounds under the condensation conditions. In a further aspect of the process at least one of the less desired types of by-products from the condensation zone is recycled to the reactor thereto together with regenerated catalyst and additional quantities of fresh feed material for effecting a condensation reaction and at the same time minimizing further production of the undesired side reaction product. The regeneration step appears to result in freeing the catalyst from tarry residue.

In the process under consideration, unsupported alkali metal catalysts, when contacted with olefinic condensable organic compounds, produce a mixture of condensation products predominating in terminal olefin dimers, in contrast to internal olefins which are produced by alkali metal catalyst supported on activated carbon. Out of this mixture usually one or several components are desired while the remainder of the product comprises either unreacted feed materials or undesirable or less desirable condensation products. As the reaction progresses, both the activity of the catalyst and its selectivity for terminal olefin production diminish. Consequently, it is important to regenerate the catalyst and to so regenerate it that the production of the desired condensation product is increased while the ratio of undesired or unreacted components is minimized.

Taking as an illustration the condensation of propylene in the presence of an unsupported alkali metal catalyst, a typical condensation product will contain C_1 to C_5 hydrocarbons, propane, C_9 olefins, heavier residues, C_6 olefins and particularly 4-methyl-1-pentene and 4-methyl-2-pentene. The former of these last two products is highly desirable since it can be utilized to great advantage in the preparation of the so-called Ziegler polymers. As the unsupported alkali metal catalyst is recycled for the condensation of further quantities of propylene, the ratio of 4-methyl-1-pentene to the corresponding -2-pentene rapidly decreases and shortly reaches an uneconomic level wherein undesirably high quantities of -2-pentene are being produced at the expense of the production of -1-pentene. Regeneration according to the process comprising heating of this used catalyst at a temperature of from 200° to 400°C. for a time from 15 minutes to 24 hours in the absence of hydrogen but in the presence of a normally liquid inert carrier substantially completely regenerates the catalyst with respect to this ratio of desired to undesired components.

In some cases, however, regeneration of total activity (i.e., total amount of conversion products formed under a given set of conditions) is not fully accomplished by this relatively mild treatment. Under some circumstances, therefore, it is economic and desirable to subject at least a portion of the catalyst periodically to harsher conditions whereby the alkali metal is vaporized from the carbonaceous residue with which it is associated prior to recombining with fresh alkali metal or alkali metal regenerated under milder conditions before recycling for the treatment of further portions of the feed organic compounds.

Figure 2 shows the type and arrangement of equipment which may be employed in the conduct of the process. Fresh alkali metal from source 1 may be introduced, if necessary together with regenerated alkali metal catalyst, and a feed hydrocarbon such as propylene from source 2 into the reactor 4, a stirred autoclave or pipeline reactor. The autoclave is preferably fitted with an agitator or stirrer or is vibrated in order to mix the contents thoroughly. Supersonic vibration may be employed for this purpose. The propylene and alkali metal catalyst (approximately 1 part by weight of catalyst for about 40 or more parts by weight of propylene) are heated with agitation in the autoclave at a temperature in the order of 200°C. for a period of 0.3 to 5 hours, the maximum pressure rising in this time to 900 to 3,000 lbs. per sq. in. Completion of the condensation between pairs of propylene molecules is indicated by a decrease in the pressure at the maximum reaction temperature to approximately half of the maximum pressure previously attained. One means of separating the catalyst so utilized comprises vaporizing the volatile components of the condensation product through line 41 to the base of a

FIGURE 2:   FLOW DIAGRAM OF PROCESS FOR DIMERIZATION OF PROPYLENE TO 4-METHYL-1-PENTENE IN THE
PRESENCE OF ALKALI METAL CATALYST

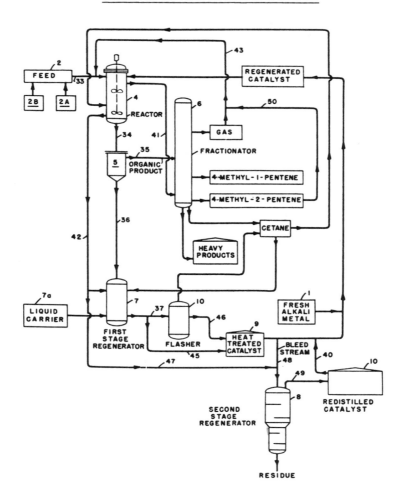

Source: C.W. Bittner and G. Holzman; U.S. Patent 3,075,027; January 22, 1963

fractionating column 6. The catalyst may then be regenerated in the reactor zone. The organic product is treated by
fractional distillation in fractionator 6 to obtain predetermined products including gas, heavy products, carrier hydrocarbons, such as cetane, and the principal product, namely, 4-methyl-1-pentene. Gases including unreacted propylene
taken from the top of the fractionating column 6 also may be recycled by means of line 43 to the reactor. Cetane or
other substantially inert carrier hydrocarbon which is recovered during the fractionation from tower 6 may be recycled to
the reactor or may be utilized also as a carrier during catalyst regeneration, being conveyed to the first stage regenerator
by means of line 44.

The catalyst, sent by means of line 36 to the first stage regenerator, is heated in the presence of an inert liquid carrier
from source 7a by a heater not shown to a temperature between 200° and 400°C., preferably 225° to 350°C., for a
period between 30 minutes and 24 hours, preferably 1/2 to 2 hours. This may be conducted under pressure such as in an
autoclave or in a vessel through which a substantially inert gas such as nitrogen is passing to provide an inert atmosphere
for the regeneration. The heat treated catalyst is then passed either directly by means of line 45 to a heat treated catalyst
storage zone 9 or by way of a flasher 10 for the partial or complete removal of carrier fluids such as cetane and thence
by line 46 to the heat treated catalyst storage zone 9.

If the total activity of the alkali metal catalyst has degenerated to such an extent that its use is unsatisfactory even after
the mild heat treatment, it may be subjected to distillation, portions at least of the catalyst being taken either by means
of line 47 or from the bleed stream line 48 to the second stage regenerator zone 8 wherein it is subjected to heating

sufficient to vaporize the catalyst and convey it by means of line 49 to a redistilled catalyst storage zone 10. This re-distilled catalyst is then recycled by means of line 40 for use in the condensation of further quantities of the feed propylene, combining it if necessary with make-up quantities of fresh alkali metal and/or further proportions of heat treated catalyst from the storage zone 10 or the first stage regenerator 7. If the catalyst is to be modified with promoters such as iron or the like this must be done prior to recycling to the reactor zone 4.

DIOLEFIN POLYMERIZATION

The polymerization of diolefins using sodium catalysts is old in the art, the World War I German Buna synthetic rubber having its name derived from the first letters of the German words butadiene and natrium (for sodium). Later emulsion polymerization proved a more practical process but interest has continued in the development of catalysts based on the low cost alkali metal, sodium.

Thus, Morton and coworkers in a series of papers in the Journal of the American Chemical Society, starting in 1947, described an organoalkali metal catalyst for the polymerization of olefins and particularly dienes which they term an alfin catalyst, Journal of the American Chemical Society 69, 161; 167; 950; 1675; 2224 (1947). The name "alfin" is taken from the use of an alcohol and an olefin in their preparation. The alcohol, a methyl n-alkyl carbinol, usually isopropanol, in the form of the sodium salt, the olefin, also in the form of the sodium salt, and an alkali metal halide, form a complex that constitutes the catalyst.

These catalysts were reported by Morten et al to cause the polymerization of butadiene, isoprene and other dienes, alone and together with other copolymerizable organic compounds, in most cases olefinic in nature. The catalyst was discovered in the course of a study of the addition of organosodium compounds to dienes. Later on, Morton summarized the work done up until 1950 in Industrial and Engineering Chemistry, 42, 1488 through 1496 (1950).

National Distillers and Chemical

A process developed by T.B. Baba; U.S. Patent 3,640,980; February 8, 1972; assigned to National Distillers and Chemical Corp. involves the preparation of alfin polymers by polymerization of the monomer by an alfin catalyst in the presence of a molecular weight moderator, treating the alfin polymer reaction mixture with a small amount of carbon dioxide to convert organometallic compounds to organic acid salts, extracting the acid salts with water and separating unreacted monomer, volatile low polymer and solvent, and thereafter purifying solvent and recycling it and optionally the monomer for reuse, and washing and drying the alfin polymer.

In accordance with the process, a continuous process for the preparation of alfin polymers is provided, comprising con-tinuously blending monomer, alfin catalyst, molecular weight moderator and solvent, continuously effecting the poly-merization of the monomer at a temperature at which the reaction proceeds by an alfin catalyst in the presence of the molecular weight moderator, treating reaction mixture with carbon dioxide to convert organometallic compounds such as acetylides and alkali metal cyclopentadiene compounds to the acid salts, continuously separating unreacted monomer, volatile low polymer and solvent from the alfin polymer reaction mixture and dissolving and extracting such acid salts in water, and steam distilling volatile materials from the resulting dispersion, and thereafter recovering solvent and optionally the monomer and recycling them for reuse, and washing and drying the alfin polymer.

The carbon dioxide treatment by converting acetylides and cyclopentadiene metal compounds into acid salts makes it pos-sible to separate these from the solvent, and prevent acetylene and cyclopentadiene which otherwise would be regenerated when their organometal derivatives are brought in contact with water, from being distilled off with the solvent. If this occurs, they can be recycled with the solvent, and can react with and destroy alfin catalyst by being converted again into their alkali metal derivatives. Moreover, with each recycling, and each fresh portion of monomer, the content thereof will be increased, until a substantial loss of catalyst can occur. This is prevented by the carbon dioxide treatment.

Figure 3 is a flow diagram of the process. The synthesis of the alfin catalyst in this system takes place in Zone A. The process shown employs sodium, which is prepared as a dispersion in a liquid diluent at a 25 to 50 weight percent sodium concentration. The sodium dispersion is fed via pump 1 to the storage tank 2 where it is stored under nitrogen. Diluent enters via line 3 and sodium (molten) via line 3 into the mixing tank 5, where it is circulated via line 7 to a Gaulin mill 8 to reduce the particle size of the sodium, and then back via line 9 to the mixing tank, to provide an intimate dispersion of sodium of a particle size of less than 10 microns average diameter in the diluent. The finished dispersion is bled off con-tinuously via line 10 to one of two storage tanks 11, 12, equipped with agitators to maintain uniformity.

To prepare the alfin catalyst, a batch technique is used. Diluent from storage 14 is charged via line 15 to the catalyst synthesis reactor 16, an agitated vessel equipped with cooling facilities. Sodium dispersion is added via line 15 from tanks 11 or 12, and isopropyl alcohol is gradually added from storage 17, via line 18 with agitation and cooling at a temperature of approximately 0° to 80°C. Since the reaction is exothermic, the alcohol addition is slow. In this way, 1/3 of the sodium is converted to sodium isopropoxide. The addition of butyl chloride from storage 19 via line 18 then converts most

FIGURE 3: FLOW DIAGRAM OF CONTINUOUS ALFIN RUBBER PROCESS

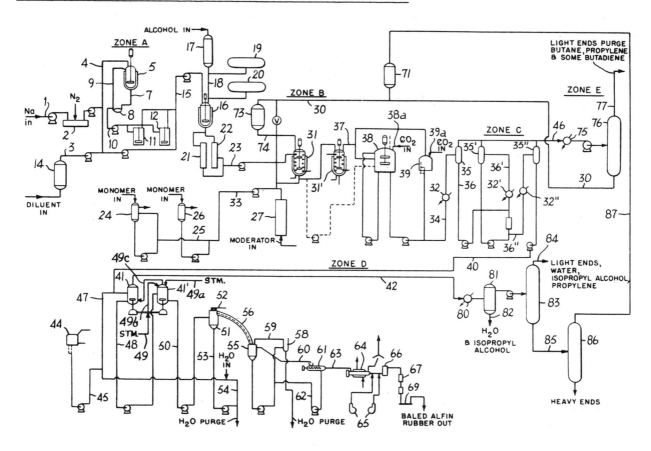

Source: T.B. Baba; U.S. Patent 3,640,980; Feburary 8, 1972

of the remaining sodium in the tank 16 to equimolar quantities of sodium butyl and sodium chloride. This also is an exother-mic reaction and cooling is required. The temperature is held within the range from 0° to 80°C. After the addition of butyl chloride is complete, the reaction is allowed to proceed to completion, with agitation. Propylene from storage 20 is then added directly via line 18 to the liquid contents of the vessel. This addition converts sodium butyl to sodium allyl, with the formation of butane as a by-product. Very little heat is evolved at this point, and the reactor is kept under the pressure of the propylene solution. The pressure at this point should be less than 40 psig. The contents of the reactor are held at this temperature for several hours, and the pressure then reduced to atmospheric by venting. Butane and excess propylene may be partially removed by heating. The contents then are transferred to one of two catalyst storage tanks 21, 22.

The catalyst preparation can be converted to a continuous operation by providing three catalyst reactors in series, in which each step of the catalyst preparation is carried out in sequence. Catalyst suspension is supplied to the polymerizers con-tinuously from one of tanks 21, 22 via line 23. The tanks are equipped with agitators to avoid settling of the solids.

The alfin monomer polymerization process takes place in Zone B. The process will be described for preparing a butadiene-isoprene rubber. Monomer feed is prepared for use in the polymerization by removing water and any inhibitor in the strippers 24, 26 from butadiene and isoprene, since these substances destroy catalyst. The preparation of a butadiene-styrene rubber is similar, except that only butadiene is dried. The dry monomers are fed via lines 25, 33 to the first polymerization reactor 31. Dry moderator is stored in tank 27.

A plurality of polymerizer reactors 31, connected in series, in this case, three, is used. These are each jacketed, and contain cooling coils. The coolant is water, or other suitable liquid. The reaction temperature is within the range from 100° to 200°F. All polymerizers are operated liquid-full. The maximum polymerization pressure is that needed to en-sure adequate pressure containment in the event of an upset, and also to ensure sufficient pressure for the reactor effluent to flow to the alfin cement blend and feed tanks. The polymerization is carried out by passing recycle stream 30 consisting essentially of isooctane and some recycled butane and butadiene to the first of the three polymerizer reactors 31, 31' (only two are shown). Dry moderator from tank 27 and fresh dry monomer in line 33 are mixed with the recycle in the desired proportions and charged together to the polymerizer. Catalyst via line 23 is injected separately. Since the reaction is exothermic, heat must be removed.

The polymerizer effluent from the last reactor, while still at reaction temperature, is then run via line 37 to the carbonation tank 38, where it is carbonylated and also blended with alfin cement to the desired Mooney value, if necessary. Carbon dioxide is admitted via line 38a in the proportion needed, and the mixture held in the tank for a sufficient time to permit carbonylation to become complete. From a few minutes to several hours, is adequate. Carbonylation can also be effected in the concentrator feed tank 39. Carbon dioxide is admitted via line 39a. Also, carbon dioxide can be added to the alfin cement in line 37, and carbonylation then takes place in course of transit in the line and in the tanks 38 or 39.

During normal operation, when product of the proper Mooney is being made, the flow will be directly to the feed tank 39. Blending to the desired Mooney level can be obtained by mixing alfin cement from various storage tanks in the blend tank 38. Both tanks 38 and 39 are kept under carbon dioxide atmosphere. The stirring and agitation therein ensure thorough mixing and complete carbonylation.

Alfin cement of the desired Mooney is charged from the tanks 38 and 39 via line 34 to the first of three flash stages of concentration in Zone C. A first stage heater 32 is provided in the line 34, in the event that the alfin cement stream is below flash temperature. In addition, a spare flash heater servicing the entire concentration system can be provided in a recycle line connected to the flashers.

The concentration system is composed of three flashers 35, 35', 35", and these are operated so that no vaporization takes place in the tubes of the heaters 32, 32', 32". This is effected by maintaining a pressure on the discharge of the heaters greater than the boiling point of the solution, preferably higher than the maximum skin temperature of the tubes. Here, steam condensing at a temperature of 350°F. is used.

The effluent from the first stage heater is at approximately 200° to 275°F., and 50 to 110 psig and flashes in 35 by reduction of pressure to 1 psig. The vapor stream is withdrawn through line 46, and the liquid stream is also withdrawn in line 36, and charged to the inlet of the second stage heater 32'. The rubber cement now contains, for example, 9.8 weight percent of alfin rubber. It issues from the heater at approximately 210° to 275°F. and 80 to 100 psig and flows into 35', the second stage flash tank, which is maintained at approximately 1 psig. The temperature after flash at this point is approximately 150° to 230°F. The vapor stream is separated and withdrawn via line 46. The liquid from the flash tank 35 containing, for example, approximately 11.5 weight percent of alfin rubber is withdrawn via line 36" and is charged to the third stage heater 32". After these two stages of concentration, most of the butadiene and the isoprene remaining unreacted has been removed as vapor.

The liquid effluent from the third stage heater 32" at approximately 150° to 275°F. and 60 to 100 psig flows to the third stage flash and surge tank 35", wherein the liquid flashes upon pressure reduction, and the vapor stream at approximately 180° to 230°F. and 1 psig is withdrawn to line 46. This vessel provides surge capacity for rubber cement of approximately a 15 weight percent alfin rubber concentration. The liquid cement is withdrawn via line 40 and is then charged to the first stage of the crumb formers 41, 41'. Approximately 1/2 of the total solvent has been removed at this stage by evaporation, and substantially all of the butadiene and isoprene have been recovered.

Alfin cement of the desired Mooney is fed via line 40 to the first of the solvent strippers or crumb formers 41, 41'. Approximately 80 to 90% of the remaining solvent is removed in the first solvent stripper. The combined vapor stream from the solvent strippers flows to a purification system and is then recycled. The crumb formation and finishing operations take place in Zone D. These are the same whether an isoprene or styrene rubber is made. The isoprene rubber case is described.

Rubber cement from flashers, for example, containing approximately 12 to 15 weight percent rubber is continuously charged via line 40 to the first of two solvent strippers 41, 41'. It is mixed with hot recycled water entering via line 47 so that a suspension of alfin cement in water results. If desired, a dilute solution of emulsifying agent from storage 44 can be added via line 45. The resultant mixture enters the first solvent stripper 41, a vessel equipped with a stirrer and overhead collection line 42 running to a condenser. The water is hot enough to flash some of the solvent. Steam and solvent vapors from 41' are injected via line 49c to effect a steam distillation, and heat the mixture to a temperature of 50° to 130°C., while the mixture is stirred. Solvent vapors escape via line 42. A slurry of alfin rubber crumb results, and the rubber slurry is removed from below and is sent via line 48 to the second stage solvent stripper 41', which is similar to the first stage, and the rubber entering the second stage has for example a solvent content of the order of 5 to 10 weight percent, based on the alfin rubber content.

Stripper 41' operates at a temperature of approximately 212°F. Steam is also injected directly into this vessel via lines 49a and 49b. An aqueous slurry of alfin rubber of the order of 2 to 6 weight percent rubber is withdrawn via line 50. The solvent content of the rubber at this point is of the order of 1 weight percent, based on the alfin rubber.

The product vapor stream in line 42 contains essentially all the hydrocarbons that were present with the exception of the rubber. In addition, it contains propylene, formed by decomposition of the catalyst with water to form sodium hydroxide. It also contains isopropyl alcohol, formed by hydrolysis of the sodium isopropoxide, alone with any excess carbon dioxide. The rubber crumb contains small quantities of the moderator dihydronaphthalene, styrene (if present), as well as a small amount of solvent. The quantity of solvent in the crumb at this stage should be kept to a minimum by appropriate adjustment of the steam stripping conditions.

The slurry from line 50 enters a separator 51 equipped with a mechanical rake 52, so that rubber crumb which floats to the surface of this vessel can be skimmed off. The water in the lower portion of this vessel, relatively free of rubber crumb, is recycled to stripper 41 via lines 53, 54, 47. In addition, to prevent buildup of salts, a proportion is purged, and replaced by make-up water which enters at line 54.

The rubber crumb which is present in the form of small particles and contains substantial water and which is raked out enters a dewatering screen separator 55 via a chute 56. In the chute, the rubber crumb is contacted with a stream of cold water which cools the crumb. The underflow from the screen consists essentially of water containing a small amount of rubber fines, and is withdrawn and pumped to a secondary fines settler 58. Rubber crumb is allowed to overflow from the upper portion of this vessel, and passes via line 59 back on to the screen separator 55. The underflow consists of water containing dissolved salts, and is purged.

The alfin rubber crumb discharged from the separator 55 is fed by conveyor 60 to an expeller 61. The expeller by means of screw compression reduces the water content to approximately 9 weight percent. The water discharged from the expeller is returned to the fines settler 58 via line 62. The rubber from the expeller passes through line 63 and enters an expander 64. Here, by compression, and jacket-heating, the rubber is heated, so that upon discharge water as steam and solvent flash off. A stream of hot purge air to carry away water vapors is provided by blowers 65. The alfin rubber at this point in the form of crumb is conveyed to a crumb conveyor and cooler 66 and subsequently to a baler 67 where it can be packaged in 75 pound bales. These are conveyed via conveyor 69 to storage. The solvent and other volatiles removed at the expander are vented.

The solvent recovery and purification system shown in Zone E is designed to (1) recover solvent and optionally monomer and (2) to purge the system of essentially all the alcohol and part of the butane and propylene. Solvent and monomer vapors in line 46 are condensed in the condenser 75, and a portion of the butane, propylene, along with some butadiene distilled in the light ends tower 76 and removed as an overhead stream via line 77. The dry liquid bottoms, consisting of solvent and most of the monomers, is recycled via line 30 to solvent storage 73, and then via line 74 to the polymerizers.

Wet solvent vapors in line 42, along with isopropyl alcohol and propylene, are condensed in condenser 80, decanted in decanter 81, and the alcohol-water layer removed at the bottom of the decanter via line 82. The wet liquid hydrocarbon layer is sent to the butylene removal tower 83, where the propylene, water, and remaining isopropyl alcohol are removed overhead in line 84 by distillation. The bottoms are sent via line 85 to a heavy ends tower 86 wherein wet purified solvent is taken overhead in line 87 and recycled to the process via dryer 71 and line 30. The bottoms (the heavy ends) are rejected. If styrene is present, it can be recovered from the heavy ends.

COPPER-CONTAINING CATALYSTS

ACRYLONITRILE MANUFACTURE

Allied Chemical

A process developed by J.C. Eck; U.S. Patent 3,186,953; June 1, 1965; assigned to Allied Chemical Corporation involves copper catalysts used in the production of acrylonitrile by the reaction of acetylene with hydrogen cyanide. The production of acrylonitrile by the reaction of acetylene and hydrogen cyanide in the presence of an aqueous cuprous chloride catalyst is described in U.S. Patents 2,322,696; 2,486,659; 2,733,259; and many others.

The preferred catalyst mixture suitable for such use consists of a mixture of cuprous chloride, ammonium chloride, copper powder, hydrochloric acid and water. Other catalyst mixtures include a mixture of potassium and sodium chloride in place of the ammonium chloride.

The active ingredient in the catalyst solution is monovalent copper. Cuprous chloride is the most convenient source of the monovalent copper and is relatively insoluble in water. It is therefore customary to add ammonium chloride to solubilize the cuprous chloride as a cuprous ammonium chloride complex. Hydrochloric acid is added in relatively small amounts to bring the catalyst completely into solution. It is, however, important to control the concentration of hydrochloric acid. While small amounts depress the formation of vinyl acetylene, larger amounts cause the formation of vinyl chloride. Hydrochloric acid also influences the equilibrium between the hydrogen-cyanide complex and the hydrogen cyanide itself. Metallic copper is added to the catalyst solution in order to reduce any divalent copper which may be present.

The reaction conditions prevailing in the system in which acrylonitrile is formed from acetylene and hydrogen cyanide in the presence of the aqueous cuprous chloride catalyst also lead to the formation of a small amount of by-product tar which deposits on suspended catalyst present and prevents this suspended catalyst from dissolving in the solution. This in turn decreases its effectiveness. Control of the decreasing efficiency of the catalyst and the necessity to regenerate the catalyst are factors which play an important part in determining the commercial success of the operation.

A number of methods have been proposed for the recovery of the copper values from the spent catalyst or for reactivating the catalyst by first driving off the water, then heating the dried mass to a temperature at which the tar contaminants are converted to solid carbonaceous material which can then be removed by filtration after the catalyst is again taken up in water. While the methods of the prior art may be effective, they are cumbersome and expensive. It is accordingly an object of this process to provide a method of removing these tarry by-products from the catalyst solution which is effective, cheap, and simple.

The process involves solvent extraction of a portion of the catalyst solution with a halogenated hydrocarbon. In general, the chlorinated solvents which are preferred to use in removing the tarry products from the catalyst suspension have the following characteristics:

(1) They should be relatively insoluble in water;
(2) should not react with hydrochloric acid;
(3) should have a boiling point above 90°C.; and
(4) should preferably have a specific gravity which is considerably higher or considerably lower than that of the catalyst solution.

It is preferable to use a solvent having a high specific gravity since the undissolved portion of the catalyst on which the tarry product is dispersed would tend to settle even with agitation. Under these circumstances, a heavy solvent would have a greater chance to extract tarry by-products from the catalyst. It is desirable that the solvent have a boiling point above 90°C., for if any of the solvent is carried back into the reactor along with the catalyst it would not be distilled out through the exit gas lines along with the acrylonitrile product but would remain in the catalyst solution where it would not be harmful. With the aid of these solvents it is possible to remove the tarry by-products which are dispersed on the suspended catalyst. By this means the efficiency of the catalyst is maintained and the life prolonged.

The conduct of the process may be illustrated by reference to the flow diagram in Figure 4. Acetylene is fed to the bottom of reaction chamber 3 through feed pipe 1. As the gaseous acetylene passes through, it comes into intimate contact with a catalyst mixture in which the components are present in the following weight ratios: 650 parts of cuprous chloride, 8 parts of copper powder, 20 parts of concentrated hydrochloric acid (specific gravity 1.18) and 560 parts of water. As the gaseous acetylene bubbles through the catalyst solution, it promotes a circulatory action causing the solution to flow downward through leg 4 of the reactor and come into contact with HCN which is being fed into the system through feed pipe 2. An intimate mixture of HCN, acetylene, and catalyst solution results.

The admixed solutions continue their circulatory motion downward through leg 5 and back to the reactor chamber 3. This causes a maximum amount of hydrogen cyanide to be present at that point where the acetylene enters the reaction chamber through feed pipe 1. Approximately 5% of the reaction mixture in the reactor 3 is continuously tapped through flow arm 6 at a rate such that the volume in packed extractor 7 is maintained at 1 to 10% of the volume in the reactor chamber 3. Tetrachloroethylene is cycled countercurrently through extractor 7 and leg 8 with the aid of pump 18 in such manner as to extract by-products produced in the reaction which took place in reactor 3 and were fed to the bottom of extractor 7 via flow arm 6.

The extracted reaction mixture leaves extractor 7 by means of take-off arm 9 and passes to agitated vessel 10 which contains solid copper metal. A bleed-off valve is provided at 16 to permit the removal of contaminated solvent which is replaced with fresh solvent at 15. Concentrated aqueous hydrochloric acid is continually added to the catalyst solution at inlet 11 in vessel 10 in order to maintain the pH of the solution at 1 to 2. The purified catalyst solution is subsequently removed via pipe 12 under the influence of pump control 13 and is then fed back to the reaction chamber

FIGURE 4: FLOW SCHEME FOR ACRYLONITRILE MANUFACTURE USING AQUEOUS CUPROUS CHLORIDE CATALYST AND CONTINUOUS SOLVENT PURIFICATION OF CATALYST

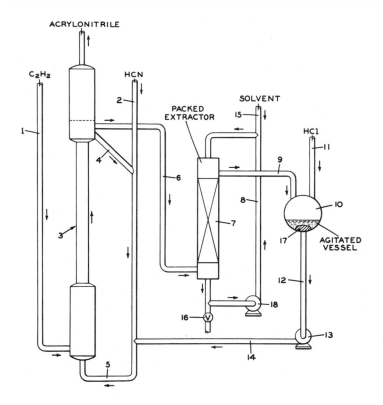

Source: J.C. Eck; U.S. Patent 3,186,953; June 1, 1965

through pipe 14 and leg 5 of the reactor 3 where it again is used in the production of acrylonitrile. It can readily be seen that this method readily lends itself to either continuous or intermittent treatment of the catalyst solution and that by the constant removal of the tarry by-product from the catalyst solution, it is possible to greatly prolong its effective life.

Du Pont

A process developed by S.N. Vines, U.S. Patent 3,092,451; June 4, 1963; assigned to E.I. du Pont de Nemours and Company relates to the treatment of spent catalyst solutions formed in the production of acrylonitrile by reaction of hydrogen cyanide with acetylene in organic liquid containing cuprous chloride, to recover the organic liquid and the copper content as cuprous chloride.

A well-known commercial process for producing acrylonitrile involves continuously passing hydrogen cyanide and acetylene through a reactor containing an aqueous solution of cuprous chloride catalyst at a temperature of 70° to 100°C. The resulting liquid phase reaction produces a variety of water-insoluble by-products in addition to the desired acrylonitrile. The more volatile by-products pass out of the reactor with the acrylonitrile and unreacted hydrogen cyanide and acetylene, but a mixture of by-products accumulates in the reactor and will soon stop the process unless it is removed. This mixture of by-products is a highly viscous, very dark, tarry material, which is reasonably fluid at the reaction temperature but solidifies at room temperature. It is a relatively easy matter to drain a sufficient amount of such water-insoluble tars from the reactor to avoid an excessive accumulation.

The use of nonaqueous organic solvent solutions of cuprous chloride instead of aqueous catalyst solutions in the above process decreases by-product formation, as by avoiding the formation of acetaldehyde by reaction of acetylene with water. Furthermore, various anhydrous organic liquid catalyst systems have been found which also provide potentially important improvements in productivity, by greatly increasing the activity and useful life of the catalyst in some unexplained manner. However, the tarry by-products formed are soluble in the organic liquids used. These by-products dilute the catalytic reaction system, increase its viscosity and cause the productivity per unit volume of catalyst to drop when the concentration reaches about 20% of tar, the operation becoming uneconomical at tar concentrations approaching 30%. It is therefore necessary to replace the catalyst medium contaminated with dissolved by-product tars, either continuously or intermittently, with fresh components at a rate which will remove undesirable amounts of tar from the system.

Since disposal of such tarry mixtures is a troublesome problem and the components of nonaqueous catalyst solutions are expensive, an effective process for recovery of values therefrom and preferably as components suitable for reuse in the catalyst medium, especially one leaving waste in a form which is readily disposed of, is highly desirable.

In accordance with this process it has been found that anhydrous organic liquid solutions of cuprous chloride catalyst containing 10 to 30% of nonvolatile tarry by-products resulting from use in the production of acrylonitrile can be effectively treated to accomplish the above objectives by (1) evaporating organic liquids which are volatile at temperatures up to 250°C. under subatmospheric pressure to form a desolvated mixture; (2) heating the desolvated mixture at 225° to 450°C., preferably in the presence of hydrogen chloride, to form a pulverulent residue of insolubilized or charred by-products containing cuprous chloride; and (3) dissolving the cuprous chloride in an organic solvent.

For highly effective synthesis of acrylonitrile from hydrogen cyanide and acetylene, it is advantageous to have present in the catalyst mixture an organic promotor to the extent of about 0.2% to not over about 15% by weight of the mixture. The promoter used should preferably boil above about 100°C. and should be volatilizable at a temperature not exceeding about 250°C. at a pressure of about 40 mm. mercury. Promoters of the amide type are especially advantageous and among the amide promoters the lower dialkyl formamides have been found to be particularly useful.

In general it is possible to recover the volatiles (solvent and organic promoter) from used catalyst to the extent of about 95% by subjecting the used catalyst to a temperature between about 125° to 210°C. under a partial vacuum between 20 to 100 mm. of mercury pressure. Most of the remainder of the volatiles can be recovered during the subsequent charring operation whereby the acrylonitrile synthesis by-product tars are rendered insoluble in the organic solvent. Vacuum distillation is preferred because recoveries are higher and corrosion of equipment lower than at atmospheric pressures.

Figure 5 shows a suitable form of apparatus for the conduct of this process. Used catalyst is continuously or periodically removed from the acrylonitrile reactor, not shown, at a suitable rate which may, for example, be 10% per 24 hours of the total catalyst mixture. This withdrawn used mixture is fed by way of line 1 into a suitable vessel such as a Dopp kettle 2 provided with jacket 3 for heat input to evaporate the volatile organic liquids (volatilize the organic solvent and promoter, specifically benzonitrile and dimethylformamide, at about 125°C. and 0.4 psia). Inert gas, such as nitrogen, may be used to flush the vessel and enters by way of line 4. The contents are stirred while the solvents are volatilized and the molten residue comprising the cuprous chloride and by-product tars is then transferred through lines 5 and 7 propelled by impeller 6 into a rotating kiln 8 which may operate at atmospheric pressure and which may be

FIGURE 5:   OVERALL FLOW DIAGRAM OF COPPER CATALYST RECOVERY FROM ORGANONITRILE SOLUTION

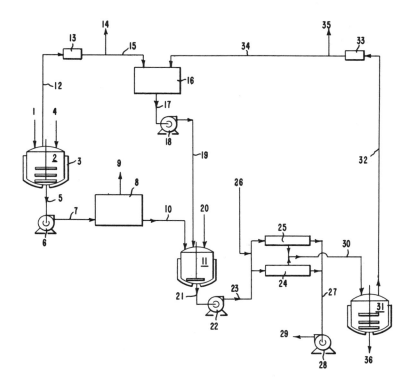

Source: S.N. Vines; U.S. Patent 3,092,451; June 4, 1963

heated by any of the usual methods, not shown, to a temperature of about 350°C. such that the contained by–product tars will be partially charred or at least rendered insoluble in the catalyst solvent. Vapors leaving the kiln by way of line 9 may be cooled and the condensibles trapped by any desired means, not shown. The insolubilized tars and cuprous chloride form a pulverulent mass which is transferred to the extraction vessel or slurry tank 11 by way of line 10, in which the cuprous chloride is dissolved by the organic solvent and thus separated from the insolubilized tars.

The volatilized solvent vapors leaving the Dopp kettle by way of line 12 are recovered by liquefication in cooler 13. Reduced pressure is applied to the kettle by a pump or other means not shown at 14 and the liquid solvent passes from the cooler 13 by way of line 15 to solvent reservoir 16 to which solvent from other sources may also be added. Solvent is transferred from reservoir 16 through 17 and 19 by way of actuating means 18 to the extraction vessel 11. As required, additional solvent and cuprous chloride may be added to the extraction vessel by way of line 20. The slurry of cuprous chloride, solvent and insolubilized tars is transferred by way of lines 21, 22 and 23 to filters 24 and 25 which may be operated in sequence according to usual chemical engineering practice. Solvent from any source such as 26 may be circulated to filters 24 and 25 to wash the cake comprising mainly insolubilized tars. The solvent extract and washings may contain up to 90% and more of the cuprous chloride and this is recycled to the acrylonitrile reactor, not shown, by way of lines 27, 28 and 29.

The extracted insolubilized tars from the filters 24 and 25 are transferred by way of line 30 to a vessel 31 similar to vessel 2 to recover the solvent adhering to the filter cake. Recovered solvent vapor is removed via line 32, cooler 33 and finally the condensed solvent may be sent to solvent reservoir 16 by way of line 34. Pressure reduction for the vaporization of solvent in vessel 31 is applied at 35 in a manner similar to 14 above. Insolubilized by–product tars finally are discharged from the system at 36 for suitable disposal by methods not further considered here.

Figure 6 shows in somewhat more detail the apparatus for desolvating used catalyst and recovering the volatile materials. Feed tank 37 is provided with heating coil 38 for maintaining the used catalyst mixture in a fluid state. The mixture passes through line 39 to metering pump 40 and is pumped continuously through line 41 into a film–type evaporator 42 having a rotor speed of 1,000 to 2,000 revolutions per minute. The evaporator is provided with a steam jacket 43 for heating the evaporating surface up to the temperature required to drive off the volatiles. The desolvated catalyst is continuously withdrawn from the evaporator through line 45 by discharge pump 46 and passed through line 47 into a

15

FIGURE 6:   APPARATUS FOR VOLATILE COMPONENT RECOVERY IN COPPER CATALYST REGENERATION

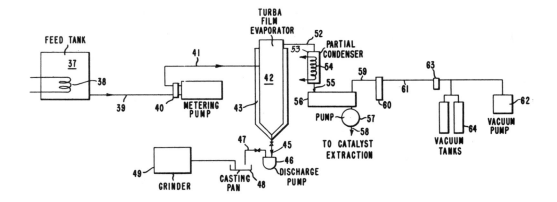

Source: S.N. Vines; U.S. Patent 3,092,451; June 4, 1963

casting pan 48. The desolvated mixture of copper salts and by-product tars solidifies upon cooling and is broken into small particles in grinder 49. The vaporized organic liquids leave the evaporator by way of line 52 and are recovered in partial condenser 53. The cooling means 54 is maintained at a temperature which will pass acrylonitrile and condense higher boiling liquids. The liquids and vapors pass through line 55 into solvent tank 56, where the condensed liquids are collected and then removed by pump 57 and line 58 to the catalyst extraction operation described in connection with the preceding figures. The uncondensed vapors, consisting principally of acrylonitrile and hydrogen cyanide, are separated in solvent tank 56 and pass through line 59 to total condenser or trap 60 for recovery. The system is exhausted through line 61 leading from the total condenser to vacuum pump 62, suitable for maintaining an absolute pressure of about 50 mm. of mercury in the evaporator. Vacuum tanks 64 and pressure regulator 63 are provided to assist the vacuum pump in maintaining the desired low pressure.

The rotating kiln 8 in the first figure used for the charring step is shown in more detail in Figure 7. The desolvated catalyst mixture to be charred will usually have the approximate composition of 55 to 70% cuprous salts, principally cuprous chloride, and 25 to 40% tars, together with a small percentage of unvolatilized solvent and ammonium chloride. Sticking during feeding into the kiln at the higher range of tar contents can be avoided by blending with recycled charred catalyst. Desolvated catalyst from the grinder is mixed with recycled char in blender 68 and is then conveyed by way of line 69 into feed hopper 70. A variable speed screw feeder 71 is used to introduce the mixture into the heated section 72 of the cylindrical kiln. This screw is provided with a water-cooled jacket 73 to prevent sticking.

The kiln is rotated slowly in the conventional manner for such equipment to mix the material and expose it uniformly to the heated surface and atmosphere. A stationary scraper bar 74 is arranged to rub against the inside of the heated

FIGURE 7:   APPARATUS USED IN CHARRING STEP OF COPPER CATALYST RECOVERY

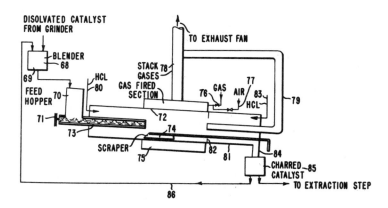

Source: S.N. Vines; U.S. Patent 3,092,451; June 4, 1963

cylinder to keep the surface free of deposits and assist in moving the material through the kiln. The kiln is surrounded with a gas-fired jacket 75 which provides the heat for charring the tars. Gas is supplied through line 76 and mixed with air supplied through line 77 for combustion. The combustion gases exhaust to the stack through conduit 78. The off-gases formed in charring the tars are exhausted from the kiln through conduit 79 and likewise pass to the stack. Conduit 79 is insulated to avoid condensation in the line.

A hydrogen chloride atmosphere is maintained in contact with the heated catalyst mixture by introducing hydrogen chloride into the feed end of the kiln through line 80. Since the hydrogen chloride is diluted by gases produced in charring the tars, a feed rate of about 15 to 40 lbs. HCl/100 lbs. of desolvated catalyst mixture is desirable in order to avoid dilution to an extent which will seriously reduce the amount of cuprous chloride subsequently recovered. The catalyst mixture is heated to about 400°C. and retained for about one-half hour in heated section 75; it then passes into a cooling section 81 of the kiln. The retention time can be controlled by adjusting the incline of the kiln and by the use of a retention ring 82 of suitable size at the exit end of the heated section.

The charred catalyst mixture is cooled sufficiently in section 81 to avoid oxidation upon exposure to air. Hydrogen chloride can be introduced into this section through line 83, if necessary, to supplement that introduced at the other end of the kiln through line 80. The catalyst passes from the kiln through exit conduit 84 into a feed hopper 85 for the cuprous chloride extraction step of the process described previously. A portion of this charred catalyst may also be recycled through line 86 to blender 68 for the purpose already described.

Another process developed by S.N. Vines; U.S. Patent 3,309,173; March 14, 1967; assigned to E.I. du Pont de Nemours and Company involves first removing the volatile organic liquids from the catalyst solution to form a desolvated mixture, heating the desolvated catalyst in the range of from 800° to 1800°C. to effect the complete combustion of organic components present in the desolvated catalyst, quenching the combustion products to a temperature between about 400° and 100°C. and recovering the solid products from the quenching step. This solid product which is a mixture of cuprous chloride and cuprous oxide may be converted substantially to pure cuprous chloride by post-treatment with HCl at a temperature of from 50° to 150°C. for reuse as catalyst.

Figure 8 shows the essential elements of this process. Used catalyst is treated in a suitable vessel such as a Dopp kettle, not shown, to remove organic liquids, i.e., organic solvent and a promoter, if present, by vaporization and form a desolvated catalyst mixture comprising the copper compounds and by-product tars. This desolvated catalyst mixture is fed through line 11 to a slurry tank 12 which is provided with a heating jacket 13 for heat input to the catalyst mixture. Water enters the slurry tank 12 by line 14 and is stirred with the catalyst mixture to form a water slurry of about 50% solids which is then transferred through lines 15 and 16 by pump 17.

After being atomized with air introduced into line 16 through line 18, the catalyst mixture is fed into a high efficiency vortex burner 19 where it is burned at a temperature ranging from 800° to 1800°C. with an excess of air which is introduced into the combustion zone 20 of burner 19 through line 21. The excess air and water assist in holding the maximum temperature of combustion in combustion zone 20 below 1800°C., thereby avoiding decomposition of the copper compounds present in the catalyst. Burner 19 is mounted at a slight downward angle from a horizontal plane and discharges

FIGURE 8: ALTERNATIVE FORM OF APPARATUS FOR CHARRING STEP OF COPPER CATALYST RECOVERY

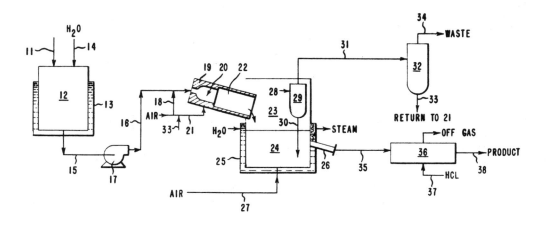

Source: S.N. Vines; U.S. Patent 3,309,173; March 14, 1967

directly into the side of quenching chamber 23 by means of conduit 22 which forms an airtight connection with burner 19 and quenching chamber 23. The combustion products entering quenching chamber 23 from combustion zone 20 involve vapors composed chiefly of combustion gases and volatilized cuprous chloride and a pulverulent mass of nonvolatile copper compounds. Quenching chamber 23, provided with a cooling jacket 25, contains a fluidized bed 24 of previously quenched copper compounds to effect rapid cooling of the combustion products. This is made feasible by the wall-to-solids heat transfer coupled with a turnover of pulverulent copper compounds in bed 24 by the continuous withdrawal of a pulverulent mass from withdrawal tube 26 for post-treatment discussed later.

The bed 24 is maintained in a uniformly fluidized condition with air introduced at the bottom of quenching chamber 23 by means of line 27. The depth of fluidized bed 24 is maintained sufficiently below the entry point of the combustion products via conduit 22 to provide a cooling surface on the walls of the quencher and also to provide sufficient free volume above the bed so that, on shutdown in a nonfluidized condition, the copper compound bed will come to rest below the entry point of the combustion products in order to avoid fusion of a portion of the bed in the hot zone adjacent conduit 22.

Operation of the fluidized bed can be readily carried out in a temperature range from 100° to 300°C. whereby the combustion products entering quenching chamber 23 are rapidly cooled with the cuprous chloride being condensed in the bed and only the combustion gases remaining in the vapor state. The latter, as off-gases, are passed by means of line 28 into an internal cyclone 29 to recover some entrained copper compound fines which are returned by standpipe 30 to bed 24. The off-gases from cyclone 29 are transferred by line 31 outside quenching chamber 23 to a second cyclone 32 for removal of further fines. These latter fines are returned by means of line 33 to line 21 for recycle through burner 19 by entrainment in the secondary combustion air present in line 21. The off-gases from cyclone 32 are exhausted to the atmosphere by means of line 34 as waste after water scrubbing to remove trace amounts of toxic materials.

The pulverulent mass of copper values, principally a mixture of cuprous chloride and cuprous oxide, is transferred from the quenching chamber 23 by means of withdrawal tube 26 and line 35 to a rotary contactor 36. The cuprous oxide portion of the mixture is converted to cuprous chloride within rotary contactor 36 by treatment with HCl gas which is introduced therein by means of line 37. The mixture within contactor 26 is maintained at a temperature of about 50° to 150°C. and the conversion of cuprous oxide to cuprous chloride readily progresses with time in the presence of water vapor formed as the product of cuprous oxide and hydrogen chloride. The final product, cuprous chloride, is removed from contactor 36 by means of line 38 and may be pneumatically conveyed to storage bins or to an anhydrous catalyst makeup tank for reuse.

AUTOMOTIVE EXHAUST TREATMENT

Du Pont

A process developed by A.B. Stiles; U.S. Patent 3,317,439; May 2, 1967; assigned to E.I. du Pont de Nemours and Company involves catalyst aggregates of crystallites of a catalytic material which are kept apart by crystallites of a refractory which melts above 1000°C., the aggregates thus constituted thereafter having crystallites of a refractory which melts above 1000°C. and added thereto to keep the former two groups of crystallites apart and thus form a catalytic aggregate which is stabilized against crystallite growth and inactivation at high temperatures. Figure 9 is an artist's representation of such a catalytic aggregate showing the association of crystallites of copper chromite with silica and chromium oxide.

FIGURE 9: ENLARGED VIEW OF CATALYST AGGREGATE USED IN CATALYTIC EXHAUST GAS TREATMENT

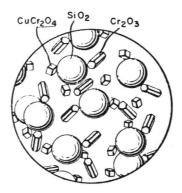

Source: A.B. Stiles; U.S. Patent 3,317,439; May 2, 1967

The process will now be further described with reference to Figure 9. A catalytic material, copper chromite, is formed in conventional manner as crystallites of small size. Ordinarily these crystallites grow rapidly when the catalyst is put into use under conditions such that the catalyst is heated. According to the process, a refractory material, such as silica, which melts above 1000°C. is formed in conventional manner as crystallites in the same size range as the copper chromite and intimately mixed with it. These are then heated to drive off any volatile components and the resulting aggregate is treated with still another refractory, chromium oxide, in the form of small crystallites. The drawing shows an association of such crystallites and it will be seen that the crystallites of refractories keep each other apart and also keep the copper chromite crystallites apart so that crystal growth is hindered.

These catalysts can be used in the same ways as prior art catalysts containing the same active catalytic materials. Thus they can be used for a wide variety of oxidations and reductions and many have especially great value for oxidation or reduction, or both, of combustion gases.

Sergeys et al

A process developed by F.J. Sergeys, P.K. Maher, W.S. Briggs and C.V. McDaniel; U.S. Patent 3,346,328; October 10, 1967 involves removing the noxious components from the exhaust gases of internal combustion engines by contacting the gases with a copper metal catalyst upon a zeolitic base where a substantial portion of the metal is present as metal oxide. The balance of the metal is contained in the zeolite structure as metal ions.

Such a catalyst which has unusual selectivity and activity for oxidation of the hydrocarbons, carbon monoxide and other harmful components of auto exhaust can be prepared using a copper catalyst on a crystalline alumino-silicate base, where the copper is present as both the exchanged cupric ion and copper oxide. This catalyst is suitable either alone or as a promoter in other catalytic combinations.

This catalyst is prepared using a crystalline zeolite as the basic building block. The zeolites are crystalline metal aluminosilicates with a three-dimensional structure of silica and alumina tetrahedra. The charge of the anionic alumino-silicate network of the zeolite is most often balanced by nonactive cations such as H^+, NH_4^+, substituted ammonium, alkali, or alkaline earth metal cations. By both cation exchange treating these zeolites with catalytically active metal ions and also depositing a portion of the metal ions as the oxide in the internal adsorption area, it is possible to produce a very active and selective oxidation catalyst. Further, by varying the amount of metal ion and the relative equivalent ratio of exchanged metal ion to deposited metal oxides, it is possible to vary the activity and selectivity of the catalyst to meet the demands of the particular exhaust gas composition to be oxidized.

Both natural and synthetic zeolites are suitable for the product. Suitable examples of natural zeolites include faujasite, erionite, chabazite and mordenite. Synthetic zeolites which have been found suitable include, for example, those zeolites designated Type A, Type X, Type Y and Type L by the Linde Division of Union Carbide Corporation. In addition, synthetic sodalite and synthetic mordenite types of zeolites also give satisfactory results. The approximate gross compositions of these synthetic zeolites and their pore sizes are set out in the table below.

Zeolite	Approximate Gross Composition	Pore Size (A.)
Type A	$MO:Al_2O_3:2SiO_2:xH_2O$	3 - 5
Type X	$M_2O:Al_2O_3:2 - 3SiO_2:xH_2O$	9 - 13
Type Y	$M_{2/n}O:Al_2O_3:3 - 6SiO_2:xH_2O$	9 - 13
Type B	$M_{2/n}O:Al_2O_3:2 - 10SiO_2:xH_2O$	2 - 3
Type L	$M_{2/n}O:Al_2O_3:2 - 10SiO_2:xH_2O$	
Mordenite	$M_{2/n}O:Al_2O_3:8 - 12SiO_2:xH_2O$	6 - 10

M is a metal cation (normally in the original prepared state Na).
n is its valence.
x is a function of the degree of dehydration and normally ranges between 0 to 9.

The active metal cations are present in the zeolite either as exchange elements or as the metal oxides. Additional desirable characteristics are imparted to the product by loading the internal adsorption area with metal oxides. The active metal ions or oxides include those of the elements copper, cobalt, nickel, chromium, manganese, vanadium, molybdenum, Group VIII metals, yttrium and the rare earths. The preferred catalyst is prepared to contain cupric ions by exchange with a cupric salt and copper oxide by conversion of the cupric salt deposited within the pores of the zeolite. The amount of metal ions and metal oxide used is determined by the particular use intended for the finished catalyst.

In the preparation of a catalyst containing copper, for instance, the characteristics for carbon dioxide and hydrocarbon conversion can be substantially altered by altering a portion of or all of copper present as Cu^{++} to CuO. In preparing these catalysts, for example, by using the copper salts as the source of the active metal or metal oxides, catalysts have been prepared in which from 5 to 90% of the sodium ions in the zeolite have been replaced with active metal ions and with varying amounts of active metal oxide interstitially held in the pores. Catalysts containing up to 16% copper ion in the zeolite structure and up to 30% CuO present in the pores have been prepared with excellent results.

Since the copper ion plays an important part in hydrocarbon conversion while copper oxide is an influential factor in carbon monoxide conversion, it is desirable to have some of the copper present as the ion and some as the oxide. This phenomenon facilitates the preparation of a catalyst selective to convert hydrocarbons and/or carbon monoxide preferentially, depending on the type of engine and fuel being used.

International Copper Research

A process developed by R.A. Baker and R.C. Doerr; U.S. Patent 3,398,101; August 20, 1968; assigned to International Copper Research Association Inc. relates to copper-containing redox catalyst which is effective for promoting the oxidation and reduction of various chemical compounds under different reaction conditions. It has been found to be especially effective in the treatment of exhaust gases from hydrocarbon combustion engines to control the emission of noxious components, carbon monoxide, hydrocarbons and oxides of nitrogen therein.

Several chemicals, particularly copper chromites, have been found to exert catalytic effects in the treatment of automobile exhaust to remove selectively its noxious components. They promote the oxidation of carbon monoxide and hydrocarbons in an oxidizing atmosphere and the reduction of oxides of nitrogen in the reducing medium in the presence of a reducing agent. These catalysts, however, undergo severe attrition after having been subject to several cycles of oxidation and reduction and lose their effectiveness. Furthermore, the catalytic activities of these compounds decrease rapidly during the treatment of the exhaust, especially the leaded exhaust. The instability of these catalysts as the result of low resistance to mechanical, chemical and thermal attritions which contribute to their short catalytic lives and their rapidly decreasing effectiveness when exposed to the leaded exhaust reduces their usefulness for catalytic treatment of automobile exhaust.

It has now been found that a catalyst having cobaltic oxide and cupric oxide provides superior catalytic performance for promoting the reduction of oxides of nitrogen in the presence of a reducing agent such as carbon monoxide. This catalyst has exceptionally high stability in resisting attritions and has a catalytic life of more than 350 hours under normal operating conditions. It has also been found to be effective in promoting the oxidation of various chemicals, particularly the carbon monoxide and hydrocarbons in the exhaust gases of internal combustion engines, rendering it particularly desirable for the treatment of automobile exhaust in a homogeneous two-stage device. This device will reduce the oxides of nitrogen in the first stage and oxidize the remaining carbon monoxide and the hydrocarbons in the second state with the assistance of external oxygen.

Broadly stated, this redox catalyst composition consists essentially of cobaltic oxide, cupric oxide and a catalytic carrier. The ratio of the oxides in the composition is one part of cobaltic oxide to about 3 to 35 parts of cupric oxide by weight. While any catalytic carrier that will not substantially interfere with the activity of the catalyst can be used in preparing this catalyst, aluminum hydroxide is eminently suitable. Other satisfactory carriers include silica, alumina and Carborundum. Conventional methods for preparing this catalyst may be used and the catalyst prepared can be a homogeneous mixture in the form of an unsupported catalyst, or a heterogeneous mixture in the form of a supported catalyst. The amount of catalytic carrier in the composition can be varied within a broad range depending on a number of variables such as the ratio of oxides in the catalyst, the carrier used, and the desired physical characteristics of the catalyst. For a homogeneous unsupported catalyst using aluminum hydroxide as a catalytic carrier, a suitable catalyst has 2 to 15% by weight of cobaltic oxide, 50 to 70% by weight of cupric oxide, and the balance being the carrier.

Monsanto

A process developed by J.F. Roth; U.S. Patent 3,493,325; February 3, 1970; assigned to Monsanto Company is a process for catalytic treatment of the exhaust gases in a hydrocarbon combustion engine which involves passing the gases over a degradation-resistant catalytic composite comprising an alumina supported copper oxide where less than 4% by weight of the catalytic composite has a diameter size of at least 50 A. The absence of secondary air will cause a reduction of reducible constituents and a presence of secondary air will cause an oxidizing of the oxidizable constituents in the exhaust gas.

Oxy-Catalyst

A process developed by W.R. Calvert; U.S. Patent 3,053,773; September 11, 1962; assigned to Oxy-Catalyst, Inc. relates to the rejuvenation of catalytic exhaust purifiers of the type employed with internal combustion engines which employ leaded gasoline. Of late years, the almost universal use of so-called "leaded" gasolines has increased the difficulty of purification of exhaust gases both by introducing another undesirable contaminant in the exhaust gases in the form of lead compounds formed in the engine and exhausted therefrom, and by increasing the difficulty of purifying gases by means of catalytic oxidation. The term "leaded gasoline" as it is commonly used refers to a gasoline to which has been added a compound of lead, most usually tetraethyl lead, for the purpose of increasing the octane rating of the gasoline. The tetraethyl lead, which is usually added together with halogenated compounds which inhibit the deposition of the lead on the cylinder walls, is added to the gasoline in relatively small amounts, such as, for example, the amount equivalent to 3 grams of metallic lead per gallon of gasoline. Under the combustion conditions prevailing in the engine, the lead oxide and other lead compounds which are formed are carried out of the engine in the exhaust gases. These compounds are for the most part lead oxide, tetraethyl lead and lead halides, such as lead chloride and lead bromide, and complexes of these compounds.

When a catalytic exhaust purifier containing a bed of oxidation catalyst is employed to catalytically oxidize carbon monoxide, hydrocarbons and the organic constituents of the exhaust gases, these lead compounds tend to deposit upon and accumulate within the catalyst bed which is advantageous from the standpoint that the bed acts as a filter to remove these undesirable contaminants and prevent their escape to the atmosphere. However, it is disadvantageous in that these compounds tend to reduce the oxidation activity of the catalyst and eventually render it unable to carry on its intended function of oxidizing the undesirable oxidizable fumes.

In accordance with the process, it has been found that it is possible in the operation of a catalytic purifier on engines using leaded gasoline to periodically rejuvenate the catalytic purifier by a mechanical removal of catalyst particles and of the accumulated loose lead compounds which are in the form of dust and then treating the catalyst with a solution of the catalyst metal. Surprisingly this process restores the catalyst to essentially its original activity and insures free passage of the exhaust gases through perforations in the structure containing the catalyst.

The method is advantageous in that it can be carried out, if desired, without removing the exhaust purifier from its permanently installed position. It is further advantageous in that it greatly improves the step of replacing catalyst metal by preventing the clumping and other deleterious effects caused by the presence of the large amounts of catalyst dust and particles.

In accordance with this process, the catalytic purifier is first vibrated to dislodge dust and particles from the walls, from the catalyst bed, and from the other internal structural parts of the catalytic purifier. Advantageously an entraining fluid, for example, a gas such as air or a liquid such as water, is passed through the interior of the catalytic purifier while it is being vibrated in order to entrain the dust and particles and remove them from the interior of the purifier. Alternatively, the vibration can be carried out first to dislodge the material to be entrained and the entraining fluid passed through the interior of the purifier subsequently to remove this material. The vibrations may be within a wide range for satisfactory results, but will have preferably an amplitude of vibration of the entire purifier housing from about 1/1,000" to about 1/3", and preferably will have a frequency of from about 60 to 9,000 vibrations per minute. Electrical, mechanical and sound vibrators may be employed. The vibration is not only of great value in cleaning the purifier and its catalyst, but also is of importance in removing dust to facilitate the later employment of liquid reagents which otherwise would form the dust into interfering wet clumps.

A liquid solution, preferably an aqueous solution, of a decomposable compound of the catalyst metal employed on the catalyst in the purifier is then flowed into the catalyst in the purifier. Thus, for example, an aqueous solution of a water-soluble salt of the appropriate metal preferably a salt of strong acid, such as salts of inorganic acids, for example, a nitrate, sulfate or chloride salt, may be employed. Thus, an aqueous solution of copper nitrate and chromium nitrate made, for example, from hydrated copper nitrate [$Cu(NO_3)_2 \cdot 3H_2O$] and hydrated chromium nitrate [$Cr(NO_3)_3 \cdot 9H_2O$] or chromium trioxide in water is useful where the catalyst is a copper chrome catalyst. Aqueous solutions of the sulfates and chloride salts of copper and chromium are further exemplary. Similarly where platinum, nickel or silver is the catalyst metal, exemplary is an aqueous solution of a water-soluble salt of the metal involved, such as a salt of a strong acid, such as nitric, hydrochloric or sulfuric acid.

This solution is permitted to remain in contact with the catalyst in the catalyst purifier until a substantial amount has been absorbed, normally from 2 to 30 minutes. The remaining solution is then drained and the catalyst is dried and the metal salt is decomposed by passing the engine exhaust through the purifier, preferably while running the automobile on the road. The drying and decomposing can, if desired, be carried out by passing any reducing gas such as a gaseous or vaporized fuel, for example, propane or city gas, through the purifier at a temperature in the range of 400° to 1000°F. Alternatively, the catalyst can be dried and partial decomposition to the metal oxide and possibly some metal accomplished by the use of hot air with the reduction carried out in a subsequent step by passing the aforementioned reducing gas through the purifier.

GASOLINE TREATMENT

Phillips Petroleum

A process developed by D.S. Joy and R.E. Dollinger; U.S. Patent 3,374,181; March 19, 1968; assigned to

Phillips Petroleum Company involves an improved technique for the preparation of a copper treater catalyst. The copper treater catalyst has heretofore been prepared by application of the selected salt solution to a mass of adsorbent in a bin or tank followed by shoveling the impregnated mass of fuller's earth much in the manner of mixing a Portland cement, sand, and water mix, followed by drying the mixture in drying pans in an oven. This method has not consistently produced a uniform distribution of the impregnating salt solution on the adsorbent. This process is directed to an improved method of incorporating the aqueous copper salt solution on the adsorbent and drying the impregnated mass of adsorbent to a selected water content in the range of about 10 to 20 weight percent.

This process comprises spraying the selected salt solution onto a mass of the selected adsorbent in a mixing and blending zone to provide the desired uniform concentration of cupric chloride in the adsorbent, passing the resulting homogeneous mixture into a fluidized bed drying zone and maintaining the particulate material in suspension in a fluidized bed in the zone by passing a fluidizing and drying gas upwardly through the bed so as to reduce the water content of the material to the range of about 10 to 20 weight percent, and preferably to about 15 weight percent, and leave the salt in a concentrated solution in the remaining liquid in the adsorbent, and recovering the partially dried adsorbent from the drying zone.

Figure 10 shows the essential equipment elements involved in the process. As shown there, a mixing tank 10 for the salt solution is connected with a spray line 12 by means of line 14 containing a gear pump 16. This gear pump is a Hastaloy pump and all of the equipment contacted with the salt solution during the process is fabricated of corrosion resistant material such as stainless steel or plastic material such as Marlex polyolefins of Phillips Petroleum Company, Teflon, etc. The lines and conduits are preferably fabricated of Marlex pipe.

A hopper 18 is provided with a vibrating feeder 20 which feeds the adsorbent into a mixing and blending device 22, preferably a ribbon blender, into which spray line 12 directs a dispersion of the salt solution. A ribbon blender simultaneously passes the adsorbent longitudinally therethrough and effects a spiral blending movement thereof. The resulting homogeneous mixture of adsorbent and salt solution is fed by screw conveyor 24 or other suitable means into a fluidized bed dryer 26.

FIGURE 10:   APPARATUS FOR THE PREPARATION OF COPPER CATALYST FOR GASOLINE TREATMENT

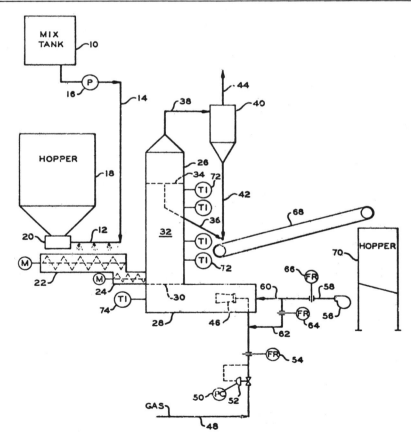

Source: D.S. Joy and R.E. Dollinger; U.S. Patent 3,374,181; March 19, 1968

Dryer 26 is positioned directly above the downstream end of furnace 28 and is separated therefrom by a gas distributor plate 30 which uniformly distributes the drying and fluidizing gas over the transverse cross section of the dryer. A fluidized bed of adsorbent 32 is maintained in the lower section of dryer 26 and extends to about the level indicated at 34 from which outlet 36 extends through the wall of the dryer. Effluent gas from bed 32 passes through conduit 38 to a cyclone separator 40 which separates any finely divided adsorbent carried overhead in the gas stream and gravitates the same through conduit 42. Effluent gas from cyclone 40 substantially free of adsorbent particles is passed through line 44 to atmosphere or any part of the gas may be recycled to furnace 28, if desired.

Furnace 28 is provided with a burner 46 which is connected by line 48 with a source of fuel gas such as natural gas. The flow of gas is controlled by a pressure controller 50 which is in operating control of motor valve 52 in conventional manner. A flow recorder 54 is also positioned in line 48. Air is supplied by blower 56 and line 58, a portion being injected as tempering air directly into the furnace through line 60 and another portion being passed to fuel line 48 through conduit 62 in which a flow recorder 64 is positioned. Likewise, a flow recorder 66 is positioned in line 58.

The partially dried reagent or catalyst passing through lines 36 and 42 is delivered onto conveyor belt 68 for delivery to sack loading hopper 70 or other storage or packaging facility. Temperature indicators 72 are positioned at different levels along fluidized bed 32 and a similar temperature indicator 74 is positioned on the outlet end of furnace 28 to record the furnace outlet gas temperature which is the inlet gas temperature to bed 32.

The process produces a finished catalyst or reagent of any selected cupric ion content and moisture content. The product is of uniform consistency and can be easily duplicated in successive runs. The manufacturing process is continuous, requires relatively inexpensive equipment, and has high capacity for the size of the equipment. There is no manual handling of the materials utilized in preparing the reagent.

HYDRATION OF ACETYLENE

Phillips Petroleum

A process developed by C.A. Wentz, Jr.; U.S. Patent 3,249,555; May 3, 1966; assigned to Phillips Petroleum Co. is a process for reactivating a calcium phosphate-copper phosphate catalyst for the hydration of acetylene. The hydration of acetylene to form acetaldehyde is a conventional process which frequently utilizes as the catalyst for the reaction calcium orthophosphate in admixture with a minor concentration of copper orthophosphate. One of the problems involved in this process is the fairly rapid decrease in the activity of the catalyst. In order for the process to be economically feasible, the catalyst must have an acetylene conversion of more than about 70 mol percent.

A fresh calcium orthophosphate-copper orthophosphate catalyst, when properly made, has an activity of more than 98 mol percent conversion of acetylene. Even with catalysts of this high activity, the conversion drops to below about 75 mol percent conversion within 18 to 20 hours of operation and is considered "dead." Activity is temporarily restored by regenerating the catalyst with a stream of air and steam at elevated temperature in the range of 700° to 850°F. The regeneration period is about 4 hours. Eventually, the catalyst cannot be restored by air-steam regeneration to an activity substantially above about 70 to 75 mol percent conversion and must be reactivated by other means or discarded.

The process is concerned with a method of restoring an apparently dead catalyst consisting essentially of copper phosphate in admixture with calcium orthophosphate which is not restorable with air and steam.

In somewhat more detail, the process comprises contacting a calcium phosphate-copper phosphate catalyst with an aqueous solution of a copper compound capable of exchanging copper for calcium in the calcium orthophosphate and producing a water-soluble calcium compound which is then washed out of the composite catalyst. The preferred copper compound is copper acetate, $Cu(C_2H_3O_2)_2$, but other copper compounds effective in the process include the nitrate, chloride, chlorate, bromide, iodide, formate, and salicylate. The catalyst is frequently utilized in pelleted form such as 1/8" x 1/8" pellets, although 1/16" x 1/16" up to 1/4" x 1/4" pellets may be utilized. In reactivating the catalyst in pelleted form, the pellets may simply be immersed in the dilute aqueous solution of the selected copper compound for an extended period such as 4 to 24 hours, or longer. However, it is preferable and more effective to crush or grind the pellets or otherwise communite them to small particle size and soak the comminuted catalyst in the copper solution. The reactivated particulate catalyst is then repelleted in any suitable manner for use.

HYDROGEN MANUFACTURE

Catalysts & Chemicals

A process developed by R.E. Reitmeier and H.W. Fleming; U.S. Patent 3,388,972; June 18, 1968; assigned to

Catalysts & Chemicals Inc. yields catalysts useful for water gas shift reactions, the reaction of carbon monoxide with steam which are favored by low temperature operations, but occasionally are subjected to abnormal temperature increases. Increases in temperature have been injurious to low temperature shift catalysts. A shift catalyst suitable under such conditions contains copper, zinc oxide and alumina. The zinc to copper weight ratio is 0.5 to 3 zinc to 1 copper, and the catalyst contains 1 to 55% alumina based on the copper oxide-zinc oxide-alumina precursor.

Imperial Chemical Industries

A process developed by G.W. Bridger, D.O. Hughes and P.W. Young; U.S. Patent 3,514,261; May 26, 1970; assigned to Imperial Chemical Industries Limited, England involves producing carbon dioxide and hydrogen by reacting carbon monoxide with steam in the presence of a catalyst comprising the product obtained by reducing the mixed oxides of copper, zinc and chromium.

The catalyst for the reaction of carbon oxides with hydrogen contains copper, zinc and chromium, preferably in a ratio falling within the area substantially defined by the following points on the triangular phase diagram:

Cu	Zn	Cr
95	4	1
15	84	1
20	70	10
20	30	50
45	5	50

and especially by a perimeter lying within the above perimeter and passing through the points:

Cu	Zn	Cr
90	8	2
25	60	15
25	45	30
90	6	4

The catalysts may contain support materials, diluents or binding materials, of types well known in catalyst-making. These however do not appear to be essential, very satisfactory results being obtained without them. The method of producing such a catalyst comprises coprecipitating from solution copper, zinc and chromium as one or more compounds readily convertible to oxides, under such conditions that at the end of the coprecipitation stage at most a minor part of the precipitated copper compound has decomposed.

The nature of the decomposition of the copper compound is illustrated by the following description. When a solution of the mixed nitrates of copper, zinc and chromium is added to a solution of sodium carbonate, a precipitate is formed which initially is flocculent and of a royal blue color. The precipitate rapidly changes to the pale bluish or greenish condition which is characteristic of the above-mentioned compound, then to a khaki color which may darken further. The change to the khaki color and then to darker colors is the result of decomposition which is to be avoided. It is believed that copper oxide is one of the products of this decomposition and that it is especially important to avoid the formation of copper oxide during the coprecipitation stage.

The coprecipitation is preferably carried out in the presence of an excess of the acid radical of the compound readily convertible to an oxide. Thus if a batchwise method is employed, the solution of the copper, zinc and chromium is preferably added to the solution of that acid radical. Conditions tending to decrease decomposition of the copper compound appear to include the following:

(a) Low temperature, for example, room temperature; and
(b) short time of contact of the copper compound with alkaline solutions, especially at pH values above about 9.5.

It is preferred to use condition (b) while keeping the temperature relatively high, for example, at between 80°C. and boiling point. The short time of contact of the copper compound with alkaline solutions is of special importance when the compound is a salt of a weak acid, for example, a hydroxide or carbonate, and is formed by adding the copper salt, without or with the zinc and/or chromium salts to an alkali metal hydroxide or carbonate.

The short times of contact with the alkaline solution may be achieved, for example, by adding the copper salt, without or with the zinc and/or chromium salt, to a vessel and at the same time adding the alkaline solution at such a rate that the alkali concentration is continuously and rapidly reduced to a low level, corresponding to a pH less than 9.5, preferably less than about 8, for example between 7.5 and 8.0.   The pH values quoted are measured at room temperature. As an alternative, the two solutions may be mixed at a flowing junction, the relative rates of flow being adjusted as for simultaneous addition to a vessel.

By making the catalyst by this method, it is found possible to increase the proportion of copper to zinc and chromium in the catalyst in such a way that the activity of the catalyst, for example in the carbon monoxide–steam reaction and in the methanol synthesis reaction, is increased. When the attempt is made to make catalysts of high copper proportion, that is in which the atomic ratio of copper to zinc + chromium is at least 1:1, for example containing 60% of copper by atom, without taking precautions to prevent the above-mentioned decomposition of the copper compound, the resulting catalysts are somewhat inhomogeneous and are no more active than catalysts of a considerably lower copper proportion. The method may also be used to make a catalyst of lower copper proportion, for example 30% by atoms of the total copper + zinc + chromium content.

After the coprecipitation stage, the precipitate should be to a large extent freed of electrolytes, for example by washing. Suitably the electrolyte content, calculated as sodium oxide equivalent, is less than 1.0%, especially less than 0.1% by weight of the dry solids present. By "dry solids" is meant solids stable at 900°C. The removal of the electrolytes is made easier if the precipitates and mother liquor are heated to a temperature higher than the precipitation temperature, before being separated. After this heating stage but before separation a final adjustment of pH may be made if desired. Alternatively, or additionally, the separated precipitate may be heated with one or more changes of washing water. These wet heating stages should be carried out with care otherwise the activity of the catalyst may be decreased to some extent.

After being washed the precipitate is dried, conveniently at such a temperature, for example 105° to 150°C., that not more than half the total copper compound present is converted to copper oxide. The dried material is calcined at, for example, 200° to 300°C. to convert it at least partly to the mixed oxides. The calcined material may be ground finely for use in a fluid bed reaction or may be formed into pieces by, for example, pelleting under pressure using graphite as lubricant. It may also be granulated or extruded and binding agents may also be added.

Before the oxide mixture can show its full activity as a catalyst it should be partly reduced. This may be conveniently effected by passing a reducing gas, for example hydrogen or carbon monoxide, preferably diluted with an inert gas such as nitrogen or steam, at atmospheric pressure over the oxide mixture at temperature preferably in the range of 120° to 250°C. When the inlet gases of the process which the catalyst is to catalyze have reducing properties, the reduction of the oxide mixture may be effected by these gases preferably suitably diluted in the plant in which the process is to be carried out.

The catalysts made by the method described are highly valuable for the reaction of carbon monoxide with steam, for the synthesis of oxygenated hydrocarbons, especially methanol from carbon oxides and hydrogen, especially as described in U.S. Patent 3,326,956 on June 20, 1967 in the name of P. Davies and F.F. Snowdon, and for organic hydrogenation and dehydrogenation reactions at temperatures up to about 300°C.

HYDROGENATION

Kao Soap

A process developed by B. Miya; U.S. Patent 3,267,157; August 16, 1966; assigned to Kao Soap Company, Ltd., Japan relates to a method of remarkably increasing the activity of copper series catalysts to be used in organic chemical reactions.

There are known methods of activating catalysts wherein a promoter is added, a mixed catalyst is made, a mutually promoted catalyst or an active catalyst is made by mixing two inactive substances, and where a carrier is used so that the surface area of the catalyst may be increased. However, it usually happens that the activity of such catalysts diminishes during use. The activity of some of such catalysts cannot be restored. The activity of other catalysts can be partially restored, for example, by a method where carbon deposited on the surface of the catalyst is oxidized and removed by feeding air thereto. However, in such cases, the initial activity also will not be completely restored. It is all the more impossible for the activity of the catalyst to attain a value higher than its initial one.

An object of the process is to provide a method of increasing the activity of copper series catalysts to be used in organic chemical reactions, specifically in high temperature, high pressure, hydrogen reduction reactions which are used primarily for converting oil and fat series carbonyl radicals into hydroxyl radicals, reducing a fatty acid ester to an alcohol and converting unsaturated bonds to saturated bonds, before or during the reactions, by a method entirely different from the methods mentioned above. In a hydrogen reduction reaction for converting an oil and fat series carbonyl radical into a hydroxyl radical or a reaction for reducing a fatty acid ester to an alcohol and converting unsaturated bonds to saturated bonds, such low boiling point compounds as water, methanol, ethanol, n- or i-propanol or n-, i-, sec- or tert-butanol will be produced during the reaction. It has been discovered that when either (1) the low boiling point compound produced during such reaction is discharged out of the reaction system together with hydrogen at the time of a high temperature in the reaction system, or (2) when a low boiling point compound (which may be identical with or different from the one produced by the reaction) is added to the reaction material in advance and is discharged out of the reaction system together with hydrogen or hydrogen and the low boiling point compound produced by the reaction at the time of a high temperature in the reaction system, the activity of the copper series catalyst used in the reaction will be remarkably increased. The low boiling point compound and the hydrogen are removed from the reaction system by flashing, that is, by opening a valve to place the pressure vessel in communication with the atmosphere.

The reaction velocity, after the activation of the catalyst according to the process is carried out, will be several times as high as would be the case if such activation were not carried out. Further, when such treatment is repeated, the reaction velocity will increase still further. When such a treatment was carried out twice, the reaction velocity became 13.3 times as high as it was before such treatments. When the amount of the low boiling point compound added to the reaction material in advance is increased, the reaction velocity will further increase. Thus, in one example, the reaction velocity reached the surprising value of 23 times as high as it was before undergoing treatment according to this process.

Nippon Oil

A process developed by T. Ohmori; U.S. Patent 3,555,106; January 12, 1971; assigned to Nippon Oil Company Ltd., Japan provides for a catalyst composition containing an inert catalyst carrier formed by calcining a mixture of alumina and silica at a temperature below 850°C. onto which carrier are dispersed copper and nickel as active metal components. The weight of the copper exceeds the weight of the nickel, the weight of the carrier exceeds the weight of the active metal components, at least 25% by weight of the active metal components being in the metallic state, and the remaining percentage being in the form of their oxides. The process also provides for the use of the aforementioned catalyst for selectively hydrogenating acetylenic hydrocarbons.

In the process of thermal cracking, catalytic cracking, or dehydrogenation of petroleum hydrocarbon fractions, hydrocarbon fractions containing a large quantity of diolefinic hydrocarbons, that is, butadiene, isoprene, piperylene, and the like, are produced. Although these fractions usually include saturated hydrocarbons, monoolefinic hydrocarbons, and acetylenic hydrocarbons besides diolefinic hydrocarbons, they are used as raw material for petrochemical products by separating or refining the diolefinic hydrocarbons in accordance with a well known process. In this case, the coexistence of acetylenic hydrocarbons with diolefinic hydrocarbons is extremely undesirable, and it is well known preliminarily to remove acetylenic hydrocarbons for the upgrade of the quality of diolefinic hydrocarbons.

Heretofore, as a process for removing acetylenic hydrocarbons from coexisting diolefinic hydrocarbons, a hydrogenation refining process had been proposed, and this process has been industrially applied. For example, a selective vapor phase hydrogenation process (U.S. Patent 2,426,604) has been known, in which a catalyst consists essentially of between 85 and 99.5% by weight of copper intimately admixed with between 15 and 0.1% of a different metal, the oxide of which is reducible to the metal with hydrogen at a temperature below 550°C., both of the metals being dispersed on an inert porous supporting material, and the process comprises heating the vapors of a hydrocarbon fraction in the presence of the catalyst at a reaction temperature below 300°C.

For example, 50 to 60% by weight of 1,3-butadiene, 38 to 45% by weight of butenes, about 2% by volume of acetylenic hydrocarbons, and about 4% by volume of hydrogen are used as a raw gas, and a binary mixed catalyst consisting of copper and another metal selected from the group consisting of nickel, silver, cadmium, titanium, iron, vanadium, zinc, etc., is used, and in this case, about 2% by volume of acetylenic hydrocarbons in the raw material are decreased to approximately 0.01 to 0.08% by volume therein after the reaction.

However, these processes as mentioned above have such disadvantages as that when the acetylenic hydrocarbons are subjected to hydrogenation, at the same time, conjugated diolefinic hydrocarbons such as butadiene are also subjected to hydrogenation, and consequently, the loss of useful materials is increased, and the activity of the catalyst is apt to decrease.

It is an object of the composition to overcome the disadvantages of many prior-art catalysts comprising copper as a principal constituent, such as inferior selectivity, and the susceptibility to decrease of the hydrogenation activity for acetylenic hydrocarbons.

In the production of the catalysts according to this process, generally copper salts and nickel salts are separately applied on the carrier in the form of an aqueous solution or aqueous ammonia solution thereof as the means for applying the active metals onto the carrier. However, other processes may be employed such as a process of impregnating a mixed solution of active metals into the carrier, a kneading process, a coprecipitation process, or a precipitation process, etc. However the employment of the impregnating process or kneading process is preferable. The surface area and pore volume of the carrier can be varied in accordance with the ratio of alumina and silica or diatomaceous earth. However, it is preferable for attaining the objects of this process to make the surface area within the range of from 10 to 250 m.^{2}/g. and the pore volume within the range of from 0.02 to 2.0 cc/g.

It is necessary to bake the catalyst in the presence of air and nitrogen or in the presence of air alone at a temperature of from that at which active metal salts or active metal hydroxides become oxides thereof, to 800°C. for 1 to 20 hours. Further, it is necessary that the copper-nickel binary mixed catalyst be subjected to a preliminary reduction with hydrogen prior to the hydrogenation reaction. The catalyst is influenced in hydrogenation activity, selectivity and durability of acetylenic hydrocarbons by the temperature of the preliminary reduction with hydrogen. It is preferable to carry out the preliminary reduction with hydrogen at a temperature in the range of from 350° to 430°C.

The hydrogenation reaction is generally carried out at a temperature of from 100° to 250°C. Referring to the quantity of hydrogen to be added to the reaction gas, it is necessary to use generally at least an equivalent of hydrogen with respect to the acetylenic hydrocarbons to be hydrogenated, and it is preferable that the reaction pressure be 5 kg./cm.^{2} g. or less. Regarding the contact time, it is necessary to vary this in accordance with the quantity of acetylenic hydrocarbons in the raw gas, and generally a gas space velocity (supplying volume at NTP of raw gas per unit time and unit volume of catalyst) is suitable in a value of from 200 to 500/hr.

OXIDATION

General Electric

A process developed by M.M. Modan; U.S. Patent 3,630,995; December 28, 1971; assigned to General Electric Co. relates to separation and recovery of copper-amine complex catalyst residues from a reaction stream in a process for the formation of polyphenylene ethers by an oxidative coupling polymerization reaction. The process comprises simultaneously terminating the reaction and extracting catalyst residue by contact of the reaction solution with an aqueous acid solution in a countercurrent, liquid-liquid extraction column. The polymer is then recovered from the reaction solution substantially free of catalyst residue. The amine is recovered from the aqueous acid solution by pH adjustment with alkali and may be reused if desired. The process is less expensive than prior art procedures and provides more effective catalyst removal than other commercially acceptable methods.

This process provides a method for terminating an oxidative coupling polymerization with an acid while simultaneously extracting catalyst residue which method uses substantially less acid per pound of polymer formed and which removes substantially all catalyst residue from the reaction solution. The method is applicable to the formation of polyphenylene ethers by an oxidative coupling reaction in the presence of a copper-amine complex catalyst. If catalyst residue were not removed from the reaction solution, it would be recovered with the polymer and contaminate the same resulting in discoloration and degradation.

Figure 11 shows a useful form of apparatus for the conduct of this process. Referring to the drawing, a reaction solution from a reactor (not shown) containing amine and copper residues along with polymer is pumped to hold tank 10 and withdrawn continuously through conduit 12 into a continuous, countercurrent, multistage liquid-liquid extraction column 14. The extraction column may be of any convenient design capable of providing a sufficient number of theoretical extraction stages to effect the desired separation of copper and amine. A conventional packed column may be used, for example, as well as a pierced-plate column, a bubble-plate column or a column containing alternate zones of quiescence and turbulence. Aqueous acid solution is fed continuously from the acid storage tank 16 through a conduit 18 into the extraction column 14.

Since the density of the acid solution exceeds the density of the polymer solution, the acid solution descends through the column dissolving the amine and copper residue in the feed while the lighter polymer solution passes countercurrently upward through the column. Polymer solution substantially free of amine and copper residue is recovered from the top of column 14 through conduit 20. The polymer may then be recovered from its reaction solution by any convenient means known in the art such as by precipitation with a nonsolvent for the polymer such as methanol.

The aqueous acid solution containing extracted copper-amine catalyst residue is withdrawn from the bottom of extraction column 14 through conduit 22 and is passed to decantation tank 24. The decantation tank is equipped with an inlet 26 for addition of an alkali, such as caustic, to bring solution pH preferably up to about 10.0.

At this pH, the amine separates out as a separate light layer that can be removed from the top of decantation tank 24 through conduit 28. In addition, copper precipitates, probably as an oxide, and the aqueous solution is passed from decantation tank 24 through conduit 30. Copper is removed in filter 32 and the acid solution may then be passed through conduit 34 to waste disposal or to an acid recovery operation.

FIGURE 11: APPARATUS FOR COPPER CATALYST RECOVERY FROM OXIDATIVE COUPLING PROCESS FOR POLYPHENYLENE ETHER MANUFACTURE

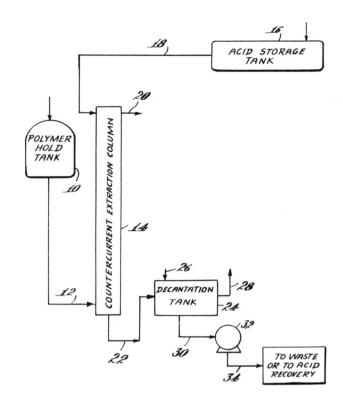

Source: M.M. Modan; U.S. Patent 3,630,995; December 28, 1971

Halcon International

A process developed by R.S. Barker; U.S. Patent 3,265,635; August 9, 1966; assigned to Halcon International Inc. involves the preparation of a lower acrolein from a corresponding lower olefin by catalytic oxidation in the vapor phase, and also to catalysts for use therein, more particularly to such a process and catalyst where the catalyst contains oxidized copper, molybdenum and chromium on a refractory support, and especially to such a process and catalyst where methacrolein is prepared from isobutylene over such a catalyst including phosphate and where the support is silicon carbide.

The lower acroleins such as acrolein itself and alpha-methacrolein are commercially important materials and several processes are known for their manufacture, however these processes leave much to be desired as to the overall yield, ease of recovery, simplicity of operation, and the like. Accordingly, the art is confronted by the problem of providing these materials from readily available low cost raw materials in an even more economical and convenient manner than is currently possible.

The process for preparing a catalyst adapted for use in the catalytic vapor phase oxidation of isobutylene to methacrolein comprises preparing an aqueous ammoniacal solution of a copper halide, a molybdate and a chromate, impregnating a refractory material therewith, drying and then activating at a temperature in the range of about 400° to 500°C. for a time in the range of about 5 to 25 hours, the proportions being such that relative to a formula weight of copper as 1.0, the formula weight of molybdenum is in the range of 0.05 to 2.0, and the formula weight of chromium present is in the range of 0.05 to 2.0. The catalyst may also contain phosphorus such that the content of phosphate in the final mixture is in the range of 5 to 25% based on the total catalyst weight exclusive of the support.

The apparatus for the effective utilization of such a catalytic material in the olefin oxidation process is shown in Figure 12. The reactor 10 consists of a vertical (4' x 1") carbon steel pipe 11 (with bottom cap 16) attached at 13 near the upper end to another (4' x 3/8") carbon steel pipe 12 to form an inverted U. The two tubes are set in an enclosed, electrically heated jacketed system 24 so that they can be heated by a liquid such as Dowtherm A, up to 360°C.

FIGURE 12:   APPARATUS USED IN OXIDIZING ISOBUTYLENE TO METHACROLEIN OVER A COPPER-CONTAINING
CATALYST

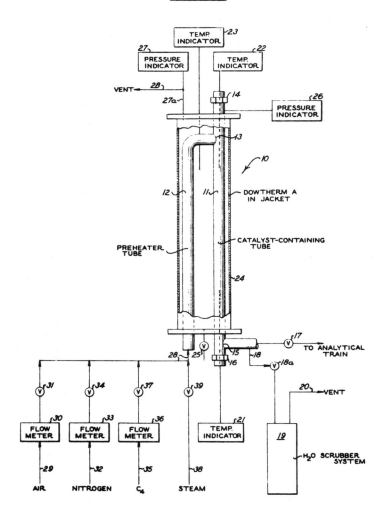

Source: R.S. Barker; U.S. Patent 3,265,635; August 9, 1966

(under about 80 psig pressure). The reactor tube is filled with catalyst (after removing cap 14) through the opening at the top. The exit tube 18 goes through valve 18a to a water scrubber 19 before venting via 20, and has a side T 17 and valve 17a for taking samples for analysis. Thermocouples 21, 22 and 23 are provided, as are pressure gauges 26 and 27 (connected via 26a and 27a, respectively). The jacket is provided with drain valve 25.

The bottom of the preheater tube is provided with manifold 28 so that different substances can be fed to the reactor. Air from a tank at 2,000 psig is fed through line 29, a flowmeter 30 and valve 31 into the tube 28. Nitrogen, when used to dilute the air and lower the oxygen concentration, is fed via line 32 through a flowmeter 33 and valve 34 to line 28. Isobutylene taken from a cylinder (via line 35) is passed through flowmeter 36 and valve 37 to the tube 28.

Steam, when used, is fed as water through a metering pump and then through heating coils to convert the water to steam, which in turn passes through line 38 and valve 39 to the tube 28. The exit side of the steam coils may have a thermo-couple to record the efficiency of the steam-producing apparatus. The manifold is held at 200°C., with the aid of an electrical heating ribbon to prevent condensation of steam in it.

A typical exit gas contains unreacted olefin, methacrolein, other carbonylic compounds, acids, carbon dioxide and carbon monoxide. Methacrolein and carbonylic materials (such as acetaldehyde and formaldehyde) and acids are caught in water scrubbers upon exiting from the reactor. Testing has shown that about 80 to 90% of all of the methacrolein is caught in the first two scrubbers and 10% of the first two amounts in the third; therefore, practically all is collected in three scrubbers. When amounts of carbonyl compounds are high, a polarographic method is used to verify amounts of methacrolein. Gas chromatographic means can also be used. Catalyst (approximately 300 cc) is prepared by impregnating water solutions of metal salts on a support, generally porous silicon carbide, then evaporating the mixture to dryness up to 120°C., then placing the impregnated support in a muffle furnace overnight (15 hours) at 400° to 500°C.

Monsanto

A process developed by R. Johnson and C.R. Campbell; U.S. Patent 3,148,210; September 8, 1964; assigned to Monsanto Company relates to the recovery of a specific mixed copper-vanadium catalyst system from the waste liquors obtained during the manufacture of adipic acid by the oxidation of cyclohexanol and/or cyclohexanone with nitric acid.

A well-known and commercial method of producing adipic acid, a valuable and widely used chemical, involves a series of steps, including: (1) the oxidation of cyclohexane in a liquid phase with air or other molecular oxygen-containing gas to a mixture of cyclohexanol and cyclohexanone at rather low conversion but high yields; (2) the separation of the unoxidized cyclohexane from the cyclohexanol and cyclohexanone intermediate reaction product; (3) the final oxidation of the intermediate material with a strong oxidizing agent such as nitric acid into adipic acid and concomitant minor amounts of other organic dibasic acids such as glutaric acid and succinic acid; and (4) isolation of the adipic acid from the by-product organic acids.

A proposed method of carrying out the nitric acid oxidation of the intermediate reaction product involves the use of a mixed catalyst system composed of vanadium and copper compounds. The adipic acid so produced is crystallized from the nitric acid oxidation product and separated from the adipic acid mother liquor. Contained in the mother liquor are the valuable catalyst compounds and soluble by-product organic dibasic acids. Heretofore economics of the process have dictated that the mother liquor be disposed of as waste, such as by burning the residual hydrocarbons, thereby losing the costly catalyst compounds.

Where vanadium compounds are used in the catalyst mixture, it has been suggested previously to recover such compounds by adjusting the pH of the waste liquor to a value in excess of 1.0 by the addition of an inorganic base such as sodium carbonate or hydroxide. The vanadium organic complex which precipitates then may be recovered by filtration and recycled to the nitric oxidation step. However, manifestly such procedure involves the addition of metal ions to the system, thereby giving rise to considerable processing difficulties. In accordance with another prior art procedure, it has been suggested to add sulfuric acid to the waste liquors and to heat the resulting mixture until substantially all the nitric acid and water have been evaporated therefrom. Thereafter the catalysts and by-products are isolated from the residue.

This latter method is not satisfactory since here again one must introduce in the system an extraneous material; and additional, expensive steps must be taken to separate the catalysts from the by-products isolated therewith. When the recovery of these valuable catalysts of copper and vanadium is attempted by first evaporating the waste liquors to dryness, a vigorous oxidation reaction occurs which renders the recovery in this manner very dangerous.

Therefore, an object of this process is to provide an improved method for recovering and reusing a copper-vanadium catalyst from waste liquor obtained in the manufacture of adipic acid by nitric acid oxidation of cyclohexanol and cyclohexanone mixtures whereby the problems normally attendant to the recovery of the catalyst are obviated.

In somewhat more detail, this improved process is carried out by subjecting the waste liquors obtained in the manufacture of adipic acid by nitric acid oxidation of cyclohexanol and cyclohexanone in the presence of copper-vanadium mixed catalyst at the point where adipic acid is crystallized and separated from other organic dibasic acids such as glutaric and succinic acids to controlled conditions inducing the nitric acid in the liquors to evaporate to such an extent that the pH thereof is raised to within the range of 1.2 and 2.2 as measured after dilution of the evaporated residue with water. The residue from the evaporation then can be diluted with water. Upon cooling the diluted residue, the vanadium component of the aforesaid mixed catalysts in the form of an organo-vanadium complex is selectively precipitated.

In a suitable manner the vanadium precipitate is isolated from the mother liquor of the diluted residue for eventual reuse in the nitric acid oxidation of cyclohexanol and cyclohexanone. Thus, the vanadium precipitate can be dissolved in dilute nitric acid, or the like, and recycled to the feed material of the aforesaid nitric acid oxidation step. Preferably the mother liquor of the vanadium precipitation and separation is further diluted with water and cooled. This cooled, diluted mother liquor is brought into intimate contact with a polymerizate having cation exchange properties to effect substantially complete removal of the ionic copper in this mother liquor. The polymerizate to which the copper ions are bound is separated from the liquid medium in a suitable manner. Finally, the copper ions are eluted from the ion exchange polymerizate, such as by contacting the polymerizate with dilute nitric acid or other suitable eluting agents. The eluted copper ions can be recycled to the feed material of the aforesaid nitric acid oxidation step.

Figure 13 is a process flow diagram of the improved recovery scheme. Adipic acid mother liquor is supplied to a liquor

FIGURE 13: FLOW SCHEME OF PROCESS FOR RECOVERY OF COPPER-VANADIUM CATALYSTS USED IN ADIPIC ACID MANUFACTURE

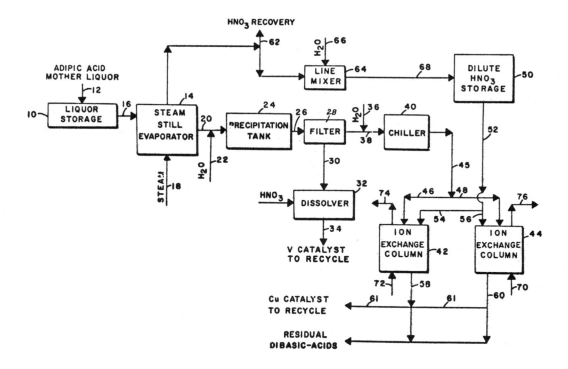

Source: R. Johnson and C.R. Campbell; U.S. Patent 3,148,210; September 8, 1964

storage zone 10 through conduit 12. The material supplied to the zone is the aqueous waste liquor obtained during the manufacture of adipic acid by the nitric acid oxidation of cyclohexanol and cyclohexanone in the presence of a copper-vanadium mixed catalyst, the copper having been added to the system as a soluble salt such as copper nitrate, or the like, and the vanadium having been added to the system as ammonium vanadate, or the like. The liquor also contains an amount of organic dibasic acids such as glutaric acid and succinic acid which are inherently produced in the afore-mentioned nitric acid oxidation procedure. Because of the differences in solubilities of adipic acid and the other homol-ogous dicarboxylic acids, adipic acid is easily isolated therefrom by simple crystallization and separation methods. The aqueous mother liquor of the adipic acid crystallization is the starting material of the process from which valuable cata-lytic compounds are recovered.

From the storage zone 10 the aqueous mother liquor is withdrawn batchwise, continuously or semicontinuously as required and passed into an evaporating means of a conventional type adapted for removing the resulting concentrate and liberated gases such as a steam still evaporator 14 via conduit 16. Steam is passed upwardly through the liquor fed to the evaporator, the steam being supplied to the evaporator by means of conduit 18. This steam sparge is continued until substantially all the nitric acid is removed from the mother liquor make.

The residue remaining after evaporation is a free-flowing liquid above 110°C. and exhibits melting points from 60° to 100°C. depending upon the dibasic acid content thereof which comprises glutaric acid and succinic acid, as well as some adipic acid. The other ingredients in the residue are the catalyst components and a small amount of water and traces amounts of nitric acid. The nitric acid may be removed from the liquor by other techniques such as by the applica-tion of vacuum and heat to the liquor, as long as the pH thereof is increased to above 1.2 as measured after dilution by the evaporization of nitric acid therefrom without the introduction of a foreign chemical. Not introducing a foreign chemical to the system for adjusting the pH provides for the recovery of the catalytic components from the adipic acid mother liquor of improved quality while, at the same time, minimizes problems normally attendant to the known catalyst recovery operations.

From the evaporator 14 the evaporator tail is passed through conduit 20. During the passage through this conduit water is added to the tail material via conduit 22, and the total stream is then passed into a precipitation zone such as provided by the precipitation tank 24. The addition of water serves to assure that the tail material composed of the organic dibasic acids, small amounts of nitric acid and copper values remain in solution if desired and the vanadium component will precipitate at a convenient temperature, thereby providing ease of material handling. In the tank or on the way thereto the temperature of the evaporated tail can be decreased to bring about precipitation of the vanadium as an organic vanadium complex. It will be apparent that the temperature to which the evaporated tail is adjusted will depend, inter alia, upon the specific composition of the diluted evaporated tail, since the optimum precipitation temperature may vary from time to time, the preferred temperature being given above. Following this precipitation, the diluted material now containing the vanadium precipitate is filtered or separated in another manner to effect removal of the vanadium bearing substance. This can be accomplished by feeding the diluted material through conduit 26 to filter 28 where the aqueous phase is separated from the vanadium-containing precipitate.

One of the several known type filters can be used for this step in the process. A rotary type of filter or one of the high speed types, such as a centrifuge, can be utilized; or a pressure-type filter can also be used. The vanadium precipitate is removed from the filter 28 in a conventional manner, and then the removed material is passed through conduit 30 to a dissolving tank 32. It is preferred to dissolve the precipitated vanadium-containing material with dilute nitric acid. The dissolved material can then be recycled by conduit 34 to the step in the production of adipic acid where cyclo-hexanol and cyclohexanone are oxidized to adipic acid with nitric acid.

The subsequent treatment of the liquid which passes through filter 28 depends upon the concentration of the dissolved material and the temperature thereof. It is usually preferred to dilute the liquid by the addition of water. This can be accomplished by supplying water through conduit 36 as the liquid is passed through conduit 38 to chiller 40. The chilling and diluting are suggested so that ordinary commercial ion exchange resins can be used conveniently, although it will be appreciated that these steps could be omitted when the resin selected is adapted to remove the copper ions from the liquid at other concentrations and temperatures.

The chilled liquid is then passed through either ion exchange columns 42 or 44 which contains a cation exchange resin in a form such as a compact bed of beads. Where two or more columns are used, of course, the recovery process may be adapted so as to be continuous. While one column can be used in a semicontinuous process, two or more columns are preferred for obvious reasons. A solid cation exchange material to which the chilled liquid is inert is employed to effect removal of the copper ions therein. The ion exchange material can be a synthetic polymerizate which will react with the copper ions in the liquid and remove same therefrom. The material should also be capable of convenient regeneration to at least part of its original activity so that it may be used over and over again and should be insoluble in the liquid containing the copper ions.

One suitable ion exchange material is a water-insoluble polymerizate of a mixture of a sulfonated polyvinyl aryl compound and a divinyl aryl compound. Such material is sold under the trademark IR-120 and DOW-50W and may be chemically identified as sulfonated polystyrene containing various levels of divinylbenzene as a cross-linking agent. The chilled liquid is passed from the chiller 40 to the ion exchange columns via conduits 45-46 or 45-48 as the case may be. The liquid which emerges from the bottom of column 42 or column 44 contains in the main the residual dibasic acids composed of glutaric acid, succinic acid and adipic acid, together with some water and nitric acid. The liquid is disposed of in some suitable manner, which disposition normally includes burning the residual acids. The copper is eluted from the ion exchange resin in a suitable manner. It is preferred that dilute nitric acid be passed downwardly through the resin in column 42 and 44 from an acid storage tank 50 by means of conduits 52-54 or 52-56.

The elution liquid containing the recovered copper catalyst component can then be recycled through conduit 58-61 or 60-61 to the step in the production of adipic acid where the mixture of cyclohexanol and cyclohexanone is oxidized with nitric acid. A convenient source of dilute nitric acid for storage tank 50 is provided from the overhead material recovered from evaporator 14. The nitric acid in the overhead material can either in whole or in part be directed to a recovery system by means of conduit 62. Alternately the overhead is diluted with water by means of a line mixer 64 provided with a water inlet line 66. The diluted acid is then passed to storage tank 50 through conduit 68. The columns 42 and 44 can be backwashed with water supplied by conduits 70 and 72, respectively, and can be withdrawn by conduits 74 and 76, respectively.

REDUCTION

Imperial Chemical Industries

A process developed by P. Davies and F.F. Snowdon; U.S. Patent 3,326,956; June 20, 1967; assigned to Imperial Chemical Industries Limited, England is a method for producing oxygenated hydrocarbons which comprises bringing a substantially sulfur-free gaseous mixture of hydrogen, carbon monoxide and carbon dioxide at elevated temperature and pressure into contact with a catalyst comprising the product of partly reducing the mixed oxides of copper, zinc and chromium. The oxygenated hydrocarbons which may be produced by the process include alcohols, aldehydes, ketones and mixtures thereof. Which oxygenated hydrocarbons are produced depends on several factors in the process operating conditions. As a general rule it may be stated that higher temperatures, higher pressures, lower ratios of hydrogen to carbon monoxide and the presence of alkali in the catalyst all appear to varying degrees to favor the formation of the higher molecular weight oxygenated hydrocarbon. The process is especially useful for the production of methanol. For producing methanol the substantially sulfur-free gas mixture preferably contains with respect to carbon monoxide at least the stoichiometric concentration and conveniently up to a five-fold excess of hydrogen, the temperature is preferably in the range of 200° to 300°C., especially 200° to 270°C., and the pressure is preferably in the range of 1 to 350 atmospheres absolute, especially 10 to 150 atmospheres, and more especially 30 to 120 atmospheres, for example, 40 to 80 atmospheres. The catalyst is substantially alkali free and preferably contains less than 1%, especially less than 0.1%, by weight of alkali metal compounds calculated as sodium oxide on the catalyst as dried at 900°C.

The space velocity at which the process is carried out is conveniently in the range of 10 to 30,000, and especially 7,000 to 20,000, hour^{-1}; these values are calculated to standard temperature 20°C. and to standard pressure 1 atmosphere and are volume space velocities, for example, liters per liter of catalyst filled space per hour. By a substantially sulfur-free gas mixture is meant a mixture containing sulfur or its compounds to the extent of less than about 10 parts per million by weight calculated as sulfur. It is preferred to have present less than 5 parts per million and more, preferably less than 1 part per million. Such gas mixtures are produced readily by modern techniques, for example, by the reaction of steam with desulfurized hydrocarbons. The ratio of carbon dioxide to carbon monoxide is preferably at least 1:200, for example in the range of 2:1 to 1:20. Higher ratios of carbon dioxide to carbon monoxide can also be used if desired. The catalyst for the reaction of carbon oxides with hydrogen contains copper, zinc and chromium, preferably in a ratio falling within the area defined by the perimeter passing through the following points on a triangular phase diagram:

Cu	Zn	Cr
95	4	1
15	84	1
20	70	10
20	30	50
45	5	50

and especially by a second perimeter lying within the above perimeter and passing through the points:

Cu	Zn	Cr
90	8	2
25	60	15
25	45	30
90	6	4

These perimeters are illustrated in the triangular phase diagram shown in Figure 14 below.

FIGURE 14: TERNARY COMPOSITION DIAGRAM FOR COPPER-ZINC-CHROMIUM OXIDATION CATALYSTS USEFUL IN METHANOL MANUFACTURE

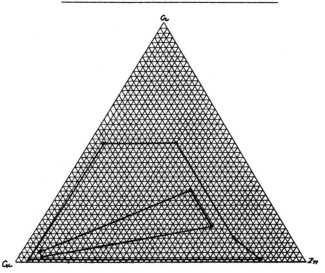

Source: P. Davies and F.F. Snowdon; U.S. Patent 3,326,956; June 20, 1967

ZINC-CONTAINING CATALYSTS

HYDROCRACKING

Consolidation Coal — U.S. Secretary of the Interior

A process developed by E. Gorin, R.T. Struck and C.W. Zielke; U.S. Patent 3,371,049; February 27, 1968; assigned jointly to Consolidation Coal Company and to the U.S. Secretary of the Interior involves regenerating spent molten zinc halide catalyst from a hydrocracking process for polynuclear aromatic feedstocks by contacting the spent catalyst with an aromatic solvent and aqueous HCl, and subsequently driving off hydrogen sulfide, water and ammonia.

It has been found that polynuclear hydrocarbons, even those which are nondistillable may be readily converted in the presence of a large quantity of molten zinc chloride or bromide, to low boiling liquids suitable for fuels such as gasoline. Some of the zinc halide is consumed by reaction with the nitrogen and sulfur compounds in the feedstock. A successful commercial process utilizing molten zinc halide catalysts must therefore provide for the regeneration of the catalyst. Figure 15 shows the overall scheme for hydrocracking and catalyst regeneration.

For ease of reference, zinc chloride, $ZnCl_2$, is used as illustrative of the catalyst, and coal extract as illustrative of a sulfur- and nitrogen-containing polynuclear hydrocarbon. Numeral 10 designates a suitable hydrocracking zone to which coal extract and hydrogen are fed through conduits 12 and 14 respectively. Regenerated molten zinc chloride is introduced through conduit 16 into the hydrocracking zone 10. The operating conditions maintained in the hydrocracking zone 10 are as follows:

Temperature, °F.	500 – 875
Pressure, psig	500 – 10,000
Liquid hourly space velocity	0.25 – 4.20
H_2/feedstock ratio, scf/lb.	5 – 50
$ZnCl_2$ catalyst (at least 15 wt. percent of hydrocarbon inventory in the hydrocracking zone)	

The hydrocracked products, i.e., low boiling hydrocarbons, are withdrawn through a conduit 18. The following reactions involving $ZnCl_2$ occur in the hydrocracking zone:

$$(1) \quad ZnCl_2 + H_2S \rightleftharpoons ZnS + 2HCl$$

$$(2) \quad ZnCl_2 + NH_3 \rightleftharpoons ZnCl_2 \cdot NH_3$$

$$(3) \quad ZnCl_2 \cdot NH_3 + HCl \rightleftharpoons ZnCl_2 \cdot NH_4Cl$$

The products of these three reactions, together with $ZnCl_2$ itself and unconverted high boiling or nondistillable coal extract, are withdrawn through a conduit 20 to a regeneration zone 22. In this zone, the products ZnS, $ZnCl_2 \cdot NH_3$ and $ZnCl_2 \cdot NH_4Cl$ must be reconverted to $ZnCl_2$. The nitrogen and sulfur portions are discharged as NH_3 and SO_2, respectively, through a conduit 24. The molten zinc chloride is returned to the hydrocracking zone through conduit 16. Figure 16 then shows in somewhat more detail the regeneration process for removing organic residue, sulfur and nitrogen compounds from the spent catalyst.

The spent catalyst withdrawn from the hydrocracking zone through conduit 20 is a single phase system for all practical purposes. In other words, the inorganic salts and the organic residue form a single phase or stable suspension. An aromatic solvent is introduced into the spent catalyst through a conduit 26. Two phases are immediately formed, an organic and an

FIGURE 15:   BLOCK FLOW DIAGRAM OF ZINC CHLORIDE CATALYZED HYDROCRACKING PROCESS

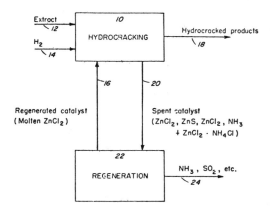

FIGURE 16:   DETAILED SCHEME FOR ZINC CHLORIDE CATALYST RECOVERY FROM HYDROCRACKING PROCESS BY SOLVENT AND ACID TREATMENT

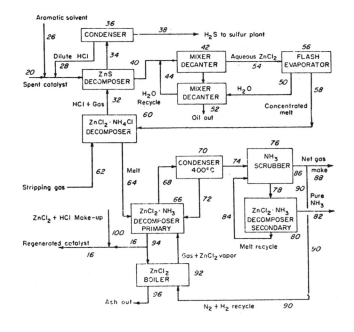

Source: E. Gorin, R.T. Struck and C.W. Zielke; U.S. Patent 3,371,049; February 27, 1968

inorganic phase. Aqueous hydrochloric acid is also introduced into conduit 20 via a conduit 28. This acid joins the inorganic phase and helps to ensure the formation of the phases, but more importantly, serves to convert the zinc sulfide in the ZnS decomposer 30 back to zinc chloride, in accordance with the equation:

$$(4) \quad ZnS + 2HCl \rightleftharpoons ZnCl_2 + H_2S$$

The temperature maintained in the decomposer is 50° to 200°C. Supplemental HCl is introduced into the decomposer through a conduit 32, as will be more fully described shortly. The product H_2S gas and any vaporous HCl are discharged through a conduit 34 to a condenser 36 where the H_2S gas is discharged separately through a conduit 38 to a sulfur plant. The vaporous HCl is condensed and returned via conduit 28 to the inlet conduit 20.

The two-phase liquid product from the ZnS decomposer is conducted through a pipe 40 to a mixer-decanter 42. Here the two phases are thoroughly washed with water introduced into conduit 40 from a conduit 44. The temperature is maintained between about 200° and 250°C. The washed phases are separated into an oil phase and an aqueous phase by decantation. The oil phase is withdrawn through a conduit 46 to a second mixer-decanter 48 where the oil phase is again washed with water introduced into the decanter from a conduit 50. The washed oil is separated from the wash water by decantation, and discharged through conduit 52. The wash water containing some zinc chloride as a result of the washing is recycled through conduit 44.

The aqueous phase from the mixer-decanter 42 is conducted via a conduit 54 to a flash evaporator 56. The water is flashed off the aqueous phase leaving a concentrated melt of $ZnCl_2$, $ZnCl_2 \cdot NH_4Cl$ and $ZnCl_2 \cdot NH_3$. The latter two adducts may not have the molar ratios indicated for the zinc chloride and the adductant. However, for convenience of reference, they are so identified. The water evaporated in the flash evaporator is conducted to the mixer-decanter via conduit 50.

The concentrated melt from the flash evaporator 56 is conducted via a conduit 58 to a suitable vessel 60 for the decomposition of the NH_4Cl adduct, in accordance with the following equation:

$$(5) \quad ZnCl_2 \cdot NH_4Cl \longrightarrow ZnCl_2 \cdot NH_3 + HCl$$

A temperature between 325° and 470°C. is maintained in the vessel 60 while a suitable inert stripping gas, introduced via a conduit 62, is passed through the melt to remove the evolved HCl. The latter, together with the stripping gas, is conducted to the ZnS decomposer via conduit 32.

The concentrated melt of $ZnCl_2$, now containing $ZnCl_2$ and $ZnCl_2 \cdot NH_3$, is withdrawn from vessel 60 via a conduit 64 and fed to a first or primary $ZnCl_2 \cdot NH_3$ decomposer 66. In this decomposer, which is held at a temperature between 525° and 600°C., the decomposition of $ZnCl_2 \cdot NH_3$ occurs as follows:

$$(6) \quad ZnCl_2 \cdot NH_3 \longrightarrow ZnCl_2 + NH_3$$

Some nitrogen and hydrogen are also produced as a result of the decomposition of NH_3. The product gases and any vaporous $ZnCl_2$ and $ZnCl_2 \cdot NH_3$ are discharged through a conduit to a condenser 70 which is maintained at a temperature in the neighborhood of 400°C. At the latter temperature, any $ZnCl_2$, and $ZnCl_2 \cdot NH_3$ are condensed and recycled through a conduit 72 to the primary decomposer 66. The ammonia and other gases are carried via a conduit 74 to an NH_3 scrubber 76 where the ammonia is separated from the gases nitrogen and hydrogen. This separation may be conveniently effected by using $ZnCl_2 \cdot NH_3$ as the scrubbing agent, in accordance with the following equation:

$$(7) \quad ZnCl_2 \cdot NH_3 + NH_3 \longrightarrow ZnCl_2 \cdot 2NH_3$$

The above reaction is effected at a temperature between 250° and 270°C. The double ammoniate $ZnCl_2 \cdot 2NH_3$ is withdrawn via a conduit 78 to a second or secondary $ZnCl_2 \cdot NH_3$ decomposer 80. The $ZnCl_2 \cdot 2NH_3$ is decomposed at a temperature of about 400°C., as follows:

$$(8) \quad ZnCl_2 \cdot 2NH_3 \longrightarrow ZnCl_2 \cdot NH_3 + NH_3$$

The gaseous ammonia, so produced, is quite pure and is discharged via a conduit 82. The ammoniate $ZnCl_2 \cdot NH_3$ in molten state is recycled through a conduit 84 to the NH_3 scrubber 76. The gas from which the ammonia has been removed consists essentially of nitrogen and hydrogen, and is removed through a conduit 86. A portion of it is discharged from the system via a conduit 88, while the remainder is recycled through a conduit 90 to a $ZnCl_2$ boiler 92.

The $ZnCl_2$ boiler 92 serves primarily to rid the system of any ash that tends to collect because of the ash content of the coal extract fed to the hydrocracking zone. Some of the $ZnCl_2$ regenerated in the primary $ZnCl_2 \cdot NH_3$ decomposer 66 is sent to the boiler via a conduit 94, while the rest is recycled through conduit 16 to the hydrocracking zone. The boiler is maintained at a temperature between 600° and 675°C. The $ZnCl_2$ is vaporized; ash drops out, and is removed through a pipe $\underline{96}$. The $ZnCl_2$ vapor is carried by the gas from conduit $\underline{90}$ back to the zone $\underline{66}$ via a conduit $\underline{98}$. Any loss of $ZnCl_2$ or of HCl in the system is made up by introduction thereof through a conduit $\overline{100}$ into the recycle line $\underline{16}$.

Another process developed by E. Gorin, R.T. Struck and C.W. Zielke; U.S. Patent 3,594,329; July 20, 1971; assigned jointly to the consolidation Coal Company and the Secretary of the Interior is one in which spent zinc chloride catalyst is regenerated by combustion in the vapor phase in the presence of a fluidized refractory solid such as silica sand. Use of a near-stoichiometric amount of air results in substantially complete removal of sulfur, nitrogen and carbon impurities, while use of about 40 to 60% of the stoichiometric amount of air results in production of a low-sulfur fuel gas. A semi-works scale apparatus for the conduct of such a catalyst recovery process is shown in Figure 17.

The melt feed is fed via line $\underline{1}$ to inlet tube $\underline{2}$, provided with rod $\underline{3}$ for mechanically clearing tube $\underline{2}$, if a plug should develop. The melt is then dropped from drip tip $\underline{4}$ in the top flange $\underline{5}$ of combustor $\underline{6}$ (3 1/2" Sch. 40 Inconel 600) into the fluo-solids bed $\underline{7}$ (bed depth: 12 to 16"). The fluo-solids are contained in mullite reactor liner $\underline{8}$ (2 3/8" I.D. x 28"), supported by mullite support $\underline{9}$. The fluidizing air supplied via line $\underline{10}$ and preheater tube $\underline{11}$ that enters the combustor at the upper flange and extends to within about 3/4" of the tip of the liner cone where it discharges into the fluo-solids bed. The top flange of the combustor is also provided with thermocouple well $\underline{12}$ and cooling air inlet $\underline{13}$ and outlet $\underline{13}'$.

Vapors from the combustor are conducted via line $\underline{14}$ to condenser $\underline{15}$ (2" Sch. 40 Inconel 600, 32" long), provided with cooling coil $\underline{16}$ and cooling air where they are cooled to a temperature of about 650° to 700°F. and where the $ZnCl_2$ condenses and HCl and ZnO, formed by hydrolysis in the oxidation zone, interact to re-form $ZnCl_2$ and H_2O. These products are collected in balance melt receiver $\underline{17}$ or in lineout receiver $\underline{18}$, both kept at temperatures of about 650°F.

Residual $ZnCl_2$ fog from the condenser is removed from the gas stream by electrostatic precipitator $\underline{19}$ or $\underline{20}$ (balance and lineout, respectively; 3" Sch. 10 x 27" long), also kept at about 650°F., i.e., above the melting point of $ZnCl_2$.

The effuent gas from the precipitators, via lines 21 and 22, is then treated according to conventional procedures, not illustrated in the figure. The effuent gas, essentially $ZnCl_2$-free, is passed through a cooler where water and some HCl are removed. A small side stream of the dry gas is then passed through an Ascarite trap to remove acid gases and then to a Beckman Model E-2 oxygen analyzer. The main stream of dry gas is passed through tandem scrubbers containing aqueous hydrogen peroxide which removes SO_2 plus HCl and aqueous sodium hydroxide which removes CO_2 from the product gas. A fraction of the offgas is diverted to a gas holder and the remainder is metered and vented.

FIGURE 17: APPARATUS FOR REGENERATION OF ZINC CHLORIDE HYDROCRACKING CATALYST BY VAPOR PHASE OXIDATION IN A FLUIDIZED SAND BED

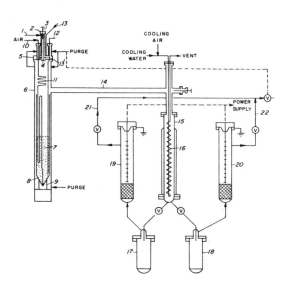

Source: E. Gorin, R.T. Struck and C.W. Zielke; U.S. Patent 3,594,329; July 20, 1971

To charge the fluo-solids bed and start a run, the thermowell of the combustor is replaced with a tube surmounted by a closed hopper containing fluo-solids. The fluidizing air and argon purge flows are then established, and the fluo-solids are charged to the reactor held at 1200 to 1400°F. After replacing the thermowell, the desired pressure is established, and the combustor is heated to about 50°F. below the desired run temperature. The feed is then started with the vapors going to the lineout train. When all temperatures are lined out and the oxygen content of the effuent gas is constant, the vapors are diverted through the balance train to start the balance period. The weight of melt fed during the balance varied from 900 to 6,700 grams, depending on the feed rate.

All products, including the fluo-solids, are collected and analyzed. Determinations are made of chlorine on the product water, chlorine and sulfur on the hydrogen peroxide scrubber effluent, and chlorine and CO_2 on the sodium hydroxide scrubber effluent to obtain the amounts of HCl, SO_2, and CO_2 collected in these materials. The scrubber gas collected in the gas holder is analyzed for H_2, CO, CO_2, SO_2, N_2 and O_2 by two-stage chromatography. The results from the water, scrubbers, and gas holder are consolidated to obtain the effluent gas composition. Material balances and elemental balances are made. The amount of ammonia decomposition is determined by the difference between inorganic nitrogens in the feed and effluent melts. The amounts of ZnS and NH_3 in the melts are determined by elemental analyses of the fractions produced by washing with water, benzene, and methyl ethyl ketone.

OXIDATION

Inventa AG

A process developed by H. Fueg; U.S. Patent 3,350,456; October 31, 1967; assigned to Inventa AG für Forschung und Patentverwertung, Switzerland is based on the discovery that satisfactory yields of cyclohexanone may be obtained by cyclohexanol oxidation when a catalyst is used which consists of zinc oxide and zinc carbonate or a mixture of zinc oxide and carbonate with alkaline earth oxides and carbonates. As alkaline earth oxides calcium oxide or magnesium oxide are preferred. An addition of graphite may be made to the catalyst which acts as a lubricant and enhances shaping.

Cyclohexanol which is obtained by oxidation from cyclohexane contains a small amount of organic acids which cannot be removed by distillation. When such cyclohexanol is dehydrogenated by prior-art processes, such as that in U.S. Patent 2,338,445, which uses a zinc-copper alloy catalyst, the conversion decreases to about 50 to 20%. When, however, dehydrogenation is carried out by this process, the conversion amounts to about 80% as is the case of dehydrogenation of cyclohexanol obtained from phenol, and the yield is about 90 to 96%. A form of apparatus which is suitable for the experimental conduct of the oxidation process is shown in Figure 18.

In the drawing, a copper tube 1 having a diameter of 30 mm. is shown surrounded by a jacket 2 of a furnace provided with heating wire coils 3 arranged along the entire length of tube 1 and capable of being supplied with up to 3000 watts. For measuring the temperature in the tube, a thermo element 4 extending into the tube is provided, whose dial for reading the temperature is shown at 4'. Liquid cyclohexanol for dehydrogenation is introduced from a container C by means of a glass centrifugal pump 5 and via a rotameter 6 into the tube 1 where the catalyst is illustrated by 15. The cyclohexanone formed in the reaction is collected in a vessel 7 surrounded by a cooling jacket 16. Hydrogen escapes from the vessel 7 and is passed through a reflux condenser 8 while entrained cyclohexanone collects in a flask 18. The effective contact space is 500 cc. The following are some specific examples of catalyst preparation for this process.

(1) Preparation of a mixed catalyst containing $ZnO + ZnCO_3 + CaO + CaCO_3$. This may be done by mixing together, with constant stirring, in sodium carbonate solution a solution of $Zn(NO_3)_2 \cdot 6H_2O$ and a solution of $Ca(NO_3)_2 \cdot 4H_2O$ in a molar ratio of 1:1. (The sodium carbonate solution is in excess.) The gel-like precipitate of $ZnCO_3$ and $CaCO_3$ is washed free of nitrate with distilled water and is filtered. The precipitate, which is then almost dry, is made into a paste, e.g. on a kneading machine. The paste is stored for some time and slowly dried. To the dry powder, a small amount of graphite is added, for instance 0.5 to 1.0% and the mass is then tableted. The tablets may have a diameter of 10 mm. and a height of 5 mm. Bulk weight 0.96. The freshly prepared catalyst contains: CO_2, 27.92%; Zn, 29.85%; Ca, 18.46%.

The $CaCo_3$ present contains $6H_2O$ bound as crystal water. As the catalyst is used in the process, it gradually loses part of its CO_2 and the carbonates of Zn and Ca are converted into oxides. At the prevailing high temperatures between about 400° and 450°C., the CO_2 content decreases after several days to 22.4% and remains fairly constant at that value. After one year in use the catalyst was found to have the following composition: CO_2, 22.46%; Zn, 32.91%; Ca, 19.70%; which corresponds to a ratio of $ZnO:ZnCO_3 = 1:0.386$ and $CaO:CaCO_3 = 1:79.5$.

(2) Preparation of a catalyst containing $ZnO + ZnCO_3$. In a manner similar to the one described under (1) above, a solution of $Zn(NO_3)_2 \cdot 6H_2O$ is introduced while stirring into a sodium carbonate solution, whereby a gel-like precipitate of $ZnCO_3$ is obtained. Washing, paste formation, drying and further processing is carried out in the manner described. The freshly prepared catalyst contains CO_2, 35.2%; Zn, 49.7%.

During the use of the catalyst at temperatures between 400° and 450°C., the CO_2 content gradually decreases so that a mixture of ZnO and $ZnCO_3$ will be present in the catalyst during the reaction in about the following ratio: CO_2, 3.3%; Zn, 79.6%.

(3) Preparation of MgO + $MgCO_3$ in combination with ZnO + $ZnCO_3$. The preparation is analogous to the one described under (1) for ZnO + $ZnCO_3$ with CaO + $CaCO_3$. Obtained is a catalyst of the original composition: CO_2, 41.0%; Zn, 30.5%; Mg, 11.4% which gradually becomes the following composition: CO_2, 2.11%; Zn, 53.1%; Mg, 19.5%.

FIGURE 18:: APPARATUS FOR PREPARATION OF CYCLOHEXANONE FROM CYCLOHEXANOL OVER ZINC-BASED
CATALYST

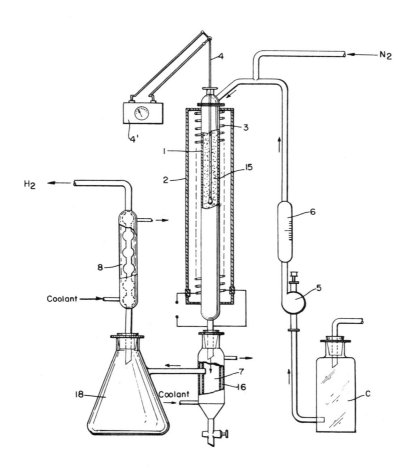

Source: H. Füeg; U.S. Patent 3,350,456; October 31, 1967

POLYMERIZATION

Esso Research and Engineering

A process developed by I. Kuntz and W.R. Kroll; U.S. Patent 3,459,721; August 5, 1969; assigned to Esso Research and Engineering Company is one in which epoxy compounds are polymerized with a catalyst consisting of (a) pure dialkyl aluminum acetylacetonate in which the alkyl group contains one to eight carbon atoms, (b) dialkyl zinc or cadmium in which the alkyl group contains one to ten carbon atoms, and (c) H_2O.

MERCURY-CONTAINING CATALYSTS

For additional details on the catalytic uses of mercury and the pollution problems attendant thereto, the reader is referred to "Mercury Pollution Control" by H.R. Jones, Noyes Data Corp., Park Ridge, New Jersey (1971).

VINYL CHLORIDE MANUFACTURE

Knapsack AG

A process developed by A. Jacobowsky; U.S. Patent 3,537,843; November 3, 1970; assigned to Knapsack AG, Germany involving the recovery of mercury from a contaminated inactive mercuric chloride/active carbon-catalyst comprises burning the active carbon with a deficiency of oxygen or air, expelling or subliming off, together with the combustion gases, a mixture of gaseous mercury, mercurous chloride and mercuric chloride, condensing the said gaseous mixture by cooling it, and, by adding a suitable reducing agent, reducing to mercury the mercurous chloride and mercuric chloride contained in the condensate.

It is known that vinyl chloride can be produced from acetylene and hydrogen chloride in contact with active carbon-catalyst generally impregnated with 10% by weight mercuric chloride. The reaction of the components is initiated at a temperature of about 80°C. which, inside a contact furnace, increases to 120° to 150°C., as a result of the exothermal reaction, and is maintained at that value by cooling. The fact that the mercury chloride is subject to sublimation gradually results in fatigue phenomena for the catalyst. These are even favored by the fact that the active carbon surface area becomes reduced by condensation and polymerization products that penetrate into the pores of the active carbon.

The catalyst will often be found to be exhausted and inactive after an operation period of 8 to 15 months, depending on the catalyst load. Such exhausted catalyst still contains between about 2.2 and 5.8% by weight mercuric chloride. This depends on the operation period and reaction conditions. The inactive catalyst is required to be removed from the contact furnace and has previously been banked out in a prepared bed of lime. The step of rejecting a mercury catalyst is disadvantageous for reasons of economy and involves the danger of ground water poisoning.

The challenge has thus been to develop a process which enables this disadvantage to be obviated and the mercury to be recovered under reasonable conditions. Attempts made earlier with the object of recovering mercuric chloride from resinified active carbon by extraction with water or acetone, have been found to be unsatisfactory from economic aspects. The reason is that no more than 0.1 or 0.7% by weight of the mercury chloride contained in the catalyst is found to have dissolved after extraction for 1 hour with the use of water or acetone as the extractant. In laboratory tests, the extraction effect has also been found to decrease as the extraction period increases, whereas, the rejected catalyst has been found to become extracted considerably more rapidly under outdoor conditions.

It has now been found that mercury in the form of a mixture comprising mercury, mercurous chloride and mercuric chloride can be recovered from inactive catalysts by burning the active carbon used as the catalyst carrier. This is a very desirable step forward in the art bearing in mind that more than 98.2% of the mercury initially contained in the catalyst is recovered and that the resulting off-gases as well as the minor amounts of slag are both free from mercury.

Figure 19 shows the apparatus which may be employed in this process catalyst recovery operation. Inactive catalyst placed in bunker 1 is allowed to travel via down pipe 2 into shaft furnace 3 provided with grate 4 and slag removing means 5. After ignition through opening 7 of the fuel gas introduced into furnace 3 through line 6 and after access of some air through line 8, the catalyst 9, which is placed on grate 4, starts glowing. The supply of fuel gas is arrested and air for combustion is supplied at increased rates. The catalyst 9 burns on the grate and steadily sinks down, fresh catalyst being supplied through down pipe 2.

With respect to the combustion process, it is possible to subdivide shaft furnace 3 into three zones comprising a combustion zone a, a sublimation zone b and a drying zone c. The combustion gases coming from combustion zone a, in which the mercuric chloride is partially reduced to mercurous chloride and mercury, travel through sublimation zone b and produce partial sublimation of the mercuric chloride contained in the catalyst. In drying zone c, the burnt catalyst is freed from water and further volatile constituents.

The combustion gases produced on burning the catalyst, which leave shaft furnace 3 through line 10, are supplied to tube condenser 11 to be cooled therein. The combustion gases contain carbon monoxide, carbon dioxide, nitrogen, mercuric chloride and mercurous chloride as well as mercury in vapor form. Both the mercury in vapor form and the mercury chlorides precipitate on the tubes and accumulate in the bottom portion of condenser 11. Condensed material depositing on the walls of condenser 11, if any, is removed from time to time by means of scraper 12. The condensates accumulating in the bottom portion of condenser 11 are supplied together with the combustion gases through line 13 to scrubbing tower 14 charged overhead with a reducing scrubbing solution 15.

In the scrubbing solution accumulating in the bottom portion of scrubbing tower 14, the mercury chlorides are reduced to mercury, which is withdrawn through outlet opening 17. The combustion gases which ascend in scrubbing tower 14 also contain mercury chloride contaminants. These are also scrubbed out and reduced to mercury. The purified combustion gases leave scrubbing tower 14 through line 16 and are burnt. The scrubbing solution is recycled through line 18 to scrubbing tower 14, preferably by means of pump 19. Consumed scrubbing liquid can be replaced with fresh scrubbing solution, which is stored in container 20, and is introduced through line 21 into the cycled scrubbing solution.

FIGURE 19:   APPARATUS FOR RECOVERING MERCURY FROM A CONTAMINATED INACTIVE MERCURY-ON-CARBON CATALYST

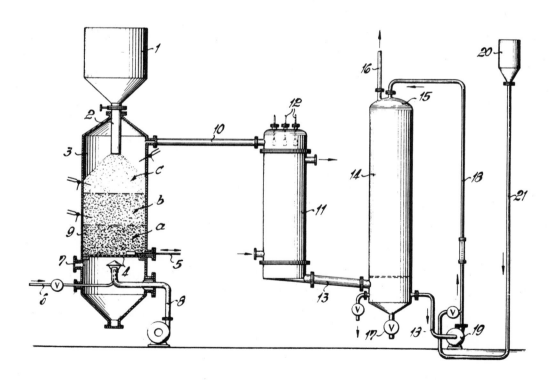

Source: A. Jacobowsky; U.S. Patent 3,537,843; November 3, 1970

ALUMINUM-CONTAINING CATALYSTS

ALKYLATION

Phillips Petroleum

A process developed by T. Hutson, Jr. and C.O. Carter; U.S. Patent 3,506,409; April 14, 1970; assigned to Phillips Petroleum Co. involves removing residual amounts of catalytic metal halides from alkylate streams by passing the alkylate through a first saturated coalescing zone and then through an active adsorption zone while the third zone is being regenerated, after which the several zones are cycled to allow the use of the regeneration zone as the active adsorber.

Figure 20 shows both the hydrocarbon alkylation and catalyst recovery steps. In the drawing, reactants and catalysts are mixed by means of stirrer 15, driven by motor 14, in reactor 13. The effluent from the reactor passes through line 16 to settler 17, wherein the bulk of the catalyst separates from the product, which could be, for example, a detergent grade alkylate. The catalyst can be recycled to the reactor via lines 18, 4, and 3, or can be drained from the settler through line 19.

FIGURE 20: FLOW SCHEME FOR ALUMINUM CHLORIDE CATALYZED ALKYLATION AND CATALYST REMOVAL FROM PRODUCT STREAM

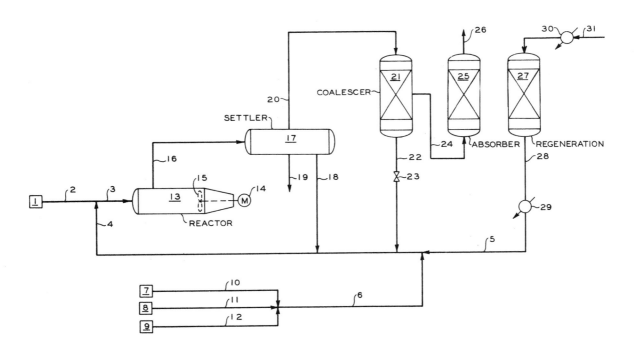

Source: T. Hutson, Jr. and C.O. Carter; U.S. Patent 3,506,409; April 14, 1970

The partially purified product material passes through line 20 to coalescing zone 21, which contains supporting material for the metallic halides having an adsorbing surface substantially loaded therewith, and is maintained under nonreactive conditions. Partially purified product passing thereto typically contains well-dispersed metallic halides which tend to coalesce and remain in the interstitial volume between the adsorbing particles in zone 21. The mixture now containing dissolved metallic halides passes via line 24 to adsorbing zone 25, which contains an adsorbing material having a non-loaded surface, and the dissolved metallic halides are adsorbed thereon. Substantially pure product passes from zone 25 through line 26.

When the adsorbing surface of the material in zone 25 becomes loaded with metallic halides, the bed of material can be switched to coalescing zone 21. When the interstitial volume of support material in zone 21 becomes plugged with metallic halides, it is necessary to remove the halides from this zone so as to regenerate the adsorbing and coalescing power of the support material. In the drawing, zone 27 is shown on the regeneration cycle, which can be accomplished, for example, by flushing a suitable solvent for the metal halides through the zone, by heating the material therein, by burning off the solvent using steam and/or air, or by any combination of these or other techniques. Heated solvent after passing through line 31, heater 30, zone 27, line 28, and cooler 29, can be recycled to reactor 13 with the entrained catalyst via lines 5, 4 and 3. Reactants, catalysts, etc., can be fed to reactor 13 from zones 1, 7, 8, and 9, via lines 2, 10, 11, 12, and 6, respectively, to initiate the operation of the system.

ISOMERIZATION

Phillips Petroleum

A process developed by T. Hutson, Jr. and C.O. Carter; U.S. Patent 3,476,825; November 4, 1969; assigned to Phillips Petroleum Company involves an alkylation process wherein AlCl3 catalyst is formed in situ, excess catalyst is used in a hydrocarbon isomerization process, the isomerized hydrocarbon is used in the alkylation process, and spent catalyst from the process is regenerated with the use of hydrogen formed in situ in the alkylation process.

Figure 21 shows the butane isomerization step of such a process. Make-up normal butane in line 80 is passed to feed dryer 82 through line 84 to feed vaporizer and superheater 86 through line 88 to vessel 90 wherein the normal butane is contacted with aluminum chloride supported on a clay bed. The isomerization vessel 90 is preferably maintained at a temperature in the range of 250° to 310°F. at a pressure of 225 psig. Under these conditions the butane conversion is about 55% with a selectivity of 91%. The aluminum chloride used is 0.092 lb. of aluminum chloride per barrel of feed. Vessels 90, 91 and 175 serve as both the isomerization reactors and the clay treating beds used in this embodiment. While vessel 90 is on stream as a reactor, vessel 91 is on stand-by and vessel 175 is on regeneration as a clay treater. In order to simplify the drawings, not all piping to vessels 90, 91 and 175 has been shown.

FIGURE 21:   FLOW SCHEME FOR BUTANE ISOMERIZATION

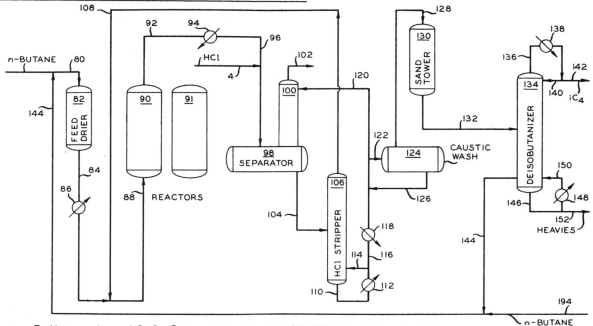

Source: T. Hutson, Jr. and C.O. Carter; U.S. Patent 3,476,825; November 4, 1969

This isomerized hydrocarbon is passed through line 92 to condenser 94 through line 96 to separator 98. Venting gases are removed through absorber 100 and through line 102. Liquid product is removed from separator 98 through line 104 and passed through HCl stripper 106 wherein HCl is removed through line 108 and recycled to line 88. Butane removed from the bottom of stripper 106 in line 110 is heated in heater 112 and a portion of the bottoms product is recycled to stripper 106 through line 114. The isomerized hydrocarbon in line 116 is cooled in heat exchanger 118 and a portion of the liquid is recycled through line 120 to absorber 100. The other portion of the cooled liquid is passed to caustic wash zone 124 and a portion of the liquid is recycled through line 126.

The wash hydrocarbon is removed from wash zone 124 through line 128 and passed to sand tower 130 wherein any caustic entrained in the hydrocarbon is removed. The hydrocarbon is passed through line 132 to deisobutanizer 134 wherein iso-butane is removed through line 136, condensed in heat exchanger 138 and returned to the deisobutanizer 134 through line 140. A portion of the condensed isobutane is passed through line 142 to be admixed with ethylene in line 14 as will be hereinafter described. Heavy ends are removed from deisobutanizer 134 in line 146 and a portion thereof is heated in heat exchanger 148 and recycled to the deisobutanizer through line 150. The other portion of the heavies are removed through line 152. Normal butane removed from the deisobutanizer 134 through line 144 is recycled to the isomerization process. Make-up hydrochloric acid for the isomerization process is added through line 4 to line 96. Then isobutane and ethylene are reacted in an alkylation step in the presence of aluminum chloride as shown in Figure 22.

The combined stream of ethylene and isobutane is passed through heat exchanger 154 to temperature condition the feed through line 156 to reactor-settler 160. Also charged to the reactor 160 is hydrochloric acid through line 16 and metallic aluminum through line 18. In the alkylation section of the reactor 160, the temperature is preferably maintained at 117°F. The reaction feed temperature is 94°F., the settler temperature is 115°F., and the reactor pressure is 100 psig. The mol ratio of isobutane to ethylene is preferably maintained at 8 to 10. Unreacted hydrocarbon is removed through line 162 from reactor-settler 160, compressed in compressors 164, passed to fractionator 168. Product alkylate is removed through line 170 and passed to coalescer 172.

FIGURE 22: FLOW SCHEME FOR ISOBUTANE-ETHYLENE ALKYLATION TO YIELD HEXANES

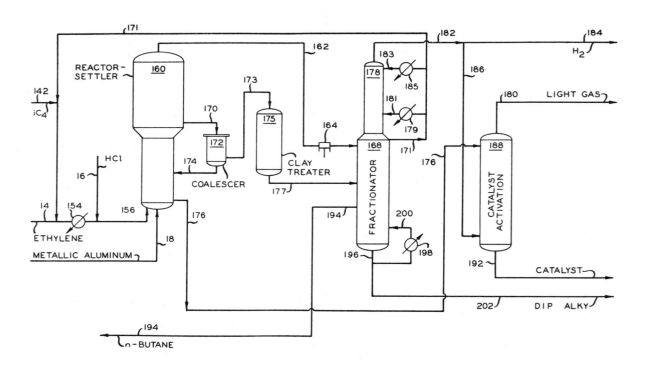

Source: T. Hutson, Jr. and C.O. Carter; U.S. Patent 3,476,825; November 4, 1969

Aluminum chloride complex, formed in situ in the reactor-settler 160 is removed from the product alkylate in coalescer 172 and returned to the reactor through line 174. Product alkylate, preferably diisopropyl, containing a small amount of aluminum chloride catalyst, is passed through line 173 to vessel 175, shown as a clay treater, wherein final traces of aluminum chloride catalyst are removed. The product, free from catalyst, is passed through line 177 to fractionator 168. In the process of this method, when vessel 175 has been saturated with aluminum chloride catalyst, it will be regenerated by being switched on stream to line 88 to be used as a butane isomerization reactor. After the aluminum chloride has been depleted from vessel 90, it is switched onto standby as vessel 91 for cooling, at which time vessel 91, depleted of aluminum chloride, is switched into line 173 as a clay treater.

A hydrogen stream is removed through absorber 178 in line 182 and a portion thereof can be removed as high purity hydrogen through line 184. Preferably, however, the hydrogen stream is passed through line 186 to catalyst activation unit 188. Isobutane removed from fractionator 168 through line 171 can be cooled in heat exchangers 179 and 185 and returned to the fractionator and absorber through lines 181 and 183, respectively. Alternately, a portion of the isobutane removed can be added to line 142 for recycle via line 171. Normal butane is removed from fractionator 168 through line 194 and is admixed with the normal butane in line 144 for recycle to the isomerization process. Alkylate product is removed from fractionator 168 through line 196. A portion of the alkylate is heated in heat exchanger 198 and returned to the fractionator 168 through line 200. The other portion is removed through line 202 as a product. Spent aluminum chloride complex from the reactor-settler 160 is removed through line 176 and passed through catalyst activation unit 188 wherein it is contacted with hydrogen as will be hereinafter described. Hydrogen off-gas containing 25% hydrogen is removed through line 180 and can be used in the hydrogenation of benzene to cyclohexane. Activated catalyst is removed from the activation unit 188 through line 192 and is passed through the hexane isomerization reactor 210 in the subsequent process step. The catalyst recovered from the alkylation step is then used in hexane isomerization as shown in Figure 23.

FIGURE 23: FLOW SCHEME FOR HEXANE ISOMERIZATION

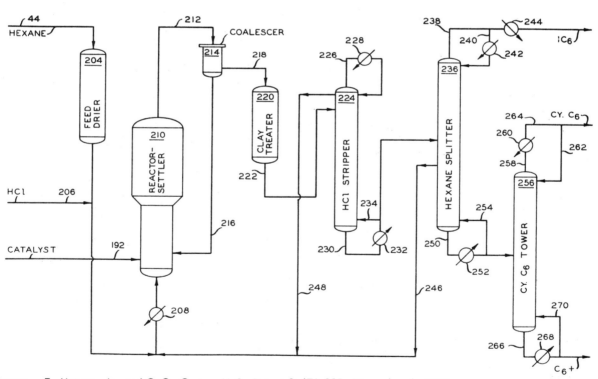

Source: T. Hutson, Jr. and C.O. Carter; U.S. Patent 3,476,825; November 4, 1969

Hexane concentrate from a benzene hydrogenation process is passed through line 44 to a feed drier 204 and admixed with HCl in line 206. The mixture is temperature conditioned in heat exchanger 208 and passed to isomerization unit 210 wherein the hexane is isomerized to produce isohexane and cyclohexane from normal hexane and methylcyclopentane, respectively. Preferably, the conditions in the hexane isomerization reactor-settler 210 include a reactor temperature of 160°F., a feed temperature of 78°F., a weight ratio of aluminum chloride to feed of 0.75 to 0.85 and a weight ratio of HCl to feed of 0.03 to 0.05, preferably 0.04. The effluent from the hexane isomerization unit 210 is passed through line 212 to coalescer 214 wherein catalyst is separated from hydrocarbon and the catalyst is recycled to reactor 210 through line 216. The hydrocarbon is passed through line 218 to a clay treater 220 for catalyst removal. The clay treater 220 can be regenerated as has been hereinbefore described with reference to vessel 175. The hydrocarbon, now free of catalyst, is passed through line 222 to HCl stripper 224 wherein HCl is removed through line 248 and recycled to the reaction. Further, the overhead stream 226 containing predominantly HCl is condensed in heat exchanger 228 and returned to the stripper as reflux therefor. Hydrocarbon is removed from HCl stripper through line 230 after heater 232 and a portion of the hydrocarbon is recycled to the stripper through line 234 to provide heat therefor. The other portion of the hydrocarbon is passed to hexane splitter 236 wherein isohexane is removed overhead through line 238. A portion of the overhead is passed through line 240 through condenser 242 and refluxed to the hexane splitter 236. The other portion is passed through heat exchanger 244 to condense the same and thereby recover isohexane.

Normal hexane is removed from hexane splitter 236 through line 246 and recycled to the isomerization process. Cyclohexane and C6+ hydrocarbons are removed through line 250, reheated in heat exchanger 252 and a portion thereof returned to hexane splitter through line 254. The other portion is passed to fractionator 256 wherein cyclohexane is removed through line 258, condensed in heat exchanger 260 and a portion thereof returned as reflux through line 262. The other portion is removed as product having 98% pure cyclohexane through line 264. C6 and heavier hydrocarbons are removed from fractionator 256 through line 266, heated in heat exchanger 268 and a portion thereof returned to fractionator 256 through line 270. The other portion is removed as a product which is useful in motor fuels. The aluminum chloride catalyst activation step using hydrogen is shown in Figure 24.

FIGURE 24: APPARATUS FOR ALUMINUM CHLORIDE CATALYST ACTIVATION USING HYDROGEN

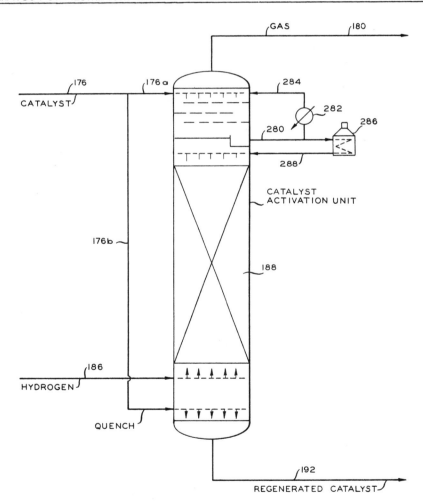

Source: T. Hutson, Jr. and C.O. Carter; U.S. Patent 3,476,825; November 4, 1969

Aluminum chloride complex which is a heavy liquid is passed through line 176a to the top portion of activation unit 188. Hydrogen is passed through line 186 to bottom portion of the reactor 188 which contains a plurality of gas-liquid contacting devices between the bottom and top thereof. A portion of the liquid complex is removed through line 280, cooled in heat exchanger 282 and returned through line 284 to the reactor as reflux. Another portion of the liquid is heated in heat exchanger 286 and passed through line 288 at a temperature of 425°F. back to the portion of the reactor below line 280. As is obvious to one skilled in the art, the hydrogen stream passes upwardly and countercurrently contacts the downwardly flowing aluminum chloride complex. The gas is removed overhead through line 180. The activated catalyst as it reaches the bottom of the reactor is contacted with a quench complex which enters the bottom of the reactor through line 176b. The regenerated complex is removed through line 192. Preferably, the temperature of the reactor is maintained at 420°F. and the pressure is maintained at 1,000 psig. The ratio of hydrogen (scf) to barrels of complex is 2,500 and the feed temperature is 100°F.

Standard Oil

A process developed by H.L. Muller and T.P. Li; U.S. Patent 3,427,254; February 11, 1969; assigned to Standard Oil Company (Indiana) is a recycle isomerization process for conversion of low octane hexane isomers into dimethyl butane, primarily neohexane (2,2-dimethyl butane), which can be used as a high octane gasoline blending component. The process is also suitable for isomerizing pentanes or mixed pentanes and hexanes. Heptanes may also be present in the feed, however, catalyst life is shortened as the concentration of heptanes in the feed is increased.

In this process the initial hydrocarbon conversion activity of a catalyst which has been regenerated is improved by treating the regenerated catalyst with hydrogen gas at an elevated temperature and pressure. The catalyst consists essentially of the reaction product of aluminum chloride with the surface hydroxyl groups of a surface hydroxyl-containing adsorbent solid. The isomerization process includes contacting the regenerated catalyst with gaseous anhydrous hydrogen chloride at a pressure in the range of 15 to 500 psi for a time sufficient to permit essentially complete association of the hydrogen chloride with the catalyst. The hydrogen chloride contacting must take place prior to contacting the regenerated catalyst with the hydrocarbon to be converted. Figure 25 shows the overall process.

FIGURE 25:   FLOW DIAGRAM FOR HEXANE ISOMERIZATION AND ALUMINUM CHLORIDE CATALYST REGENERATION

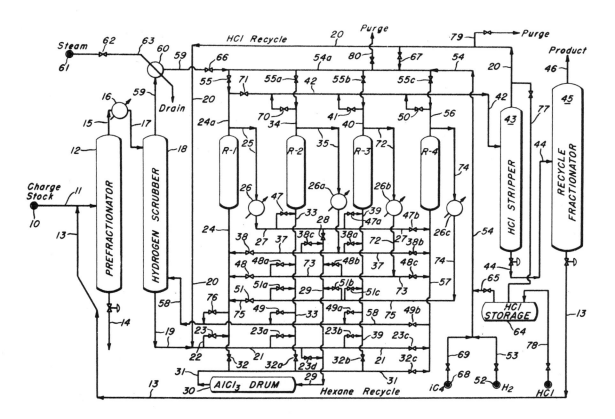

Source: H.L. Muller and T.P. Li; U.S. Patent 3,427,254; February 11, 1969

Charge stock from source 10, which has been pretreated to reduce sulfur, olefin, and aromatic concentrations to acceptable levels, is charged via line 11 to prefractionator 12 along with hexane recycle from line 13. A bottom stream containing naphthenes not required as cracking inhibitor, any heptanes and heavier portion from the fresh feed, and a small quantity of hexanes are withdrawn via valved line 14. This bottom stream can be reformed or blended directly into gasoline. The prefractionator overhead, which contains most of the hexane isomers and sufficient naphthenes to inhibit cracking in the isomerization section, is passed via line 15, through cooler 16 wherein the temperature of the overhead stream is reduced to 50°F. The cooled feed stream is then passed via line 17 into the upper portion of a hydrogen scrubber 18. The feed stream flows downward through the hydrogen scrubber countercurrently to a rising hydrogen stream, scrubbing from the hydrogen stream any hydrocarbons contained therein which boil above butane. This hydrogen stream is used for regeneration as discussed below. The feed stream passes from the hydrogen scrubber through line 19, is joined by HCl recycle from line 20, and is passed to the reaction section.

For the purpose of illustration, a process employing four reactors is shown and flow through the reaction section will be discussed as though reactor R-4 is being regenerated, reactor R-1 is the next reactor to be regenerated and thus is in the first position, reactor R-2 is the freshly regenerated reactor and thus is in the second position, and as though reactor R-3 was regenerated prior to reactor R-2, following the last regeneration of reactor R-1, and thus reactor R-3 occupies the third and last position in the series of reactors on process. The combined fresh feed-recycle HCl stream is passed into the feed manifold 21, then through line 22 and valve 23 is opened and valves 23a, 23b, 23c, and 23d are closed. The process stream passes upwardly through line 24 to reactor R-1, leaving through lines 24a and 25 into cooler 26 wherein the exothermic heat of reaction is removed. The cooled reactor R-1 effluent then passes into R-1 effluent manifold line 27, then through valve 28 into aluminum chloride drum inlet line 29. The process stream passes through the aluminum chloride drum 30 into the aluminum chloride drum outlet manifold 31, then through valve 32a into reactor R-2 inlet line 33 and then into reactor R-2.

The process stream then flows from reactor R-2 via lines 34 and 35 through cooler 26a into reactor R-2 outlet manifold line 37, valve 38a and reactor R-3 inlet line 39 into reactor R-3. The process stream is passed from reactor R-3 via line 40, and valve 41 into reaction section outlet manifold line 42, thence into HCl stripper 43. HCl is stripped from the reaction section effluent and is recycled via line 20 to reaction section inlet line 21. HCl-free effluent is then passed via valved line 44 into the recycle fractionator 45 from which dimethylbutane product is removed via line 46. The hexane recycle stream removed from the bottom of the recycle fractionator contains diisopropyl, methylpentanes, normal hexane and cycloparaffins as well as a trace of C_7+ material. This bottom stream is recycled via valved line 13 to prefractionator feed line 11.

Reactor R-4 is isolated from the process stream for regeneration by closed valves 47b, 38b, 48c, 23c, 32c, 50, 51, 51a, 51b, and 51c. Hydrogen from source 52 is introduced to reactor R-4 via valved line 53, line 54, hydrogen inlet manifold 54a, valve 55c and line 56. Hydrocarbon is drained from reactor R-4 through line 57, valve 49b and line 58 into the hydrogen scrubber 18. The hydrocarbon from reactor R-4 joins the feed stream in the hydrogen scrubber 18 and enters the reaction section via lines 19 and 21. Alternatively, a surge drum may be provided into which reactor R-4, or any other reactor, may be drained prior to regeneration, and from which the reactor may be refilled prior to being returned to process.

The use of such a surge drum will prevent upsetting the operation of the HCl stripper by intermittent variation in load when a reactor is drained or refilled. Scrubbed hydrogen passes from the hydrogen scrubber via lines 59 through steam heater 60 and is recycled by a compressor (not shown) into line 54a and into reactor R-4 via valve 55c and line 56. The recycle hydrogen stream is heated by introducing steam from source 61 through valve 62 and line 63 into the steam heater 60. The hot recycle hydrogen stream in turn heats the catalyst in reactor R-4 to the preferred regeneration temperature of 250° to 300°F. When the catalyst in reactor R-4 reaches the desired temperature, steam valve 62 is closed and the hydrogen flow is stopped by stopping the recycle compressor and closing valves 55c and 49b.

The hot catalyst is allowed to stand in the presence of hydrogen under pressure for a time sufficient to complete regeneration of the catalyst. If desired a small amount of HCl may be present with the hydrogen. Alternatively isobutane from source 68 may be introduced with the hydrogen via valved line 69 and line 54 to aid in removing olefinic contaminants from the catalyst by alkylating them. It has been found best not to include HCl and isobutane simultaneously. The time required for the regeneration is normally in the range of 6 to 72 hours. After the catalyst regeneration is complete, valves 55c and 49b are again opened, the recycle compressor started and the catalyst cooled to 100° to 150°F. by recycling cool hydrogen via lines 59, 54a, 56, 57, and 58. Although not necessary, it is preferred that the recycle hydrogen used for cooling the catalyst contain a small amount of HCl to initiate HCl treatment. The HCl can be introduced into the recycle hydrogen stream from HCl storage drum 64 via valved line 65 and line 54 into the hydrogen line 54a.

After the catalyst in reactor R-4 is cooled to the desired temperature, usually 75° to 125°F., the reactor is depressured by stopping the recycle hydrogen compressor, closing valve 55c and releasing the hydrogen from the reactor via valve 49b and line 58 into the hydrogen scrubber and then through lines 59, 54a and valve 80 to vent. Valve 66 is then closed and the reactor is pressured to 150 to 250 psi with HCl from recycle line 20 via valved line 67. The catalyst in reactor R-4 is allowed to stand in the presence of HCl under pressure. A time of 5 to 15 hours, more or less, is normally sufficient to complete the HCl treatment. When the HCl treatment is completed, valved line 67 is closed and the reactor is again depressured via line 57, valve 49b and line 58 into hydrogen scrubber 18 where the HCl is absorbed in reaction section feed. Valve 49b is then closed and the reactor R-4 is then filled with liquid by opening valves 48c and 50. When the reactor is filled with liquid, valves 48c and 50 are closed and then R-4 is put on stream and another reactor is removed from process for regeneration of the catalyst therein.

As has been pointed out above, HCl is released from the catalyst during the heating step of the regeneration sequence. If it is desired to prevent the HCl concentration from building up excessively in the process stream during the time a reactor is undergoing hydrogen treatment, HCl may be withdrawn from the system via valved line 77 into HCl storage drum 64. Make-up HCl is added to HCl storage drum 64 via valved line 78. If contaminants build up excessively in the HCl recycle stream, a portion of this stream can be purged from the system via valved line 79. Likewise hydrogen can be purged from the system via valved line 80.

CATALYST REGENERATION

The use of Friedel–Crafts type catalyst such as aluminum halide for the conversion of hydrocarbons either alone or in the presence of such added promoters as hydrogen halide, organic halide etc., is well known. Moreover, it is often desirable to modify the catalytic activity of the aluminum halide by interacting the catalyst with a hydrocarbon to form a light complex. In practically all of these processes, the aluminum halide catalyst is gradually converted to heavy aluminum halide-hydrocarbon sludge. The catalyst apparently forms complex compounds with the hydrocarbons undergoing treatment, and in doing so its catalytic activity is diminished or eliminated. The sludge is an exceedingly complex mixture of highly olefinic, conjugated and cyclic hydrocarbons (the hydrocarbon is sometimes described as a conjunct polymer) formed by a combination of reactions such as polymerization, hydrogen transfer and cyclization. Throughout this specification and claims, the terms "aluminum halide-hydrocarbon sludge," "aluminum halide sludge" and "sludge" are all intended to designate the reaction product of an aluminum halide catalyst with a hydrocarbon or hydrocarbon mixture in which the activity of the catalyst is substantially diminished or eliminated.

In commercial processes, the sludge is discarded after dilution with water to render it innocuous. Of course, the catalyst lost in the sludge and the cost of disposing of the sludge adds materially to the cost of such processes. In many cases, these high costs have hindered the commercial exploitation of Friedel–Crafts catalyst and of processes utilizing these catalysts.

Many methods have been suggested for the recovery of active catalytic material from aluminum halide sludge. Suggested methods include distillation, coking, destructive hydrogenation of the hydrocarbon complex, decomposition of the aluminum halide to aluminum oxide and hydrogen halide, etc. One of the more attractive processes is the hydrogenation of the hydrocarbon complex.

Esso Research and Engineering

A process developed by C.W. Tyson, R.J. DeFeo and W.F. Arey, Jr.; U.S. Patent 3,026,176; March 20, 1962; assigned to Esso Research and Engineering Company relates to the disposal and recovery of supported Friedel–Crafts catalysts. In well-known processes such as paraffin isomerization in which aluminum halide catalysts such as aluminum bromide or aluminum chloride are used in conjunction with a support such as bauxite, alumina, molybdenum oxide, clays, and the like, the catalyst gradually becomes deactivated with use. During the isomerization the catalyst probably forms hydrocarbon complexes until it is no longer effective for the intended purpose, and must be replaced.

Depending upon the economics, it may be desirable to recover the catalyst or, under other circumstances, to dispose of it. In either case, substantial difficulties are encountered. The fouled catalyst is extremely corrosive, and such disposal means as dumping would involve exposure to air and moisture, with resultant liberation of toxic hydrogen halide and air contamination. Disposal by dumping in water further creates a pollution problem. It is one of the objects of the process to provide a method of removing fouled and spent supported Friedel–Crafts type catalyst from a reaction zone in a safe and effective manner.

In accordance with the process, it has been found that the spent fouled catalyst may be converted to a substantially neutral product by charging the reactor vessel containing the catalyst with gaseous anhydrous ammonia. The latter has been found to form a neutral, difficultly hydrolyzable complex with the catalyst, thus making it safe for disposal by having it in a form that does not readily produce hydrogen halide gas on contact with the moisture in the air.

The spent catalyst in this case does not contain $AlBr_3$ in its original form as is evidenced by the lack of paraffin conversion. This $AlBr_3$ is probably tied up as a hydrocarbon complex, and in this way is deactivated. It may, however, have entered the crystal lattice of the support, or have undergone some other change. The $AlBr_3$, in whatever form it exists, cannot be removed by simple hydrocarbon washing, whereas pure $AlBr_3$ is quite soluble in hydrocarbons. The anhydrous ammonia, therefore, reacts with this mixture of $AlBr_3$ and hydrocarbon, forming a neutral complex.

In accordance with another embodiment of the process, the ammonia-treated spent catalyst may be effectively recovered and converted to the active species. Figure 26 illustrates the process as applied to the isomerization of a light virgin C_5/C_6 naphtha to form high octane motor fuel. To carry out this reaction, reactor 2 has been packed with a support, preferably alumina or a calcined bauxite Porocel, and a stream of C_5/C_6 naphtha in which 0.1 to 10% $AlBr_3$ has been passed over the bed at a temperature of 80° to 300°F. HBr had been used as a reaction activator, and $AlBr_3$ was deposited upon the support until 10 to 75% by weight thereof comprised $AlBr_3$, and a substantial equilibrium had been attained. Gradually reaction yields, which had been up to 92% of feed, decreased, due to the catalyst deactivation referred to above.

FIGURE 26:   APPARATUS FOR ALUMINUM CHLORIDE RECOVERY FROM SPENT SUPPORTED CATALYST

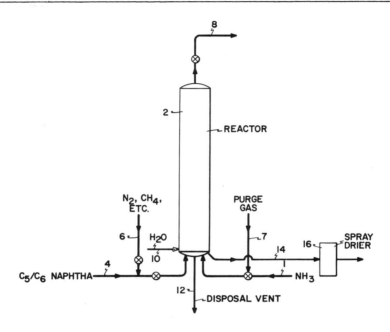

Source: C.W. Tyson, R.J. DeFeo and W.F. Arey, Jr.; U.S. Patent 3,026,176; March 20, 1962

In accordance with the process, the flow of feed hydrocarbons through line 4 is now interrupted, and preferably a purge gas, such as a light hydrocarbon, N_2, methane and the like, is passed through lines 6 and 4 to remove traces of feed and any moisture. Purge gases may be removed through line 8, and, if desired, purging may be followed by evacuation. Thereupon, gaseous ammonia is passed into vessel 2 through line 1 and the pressure is allowed to reach at least atmospheric and up to 1,000 psig.

The reactor is now allowed to stand under ammonia pressure for at least two hours, and up to 10 days to assure complete reaction of the ammonia and $AlBr_3$-hydrocarbon complex. The reaction may be followed with the temperature of the reactor, as the heat of reaction will diminish as the reaction goes to completion. The pressure is released, and the reactor and lines 1 and 8 are purged through line 7 with N_2, methane, etc. This eliminates the noxious vapors of excess ammonia upon dumping the neutralized catalyst. Suitable containers are now placed under the reactor, and the neutralized catalyst is released through the disposal vent to the containers. This neutral material may now be safely dumped, or treated as below to recover the active acidic component.

The recovery of the catalyst may be carried out in several ways. The solid neutral complex of the ammonia and $AlBr_3$-hydrocarbon mixture supported on Porocel may be slurry washed with water. The complex breaks down, the $AlBr_3$ is hydrolyzed to give free HBr which immediately combines with the ammonia present to give a neutral solution of NH_4Br. The solution may be evaporated or spray dried to give the solid halogen values which may be stored until required, and may be converted to HBr by treatment with an inorganic acid such as H_2SO_4. The HBr may be used to react with Al metal and form $AlBr_3$ for reuse in the catalyst system.

An alternate procedure involves treatment of the NH_4Br solution with elemental Cl_2, which liberates free elemental Br_2 from the solution. This Br_2 may then be used to form $AlBr_3$. A third method involves the heating of the $AlBr_3$-hydrocarbon-ammonia complex to break it down. Free ammonia is liberated, which may then be reused to deactivate another reactor.

The $AlBr_3$ may then be distilled and recovered. In one of the modifications of the process wherein the catalyst is to be recovered, water may be admitted through line 10 into reactor 2 and allowed to circulate either upflow or downflow over the NH_3-treated catalyst bed while the ammonia pressure on the system is maintained constant. Since the neutralization step is exothermic, a cooler (not shown) is inserted in the water recycle stream to control the temperature to any predesired maximum temperature, i.e. 100° to 250°F. For the same reason the initial addition of ammonia is at a low pressure of 5 lbs. absolute or below; pressure is raised after the water circulation is established for temperature control.

The halide content of the bed is leached out as the soluble neutral ammonia additive product. Similarly, any residual HBr is converted into ammonium bromide. The presence of gaseous ammonia insures that all parts of the system are maintained in an alkaline condition so that corrosion of steel and other parts is eliminated.

When solution of the halogen-containing compound is complete, a concentrated aqueous solution is withdrawn through line 14 and passes spray drier 16 to recover the solid halogen values which may be stored until required and may be converted into HBr by treatment with an inorganic acid, such as H_2SO_4, in a manner known per se. The reactor vessel is now purged with steam, passed through line 6 and finally swept clean of vapors by air. The catalyst support may now be removed from vessel 2 via line 12 for discard or for revivification without the hazards mentioned heretofore.

The spray dried ammonium halide complex may be used as a source of hydrogen halide to react with aluminum metal and form the aluminum halide again with evolution of hydrogen. This may be done in a solution of oil and the product after removing the hydrogen sent to the reactor to saturate a fresh charge of catalyst support. The regeneration of the hydrogen halide may be accomplished by using an acid such as sulfuric acid to free the hydrogen halide from the ammonium halide. This, of course, could be done either with a solution before spray drying or by the solid ammonium halide after spray drying. The ammonium sulfate resulting from the reaction above may be either sold as a chemical or the ammonia may be regenerated by reaction with sodium hydroxide, calcium hydroxide, or other alkali.

A process developed by R.F. Stringer and J.P. Bilisoly; U.S. Patent 3,121,695; February 18, 1964; assigned to Esso Research and Engineering Company relates to the removal of impurities from and the recovery of support material which may be reused with a Friedel-Crafts catalyst.

In accordance with the process, it has been found that support material from the spent supported Friedel-Crafts catalyst may be recovered by: (1) contacting the spent catalyst with anhydrous ammonia, (2) water washing the ammonia-neutralized catalyst support, and (3) drying the washed catalyst at temperature in the range of 600° to 1000°F., preferably 700° to 1000°F., for complete reactivation.

It is known that anhydrous ammonia neutralizes the catalyst mass. It has been thought that it reacts with the aluminum halide-hydrocarbon complex to form a neutral ammonia-aluminum halide-hydrocarbon complex. Though the actual reaction product is not material for the purpose of this process, for clarity, the ammonia neutralized material will be referred to as an "ammonia-aluminum halide-hydrocarbon complex." Water washing has been found to hydrolyze this complex causing its breakdown and release from the support. The critical drying step of this process removes the water and any residual oil and aluminum hydroxide remaining on the support and thereby reactivates the support for reuse in the reaction zone as support for the Friedel-Crafts catalyst.

Figure 27 shows the application of the process to the aluminum bromide isomerization of a light virgin C_5/C_6 naphtha to a high octane motor fuel. Reactor 2 is packed with a support, preferably alumina or a calcined bauxite (Porocel), and a stream of C_5/C_6 naphtha containing 0.1 to 10 weight percent dissolved aluminum bromide is passed over the bed at a temperature of 80° to 300°F. Hydrogen bromide is used as a reaction activator. Aluminum bromide is deposited upon the support until it comprises 10 to 75 weight percent of the total catalyst mass, and a substantial equilibrium is attained. Gradually the conversion to isomers, which in the case of C_6 paraffins may go up to 92% isohexanes based on paraffinic hexanes, decreases due to catalyst deactivation. Regeneration of the bed is then necessary.

Regeneration of the catalyst bed is initiated by recovering the hydrogen bromide in the bed by passing fresh feed via line 4, or recycle isomerized product by lines 13 and 4, through the catalyst bed. Where the isomerized product has any hydrogen bromide therein, it must be removed prior to its use as a wash.

After stripping to remove hydrogen bromide, anhydrous ammonia is passed into vessel 2 through lines 3, 6 and 4. The reaction of ammonia and aluminum bromide-hydrocarbon complex on the support is extremely exothermic, e.g., to the extent of 140 calories per gram of total catalyst when the catalyst contained 36 weight percent $AlBr_3$ based on weight of Porocel support. This large heat release may require special temperature control. The anhydrous ammonia may be withdrawn by line 8.

The preferred technique for controlling temperature during the ammonia neutralization step is to circulate naphtha by means of lines 5 and 6, pump 14 and line 4 through an interstage cooler 15 normally used in the isomerization step. The naphtha may be supplied by line 4. The naphtha circulation rate and naphtha temperature is controlled so that temperatures in

FIGURE 27: ALTERNATIVE APPARATUS FOR ALUMINUM CHLORIDE RECOVERY FROM SPENT SUPPORTED CATALYST

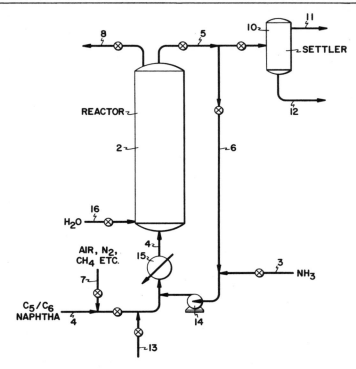

Source: R.F. Stringer and J.P. Bilisoly; U.S. Patent 3,121,695; February 18, 1964

reaction zone 2 during ammonia neutralization are maintained below 300°F. Anhydrous ammonia in amounts of 1 to 10 weight percent is blended into the circulating naphtha stream using stoichiometric amounts for complete neutralization of the aluminum bromide–hydrocarbon complex.

After ammonia neutralization, the reactor is purged by gas entering from lines 7 and 4 and being removed therefrom through line 8. Gases, such as air, nitrogen, methane and the like, are suitable. This serves to remove the last traces of ammonia and naphtha. If preferred, the gases may be heated to aid in vaporization of the naphtha and anhydrous ammonia remaining in the reactor.

The ammonia neutralized catalyst is hydrolyzed with water entering line 16 to remove from the support the ammonia–aluminum halide–hydrocarbon complex. The water, which may be at a temperature in the range of 50 to 150°F., appears to break the complex into NH_4Br, an aluminum hydroxide in gel form (which suspends in the water) and a small quantity of hydrocarbon oil. The aqueous mixture of these fragmentary products leaves the reactor through line 5 and is directed thereby to settler 10 wherein the hydrocarbon oil is separated from the aqueous mixture of NH_4Br and aluminum hydroxide gel. The hydrocarbon oil phase is withdrawn through line 11. The aqueous mixture is withdrawn by line 12 for treatment discussed in more detail hereinafter. Water washing is continued until aluminum hydroxide has been essentially removed from the support, which may be determined by a visual inspection of clarity of the wash water. A minimum of three volumes of wash water per volume of catalyst is necessary for efficient washing.

Ammonia neutralization prior to water washing is necessary for two reasons. First, the ammonia–aluminum halide–hydrocarbon complex has a pH of about 7 to 8 in aqueous solution and thus does not promote corrosion. Corrosion rates of this solution on carbon steel are very low, e.g., about 1.4 mils/year. On the other hand, directly water washing the spent supported aluminum bromide catalyst causes excessive corrosion because of the formation of hot aqueous HBr. Corrosion rates of aqueous HBr on carbon steel in excess of 500 mils/year are well known. Secondly, the combination sequence of ammonia neutralization and water washing is more effective in cleaning the support surface than water washing alone.

Catalyst wash rate must be controlled for two reasons. First, the wash rate must be sufficiently low to avoid loss of support due to entrainment of the smaller support particles and to avoid attrition. Secondly, the wash rate must be sufficiently high during the initial part of the washing to avoid formation of highly concentrated aluminum hydroxide gel. This concentrated gel, because of its high viscosity, entrains large particles of support from the bed and would carry them from the reactor.

The washed support is then dried at temperatures in the range of 600° to 1000°F. by passing a heated gas over the support by means of lines 7 and 4. The gas is withdrawn by line 8. The gas may be any inert gas, such as nitrogen, or carbon dioxide etc., air or even flue gas from the combustion of hydrocarbons with air. Drying with gases at temperatures in the range of 600° to 1000°F. is continued until the water content of the effluent gas leaving the reactor by line 8 is equivalent to the water content of the gas entering vessel 2. It has been established that the washed support must be dried to a low water content before reuse in the isomerization process and operation within this drying temperature range effects this result.

Upon completion of the drying step, the support may then be again saturated with AlBr3 by contacting the support with a hydrocarbon stream containing dissolved aluminum bromide. Of course, hydrogen bromide may also be present in the hydrocarbon solution. The hydrocarbon conversion process may then proceed again.

It is clear that the use of this process eliminates the necessity of frequent charging and discharging of the catalyst and reduces the catalyst operation cost, since the same support may be used over and over again. As by-products of the process, stream 12 may be treated in a variety of manners to recover therefrom constituents which may be reused in this process. For example, the aqueous solution of NH4Br and suspended aluminum hydroxide may be evaporated or spray dried to give the solid halogen salts which may be used for many purposes, e.g., they may be converted into HBr by treatment with an inorganic acid, such as H2SO4. The HBr may be used to react with aluminum metal and form aluminum bromide for reuse in the catalyst system. An alternate procedure involves treatment of the NH4Br solution with elemental chlorine, which liberates free elemental bromine from the solution. This bromine may be used to form aluminum bromide by reaction with aluminum.

Phillips Petroleum

A process developed by H.J. Hepp and L.E. Drehman; U.S. Patent 3,425,955; February 4, 1969; assigned to Phillips Petroleum Company involves regenerating aluminum halide containing catalysts by employing a cooling section and a scrubbing section disposed about an intermediate regeneration section, the cooling section being cooled with catalyst to be regenerated, catalyst being removed from the scrubbing zone and heated and introduced into the regeneration zone, and catalyst being removed from the scrubbing zone and cooled and reintroduced into the scrubbing zone. Figure 28 shows the apparatus which may be employed in the conduct of such a process.

FIGURE 28: APPARATUS FOR REGENERATION OF ALUMINUM CHLORIDE CATALYSTS

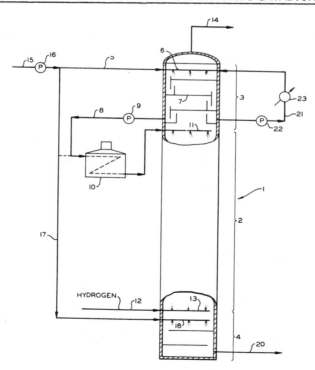

Source: H.J. Hepp and L.E. Drehman; U.S. Patent 3,425,955; February 4, 1969

In the drawing there is shown a reactor column generally denoted as 1 which has an intermediate regeneration section or zone 2, an upper scrubbing section 3, and a lower cooling section 4. Conduit 5 carries relatively cool, liquid complex into the top of scrubbing section 3 to be sprayed thereonto through sprayer 6. The liquid complex then passes over a series of conventional gas-liquid contactors 7 thereby causing that liquid to intimately contact gases rising from regeneration section 2 and scrubbing aluminum chloride carried thereby therefrom. Gas-liquid contactors can be well-known sieves, Koch trays, packing, saddles and the like or combinations of same which are well-known in the art.

Liquid complex can then pass, at least in part, out the bottom of scrubbing section 3 directly into regeneration section 2 to be contacted with hydrogen and regenerated in the conventional manner. However, it is preferred that at least some of the liquid complex in scrubbing section 3 is removed, preferably from one or more of the lowest gas-liquid contactors, through conduit 8 by pump 9, heated by furnace 10 and then introduced into regeneration section 2, preferably through sprayer 11. A still more preferred mode is to remove through 8 an amount of complex substantially equal to that amount being fed into 3 through 5, thereby preventing any appreciable, or any at all, flow of complex from 3 directly into 2 without first being heated in 10.

Also, if desired, at least some of the liquid complex in section 3 can be removed in the same manner as with conduit 8 but through conduit 21 by pump 22, cooled by heat exchanger 23 and returned to section 3 through spray 6. Liquid can be removed through conduits 8 and 21 simultaneously, consecutively or entirely separately and/or unrelatedly. The cooled liquid can be introduced to section 3 in any known manner besides spray 6 including a separate spray means.

In the regeneration section hydrogen is introduced into a lower portion thereof through conduit 12 and gas disperser 13, which hydrogen then rises through sections 2 and 3 and out of column 1 through conduit 14 for recovery, treatment, reuse, and the like as desired.

If desired, at least part of the complex pumped through conduit 15 by pump 16 passes through conduit 17 and is sprayed into the upper portion of cooling section 4 by sprayer 18 to effect cooling in that section. It should be noted that in lieu of the use of complex which is to be regenerated, regenerated complex or hydrocarbon to be treated by the regenerated complex can be used in sprayer 18. Further, in lieu of direct cooling, indirect cooling such as by use of an outer cooling jacket around the periphery of column 1 in the area of section 4 can be employed in which case a cooling fluid such as water would be used to cool zone 4 through the walls of column 1. Regenerated complex is removed from section 4 through conduit 20 for subsequent storage, further treatment, or reuse as catalysts as desired.

Regeneration section 2 and cooling section 4 can contain gas-liquid contacting devices similar to those employed in section 3. If desired, all or any part of the complex in 17 can be passed directly to furnace 10 and then directly in section 2 for regeneration thereof.

As a specific example of operation, aluminum chloride complex containing 58 weight percent aluminum chloride, the balance being substantially all hydrocarbons, is fed through conduit 15 at the rate of 88,000 pounds per day at a temperature of 100°F. Approximately 44,000 pounds per day are passed through conduit 5 and sprayed onto section 3 through sprayer 6. Similarly, 44,000 pounds per day of this complex is passed through conduit 17 and sprayed into section 4 through sprayer 18 thereby being utilized as a diluent and quench agent, which in turn adjusts the viscosities of the re-generated complex being removed through conduit 20 to any desired degree.

In a scrubbing section 3 the liquid complex to be regenerated is heated to 250°F. by contact with the effluent gas from regeneration section 2 before it passes into regeneration section 2. Approximately 44,000 pounds per day of this liquid complex is removed through conduit 8, further heated to 425°F. by furnace 10 and introduced into regeneration section 2 through sprayer 11. Similar amounts, sufficient to effect the desired cooling, are withdrawn through conduit 21 for cooling and return to section 3. About 1,200 pounds per day (2,280 cubic feet per barrel of complex treated) of sub-stantially pure hydrogen is introduced into the bottom of section 2 through gas disperser 13.

The complex which is passed through spray 18 to cooling section 4 maintains that section at a temperature below 300°F. so that the regenerated complex removed through conduit 20 is at a temperature of about 260°F. The regenerated complex is removed through conduit 20 at a rate of 77,000 pounds per day and is preferably cooled to a temperature of 120°F. before subsequent disposition. The regenerated complex in conduit 20 has an aluminum chloride content of 65 weight percent, the remainder being substantially all hydrocarbons.

About 12,250 pounds per day of overhead gas is removed from column 1 through conduit 14. This gas has an approximate composition in mol percent of 25.4 hydrogen, 27.9 methane, 22.7 ethane, 18.7 propane, 4.7 butane and 0.6 five carbon atom-containing molecules and heavier.

Regeneration zone 2 is operated during the process of the example at a temperature of 420°F. and a pressure of 1,000 psig. The complex to be regenerated which is used as the feed for column 1 can contain widely varying amounts of aluminum chloride but will generally contain no more than 60 weight percent aluminum chloride.

Although the operating conditions for the regeneration process itself can vary widely depending upon materials and equipment employed, preferred conditions are temperatures of at least 400°F., preferably from 400° to 500°F., and pressure of at least 500 psig, preferably from 500 to 1,500 psig. An amount of hydrogen equivalent to 500 to 5,000 cubic feet per barrel of complex treated is used. By use of this process and the above described procedures a regenerated complex containing from 65 to 85 weight percent aluminum chloride can be formed depending primarily upon specific conditions of operation chosen.

A process developed by F.H. Thorn; U.S. Patent 3,474,036; October 21, 1969; assigned to Phillips Petroleum Company is one in which a spent aluminum chloride catalyst is regenerated by preheating the same and introducing it into indirect heat exchange contact with a portion of catalyst already undergoing regeneration in a reactor section of a regeneration zone. The zone comprises an absorber section, a trap-out section, a reactor section, and a bottom section. Preheated catalyst is passed through a heat exchanger located within the reactor section and discharged into the reactor section. Another portion of catalyst is passed to an upper portion of the absorber section and passes downwardly into the trap-out section from which at least a portion of catalyst now contacted with vapors rising from the reactor section is removed as for reuse. Catalyst from the reactor section having been largely regenerated is collected in the bottom section of the zone and removed for reuse together with a further portion of catalyst which can be introduced into the bottoms section to cool or quench regenerated catalyst. A gas such as hydrogen is used for removing catalyst and at least a portion thereof can be introduced to the regeneration zone, for example, to the foot of the reaction section, to act as a regeneration gas and to assist in stripping vapors from the catalyst undergoing regeneration. Figure 29 shows the essentials of the process.

FIGURE 29: ALTERNATIVE APPARATUS FOR REGENERATION OF ALUMINUM CHLORIDE CATALYSTS

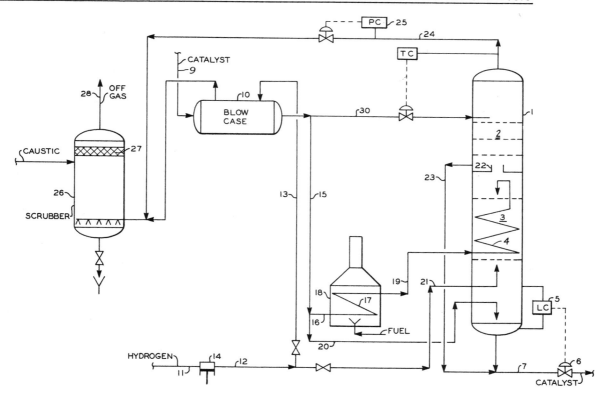

Source: F.H. Thorn; U.S. Patent 3,474,036; October 21, 1969

Referring now to the drawing, there is provided a regeneration zone having an absorber section and a reactor section, respectively identified as 1, 2 and 3. Reactor section 3 is equipped with coil 4 and is packed with contact material such as carbon Raschig rings, properly retained on a porous support means. The regeneration is provided with a liquid level control 5 operating upon valve 6 located in draw-off pipe 7.

In operation, a batch of catalyst complex from a settler as in a diisopropyl production unit in connection with which the process is now being described, is passed via conduit 9 to blow case 10 to which is supplied hydrogen by means of pipes 11, 12 and 13, and compressor 14. Under the pressure of the hydrogen, the batch of catalyst is passed by pipe 30 into the top of the regenerator or tower 1 and discharged immediately above absorber section 2. This section can contain trays, for example, cartridge sieve trays, to provide good contact between rising vapors or gases and downwardly moving liquid catalyst which has been introduced. At least a portion of the catalyst is passed by 15 and 16, through coil 17 in heater 18, and by 19 into coil 4 in section 3 of regenerator 1. The preheated catalyst is discharged from the end of coil 4 into the reactor section 3.

A further portion of the catalyst complex is passed as quench liquid by 20 into the bottom of regenerator 1. Further hydrogen is passed into the bottom of the regenerator by pipe 21. The pressure in this specific example is approximately 1,050 psia. The temperature of the catalyst in blow case is 125°F., and after preheat in coil 17 is 450° to 500°F. The distribution of the catalyst in the embodiment being described is approximately 65 barrels per day into the top, 113 barrels per day into the reactor section and 66 barrels per day into the bottom of the regenerator 1. Hydrogen passed into the regenerator system totals approximately 363,000 standard cubic feet per day of which 276,000 standard cubic feet per day passed to the bottom section of the regenerator 1. The remainder of the hydrogen passes by 13 into blow case 10. Some of this hydrogen passes, together with the catalyst, into the respective sections of the regenerator. However, some of the hydrogen also passes together with vapors from the blow case, as will appear below.

Catalyst which is passed downwardly through the absorber section is at least in part trapped out at trap-out tray 22 and passed by 23 into 7 and from the regeneration back to the catalyst settler of the diisopropyl production unit. Any overflow from trap-out tray 22 will pass downwardly through the reactor section together with preheated catalyst introduced through coil 4. In the absorber section, the catalyst introduced by pipe 30 will absorb any upwardly passing catalyst. The temperature in the reactor section is maintained at approximately 435°F. Overhead gases and any vapor are taken from regenerator 1 by way of 24 equipped with pressure controller control valve 25 to caustic treater or scrubber 26. There is also passed to this scrubber overhead gases removed from the blow case. In the caustic scrubber, the gases are treated with a solution of sodium hydroxide of approximately 15 percent concentration. The treated gases are passed upwardly through a mist extractor 27 and from the unit by 28.

Shell Oil

A process developed by H.D. Evans and R.J. Schoofs; U.S. Patent 3,210,292; October 5, 1965; assigned to Shell Oil Company involves the recovery of active catalytic material from an aluminum halide-hydrocarbon sludge. In the past, the sludge has been hydrogenated in a reactor such as a stirred reactor. There are certain operational disadvantages associated with the hydrogenation of sludge when the active catalyst material is recovered as substantially a liquid. For example, there are appreciable amounts of entrained and dissolved hydrocarbon in the sludge rejected from the conversion process. This free hydrocarbon is, to a large extent, a product of the conversion process. The hydrocracking of these hydrocarbons during catalyst reactivation degrades the value of the hydrocarbon from product such as isomerizate to light hydrocarbon gases such as methane and ethane which generally can be used only as refinery fuel.

Another disadvantage in using these reactors is that during the sludge hydrogenation reaction, certain amounts of catalyst salts are vaporized and passed from the reactor in the effluent gas. These salts condense in the piping in such critical locations as control valves and the like. The vaporized salts not only cause operational difficulties and increase maintenance cost, but also result in the loss of valuable catalysts.

It has been found in accordance with the process that when the hydrogenation reactor contains a stripping-absorption zone as well as the reaction zone, vaporized salts in the effluent gas are absorbed by the reject sludge from a conversion process and the entrained and dissolved hydrocarbon in the reject sludge is stripped by the effluent gas before possible degradation (cracking) in the reaction zone of the reactor. Figure 30 shows the form of apparatus which may be used in the conduct of the process.

Referring now to the drawing: an aluminum halide-hydrocarbon sludge from a conversion process and, if desired, antimony trihalide, enter the hydrogenation reactor 12 through line 14. The sludge passes countercurrently to the effluent gas stream through the stripping-absorption zone 16 to strip free hydrocarbon from the sludge and absorb vaporous salts from the effluent gas, and then the sludge passes into the reaction zone 18. If desired, a portion of the sludge bypasses the stripping-absorption zone, is heated in heat exchanger 20 and enters reaction zone 18 through bypass line 22. The reaction gas, hydrogen and if desired hydrogen halide, is introduced into the reaction zone 18 through line 24 to convert the sludge to active catalytic material and hydrocarbon. The reaction zone 18 can be surrounded by heat transfer means 26 to control the temperature of the hydrogenation reaction.

The effluent gas from the reaction zone, containing hydrogen, hydrogen chloride and some vaporized catalyst salts, passes upwardly through the stripping-absorption zone 16 in which it is scrubbed with the sludge. The scrubbed effluent gas passes through a final disengaging zone 28 and is withdrawn from the hydrogenation reactor through line 30. The active catalyst material is withdrawn from the hydrogenation reactor through line 32 and can be returned to the conversion process. The hydrogenation reactor is of appropriate design to insure intimate contact between the sludge and the reaction gases. A particularly suitable design is a vertical tower reactor type wherein the ratio of the length of the vessel to the

FIGURE 30:   APPARATUS FOR RECOVERY OF ALUMINUM CHLORIDE CATALYST FROM HYDROCARBON SLUDGE

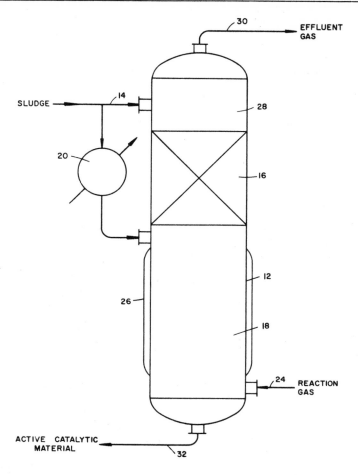

Source: H.D. Evans and R.J. Schoofs; U.S. Patent 3,210,292; October 5, 1965

diameter of the vessel is from 10 to 1 to about 70 to 1. The preferred ratio of length to diameter is 40 to 1 to about 60 to 1. The vessel can be constructed of any suitable material such as steel, nickel and various alloys or it can be for example a steel vessel lined with nickel.

While it is not necessary it is preferred to pass the liquid aluminum halide-hydrocarbon sludge through the stripping-absorption zone countercurrently to the effluent gas from the reaction zone. This stripping-absorption zone can be contained in an external vessel. However, it is preferred that the stripping absorption zone is an integral part of the hydrogenation reactor vessel. The entrained and dissolved hydrocarbon in the catalyst sludge is effectively stripped from the sludge stream by the effluent gas before possible degradation in the reaction zone of the reactor. The high efficiency of this stripping zone is attributable to the large activity coefficient of the hydrocarbon in the sludge. Moreover, the vaporized salts in the effluent gas are absorbed by the reject sludge. Several other advantages of the stripping-absorption zone are that the zone provides intimate heat transfer between the hot effluent gas and the cold sludge stream and the sludge stream becomes hydrogen saturated.

This multifunctional zone has from 1 to 10 theoretical stages, preferably 1 to 4 theoretical stages. The zone is usually designed so as to operate at from 50 to 80% of capacity. The zone generally consists either of fractionating trays such as the disc and donut arrangement or a packed bed of inert particulate solids. In a preferred embodiment, the materials of the bed are chosen so that there is a large free space to surface area with few liquid holding pockets to accumulate the sludge, metal corrosion products or other contaminants. It is naturally preferred that the materials not be subject to corrosion by the sludge or gaseous streams. Examples of suitable materials are Lessing ring, Raschig rings, packing cones, ceramic balls, spiral rings, cross-partition rings, Berl saddles, interlock saddles, and irregularly-shaped materials such as crushed rock. The length to diameter ratio of the zone is from 2:1 to 20:1, preferably 2:1 to 15:1. The superficial gas velocity is from 0.01 to 0.3 feet per second, preferably 0.05 to 0.2 feet per second.

In one preferred embodiment of the process, a bypass with a heat exchanger is provided around the stripping-absorption zone. In this manner, it is possible to use a cool sludge stream to scrub more efficiently the effluent gas while bypassing the remainder of the sludge through the heat exchanger, resulting in an overall hotter sludge stream to the reaction zone.

The reaction zone is generally operated substantially liquid full. It is preferred to maintain the liquid level in this reaction zone at 1 to 3 feet below the stripping-absorption zone. Reaction contact time varies from about 5 minutes to 10 hours depending upon the type of sludge, reactor design, contacting efficiency, temperature, etc. The preferred contacting periods are from 5 minutes to 3 hours. The reaction zone may be agitated by suitable means such as a mixer. However, it is preferred that there is no stirring in order to get the benefit of staging in the reaction zone.

The hydrogenation conditions for converting the sludge to active catalytic material are a temperature range from 100° to 250°C. and a hydrogen partial pressure from 200 psi to 2,500 psi. The preferred hydrogenation conditions are a temperature range from 150° to 225°C. and a hydrogen partial pressure from 400 psi to 1,400 psi.

The hydrogen should be essentially dry and is desirably free from hydrogen sulfide. The hydrogen consumption varies from 5 standard cubic feet per pound of hydrocarbon in the sludge to 36 standard cubic feet per pound hydrocarbon in the sludge. It is desirable to use excess hydrogen, i.e., a hydrogen feed rate to the reactor greater than the hydrogen consumption, for example, up to 10 times the hydrogen consumption, in order to maintain high partial pressures over the reactor length. Preferably the hydrogen feed rate of the reactor is 2 to 6 times the hydrogen consumption.

Standard Oil

A process developed by J.A. Ridgway, Jr.; U.S. Patent 3,094,572; June 18, 1963; assigned to Standard Oil Company (Indiana) involves an improved technique for preparing and/or regenerating an aluminum chloride-on-adsorbent catalyst. It has been found that certain aluminum alkyls, particularly aluminum triisobutyl will readily react with or be absorbed on many solids in presence of readily available hydroxyl radicals. This phenomenon may then be utilized for activating isomerization catalysts during the course of their preparation and/or regeneration. Aluminum alkyls and aluminum hydroalkyls which are effective for this process can be prepared from aluminum, hydrogen, and olefins by the K. Ziegler technique, Ang. Chem 67, p. 424 (1955).

$$Al + 1\ 1/2H_2 + 2Al(C_nH_{2n+1})_3 \longrightarrow 3HAl(C_nH_{2n+1})_2$$

$$\frac{3HAl(C_nH_{2n+1})_2 + 3C_nH_{2n} \longrightarrow 3Al(C_nH_{2n+1})_3}{Al + 1\ 1/2H_2 + 3C_nH_{2n} \longrightarrow Al(C_nH_{2n+1})_3}$$

A superactive isomerization catalyst may be prepared by treating an adsorbent such as alumina or silica in the presence of 0.2 to 2% water with a hydrocarbon solution of aluminum triisobutyl wherein the contacting is continued at 150° to 200°F. until gas evolution substantially ceases and wherein the resulting solid is thereafter saturated with hydrogen chloride. Also an aluminum chloride-on-adsorbent catalyst may be activated by contacting with a hydrocarbon solution of aluminum trisobutyl under these same conditions. When an aluminum chloride-on-adsorbent catalyst prepared by either of these methods (or prepared by any other technique) has become partially deactivated by on-stream conversion of light paraffinic hydrocarbons to more highly branched hydrocarbons, the deactivated catalyst may be regenerated by first removing hydrogen chloride and catalyst deposits which are soluble in hot hydrocarbons and then treating at about 150° to 250°F. with a hot hydrocarbon solution containing 0.5 to 10 preferably 1 to 2 weight percent aluminum triisobutyl based on catalyst undergoing treatment.

The process will now be further illustrated by citation of a specific example and by reference to Figure 31. In this example a charging stock is employed having the following composition:

	Volume %
Cyclopentane	1.7
2,2-dimethylbutane	0.9
2,3-dimethylbutane	4.0
2-methylpentane	26.5
3-methylpentane	25.0
Normal hexane	31.5
Methylcyclopentane ⎫ Cyclohexane ⎭	10.4

Since this process utilizes a catalyst which is extremely susceptible to poisoning by aromatic hydrocarbons, the maximum aromatic hydrocarbon content tolerable in the feed is dependent somewhat on the particular catalyst and the temperature of contacting, but in general, not more than 0.5 volume percent of aromatic hydrocarbons should be present in the feed. It is preferred to operate with a feed which is virtually free of aromatic hydrocarbons, i.e., contains less than 0.1 volume percent of aromatic hydrocarbons. In addition the feed should be dry and substantially free from sulfur and other materials

FIGURE 31:   APPARATUS FOR MANUFACTURE AND/OR REGENERATION OF SUPPORTED AlCl₃ CATALYST

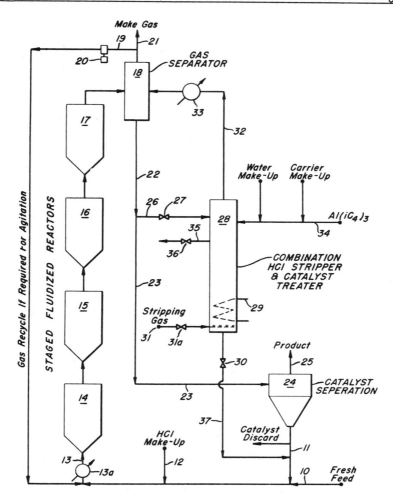

Source: J.A. Ridgway Jr.; U.S. Patent 3,094,572; June 18, 1963

which would be deleterious to aluminum chloride catalysts. The catalyst employed in this example is prepared by charging to a 500 ml. flask, equipped with a reflux condenser attached to a manometer and a stopcock for venting, 250 ml. of 10% cyclohexane + 90% 2-methylpentane, 50 ml. (3.4 g.) of finely divided silica (Santocel) previously calcined at 1000°F., and 0.25 ml. of water is charged to the flask. The contents of the flask are next agitated and purged with hydrogen after which 25 ml. of a 20% solution of aluminum triisobutyl in 2-methylpentane is added thereto. The resulting mixture is heated to 160°F. while stirring. The pressure is maintained at 960 mm. by venting until gas evolution ceases. Thereafter, the contents of the flask are cooled to 100°F. and saturated with anhydrous hydrogen chloride.

The catalyst thus prepared is contacted with the defined charging stock for a period of 0.5 to 5 hours or more at a temperature in the range of 50° to 180°F., preferably at 80° to 120°F., with a catalyst to hydrocarbon weight ratio, in the reaction zone, of about 0.05 to 1.0, preferably 0.1 to 0.5. Hydrogen may be present in amounts which are soluble in the liquid under conversion conditions and at a pressure of 25 psig but less hydrogen pressure is required than was heretofore deemed necessary (under U.S. Patent 2,443,608).

Referring to the drawing, the described charging stock, which may contain 0.1 to 1% of hydrogen chloride and dissolved hydrogen picked up in an absorber (not shown), is introduced by line 10. It picks up catalyst from line 11 and any required make-up hydrogen chloride from line 12 and introduces the catalyst slurry through line 13 through heater-cooler 13a to hopper-bottom reactor 14 wherein the residence time of the catalyst is longer than that of the liquid. The slurry may pass through a series of reaction zones 15, 16, and 17 before being introduced into gas separation zone 18. Separated gas may be recycled by line 19 and circulating compressor 20 at a rate for obtaining desired agitation in the reactors, net make-gas being vented from the system through line 21.

The liquid slurry is passed from gas separator 18 by lines 22 and 23 to catalyst separator 24 from which the product stream is withdrawn through line 25 and the catalyst slurry is withdrawn through line 11 to be reused. During continued use the catalyst gradually looses activity making periodic regeneration or replacement necessary in order to maintain a high level of conversion. To regenerate the catalyst valve 30 is closed and valve 27 is opened and a portion of the liquid slurry from gas separator 18 is passed by lines 22 and 26 through valve 27 to vessel 28 which is provided with means for heating or cooling such as coil 29 and means for agitating the slurry such as a stirrer (not shown). When vessel 28 is sufficiently full valve 27 is closed. Since the presence of HCl in vessel 28 causes increased consumption of aluminum alkyl during regeneration most of the HCl is preferably removed. This is accomplished by introducing a stripping gas through distributor 31. A suitable gas for this purpose is make-gas from gas separator 18 which has had HCl removed by absorbing in fresh feed.

The HCl and light gases removed from vessel 28 are passed by line 32 through cooler 33 back to separator 18. When the desired amount of HCl has been removed from the slurry about 1 to 10 and preferably 2 to 4 weight percent of aluminum alkyl based on total catalyst is introduced to vessel 28 through line 34 as a solution in a hydrocarbon which is preferably cyclohexane produced in the system but which may be an aliquot part of the isomerization product or a part of the charge or recycle stream. This solution is intimately mixed with the slurry by stirring or circulation through vessel 28 at a temperature of 160°F. until gas evolution ceases. The slurry is then cooled to 100°F. and the excess hydrocarbon liquid may be separated from the catalyst by withdrawing through line 35 and valve 36, although such separation step is not always necessary. The catalyst slurry in line 37 is then introduced into the incoming fresh feed stock in line 10 via line 11 where it joins the catalyst slurry recovered in catalyst separator 24. Any make-up HCl needed to activate the regenerated catalyst is introduced through line 12.

BORON FLUORIDE-CONTAINING CATALYSTS

HYDROCRACKING

Hydrocarbon Research

A process developed by P.M. Koppel and P. Maruhnic; U.S. Patent 3,525,699; August 25, 1970; assigned to Hydrocarbon Research Inc. involves the preparation of an acid site catalyst for use in catalytic hydrocracking of hydrocarbons from a typical alumina based hydrogenation catalyst by reacting the hydrogenation catalyst with a solution of boron-trifluoride etherate complex (BTFE) in benzene at reflux followed by a high temperature heating step under an inert atmosphere.

Alumina base catalysts, usually impregnated with cobalt, or molybdenum or nickel or combinations thereof, are well known for hydrogenation of hydrocarbons and especially for hydrodesulfurization. Such catalysts have little effectiveness for hydrocracking however and usually an acid site catalyst such as silica-alumina is used when some cracking is desired. As compared to the single function hydrogenation (alumina base) catalysts, the hydrocracking catalysts are far more expensive and considerably shorter lived, being readily poisoned by contaminants in the hydrocarbons.

The use of boron trifluoride complexes for hydrodesulfurization has also been suggested. A complex of boron trifluoride with water, ether, etc., as suggested in U.S. Patent 2,657,175, is formed in a reaction zone as a "homogeneous catalyst." However, this is not used to deposit Lewis acid sites on an otherwise inexpensive solid catalyst.

U.S. Patent 3,128,243 has shown the use of boron trifluoride (BF_3) gas to prepare a fluorided nickel sulfide on alumina catalyst. The BF_3 is contacted with the unfluorided catalyst under pressure, preferably in situ, by passing the BF_3 gas into the hydrocracking zone simultaneously with the feed material to be treated.

It has been found, however, that use of any of the previously known methods for fluoriding hydrogenation catalysts results in a relatively poor uniformity of fluoride distribution through the catalysts, especially when it is desired to obtain a relatively low level of fluoride on the catalyst, e.g., about 2.0% by weight fluoride. This process is directed towards a method for preparation of a fluorided catalyst which allows one to achieve a highly uniform fluoride distribution and more importantly to produce a highly uniform fluoride content on the catalyst from batch to batch.

This process consists of an upflow hydrocracking operation with such a unique catalyst. As an unexpected result of the activity of the impregnated catalyst, conversions are carried out at lower than usual temperatures with a material reduction in pour point and improved Diesel Index of the products as well as a higher ratio of furnace oil to naphtha.

Figure 32 shows the apparatus which may be used in this process. A feed stock 10 is adapted to be hydrocracked with hydrogen at 12 in an upflow reactor 14 in the presence of catalyst. By control of catalyst size, and liquid and gas velocities, the catalyst particles will be kept in random motion in the liquid and thereby substantially isothermal conditions can be maintained in the reactor.

Product (effluent) is removed overhead at 16 and where recycle is desired, it can be accomplished either by external piping (not shown) or by the internal recycle conduit 18 which extends below the distributor deck 20 and is provided with a circulating pump 22.

This process is based on the production and use of a catalyst in such a system. As a base material is used the well known and relatively inexpensive single function hydrogenation catalyst having an alumina base on which are deposited metals from the class consisting of cobalt, molybdenum, nickel molybdenum and the customary variants thereof. Such a catalyst in a close size range with a maximum dimension in the range of 3 mesh to 300 mesh (Tyler) enters catalyst pretreater 30 through line 32 and is impregnated by a boron trifluoride complex entering at 34. The impregnated catalyst removed at

FIGURE 32:   FLOW SCHEME OF PROCESS FOR HYDROCRACKING WITH FLUORIDED CATALYST

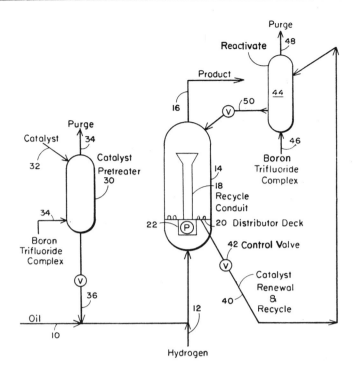

Source: P.M. Koppel and P. Maruhnic; U.S. Patent 3,525,699; August 25, 1970

36 thus becomes the catalyst for the reaction. A typical example of fluoriding an alumina base catalyst with boron tri-fluoride is to use the etherate suitably dissolved (diluted) in benzene. The catalyst base, after drying as for example, at about 300°F. at atmospheric pressure, is processed through the following (preferred) steps:

Step A consists in diluting boron trifluoride etherate in benzene in a ratio of about 1:20 and then mixing about equal parts of the catalyst base with the diluted etherate. There is an exothermic reaction between the boron trifluoride and aluminum oxide, the exact nature of which is not known. This step is preferably continued for about 30 minutes and at about 180°F. Reflux temperature controls the exothermic reaction and is high enough at atmospheric pressure to break down the etherate and release boron trifluoride for reaction.

Step B consists in heating the catalyst and liquid up to about 600°F., in a suitably slow time period of about 2 hours, in the presence of nitrogen gas as a purge. It is found that the boron trifluoride is completely absorbed by and fixed onto the normally porous catalyst base. The temperature and purge drive off the ethyl ether which may be recovered.

Step C is a cooling step which accomplishes cooling to a suitable handling or storage temperature as for example, about 75°F. and this is effectively accomplished also by using a nitrogen gas purge which restrains any oxidation. Under these conditions the catalyst gained in weight approximately 3.2%. Fluoride analysis by the steam pyrolysis method was 2.3%. Analysis for boron spectographically indicated it was present in the atomic ratio of 1B:3F.

POLYMERIZATION

Nippon Petrochemicals

A process described by T. Horie; U.S. Patent 3,631,013; December 28, 1971; assigned to Nippon Petrochemicals Co., Limited, Japan involves making solid rubbery materials by copolymerizing an isoolefin and a diolefin in an alkyl halide as solvent with boron trifluoride and metal compounds $M(OR)_m X_n$ as catalyst wherein M is Al or Ti, R is a hydrocarbon radical or halogenated derivatives thereof, and X is a halogen. The process comprises introducing monomers, recycled solvent metal compound $M(OR)_m X_n$ and boron trifluoride into the polymerization reactor and carrying out polymerization reaction at a temperature below 0°C., preferably from −50° to −110°C. The polymerized mixture is then withdrawn continuously from the polymerization reactor, bringing the mixture into contact with a heating medium to evaporate and separate low boiling fraction mainly containing the solvent and unreacted monomers from the polymer produced, drying and fractionating the low boiling fraction to distill the solvent containing the unreacted monomer, recycling the solvent, wherein the boron trifluoride introduced into the reaction system is not previously mixed with the recycled solvent or the monomers. The process flow diagram is shown in Figure 33.

The isoolefin monomer is supplied by line 1, the diolefin monomer by line 2, and part of the recovered solvent by line 20 respectively into tank 5 and mixed therein. The resultant mixture is cooled by cooler 6 to from −50° to −110°C. and supplied to polymerization reactor 7. The metal compound $M(OR)_mX_n$ supplied by line 3 is mixed with another part of the recovered solvent coming by way of line 19 and the mixture is supplied to reactor 7. The mixture may be cooled before introduction into the reactor. Gaseous or liquid boron trifluoride is directly introduced by line 4 into reactor 7. The reacted mixture is sent from reactor 7 by way of line 8 to flashing apparatus 9. The heat transfer medium is sent by line 27 into flashing apparatus 9 and the solvent, the unreacted monomers, and low-boiling by-products are evaporated off and separated from the polymerization product.

The polymers produced are taken out dispersed in a portion of the heating medium from line 21, go through the post-treatment steps 22 such as dehydration and drying and emerge as the final product from 23. The gaseous products evaporated and separated in flashing apparatus 9 are sent to drying step 11 by way of line 10 and then to fractionating tower 13. The vapor coming out from the top of the fractionating tower through 14 is cooled in cooler 15 and stored in tank 16 as liquid.

A portion of the liquid in tank 16 is returned to the fractionating tower as a reflux and another portion is sent by way of line 18 as the recovered solvent. Part of the unreacted monomers is removed from the bottom of the fractionating tower together with the by-products and the impurities accompanying the monomers by way of line 26. Part of the bottom oil is heated by heater 24 and recycled to the fractionating tower by line 25. The following is a specific example of the conduct of the process.

Example: Isobutylene was introduced at a rate of 51.8 kg./hr. by way of line 1, and isoprene at a rate of 1.5 kg./hr. by way of line 2. The recycled recovered solvent which was supplied to the reactor by way of lines 18, 19, and 20 contained methyl chloride (204 kg./hr.) and the unreacted isobutylene (9 kg./hr.). Aluminum sec-butoxide $Al(O-secC_4H_9)_3$ was introduced by line 3 and the gaseous boron trifluoride by line 4. The polymerization was carried out at −78°C., and the reaction mixture withdrawn from the reactor contained 204 kg./hr. of methyl chloride, 16 kg./hr. of the unreacted isobutylene, 0.2 kg./hr. of the unreacted isoprene, about 0.002 kg./hr. of by-products having more than 5 carbon atoms, and 45.5 kg./hr. of the polymers.

The mixture was flashed with warm water to separate the polymers and the vapors. The vapors thus evaporated and separated were first sent to the drying step and then introduced into the middle zone of fractionating tower 13. The mixture thus introduced contained 204 kg./hr. of methyl chloride, 16.7 kg./hr. of the unreacted isobutylene, 0.2 kg./hr. of the unreacted isoprene, and about 0.002 kg./hr. of the impurities. The fractionating tower was run with the reflux ratio of 1:5 and the recovered solvent was taken out at a rate of 213 kg./hr. by line 18. The recovered solvent contained 9 kg./hr. of the unreacted isobutylene. The C_4 fraction mainly consisting of isobutylene and the C_5 and higher fraction mainly containing isoprene were recovered from the bottom of the fractionating tower at rates of 7.7 kg./hr. and 0.2 kg./hr. respectively. A water slurry containing 3% of the polymers was obtained from the flashing apparatus at a rate of 1,513 kilograms per hour. This slurry was subjected to the post-treatment steps such as dehydration and drying and the solid rubbery isobutyleneisoprene copolymers were obtained.

FIGURE 33:   FLOW SCHEME FOR SOLID RUBBERY COPOLYMERS OF ISOOLEFINS AND A DIOLEFIN

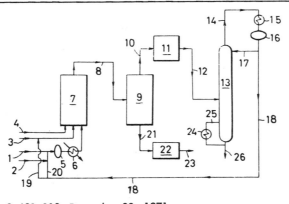

Source: T. Horie; U.S. Patent 3,631,013; December 28, 1971

Texaco

A process developed by E.T. Child and W.L. Lafferty, Jr.; U.S. Patent 3,190,936; June 22, 1965; assigned to Texaco Inc. relates to a method for regenerating exhausted or spent adsorbent and catalyst support compositions employed in the polymerization of olefins by treating the compositions with a particular solvent mixture and then drying the compositions at an elevated temperature.

The prior art has disclosed the use of adsorbents and of supported Friedel-Crafts catalysts that are highly effective for the polymerization of monoolefins. The adsorbents and the supports in the supported Friedel-Crafts catalyst are from the same classes of materials, namely activated carbon, silica, alumina and mixtures thereof. It has been found that the activity of the adsorbent and the effectiveness of the catalyst are gradually reduced during polymerization until both become substantially ineffective. Since the basic composition used as the adsorbent and as the support constitutes a significant part of the costs in such polymerization processes, it was realized that important economies could be effected if either the adsorbent or support material or both could be salvaged and regenerated for one or both of the above noted functions.

The changes in the catalyst and in the adsorbent which result in compositions exhibiting reduced effectiveness or which are totally ineffective are not fully understood. It is postulated that the active Friedel-Crafts halide is changed at least in part to relatively ineffective complexes during polymerization and that some of this remains in the support while some is adsorbed by the adsorbents. It is further believed that a certain amount of polymer is occluded both on the catalyst and on the adsorbent during the polymerization reaction. In any event, neither the supported catalyst nor the adsorbent are effective for their respective functions.

Attempts were made to regenerate the compositions rendered ineffective due to the formation of inactive halide complexes and occluded polymer by a conventional regeneration procedure. An ineffective composition was heated in a dryer at 400°F. to remove flammable material and then heated to 1000°F. in a muffle furnace. This procedure was unsatisfactory. The adsorbed polymer was not removed in the dryer and caused ignition in the muffle furnace. In addition, the higher temperature employed in the muffle furnace appeared to impair the expected life of the composition.

The removal of catalyst complexes and occluded polymer was also attempted by washing the catalyst and the adsorbent with various hydrocarbons and polar solvents. No solvent was effective for removing both the polymer and the catalyst complexes and regeneration of the support and the adsorbent could not be effected.

A method has been found for effectively removing the complexes and occluded polymer from the adsorbent and the support compositions thereby permitting the regeneration and reuse of these compositions in a polymerization process. In accordance with this method, a supported catalyst, comprising a Friedel-Crafts halide on a support, or an adsorbent either of which has become ineffective due to the formation of complexes and the deposition of occluded polymer on the surface of the composition, is contacted with a solvent mixture that is effective to remove both the complexes and polymer from the composition. Solvent mixtures which are effective for this purpose are comprised of a liquid hydrocarbon and of a polar compound.

More particularly, a solvent mixture for this process consists of a hydrocarbon and of a polar compound within certain proportions. An effective mixture is one in which the proportion of hydrocarbon to polar solvent on a weight basis ranges from 1:4 to 4:1. One or more compounds of each type may be employed so long as the total amount for each class of solvent falls within the specified proportions.

The compositions which can be beneficially treated by this process are catalysts comprising a Friedel-Crafts halide on a support and adsorbent materials. The support material and the adsorbent are characterized by being comprised of substances which will adsorb polar compounds and at the same time will not react with a Friedel-Crafts halide. Effective support materials and adsorbent materials include activated carbon, silica, alumina and mixtures thereof. Silica gel is preferred both as the adsorbent and as the support material for the catalyst.

The catalyst composition employed in olefin polymerization processes of interest consists of a Friedel-Crafts halide on a support, the latter defined hereinabove. These catalysts generally consist of about 2 to 20% by weight of the Friedel-Crafts halide with the balance or 80 to 98% consisting of the support material. Particularly effective catalysts comprising a Friedel-Crafts halide on a support are titanium tetrachloride, boron trifluoride and aluminum trichloride.

The hydrocarbons which are employed in making up the solvent mixture are the normally liquid aliphatic hydrocarbons, particularly those having from 2 to 12 carbon atoms. Effective hydrocarbons are n-octane, n-pentane, isopentane, heptane, ethane, butane, propane and the like. The polar solvents which can be employed include the lower aliphatic ketones and the lower aliphatic alcohols. These solvents are represented by the following formulas: R—CO—R and R—OH respectively in which R is an aliphatic radical having from 1 to 6 carbon atoms. Suitable polar solvents include acetone, diethyl ketone, ethyl alcohol, propyl alcohol and the like. The process is further illustrated by reference to Figure 34.

FIGURE 34:   PROCESS FOR PRODUCING POLYISOBUTYLENE USING A BORON FLUORIDE CATALYST

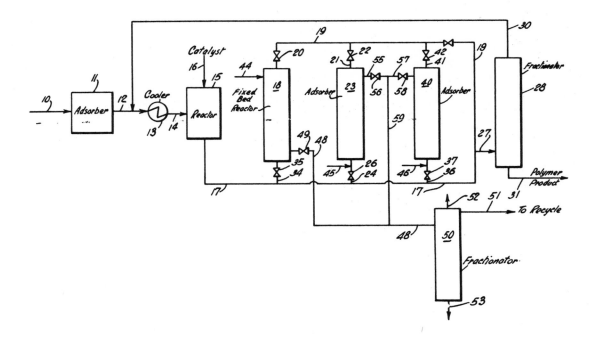

Source: E.T. Child and W.L. Lafferty, Jr.; U.S. Patent 3,190,936; June 22, 1965

An isobutylene-containing charge stock from any suitable source is charged through pipe 10 into adsorber unit 11. The adsorber unit is filled with an adsorbing medium effective for removing water or moisture and other undesirable components in the isobutylene feed. Silica gel has been found most effective for this purpose, although other well known adsorbing mediums may also be employed. The dehydrated and purified isobutylene stream is passed through line 12 into heat ex-changer 13 where it is adjusted to the temperature range at which the polymerization reaction is to be conducted depend-ing on the catalyst being employed. Generally, the feed is cooled, although polymerization may be effected at a tem-perature in the range of –100° to about 150°C. The dehydrated isobutylene feed is passed through line 14 into reactor 15. The polymerization catalyst, a mixture of a Friedel-Crafts halide in an inert hydrocarbon solvent, such as pentane or the like, is added to the reactor through line 16.

Any conventional reactor may be employed in the first stage reaction of this process. A reaction vessel or tank having a mechanical mixing means or a stirrer is preferred. However, a tubular reactor in which polymerization takes place as the feed stream and catalyst is passed through the tube is also suitable. Both basic types of reactors, those designed for back-mixing of the reaction product and those preventing back-mixing, may be employed in the first stage of the process. Polymerization in the reactor is effected at a temperature in the range of –100° to 150°C. under a pressure in the range of 25 to 500 lbs. per square inch. Under the preferred operating conditions, isobutylene is converted to polyisobutylene in the amount of 40 to 95% generally after a residence time of about 1/2 to 3 hours.

The product from the reactor containing between 40 and 95% of polyisobutylene polymer is passed through line 17 and line 34 into fixed-bed reactor 18. This reactor is preferably a tower containing a fixed-bed of a supported catalyst. The catalyst employed is a Friedel-Crafts halide on a support material. This catalyst can be a freshly prepared lot of active metal halide on support material, or more preferably, may be formed by the adsorption of the active halide re-maining in the polymer effluent in an adsorption step as explained hereinbelow.

Polymerization of substantially all of the isobutylene monomer present in the effluent from the first stage reaction is ac-complished in the fixed-bed reactor on contact with the supported Friedel-Crafts catalyst at a temperature in the range of –100° to 150°C. The residence time in this reactor is of short duration, generally in the order of about 10 to 30 min-utes. The reaction product from the fixed-bed reactor comprises the polyisobutylene polymer, hydrocarbon solvent and a minor amount of entrained and/or dissolved Friedel-Crafts halide. The reaction product from the fixed-bed reactor is passed through lines 19 and 21 into adsorber unit 23. This unit is desirably in the form of a tower and contains an ad-sorbent effective to remove the catalyst from the reaction product. The reaction product is contacted with the adsorbent and the purified solution of polymer and hydrocarbon is passed through lines 24, 17 and 27 into fractionating column 28. In the fractionator, the solvent is separated and taken off through line 30 while the polymer product is recovered through line 31. The hydrocarbon solvent is recycled through line 30 and is re-employed by combining same with isobutylene feed upstream from the stirred reactor preferably before the feed enters the heat exchanger.

When the process is started up, tower 40 is a standby adsorber unit filled with adsorbent material. This unit provides the necessary flexibility for continuously processing the incompletely polymerized isobutylene from the first stage reactor after the catalyst in the second stage fixed-bed reactor has lost its effectiveness.

As noted above, the second stage reactor containing a supported catalyst becomes ineffective over extended use due to the buildup of polymer and the formation of nonactive catalyst complexes. At the same time, the adsorber in adsorber unit 23 is adsorbing catalyst and as the catalyst builds up it is converted into a unit exhibiting substantial catalytic activity for polymerizing olefin monomers. At this juncture, it is feasible to stop the flow of the reaction product from the first stage of the process to fixed-bed reactor 18 and to divert this reaction product to unit 23 which has developed the function of a fixed-bed reactor. The isobutylene monomer remaining in the diverted stream is substantially completely polymerized to polyisobutylene polymer in unit 23 and the reaction product is passed through lines 21, 19 and 41 into adsorber unit 40 which contains an adsorbent effective for removing any catalyst remaining in the reaction product effluent. The catalyst-free reaction product from tower 40 is passed through lines 36, 17 and 27 into fractionator 28 wherein the polyisobutylene polymer is separated and recovered while the hydrocarbon solvent is recycled.

The catalytically ineffective second stage reactor is taken off-stream by interrupting the flow through lines 35 and 20. A mixed solvent, such as acetone and pentane, is introduced into this reactor through line 44. This solvent mixture removes occluded polymer and the inactive Friedel-Crafts halide complexes from the support material thereby leaving a relatively clean support material. The solvent is removed from the support material by washing with a hydrocarbon or by blowing with an inert gas, such as nitrogen or air as disclosed above. This material is suitable for use as an adsorbent in the process and is retained in a standby capacity until the adsorbent in tower 40 becomes saturated and is no longer effective to remove entrained catalyst from the product stream.

The solvent mixture containing dissolved polymer and Friedel-Crafts halide complexes is passed through line 48 into fractionator 50. The solvent is separated from the polymer and metal halide and is retained for recycle in this step for salvaging and activating catalyst support material.

A silica gel adsorbent, which had been used in a process for producing polyisobutylene by the polymerization of isobutylene in the presence of boron trifluoride and which was saturated with catalyst complexes and occluded polymer and no longer effective in the polymerization process, was contacted with a solvent mixture consisting of 50:50 weight percent of acetone and pentane at 25°C. for approximately 5 hours. After substantially all of the catalyst complexes and occluded polymer were removed, the solvent was separated from the silica gel. The rinsed silica gel was then dried and regenerated with nitrogen at an inlet temperature of 350°F. for 6 hours after the outlet temperature had risen above 100°F. The regenerated silica gel had adsorbent properties equal to the original material.

In contrast to the foregoing, acetone or pentane, when employed separately, were ineffective for dissolving catalyst complexes and occluded polymer from silica gel and did not provide a way for regenerating the spent composition.

ALUMINUM OXIDE CATALYSTS AND SUPPORTS

DESULFURIZATION

American Cyanamid

A process developed by J.D. Colgan and N. Ostroff; U.S. Patent 3,403,111; September 24, 1968; assigned to American Cyanamid Company involves the preparation of a molybdenum promoted alumina base extruded catalyst having improved crush strength after regeneration which comprises treating a hydrous alumina slurry with nitric acid, forming an extrusion mixture of the treated alumina and a molybdenum promoter and extruding the mixture. Typical fixed-bed catalysts used for hydrodesulfurization initially possess high catalytic activity and relatively high mechanical strength. Catalytic activity and catalyst strength, however, gradually deteriorate in use and for these reasons catalysts must be discarded after relatively short times. Deterioration in catalytic activity is known to be due in part to coke deposition on the catalyst during use. As a result, it is common practice to periodically regenerate the catalyst. This is usually accomplished by steam stripping the residual oil and, then, burning off the deposited coke.

The reasons for loss in mechanical strength are not well understood. It is known that the strength loss takes place in use but more particularly under the high temperature conditions used in regeneration, and further that this strength loss is catalyzed by the presence of MoO_3. Loss of strength is undesirable and frequently necessitates replacement of the catalyst because large quantities of fines are produced which are detrimental to the proper functioning of the catalyst bed, as evidenced by excessive pressure drop or poor flow distribution in the bed. Many methods have been proposed to strengthen or harden the fixed bed catalysts, pellets or extrudates, but these methods, for the most part, deal only with the strength of the fresh catalyst. While an improvement in the strength of the fresh catalyst will generally improve the strength of the used and regenerated catalyst to some extent, the improvement is not appreciable.

One method which has been proposed to improve the strength of used and regenerated hydrodesulfurization catalysts involves the addition of 1 to 15% silica. While this method imparts the desired increase in strength of the used and regenerated catalyst, it also imparts some cracking activity to the catalyst which may be undesirable. This process relates to an improvement in the strength stability of alumina-supported, formed catalysts used for hydrodesulfurization. It has been discovered that these catalysts, containing a compound of the metals Ni, Co, and particularly Mo, and chemical compositions of such compounds, have improved strength stability when nitric acid is added to the alumina support in a particular method and at a specific concentration. The specific steps involved in the process comprise (1) treating a precipitated alumina slurry with nitric acid, (2) preferably spray-drying, (3) followed by subsequently preparing (admixing) an extrusion feed of the treated alumina, (4) extruding, and (5) drying and calcining the catalyst. Alternatively, in a less effective method, the treating nitric acid may be added during the extrusion-feed-admixing (mulling) step of preparing the extrusion feed from the precipitated alumina and water, for example.

The preferred catalysts characterized by a high degree of retention of crush strength after regeneration, including steam treatment are obtained by the process as follows. Hydrated alumina is divided into a first and a separate second portion. The first portion is treated with an alkali such as sodium hydroxide or other typical alkali, sufficiently to form an aluminate such as sodium aluminate. The second portion is treated with a sulfate anion, such as with sulfuric acid, sufficiently to form alum, i.e., aluminum sulfate. The aluminate portion and the alum portion are then subsequently admixed and maintained substantially simultaneously (during the admixing) at a preferred pH 7.5 to about pH 8.5. The pH in any event should be maintained on the alkaline side. The precipitated alumina is preferably washed and preferably is admixed with a sufficient amount of diluent, preferably water, to obtain desired flow properties; and nitric acid is admixed therewith to treat the precipitated alumina. The treated catalyst is thereafter reduced in moisture (water) content (preferably spray-drying at least a portion thereof), mulled with sufficient diluent and/or precipitated alumina to form an extrusion feed, extruded, and dried and calcined. The nitric acid employed in this process is typically any commercially available aqueous nitric acid, normally the conventional (about 69 to 71%) nitric acid solution.

The nitric acid treatment employs from any minimal effective amount, about 1%, up to about 8% by weight of HNO_3 to alumina. If too high a percentage is employed, extrusion becomes difficult if not impossible because of the sticky mass which results. The preferred range of nitric acid (as HNO_3) is from about 2 to about 5%, based on total alumina. After the precipitated alumina slurry has been treated with nitric acid (in the preferred embodiment), the flow properties are adjusted, and optionally promoters such as nickel-molybdenum or cobalt-molybdenum, or other conventional promoters are admixed therewith. Treated alumina is then extruded and thereafter the extrudate is dried and calcined.

Hydrocarbon Research

A process developed by S.B. Alpert, R.H. Wolk, P. Maruhnic and M.C. Chervenak; U.S. Patent 3,630,888; Dec. 28, 1971; assigned to Hydrocarbon Research, Inc. involves carrying out catalytic reactions, such as, desulfurization in fixed, slurried, fluidized and ebullated beds utilizing a catalyst having micropores and access channels; and wherein the access channels are interstitially spaced throughout the micropores; and wherein 10 to 40% of the total pore volume is composed of access channels having diameters greater than 1000 A.; and wherein 10 to 40% of the total pore volume is composed of access channels having diameters between about 100 and 1000 A.; and wherein these access channels are substantially uniform as to their parameters and are relatively straight with minimum bending and constrictions; and wherein the remainder of the catalyst pore volume comprises micropores with diameters less than 100 A. with the remainder being 20 to 80% of the total pore volume.

The prolonged, sustained extent of diffusion of feed material into a catalyst is the controlling factor with respect to catalyst life, efficiency and process operability according to the developers of the process, and the extent of this diffusion is determined, not by the total or overall porosity of a given catalyst, but rather by the distribution and structure of large sized access channels in the catalyst. By controlling these access channels in size, shape, percentage and/or orientation in a given catalyst, one can significantly improve and prolong catalyst life and efficiency, and provide sustained access of feed material to the vast majority of all interior micropore catalytic surfaces. These channels must be interstitially spaced throughout the micropore structure of the catalyst. In order to have both sufficient strength to avoid significant mechanical attrition and yet still possess substantial reactive surface area, a catalyst must have about one access channel with diameter greater than 100 A. for about every 80,000 micropores.

One method of preparing such unique catalytic materials is a process in which there is included within the catalytic nuclei a material which is capable of radial escape during fabrication leaving the radial relatively uniform channels. For example, aluminum hydrogel is precipitated in accordance with the teachings of Example 1 of U.S. Patent 2,162,607, except that the gel has, during its formation, added thereto about 1% by volume of aluminum trihydrate seed of 3.2 M strength, consisting of aluminum hydrate covered particles of carbonaceous material having a preferred range in size of 600 to 800 mesh, and being formed during a typical Bayer process, as taught in the prior art, each seed having general dimensions in the range of 400 mesh and being, as stated, externally covered with aluminum trihydrate. Such seeds are added in sufficient quantities so as to produce a final ratio of one carbonaceous particle to each 1,000 gellated nuclei in the finished mix. Thus, for example, a specific gel after seeding in this fashion will contain approximately 10 carbonaceous particles for each milligram of precipitated gel. This gel, after filtering and drying, will be found to consist of basic catalytic carrier particles of aluminum hydrate, regardless of its designation on the Edwards and Frary scale, each having therewithin a carbonaceous particle.

Such dried catalytic particles are then directly impregnated with a solution containing, e.g., 0.5 molar quantum of salts of promoters, such as, the soluble salts of platinum, niobium, tungsten, cobalt, molybdenum or the like promoter metals. Preferably, however, these catalytic nuclei are shot in discrete stream form into a furnace operated under a pressure of about 1/10 atmosphere and at a temperature in the range of 600° to 800°F. The materials so shot are collected in a continuous, slow moving, e.g., 1 ft./sec. conveyor belt to remove them from the furnace. Examination of the particles shows that each individual nucleus has relatively uniformly radially dispersed access channels, such channels spreading from the interior of each individual nucleus to the exterior thereof, such channels occupying between about 10 and about 20% of the overall pore volume. These particles may then be impregnated with a suitable, soluble salt of a catalytic promoter metal. Following this they are further dried and calcined to produce the desired, activated catalytic structure.

Among the carbonaceous substances which may be utilized are, for example, sawdust, graphite, cellulosic materials and the like. It is to be understood that other materials capable of producing explosive, radial, gaseous escape from the nuclei may be utilized. Thus, as an illustration, there is given the following example. Included within the aluminum hydrate nuclei are centrally located crystalline aromatic hydrocarbon particles having a size in the range of 600 to 1,800 mesh. This example uses naphthalene, the remainder of this example following either alternative just described above, i.e., the naphthalene is incorporated into the alumina trihydrate nuclei, dried, exploded and then impregnated, or the nuclei incorporating the naphthalene is impregnated and then dried and exploded, it being understood that in either alternative embodiment the catalytic material is, after the above treatment, calcined in any manner known to the art so as to properly activate the carrier and/or promoter. The access channels are also capable of production by various methods which produce channels, radially directed and similar to those produced above, by a further alternative procedure. As exemplary thereof, there is given the following example.

Incorporated within the feed of Bayer process alumina seed material, or within the feed of seed material for alumina silica coprecipitation is a material such as p-secondary butyl phenol which will acicularly grow radially outward during formation of the catalytic nuclei. In this specific illustration 0.1 M strength p-secondary butyl phenol in ethyl alcohol solution is added to 2.6 M strength aluminum hydrate aqueous solution having simultaneously added thereto 0.1 M aqueous sodium silicate solution, the proportions of the additives being of substantially equal volume. The resultant mixture is allowed to stand in a relatively quiescent state, with gentle stirring, for approximately 20 minutes at ambient temperature. The resultant precipitate is then filtered and dried. The dried material is then leached with ether, following which it is further dried and impregnated, if desired, with a suitable solution of soluble catalyst promoter. The remaining treatment follows the teachings of the prior art. Alternatively, to produce a similar structure there is given the following example.

Any of the catalytic gel materials listed above, or those others known to the art, have incorporated therein, linear, re-movable material. Thus, specifically, an aluminum hydrate, in the form of partially calcined gel, and having a weight of about 1.3 kg. is mixed with, at ambient temperature and pressure, approximately 37 g. of reconstituted, cellulosic fiber particles falling within the generic class termed rayon. The mixture is homogenized in a Banbury mixer and is then fed to a screw-type plastic extruder, wherein it is further preliminarily mixed. The mixture is then further extruded under high pressure, orienting the fibers along the longitudinal axis of the extrudate. The rough, plastic extrudates resulting from this procedure are then given a preliminary calcination treatment at about 500°F. for 2 hours. Thereafter, the fibers are removed by leaching, solvent extraction or oxidation to produce the catalysts. In this specific example, the cellulosic fibers are completely removed by an oxidation process at ambient pressure and at 800°F. for a period of 1/2 hour.

Thereafter, the catalytic material having the oriented and radial channels of the process is impregnated with 0.1 M strength solutions of nickel chloride and tungsten chloride, dried at 800°F. for 3 hours in an oxygen containing atmosphere, finally ground and calcined at activating temperatures and periods. While cellulosic fibers have been specified above, there can be successfully produced catalytic structures using other natural and synthetic fibers such as: linen, cotton, silk, poly-amides, polyvinyl and polyvinylidene chlorides, polyesters, and mixtures thereof. As an observation, reference should be made to the fact that the most successful results of this procedure occur surprisingly, in those cases where the fibers are not completely uniform in diameter, but incorporate minuscule follicles or are themselves deliberately extruded in cross-section lobular form. Catalysts having access channels produced from such fibers have more uniform parameter access channels than those produced from diametrically uniform circular fibers, apparently due to some molecular or electrical affinity oper-ative during formative processes.

Simon-Carves

A process developed by J.W. Abson, M. Landau, M.C. Langton and A. Molyneux; U.S. Patent 3,414,524; Dec. 3, 1968; assigned to Simon-Carves Limited, England is one in which inactive catalytic material is activated or reactivated by contacting the material with a bacteria-containing solution. Particularly useful are the sulfate-reducing bacteria which reduce the metal sulfate content of catalytic material containing an excess of metallic sulfate. It appears that this bac-terial method of activation is gentle and the risk of damage to the material is minimized; also it appears that the activated material shows good catalytic activity. The bacterial liquor may, for example, be passed continuously through a fixed bed of the material. Alternatively an ebbing and flowing system may be used in which a fixed bed of the material is alterna-tively completely immersed in bacterial liquor and then drained.

The method of activation is also applicable to the reactivation of poisoned catalytic material. Examples of bacteria which may be used include sulfate-reducing bacteria, e.g., Desulfovibrio desulfuricans, sulfide-oxidizing bacteria, e.g., Thiobacillus thiooxidans and iron-oxidizing bacteria, e.g., Ferrobacillus ferrooxidans. These bacteria all belong to the order Pseudomonadales. In a previously proposed method of preparing a catalyst for use in removing organic sulfur com-pounds from gases, a metal oxide, e.g. NiO is prepared and this is then "sulfided" by heating in contact with hydrogen sulfide; there is a danger of catalytic activity being destroyed during the sulfiding process. It will be realized that using the bacterial method a "sulfided" catalyst may be prepared directly from material comprising the metal sulfate without the necessity for first preparing the metal oxide and heating in contact with hydrogen sulfide.

Also, for example, an active alumina catalyst may be prepared by contacting aluminum sulfate with the liquor to give aluminum sulfide which is immediately hydrolyzed to aluminum hydroxide, separating the liquor, and drying and heating the aluminum hydroxide to give alumina. Certain of the catalysts described hereinabove, the nickel and alumina catalysts for example, may be used as guard catalysts in reducing the concentration in gases of organic sulfur compounds, unsaturated hydrocarbons and other compounds to minimize the rate of poisoning by the compounds of a main catalyst with which the gases are subsequently contacted. For example, the catalysts may be used for this purpose in the detoxification of fuel gas by the "shift" reaction, the gasification of petroleum reforming and the hydrogenation of unsaturated compounds. The alumina catalysts may also be used in the Claus sulfur recovery process, viz. oxidation of hydrogen sulfide in gases to sulfur dioxide, and recovery of elemental sulfur by reaction between the sulfur dioxide so produced and further hydrogen sulfide in the gases:

$$(1) \qquad 2H_2S + 3O_2 \longrightarrow 2SO_2 + 2H_2O$$
$$(2) \qquad SO_2 + 2H_2S \longrightarrow 3S + 2H_2O$$

For example, the catalysts may be used in treating a waste gas comprising between 95 and 100% of an inert gas such as CO_2 and between 0 and 5% of H_2S. The reactivation method using sulfate-reducing bacteria is applicable, for example, to the reactivation of poisoned catalysts in which metallic sulfate is present as a result of exposure to sulfur, for example, in the form of gaseous sulfur dioxide or hydrogen sulfide, under oxidizing conditions. The poisoned catalyst may have been originally prepared either by thermal activation and/or by the above bacterial method of preparation. In this case, after separation of the liquor, the catalyst may be heated to provide further activity; this may result in the conversion of metallic sulfide to oxide and the removal of bound water. However, often the activity of the catalyst after the separation is high enough to render unnecessary heating to a temperature above that at which the reaction to be catalyzed is carried out.

An example of a poisoned catalyst in which metallic sulfate is present is a bauxite used as a catalyst in the Claus sulfur recovery process. In reactivating this catalyst it is contacted with the liquor comprising sulfate-reducing bacteria which causes aluminum sulfate to be reduced to aluminum sulfide. The aluminum sulfide is immediately hydrolyzed to aluminum hydroxide. After removal of the liquor it is often not necessary to activate the catalyst further by heating to a temperature above that at which the reaction to be catalyzed is carried out. However, the catalyst may be heated to a temperature between 400° and 530°C. to convert the aluminum hydroxide to oxide and to remove any bound water.

OLEFIN ISOMERIZATION

Phillips Petroleum

A process developed by J.W. Myers; U.S. Patent 3,631,219; December 28, 1971; assigned to Phillips Petroleum Company is one in which olefin hydrocarbons undergo double bond isomerization by contacting the olefin with a catalyst of ammonia-treated alumina. In a further embodiment, olefin hydrocarbons are alternatingly isomerized with respect to the position of the double bond and then with respect to the skeletal arrangement of carbon atoms within a single catalytic reactor. The catalyst employed for the skeletal isomerization reaction is alumina, and the catalyst used for the double bond isomerization reaction is ammonia-treated alumina.

In detail, the process comprises a series of steps wherein olefins are alternatingly subjected to both skeletal and double bond isomerization within a single catalytic reactor. Olefin hydrocarbons are contacted with a catalyst comprising catalytic alumina in the absence of ammonia. A feed olefin hydrocarbon which is to undergo double bond isomerization is contacted with the same catalytic alumina used in the first step, but which has been treated with an amount of ammonia corresponding to 0.02 to 2.0 weight percent of the weight of the catalyst. Thereafter, the catalyst is further treated to remove the modifying effects of the ammonia treatment and is once again used for the skeletal isomerization of a suitable feed olefin.

ALUMINA SOL PRODUCTION

Continental Oil

A process developed by J.W. McCarthy and S.V. Stern; U.S. Patent 3,384,458; May 21, 1968; assigned to Continental Oil Company relates to a reactor for making a finely divided solid product through chemical reaction in a liquid medium, e.g., the reaction of aluminum alkoxide and water to make alumina. Figure 35 shows a suitable reactor design. The reaction vessel 10 comprises an elongated vertical shell 12, desirably cylindrical. Shell 12 contains in superposed relation four zones, namely, (1) at the top portion of shell 12 a reaction zone 14 wherein a reaction takes place in a liquid reaction medium to form a solid precipitate, (2) extraction zone 16 immediately below reaction zone 14 wherein the solid precipitate is extracted (washed) to remove adhered liquid from zone 14, (3) a solids disengaging zone 18 immediately below extraction zone 16 where the solid is able to settle out of the extraction medium, and (4) in the bottom portion of shell 12 immediately below zone 18 a solids settling zone 20 which is shaped, preferably an approximate conical shape 22, to provide a thick slurry (body compact) of solids therein.

Reaction zone 14 has positioned therein agitator means 24 which afford vigorous stirring of the liquid reaction medium filling zone 14. Agitator means 24 provides stirring vigorous enough at the particular reaction temperature to give a rapid reaction rate. Desirably agitator means 24 comprises two propeller agitators 26 and 28 arranged to direct liquid flow therein. Preferably agitator means 24 comprises, as shown in the figure, dual propeller blades 26 and 28 arranged on a common shaft 30; the propellor blades are pitched to pull liquid into the space between the blades, as shown in Figure 35.

Shell 12 in reaction zone 14 is provided with ingress means for reactants and egress means for liquid product; herein, aluminum alkoxide is introduced by way of conduit 40, water is introduced by way of conduit 42 and alcohol product is withdrawn by way of conduit 44. Preferably conduits 40 and 42 are positioned so that the reactants are directed toward the space between blades 26 and 28 in order to shorten the hydrolysis time and to avoid entrainment of the fluffy alumina precipitate into the alcohol product stream, by giving the precipitate a downward "push". Reaction zone 14 has a substantially larger cross-sectional area than does extraction zone 16. This enlargement decreases the fluid flow rate in zone 14 and aids in decreasing the entrainment of solids into the product stream 44. Extraction zone 16 is immediately below and in direct communication with, i.e., opens directly into reaction zone 14. Zone 16 includes a plurality of annular plates 50, 52, 54, etc.; preferably these annular plates are fixed to the periphery of the shell 12 and are inclined downwardly toward the vertical axis 56 of vessel 10. The slope is sufficient to aid the flow downwardly of the solids forced into the space between immediately adjacent annular plates, e.g., 52 and 54. The annular plates may be perforate or imperforate, desirably the latter.

FIGURE 35: HYDROLYSIS REACTOR FOR PRODUCING ALUMINA SOLS FROM ALUMINUM ALKOXIDE AND WATER

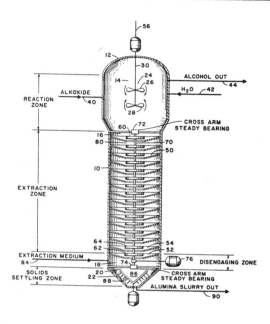

Source: J.W. McCarthy and S.V. Stern; U.S. Patent 3,384,458; May 21, 1968

Rotatable dispersers 60, 62, 64, etc., are positioned above each annular plate 50, 52, 54, etc., respectively. In this embodiment, all the dispersers except 60 are positioned intermediately of adjacent annular plates. The dispersers are in coaxial relation with shell 12; herein dispersers 60, 62, 64, etc., are positioned on a shaft 70, held in cross-arm steady bearings 72 and 74, which is driven by motor 76. Preferably, the dispersers are blades, e.g., turbine, paddle or bar, and have an effective length slightly larger than center opening 80 of the annular plates. Disengaging zone 18 is immediately below and in direct communication with i.e., opens directly into, extraction zone 16. Ingress means 84 provide an entry for liquid extraction medium. Desirably the entry point is near the upper portion of zone 18 to afford a longer disengaging distance for the solids to concentrate as they fall into solids settling zone 20 which is immediately below and in direct communication with zone 18.

Settling zone 20 is shaped to provide a thick slurry body (compact) of solids which acts to prevent by-passing of liquid through the solids discharge means 86. Solids discharge means 86 includes means for withdrawing slurry while maintaining a thick slurry body of solids in zone 20. Herein a rake 88 rotates against conical sides 22 to move the compacted solids out into the discharge conduit 90. The speed of rotation of rake 88 is such that the compact is maintained at the amount needed to avoid liquid by-passing. The operation of vessel 10 is described in connection with the manufacture of alumina from aluminum alkoxide having 11 to 14 carbon atoms in each alkoxy group, in a vessel having an internal diameter of the extraction zone 16 of about 7 feet. Water and alkoxide are introduced about midway dual contra-pitched propellors 26 and 28; reaction zone 14 is maintained at a temperature in the range of 140° to 230°F. and a pressure in the range of about 15 to 90 psia. The particular temperature is determined by the alkoxide, the rate of hydrolysis desired and the alumina crystal size desired; the pressure is determined by the need to maintain the extraction solvent liquid at the particular temperature. In general shaft 30 is rotated at about 5 to 10 rpm to provide the vigorous agitation and downward push on the alumina particles formed by the alkoxide-water reaction.

Annular plates 50 etc., are affixed to the periphery of shell 12, are imperforate, are inclined downwardly and have a central aperture of about 2.5 feet. Dispersers 60, etc., are single bars about 2.5 feet long positioned midway between immediately adjacent plates. A gentle dispersing movement is imparted to the thin slurry in zone 16 by having dispersers

60 rotate at about 5 to 20 rpm; the alumina solids and extraction liquid, e.g., butanol, are thoroughly intermingled as the bars force the thin slurry between the plates and the solids settle and slide from the inclined plate to the center opening. The solids settle out of the extraction liquid and form a thick slurry essentially free of product alcohol, e.g., 1% or less; this compact is raked at a slow speed to discharge thick slurry at about the rate at which this builds up in order to maintain a condition wherein by-passing of solvent butanol is essentially eliminated.

Such a reactor provides the following advantages and benefits:

(1) Initial investment and operating costs are substantially reduced by eliminating the extra foundations, pumps, piping, valves, and instruments required for multivessel designs.

(2) Alumina carry-over in the decanted product alcohol stream is reduced.

(3) Complete conversion of aluminum alkoxides is insured by providing high water concentrations and efficient mixing in the region of alkoxide injection.

(4) High product alcohol recoveries are provided by efficient intimate countercurrent contacting of the alumina with the extraction medium using stagewise, agitated contacting compartments.

(5) Alumina buildup in the compartments and the possibility of plugging during operation is prevented by the agitation in the extraction section; the sloping interstate baffles allow disengagement of the alumina and solvent and at the same time further minimize plugging tendencies.

(6) The conical disengaging section at the bottom of the column allows the alumina to settle more readily and to compact thus permitting a more concentrated slurry to be withdrawn with subsequent drying and/or filtration savings. A slowly revolving rake in the bottom of the disengaging section prevents liquid by-passing through the thickened slurry.

Engelhard Industries

A process developed by C.D. Keith and K.W. Cornely; U.S. Patent 3,488,147; January 6, 1970; assigned to Engelhard Industries, Inc. is one in which hydrous alumina is made by reaction of water and aluminum of about 2 to 100 microns in size. In order to carry the reaction to near completion the reaction mixture is scoured until substantially all particles have a size less than 50 microns. The reaction mixture scoured preferably has at least 5% of aluminum-containing particles larger than 50 microns in size, and the particle size reduction is made by withdrawing a portion of the reaction mixture from the reaction zone, subjecting the withdrawn material to scouring and subsequently returning it to the reaction zone. A colloid mill is advantageously employed to scour, and the volume of material milled is at least about 10 times the volume of the reaction mixture. The aluminum starting material can have a surface area of about 75,000 to 1,000,000 square mm. per gram, and the reaction can be conducted at either an acid or basic pH. The process may be conducted using the apparatus shown in Figure 36.

FIGURE 36: APPARATUS FOR THE PREPARATION OF ALUMINA SOLS BY HYDROLYSIS OF ALUMINUM METAL

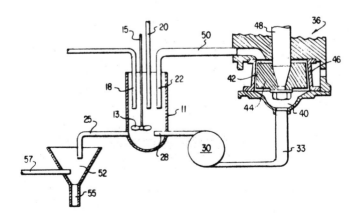

Source: C.D. Keith and K.W. Cornely; U.S. Patent 3,488,147; January 6, 1970

In the drawing, 11 represents a reactor tank having the agitator 13 fastened to the rotatable shaft 15. The reactor tank is provided with the upper inlets 18, 20 and 22 and the lower outlets 25 and 28. Outlet 28 leads to a pump, preferably the centrifugal pump 30 which has the outlet 33, which in turn leads to the colloid mill, indicated generally by 36. The colloid mill used in this process may advantageously be a Charlotte-type colloid mill. Such a mill has the inlet chamber 40 leading to an adjustably narrow annular passage 42 between a conical rotor 44 and a stator 46. The rotor 44 is operated by the shaft 48 and the annular passage 42 leads to the mill outlet 50 connected to the reactor inlet 22.

The reactor vessel 11 may be provided with an additional lower outlet where it is desired to allow settling in the reactor itself at the end of the reaction. This outlet provides for removal of the heavier layer. Preferably, the product outlet 25 from the reactor leads to a separate decantation or settling vessel 52 which can have a lower outlet 55 for the heavy unreacted aluminum powder and an upper outlet 57 for product colloidal alumina hydrate. In the process, the inlets 18 and 20 may be used, interchangeably if desired, for the induction to the reactor vessel of water, metallic aluminum and other materials used in the reaction. The aluminum is often fed to the reaction as a slurry in water. During, or at the end of the reaction period, pump 30 may be operated intermittently or continuously to send portions of the mixture in the reactor to and through the colloid mill and back to the reactor through the inlet 22. The following example illustrates the process. The aluminum used had a sieve analysis as follows:

	Percent
+200	0
200 to 230	14
230 to 270	8
270 to 325	14
–325	64

It will be noted that about 36% of the particles fail to pass a 325 mesh sieve. The colloid mill clearance was set at 0.001 to 0.003 inch equivalent to a particle size of about 25 to 75 microns. To a 50 gallon reactor fitted with a high-speed two-bladed agitator, a reflux condenser and a thermo-regulator was added 45 gallons of deionized water and, over a period of about 8 hours, 1,700 cc of 88% formic acid and 14.25 pounds of atomized aluminum metal of the size range described above as about a 20% slurry in water. The formic acid solution was added continuously at a rate of about 3.5 cc/min. The reaction started at ambient temperature, the temperature increasing to about 100°C. About 2 hours after the last addition of aluminum, a conversion of about 92.4% had been attained and the reaction appeared to stop. More than 1% of aluminum-containing particles larger than 50 microns in size remained in the slurry.

The centrifugal pump and colloid mill were then started and run at 5 gallons per minute for 30 minutes. The batch turnover, therefore, was about 3 during this run. This procedure was followed for a total milling period of about 5 hours. The initial mill clearance of about 3 mils was gradually reduced to about 1 mil as the maximum particle size diminished. Besides giving a cleaner separation by decantation, that is, giving a hydrate product containing little aluminum and little loss of hydrate in the aluminum layer, this procedure served to increase the conversion level. At the end of the run, 99.2% of the aluminum had been converted to hydrous oxide and substantially no particles greater than 44 microns remained.

Universal Oil Products

A process developed by J.C. Hayes; U.S. Patent 3,535,268; October 20, 1970; assigned to Universal Oil Products Co. is a process for the continuous manufacture of an alumina sol. Aluminum particles are continuously charged to an aluminum digestion zone and digested in contact with hydrochloric acid admixed with recycled sol. The hydrochloric acid-sol mixture is processed in contact with the aluminum particles at conditions to obviate formation of undesirable high molecular weight sol polymers. The apparatus which may be used in the process is shown in Figure 37.

The schematic drawing shows an aluminum digester 1, a sol receiver 2, and an acid charge tank 3. In a lined-out operation, the digester will contain an inventory of unreacted aluminum pellets 16, the inventory being maintained by a flow of aluminum pellets through line 4 from a feed hopper not shown. The aluminum pellets are charged to the digester at a rate substantially equivalent to the rate of digestion therein. The digester also contains an upwardly flowing acidic mixture which exits from the digester as an alumina sol through line 5 to be transferred to the sol receiver 2.

A liquid level is maintained in the sol receiver by means of a level controller 6 with alumina sol product being recovered through line 7 at a rate substantially equivalent to product make. An overhead line 8 is provided to vent any gaseous product carried to the receiver 2 by the sol stream. Alumina sol is continuously withdrawn from the receiver 2 by way of line 9 to be recycled to the digester. Dilute hydrochloric acid is charged to the process from the acid charge tank 3, water being charged to the tank through line 10, and hydrochloric acid through line 11 and admixed therein. The diluted acid is then charged through line 12 to be commingled with recycled alumina sol in line 9. The acidic mixture is then continued through line 9 by means of a recycle pump 13 and charged through 14 to the digester 1 to pass upwardly therethrough. Excess hydrogen formed in the digester is recovered overhead through line 15 at a rate to maintain the desired pressure in the digester.

FIGURE 37: CONTINUOUS PROCESS FOR THE MANUFACTURE OF ALUMINA SOLS BY THE HYDROLYSIS OF ALUMINUM METAL

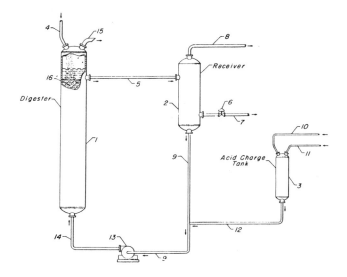

Source: J.C. Hayes; U.S. Patent 3,535,268; October 20, 1970

PREPARATION OF ALUMINA SPHERES

Inasmuch as the catalyst particles in moving-bed processes in particular are continuously colliding and rubbing over each other, the exposed edges and nonuniform surfaces of the catalyst particles are frequently fragmented with the formation of fines which are carried off by the gas stream. Moreover, while molded and extruded pellets, broken pellets and the like are customarily used in many processes, such pellets are not suitable in certain operations because of the mechanical nonuniformity, because of uneven packing in the bed and because of the extensive attrition of particles due to exposed rough edges. It has been well recognized that heavy losses in a reactor due to the creation or presence of excessive fines therein may be a function of the shape and hardness of the catalyst particles employed. For this reason, it is evident that in fixed and moving bed processes the use of stronger, substantially spherical or bead shaped catalysts will markedly reduce losses due to attrition. Moreover, these bead shaped particles are capable of packing well to form a homogeneous bed which minimizes channeling of fluids and promotes uniformity in contacting.

In the past, numerous methods have been devised and employed for the preparation of spherical catalyst particles. Numerous refinements of the basic process for their preparation have been advanced. In general, catalysts of this type are produced by suspending drops of aqueous gel-forming liquid in a water immiscible liquid for a sufficient time so that the drops set. The setting is usually accelerated by heating and by a change of pH. The aqueous liquid in addition to containing soluble materials may additionally include insoluble materials such as initially produced gelatinous precipitate or a finely divided powder of any desired composition. After setting, the beads are withdrawn from the aqueous phase and are slowly dried until free of excess moisture. The dried particles are then calcined.

American Cyanamid

A process developed by J.M. Witheford and L.A. Cullo; U.S. Patent 3,154,603; October 27, 1964; assigned to American Cyanamid Company involves the production of spherical contact particles having improved attrition resistance and controlled physical properties. In work leading to the process it was discovered that spherical contact masses having superior attrition resistance and controlled physical properties may be obtained in a surprisingly straight-forward, continuous method which may be carried out in simple equipment thereby reducing the manufacturing cost of spherical or beaded catalyst particles. In accordance with this process particles of desired shape, size and hardness are prepared by expressing a deformable plastic contact mass through an orifice to form a deformable plastic extrudate; successively breaking off the expressed extrudate to form deformable plastic cylinders; shaping the deformable plastic cylinders in a fluid medium by means of centrifugal forces to form substantially spherical bodies and drying the bodies to obtain spherical contact particles.

This process enables one to control physical properties such as surface area, pore volume, pore diameter, etc. of beaded catalysts by controlling not only the variables in the gel manufacturing stage in well-known manners but also by suitable variation of the plastic mass prior to extrusion. It has also been found that not only does the spherical shape of the particles impart improved attrition characteristics but also because the particles are formed by expressing, shaping and drying the deformable plastic contact mass while in a hydrous state, the bonding forces between the microscopic particles making up the mass are increased thereby resulting in higher intrinsic hardness. Figure 38 shows the sequence of operations in the process.

FIGURE 38: APPARATUS FOR PRODUCTION OF ALUMINA BEADS BY SHAPING IN A CYCLONE

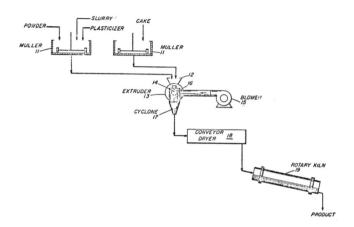

Source: J.M. Witheford and L.A. Cullo; U.S. Patent 3,154,603; October 27, 1964

As seen in the figure, a powder and a slurry of suitable composition are blended with a plasticizer in a muller 11 while a cake of suitable composition is passed through a second muller 11 and the mulled materials are admitted to the hopper 12 of a double worm screw extruder 13 supplied by the Welding Engineers Company, series 2,000, driven by a 7.5 horsepower motor at a selected rpm of either 50 or 70. The main screw of the extruder was a constant depth, deep screw which was coated with wax or with a polyfluorohydrocarbon resin and which had a true displacement of 2.3 inches per revolution. The die 14 was machined from a 2 inch diameter polyfluorohydrocarbon resin rod. The die depth was 1 inch and the die holes were drilled either as a single row across a diameter of the face (1/4 and 3/16 inch) or over the entire face of the die on a 60° triangular pitch (3/32 and 1/16 inch). The die was inserted into a steel collar which was threaded on each end for connection to the extruder and the shaping apparatus.

The shaping apparatus was a cyclone 17 ordinarily employed for the separation of solid particles from fluid streams. A turbo-blower 15 capable of developing 4 psig at 300 cfm supplied air for cutting and beading. The air was metered through an orifice and introduced into the cyclone by a flat nozzle 16 with an opening 3 x 5/16 inch. The cyclone was connected to the extruder by a coupling welded into the entrance duct, into which was screwed the steel collar containing the poly-fluorohydrocarbon die having an exposed die face 1.5 inch diameter, and brought flush to the inside of the entrance duct wall. The air nozzle was then adjusted so that the air would be directed across the exposed die face.

The cyclone was constructed of sheet stainless steel having a fairly smooth finish. The bottom discharge was modified by a 1/4 to 1/2 inch shoulder which substantially increased the particle residence time in the system. The product discharged from the cyclone was collected continuously, either by trays or by a gyratory screen which also served to classify the material. The beaded product was dried by employing forced circulation tray dryers 18. However, a rotary dryer may also be used with advantage. For calcination of the dried spherical contact particles, an indirectly fired rotary calciner of kiln 19 was used. The following is a specific example of the conduct of the process.

A hydrated alumina for use as a catalyst base is obtained by the following procedure. A heel of water (140 lbs.) is charged into a suitable reaction vessel and is heated to approximately 95°F. The agitator is turned on. An aluminum sulfate solution (36 lbs. of 8% Al_2O_3) is then added simultaneously with 68 lbs. of 15% aqueous ammonia over a period of 30 minutes. Temperature during reaction is maintained between 95° to 105°F. The precipitated alumina is washed free from salts and other impurities. The alumina is then washed over two stages of filters. After the first stage of washing, the filter cake is repulped with the addition of 1% polyacrylamide, based on alumina solids, filtered and washed over a second stage filter. The second stage filter cake is then dried to approximately 22% solids. Based on a preparation involving 25 lbs. of alumina cake, to this cake is added 1.5 lbs. of finely divided α-cellulose. The mixture is thoroughly blended and mulled. The mulling step is carried out by preextruding the mass through holes of larger diameter than the final product. The thoroughly blended, preextruded mixture is fed to the extruder with a solids content of about 24% and extruded through a die having orifices of 1/16 inch in diameter.

The extrusion rate is varied from 2.3 lbs./min. to 2.7 lbs./min. The extrudates emerging from the die are cut in short lengths, approximately 1/16 to 1/4 inch, by air which is supplied at the rate of 80 to 100 cu. ft./min. The cylindrical extrudates after cutting are quite plastic and deformable and while in this state are conveyed to a cyclone device whereby the extrudates are forcibly rotated around the containing walls of the cyclone. The extrudates are formed into balls by this action and under the force of gravity are discharged from the bottom of the cyclone. The alumina balls are then dried at 250°F. for 8 hours following which they are dried at 1100°F. for 1 hour.

Engelhard Industries

A process developed by C.D. Keith and K.W. Cornely; U.S. Patent 3,558,508; January 26, 1971; assigned to Engelhard Industries, Inc. is one whereby spherical aluminum oxide gels are made by feeding drops of a coagulable aqueous slurry of hydrous alumina into a column of a water-immiscible liquid, e.g., mineral oil, which can be maintained at close to ambient temperature. The hydrous alumina is prepared by hydrolysis of finely divided aluminum having a surface area of about 75,000 to 1,000,000 square mm./g., and the hydrolysis medium is acidic from the presence of a nonoxidizing acid for instance, formic acid. The hydrous alumina feed has a ratio of alumina monohydrate to amorphous hydrous alumina of at least 0.5/1 and the alumina monohydrate has a crystallite size of less than 65 A. Ammonia can be added to the column to aid the coagulation. Other solids, for instance, calcined alumina, alumina trihydrate, silica or carbon can be incorporated in the aqueous slurry as can aluminum oxychloride. Drying and calcination of the coagulated hydrous alumina provides adsorptive solids with advantageous characteristics when used, for instance, as catalyst components.

As shown in the flow diagram in Figure 39, a hydrous alumina slurry can be passed from slurry reservoir 1 through pipe 2, slurry pump 3 and valve 4 to the multiple orifice feed distributor head 5 located at the top of oil column 9 and containing a multiplicity of hypodermic syringe discharge needles 6 which extend through column head 7. The column head contains ammonia takeoff line 8 which is used to bleed excess ammonia from the system. The syringe needles employed can vary in size so as to give spheroidal particles of desired diameter, for instance, from about 0.007 inch to about 0.25 inch on the the basis of a calcined product, preferably about 0.04 to 0.20 inch. The slurry is then dispersed as drops from the syringe needle tips and the drops fall into a column of water immiscible liquid. The water-immiscible liquid is treated with ammonia which is introduced at the bottom portion of column 9 through feed line 14, valve 15, flow meter 16, and a porous sparger 17. When the drops of slurry initially contact the immiscible liquid they are usually lens-shaped.

However, as the drops of alumina slurry descend through the column of ammonia-treated immiscible fluid, they gradually age to produce spherical slurry particles of desired firmness. The spheroidal particles pass from the oil column to the collection vessel 19. When desired, the valve 18 can be closed and the collection vessel containing the spheroidal particles removed and replaced with a new oil-filled vessel. The valve is then opened and the process continued. The oil can be drained from the collection vessel containing the spheroidal particles through sieve 20, removed through line 21 by the operation of valve 22, passed through centrifuge 23 and recycled to oil sump 11. Oil overflow line 10 permits oil to be returned to the sump for subsequent charging to the column via pump 12 and valve 13.

FIGURE 39: APPARATUS FOR PRODUCTION OF ALUMINA SPHERES IN AN OIL-FILLED TOWER

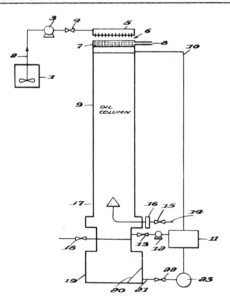

Source: C.D. Keith and K.W. Cornely; U.S. Patent 3,558,508; January 26, 1971

Figure 40 shows an alternative mode of operation. In this operation an amorphous alumina/boehmite slurry prepared by the acid hydrolysis of aluminum metal having a composition of 47% amorphous alumina/53% boehmite (20 A. boehmite crystallite size), and representing 55% by weight of the total alumina desired was intimately mixed with a quantity of calcined alumina (gamma form) which represented 45% by weight of the total alumina present. The gamma alumina used had a particle size distribution: 18% (by weight) 1 to 5μ; 43% (by weight) 5 to 10μ; 28% (by weight) 10 to 15μ; 11% (by weight) 15 to 23μ. To this slurry was added 2% by weight of carbon powder (100% through 270 mesh, 53 micron openings) based on the total calcined products present. The total amount of solids, exclusive of the carbon, in the slurry was 18.6%. The slurry was then further mixed and spheres were formed by pumping the resulting slurry by means of a gear pump to a multiple-orifice head 101.

The thusly formed spheres were passed through a column of mineral oil 102 which had been saturated with gaseous ammonia by means of sparger 103. The ammonia was added at the rate of 500 to 700 cc of NH3/min. The column of oil was 12 ft. high and 18 in. in diameter and had a viscosity of 65/75 SUS at 100°F. As the droplets contacted the oil surface they immediately began to gel, and assumed a spheroidal shape which was retained and became more firm as they descended through the oil in the column. The travel time for the spheres through the oil column was about 40 seconds. The spheroidal particles passed through the oil/NH3 interface 104 and were transported by means of liquid flow induced at point 105 by pump 106 throughout the piping indicated by 107.

The transport liquid from pump 106 was 6% aqueous ammonia which is heavier than the oil phase and also serves as an aging medium. The time of transport from interface 104 to hold tank 108 varied from 30 seconds to 1.5 minutes, often being about 35 seconds. The thusly formed spheres were held in tank 108 for from 2 to 6 minutes and discharged to container 109 by opening valve 110 after which tank 108 was again isolated by closing valve 110. The spheres were then discharged from tank 109 to a suitable drainage system 111 from which the aqueous ammonia was drained to tank 112 which served as a reservoir for pump 106. Pipes 113 and 114 were utilized to handle any ammonia and oil overflow, respectively. The drained spheres were then washed, dried at 240°F. for 16 hours and calcined at 1022°F. for 4 hours to give the finished spheroidal particles.

FIGURE 40: ALTERNATIVE APPARATUS FOR PRODUCTION OF ALUMINA SPHERES ON AN OIL-FILLED TOWER

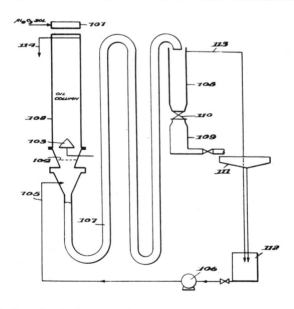

Source: C.D. Keith and K.W. Cornely; U.S. Patent 3,558,508; January 26, 1971

Koninklijke Zwavelsuurfabrieken Voorheen Ketjen

A process developed by H. Mathies; U.S. Patent 3,464,928; September 2, 1969; assigned to Koninklijke Zwavelsuurfabrieken Voorheen Ketjen, NV, Netherlands involves producing hydrogel beads by forming drops of an inorganic oxide in air above a body of water-immiscible liquid through which the drops pass so as to be formed into spheroidal globules before passing into an underlying body of an aqueous coagulating medium where the sol globules are coagulated to firm hydrogel beads and which may contain a surfactant to reduce the surface tension of the interface between the water-immiscible liquid and the aqueous coagulating medium. A water-immiscible liquid, such as, kerosene, benzene, carbon tetrachloride, vaporizing oil or mixtures thereof, is dispersed in the hydrosol with the aid of a substantially water-insoluble emulsifier consisting of an ethoxylated organic compound containing a number of polyoxyethylene groups equal to 1/2 to 1/6 the number of carbon atoms in the chain of such organic compound, whereby to obtain relatively low-density hydrogel beads with a very uniform distribution of pores therein and with a relatively high crushing strength. The process can be modified to include catalytic agents in the beads. For example, finely divided molybdenum oxide can be added to the hydrogel along with the water-immiscible liquid and emulsifier and the firm hydrogel impregnated with cobalt salt to form a disulphurizing catalyst. Also, the firm hydrogel can be subjected to acid extraction followed by impregnation with platinum and a halogen to form a reforming catalyst.

Mizusawa Kagaku Kogyo

A process developed by D. Kunii, Y. Sugahara, K. Sato, M. Saito and M. Ogawa; U.S. Patent 3,290,790; Dec. 13, 1966; assigned to Mizusawa Kagaku Kogyo Kabushiki Kaisha, Japan relates to a method of drying hydrogels characterized in that the hydrogels are heated together with porous particles, the average apparent specific gravity and average diameter of which are smaller than the average apparent specific gravity and average particle diameter of the dried gel. In the known method of drying hydrogels containing more than 90% of water such as silicahydrogel, aluminahydrogel and silica-alumina hydrogel, on account of the hydrogel's volume being 20 times as large as that in dried condition, much labor and sizable equipment have been required. Furthermore, in the conventional methods, more than 10 to 30% of the product was either cracked or damaged resulting in low yield rate. Accordingly a solution to the drying method has been a critical problem from the standpoint of thermal efficiency, labor consumption and yield rate of products, without any satisfactory method having been found heretofor.

In particular, extreme difficulties have been encountered in drying spherical hydrogels, wherein they were dried at 110°C. or higher for 30 to 40 hours. This is because when the gels were dried while flowing the air according to the general principle for the drying, cracking or destruction occurred on nearly all gels, and therefore it was necessary to dry the gels in such condition that the steam evolves slowly, without permitting air going into the dryer. More particularly the hydrogels had been dried incompletely or unevenly, and accordingly, an unnecessarily long time was required for the drying. In addition, the spherical hydrogels are fragile spherical substances which crack easily with a small amount of impact or break up with slight pressure. Further, the spherical gels are apt to stick to each other during the drying process and also thermal conduction among the spherical gels is excessively poor during the process. Syneresis fluid is discharged from the hydrogels as the gels shrink when heated, and silica or alumina contained in the fluid binds the spherical gels together, as a result of which the gels, for example stick to the wall surface of the cylindrical drying tower forming a mass of gels without dropping downward, thus presenting a cause of uneven drying.

The hydrogels having such characteristics as described above, when the gel spheres are dried in, for example, a glass tube, some spheres shrink too small while some shrink little on account of the interspace among them. This presents a strong tendency of uneven drying which is a cause of cracking. In contrast with the above, this process shows that crack free dried gels can be produced industrially easily by heating the hard-handling and easy-cracking hydrogels together with porous particles, the average apparent specific gravity of which is smaller than the average apparent specific gravity of the dried hydrogels finally obtained after drying and the average particle diameter of which is smaller than that of the dried hydrogels. Figure 41 shows the essential component of the process.

FIGURE 41: APPARATUS FOR DRYING HYDROGELS

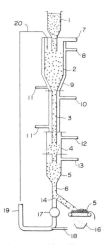

Source: D. Kunii, Y. Sugahara, K. Sato, M. Saito and M. Ogawa; U.S. Patent 3,290,790; December 13, 1966

Gel-like spheres, for example, finished spherical alumina-gel or activated clay, etc., having been mixed uniformly in advance are fed into the drying tower 2 through a hopper 1. It is usually preferable to use a powdered or granular substance of 10 to 150 mesh normally in an appropriate rate of more than 0.3 part per 1 part of the gel by weight. An intensive evaporation from the gel-like granular material occurs when heated through a metal plate or mixed with the heated powdered or granular substance circulating from a circulation tube 20 toward the top of the tower and the vapor generated while ascending is accompanied by or mixed with the fine powdered or granular substance, thereby causing intensive motions of the substance in the interspaces among the gel-like granules to carry the heat.

As can be seen in the above, the fine powdered or granular substance acts as a medium for heating, thereby effecting the tower interior being heated evenly. As previously stated, the surface of the gel-like granule is very vulnerable but the powdered or granular substance is prevented from friction or collision with the tower wall surface or other granules owing to the intervention of the powdery or granular substance and hence, the possible damage or destruction thereof is eliminated. The gel-like granules, a large part of water of which has been vaporized in the drying tower 2 then enter into a dry finishing tower 3, where they are subjected to the heat through the metal plate or the heat of the powdered or granular substance dropping through the interspace of the granules, thereby having the free water therein further being evaporated. The gel-like granules, after the residual water has been evaporated, enter into a calcination area 4, where they are fully calcined by the heated air blown in through and discharged from pipe 12, 13 while maintained at, for example, 140° to 160°C.

A mixture of dried granule product and the powdered and granular substance moves to a separation plate 14, passing through the hopper 5 and a pipe 6, where the dried granules are separated and removed continuously by means of a discharging apparatus 15, and the fine powdered or granular substance remaining in the dried granules is further separated into a receptacle 16. On the other hand, the powdered or granular substance descends through the separation plate 14, and while maintained at a predetermined flow rate by means of a powdered or granular substance flow regulator 17, circulated toward the top of the tower 2 through a lift tube 19, then through the circulation tube 20 being propelled by air or heated gas discharged from a nozzle 18. However, it is apparent that the circulation of the powdered or granular substance is not always necessary. Whereas a heating source can be either aqueous vapor, heated gas or electric heater, in the instance of the apparatus shown in the drawing it is an aqueous vapor or heated gas which is sent from 8 and 11, and discharged from 11.

It is also possible to compress the aqueous vapor discharged from 7 by means of an aqueous vapor blower or the like to raise the temperature and send the same from 8 and 10. As described hereinbefore the process makes it possible to effect drying of the hydrogels industrially in a short period of time without cracking or destruction thereof. Furthermore, more uniform products and products excelling in adsorption efficiency are obtained.

Universal Oil Products

A process developed by K.D. Vesely; U.S. Patent 3,496,115; Febraury 17, 1970; assigned to Universal Oil Products Company relates in general to the manufacture of spheroidal inorganic oxide particles and, in particular, to the manufacture of spheroidal alumina particles. Figure 42 is a flow diagram of the overall process.

FIGURE 42: APPARATUS FOR FORMING AND AGING SPHERICAL ALUMINA PARTICLES

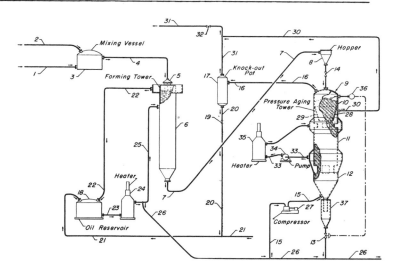

Source: K.D. Vesely; U.S. Patent 3,496,115; February 17, 1970

There is shown a blending tank 3 wherein an alumina hydrosol, introduced through line 1 from storage, and hexamethyl-enetetramine, utilized as the weak base and introduced through line 2 from an external source, are blended in the desired ratio. The alumina hydrosol is then charged through line 4 to a dropping head 5 and dispersed as droplets in the hot oil suspending medium continuously circulated through the forming tower 6 at about 200°F. The alumina hydrosol droplets, while traversing the hot oil in the forming tower are formed into semisolid spheroidal particles which leave the bottom of the forming tower and are transported in the oil through line 7 and deposited in a hopper 8 atop the pressure aging tower 9.

In accordance with the process, the spheroidal particles in traversing the pressure aging tower 9 from top to bottom, pass through three aging zones labeled 10, 11 and 12 at a pressure to maintain the water content of the particles in substantially liquid phase, the particles being collected in a receiver 37 and thereafter discharged from the pressure aging tower 9 through a valve 13 at about atmospheric pressure. The spheroidal particles are admitted to the first aging zone 10 by means of a valve 14, and are initially contacted therein with a counterflow of oil at a temperature of from about 120° to 220°F., the optimum temperature being selected to correspond with the temperature employed in the aforesaid forming tower 6. Thus, the alumina spheroidal particles of the example are contacted with the counterflow of oil at an initial contact temperature of about 200°F.

The oil utilized in the pressure aging tower 9 is charged thereto from line 15 as will hereinafter appear. The oil is processed through the three aging zones in the manner and under conditions hereinafter described, the oil being withdrawn overhead from the pressure aging tower 9 through line 16. The oil thus withdrawn is processed through a knock-out pot 17 and thereafter returned to an oil reservoir 18 passing through a pressure reducing valve 19 situated in line 20 and then by way of line 21 to the reservoir 18. In the oil reservoir 18, the oil from the pressure aging tower 9 is combined with oil recovered from the forming tower 6 through line 22. Oil from the reservoir 18 is continuously withdrawn through line 23 and passed through a heating means 24 to be recycled to the forming tower 6 and the pressure aging tower 9, and to serve as a transfer assist in carrying the pressure aged particles from the pressure aging tower 9 to conventional sphere washing means not shown. Thus, the heated oil is charged in part through line 25 as recycle oil to the forming tower 6 and in part through line 26 as recycle oil to the pressure aging tower 9, passing through line 15 including compressor 27 in the process. Hot oil is continued through line 26 to assist in transferring aged particles, discharged from the pressure aging tower through valve 13, the spheres being carried in the oil to the washing section which is not shown.

Also, line 21 is utilized to return the oil from the washing process, the oil being combined with return oil from line 20 and returned to the reservoir 18. Referring again to the first aging zone 10 of the pressure aging tower 9, the spheroidal particles are brought into contact with a counterflow of oil at the described temperature conditions and processed downwardly through the oil in a dense phase passing to a second aging zone 11 through a frustracone section 28 shown within the first aging zone 10. The first aging zone 10 is a heatup zone, its principal function being to effect a gradual heating of the spheroidal particles to a temperature of from about 240° to about 500°F., preferably not exceeding a temperature of about 350°F. A temperature in the preferred range is suitably employed at pressure of from about 40 psig to about 150 psig and sufficient to maintain the water content of the particles in a substantially liquid phase. The heatup is accomplished by heating the oil to the desired temperature in the second aging zone 11 and processing the same upwardly through the first aging zone 10 at a rate to establish the desired initial contact temperature therein.

This in effect produces a temperature gradient in the first aging zone 10 progressing from a relatively high temperature at the bottom of the zone to a relatively low temperature at the top of the zone. The spheroidal particles are processed downwardly through the first aging zone 10 at a rate to effect the desired gradual heatup. The spheroidal particles thus heated in the first aging zone 10 are continued downwardly in a dense phase and pass through the aforesaid frustracone section into the second aging zone 11. As a result of the spheroidal particle treatment in the first aging zone 10, the particles can now be processed at relatively high temperatures without the creation of stress cracks or sphere rupture. In aging zone 11 the aging process is virtually completed. The weak base contained in the spheroidal particles, in this case hexamethyl-enetetramine, is substantially completely hydrolyzed at a temperature of from about 240° to about 500°F., a temperature of from about 240° to about 350°F. being more suitable, to form ammonia and carbon dioxide.

In the schematic flow diagram provision is made for the collection of carbon dioxide and discharge of the same from the second aging zone 11. Thus, the carbon dioxide passes upwardly as formed into the void space 29 created by the frustracone 28 and is discharged through line 30 and line 31 passing through a pressure control valve 32 to the atmosphere. The carbon dioxide is separated from the process in the described manner to alleviate the turbulence which would otherwise occur in the first aging zone 10 with the constant disruption of the temperature gradient therein. In a preferred embodiment, the oil is circulated downwardly in the second aging zone 11 in concurrent flow with the spheroidal particles passing therethrough. In the flow diagram, the oil is withdrawn from the bottom portion of the second aging zone 11 through line 33 and recycles by means of a pump 34 through a heating means 35 to the top portion of the second aging zone 11. It has been found that by processing the oil downwardly in the second aging zone 11 as herein described, a better heat distribution is effected.

Included with the oil withdrawn through line 33 is oil processed upwardly through the hereinafter described third aging zone 12. The net effect of this arrangement is to establish the desired circulation of hot oil downwardly through the second aging zone 11 with excess recycle oil passing upwardly through the first aging zone 10 to effect the desired temperature gradient therein. The third aging zone 12 is provided to cool the spheroidal particles to a temperature of from about 120° to about 220°F. prior to discharge from the pressure aging tower 9 and subsequent washing procedures. Thus, the spheroidal particles pass downwardly into the third aging zone 12 still in a dense phase with the oil contained therein. As has been stated, the oil is charged to the third aging zone 12 through line 15 at a temperature of from about 120° to about 220°F. The oil passing upwardly to admix with the hot oil from the second aging zone 11 in effect creates a temperature gradient in the third aging zone 12 so that the spheroidal particles are gradually cooled to the desired temperature. The flow of spheroidal particles through the pressure aging tower 9 is regulated by a level controller 36, which activates the particle discharge valve 13 whereby the spheroids are discharged into line 26 at atmospheric pressure.

After the aging treatment, the spheres are washed in any suitable manner. A particularly satisfactory method is to wash the spheres by percolation, either with upward or downward flow of water, and preferably with water containing a small amount of ammonium hydroxide and/or ammonium nitrate. After washing, the spheres may be dried at a temperature of from about 200° to about 600°F. for 6 to 24 hours or more, or dried at this temperature and then calcined at a temperature of from about 800° to about 1400°F. for 2 to 12 hours or more, and then utilized as such or composited with other catalytic components. It is preferred that the spheres be dried slowly and also that the drying be effected in a humid atmosphere because this has been found to result in less breakage of the spheres. The alumina spheres may be used as an adsorbent, or refining agent to treat organic compounds, and are also particularly satisfactory for use as a component of a catalyst.

PREPARATION OF OTHER ALUMINA SHAPES

Du Pont

A process developed by H. Talsma; U.S. Patent 3,397,154; August 13, 1968; assigned to E.I. du Pont de Nemours and Company yields a catalytic structure comprising a catalytically active metal component (preferably at least 0.01% by weight) deposited on a porous refractory body comprising a skeletal structure of crystalline interconnected walls which define cells with an average diameter of 0.5 to 200 mils, the walls being dense and having a thickness of between 0.3 and 200 mils and containing at least 30% by weight of Al_2O_3. The wall material may be alpha-alumina or it can be constituted by compounds and solid solutions of alumina and at least one other oxide or solid solutions of at least one oxide in the compounds of alumina.

The catalytic structure of the process is prepared by heating aluminum particles and at least 0.02% by weight of the aluminum of a metal oxide fluxing agent, in the presence of oxygen and at a temperature between 400° and 1500°C., until at least 10% (preferably 90%) has been oxidized and at least 1% of the aluminum is present in a state of oxidation below a valence of 3, and depositing thereon the aforementioned catalytically active metal component. The preparation of the catalyst support itself may be carried out in two general ways. In the preferred method aluminum particles are oxidized in situ, in a porous body having a porosity of at least 20% so that an alumina-containing skeletal structure is produced. This process is described in detail in U.S. Patent 3,255,027. The unfired body can contain from 0 to 8 parts of a particulate filler refractory per part of aluminum. In order to gain the greatest amount of catalytic activity the unfired body should contain less than 33% of the filler refractory.

The catalyst support can also be prepared by oxidizing aluminum particles so separated from one another that discrete particles are produced rather than the unitary skeletal structure. The aluminum particles, for example, can be conveniently kept apart during oxidation by mixing them with a powdered refractory which does not react with aluminum or alumina to give an alumina-containing compound under the firing conditions. The porous mixture is then fired as above. The discrete particles of alumina may be separated in any convenient manner as by screening.

In the preparation of such catalyst supports, aluminum particles and particles of alloys of aluminum with other metals in which aluminum constitutes the major component may be used. The metal used is preferably clean and free from grease and oil. The aluminum particles should have one dimension of at least about 7 mils (preferably 10 mils), a second dimension, of at least 0.5 mil and a third minor dimension, i.e., between about 0.5 mil and 200 mils. For example, if spheres of aluminum are used, they must have a diameter between 7 and 200 mils (between about 3.5 and 80 mesh). Cylindrical rods as fibers must have a diameter of between 0.5 and 200 mils and a length of at least 7 mils. The length is not critical and can vary from short staple to continuous filaments. The aluminum particles can be organized into ordered structures such as a honeycomb, woven screen, etc.

Suitable fluxing agents for use in forming the catalyst supports include an oxide of a metal from the class consisting of the alkali metals, the alkaline earth metals, vanadium, chromium, molybdenum, tungsten, copper, silver, zinc, antimony, and bismuth, precursors of these oxides and hydroxides of the alkali metals. The oxides and hydroxides of the alkali metals, magnesium, strontium and barium are preferred. Figure 43 shows a longitudinal section and a transverse section of an automotive exhaust gas reactor using a catalyst produced by this process. In the drawing, a catalyst-impregnated honeycomb structure 2 is located within the enlarged portion of stainless steel reactor shell 1 in sections a through e. Opening 3 of the shell receives the gases to be oxidized and in the case of automobile exhaust gases would be attached to the exhaust ports of the cylinders. Opening 4 provides means for diluting the exhaust gases with air, while opening 5 provides an exit for the gases from the system.

FIGURE 43:   HONEYCOMB CATALYST STRUCTURE FOR OXIDATION REACTOR

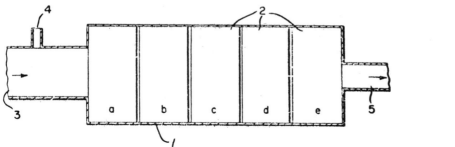
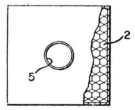

Source: H. Talsma; U.S. Patent 3,397,154; August 13, 1968

Universal Oil Products

A process developed by G.L. Hervert; U.S. Patent 3,492,148; January 27, 1970; assigned to Universal Oil Products Company yields an alumina coated catalyst-carrying composite element which comprises a metallic element with an intermediate layer of porcelain and an outer layer of porous gamma-alumina. A preferred form of the composite has the alumina held to and partially embedded into the porcelain layer as a result of the forming of the alumina in situ from aluminum particles suspended in a volatile liquid vehicle being applied to the intermediate procelain coating and then oxidizing at a temperature sufficient to soften the porcelain layer to in turn tenaciously hold a resulting gamma-alumina layer.

URANIUM, THORIUM AND RARE EARTH CATALYSTS

CONDENSATION

Rutgerswerke und Teerverwertung

A process developed by M. Froitzheim, K.F. Lang, L. Rappen and J. Turowski; U.S. Patent 3,347,936; October 17, 1967 assigned to Rutgerswerke und Teerverwertung AG, Germany relates to the condensation of phenol (C_6H_5OH) with methanol and has particular relation to such condensation in the presence of a catalyst containing magnesium oxide and uranium oxide and, if desired, boron oxide, with the formation of 2,6-dimethylphenol and, as by-product, o-cresol.

Preparation of 2,6-dimethylphenol has been the subject of various research studies, because this compound is a demanded starting material for the preparation of synthetic plastic materials, particularly polyethers, such as polyphenylene oxide. The o-cresol which is obtained as by-product in varying amounts depending on the conditions of the process is being used in increasing amounts as pesticide. Figure 44 shows the arrangement of apparatus which may be employed in the conduct of the process.

The catalyst may contain a high proportion of magnesium oxide and a lower proportion of uranium oxide, or a high proportion of uranium oxide and a lower proportion of magnesium oxide. Catalysts which contain, in addition to these components, other metal oxides in minor proportions are also effective. In the following examples 1 mol of phenol was condensed with 4 mols of methanol at a temperature of 440°C. by passing the reaction mixture through a reaction tube. Examination of the reaction products was carried out by chromatographic processing. Formation of residues was examined by fractional distillation.

The process can be carried out on a commercial scale as follows. A tube reactor, the tubes of which have a diameter of 80 mm., are filled with strands of 3 mm. diameter of a catalyst of UO_3—MgO—B_2O_3. In order to prepare a catalyst consisting of 70% MgO, 17% B_2O_3 and 13% UO_3, 600 grams of solid boric acid are suspended in 6 liters of water and into the resulting suspension 2,400 grams of MgO are stirred in and thoroughly mixed with the suspension.

In order to attain this, the addition of further 3 liters of water is necessary. The uniform mixture thus obtained is formed to strands, dried at 120°C. and is calcined in the tube furnace, while passing air through the furnace, at 600°C. After discharge from the furnace, the material is ground, mashed with a solution of 683 grams of uranyl-nitrate hexahydrate dissolved in 2 liters of water, shaped and dried at 120°C. Decomposition of the nitrate is brought about in a tube furnace by heating with passing air through the furnace and finally the material is subjected to calcination at 650°C. After discharge and cooling, the material is comminuted and shaped to 3 mm. moldings. 3 liters of the catalyst are thus obtained. The tubes of the furnace are filled with the molded catalyst and by means of a pump and a vaporizer into the furnace, 1 mol equivalent of phenol is introduced, after being heated in superheater to 440°C.

In the same manner 4 mol equivalent of methanol are vaporized and superheated and mixed with the stream of phenol prior to entering the reactor. The reaction of phenol with methanol yields 2,6-dimethylphenol and water and is exothermic. The reaction heat is utilized for the production of steam of 20 atmospheres. The reaction takes place at 450°C. No waste gas is formed and the loss of methanol amounts to 8%, based on the weight of the charge introduced into the reaction. The reaction product thus formed is processed by distillation. In a first distillation column, in which distillation is preferably carried out under pressure, the unreacted methanol is recovered and is reintroduced into the process.

In a second distillation column the reaction water is separated, which contains the phenol still present. In a third distillation column o-cresol is continuously obtained and can be reintroduced, if desired, into the process. In a fourth column, at the head of the column, a technically pure 2,6-dimethylphenol (of 95%) is obtained in addition to a residue consisting of 2,4-dimethylphenol and mesitol. 3-methylphenol and 4-methylphenol are absent in the reaction product.

FIGURE 44: PROCESS FOR ALKYLATING PHENOL WITH METHANOL USING URANIUM-BASED CATALYST

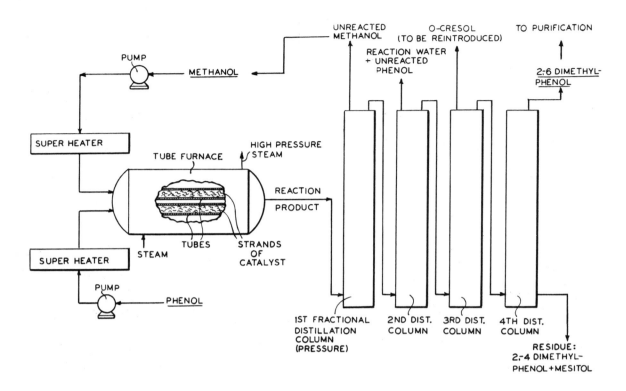

Source: M. Froitzheim, K.F. Lang, L. Rappen and J. Turowski; U.S. Patent 3,347,936; October 17, 1967

Further purification of the 2,6-dimethylphenol takes place by redistillation in a manner known by itself. The 2,6-dimethylphenol thus prepared has a solidification point of 45°C. and a concentration of at least 99.8%. If the catalyst used in the tube furnace shows reduction of its activity, the proportion of methanol is increased for a short period of time, whereupon the original activity of the catalyst will be obtained again.

DESULFURIZATION

Esso Research and Engineering

A process developed by C.L. Aldridge; U.S. Patent 3,494,860; February 10, 1970; assigned to Esso Research and Engineering Company is a continuous process for the desulfurization of heavy petroleum oils in the presence of steam and a catalyst comprising a supported metal salt in which the metal is selected from the group consisting of metals of Groups I-A, III-B, V-B, VI-B, VII-B, VIII-B, thorium and the Lanthanide Series of the Periodic Table.

The most effective method for removal of sulfur from heavy oils without substantial conversion of the oil to light hydrocarbons is catalytic hydrodesulfurization. Obviously the high cost of hydrogen places severe economic debits on this process. For this reason, desulfurization with steam is of interest because of the relatively low cost of steam. The principal object of the process is to provide a process for the desulfurization of heavy petroleum fractions with steam in the presence of an active catalyst. The process is centered on nondestructive desulfurization as distinguished from destructive desulfurization, e.g., steam cracking. Thus conversion to gas and light ends is minimized. Another object of this process is to disclose a continuous catalytic steam desulfurization process featuring a durable, regenerable catalyst.

The most preferred catalysts for this process are hydrated mixed rare earth oxides, hydrated lanthanum oxide, and thorium oxide. The catalyst support is an important aspect of the process. The support must be able to withstand continuous intimate contact with steam during the desulfurization reaction and frequent contact with hot gases during regeneration to remove carbonaceous deposits laid down during the reaction. Suitable support materials include alumina, silica alumina, bauxite, kieselguhr, molecular sieves, natural and synthetic zeolites, magnesia, charcoal. Alpha alumina is the most preferred support. The catalyst should have a surface area of at least about 1.0 m.^{2}/g. The finished catalyst can be in the form of extrudates, pills, spheres or any other attrition resistant form. Figure 45 shows the essential apparatus components involved in the conduct of the process.

FIGURE 45:   FLOW DIAGRAM OF DESULFURIZATION USING RARE EARTH CATALYST

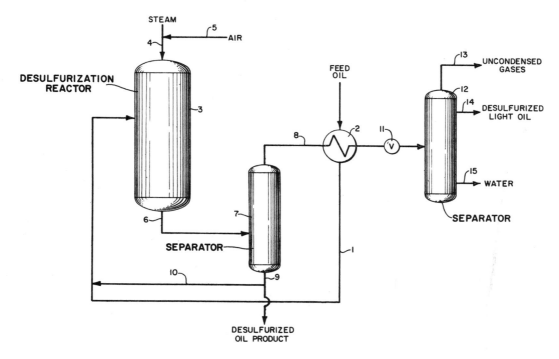

Source: C.L. Aldridge; U.S. Patent 3,494,860; February 10, 1970

Referring to the drawing, an oil feed is passed by line 1 through heat exchanger 2 to the upper section of reactor 3. The oil is preferably heated to a temperature near the desired temperature of the reactor. Steam enters the reactor via line 4. Air for in situ feed oxidation and/or regeneration of the catalyst is fed into the reactor via lines 5 and 4. In this embodiment the reactants pass cocurrently down through the fixed catalyst beds. If desired the reactor can be operated with cocurrent flow of steam, air and oil upwardly through the reactor or the steam and air can be passed upwardly, countercurrent to the downwardly moving oil. Typical steam desulfurization reaction conditions are as follows:

	Broad range	Preferred range (Residuum feed)
Temperature, °F	400–1,200	650–900
Pressure, p.s.i.a	5–1,000	15–200
Fresh feed rate, w./hr./w. on cat	0.05–5.0	0.2–1.0
Steam rate, wt./wt. on feed	0.02–3.0	0.1–0.4
Air Rate:		
Oxidation (Wt. percent O₂ on oil)	0.1–5.0	0.5–1.3
Regeneration (vol. percent air on steam)	1.0–75.0	10–30

In this embodiment a fixed bed of catalyst is used, however alternate means of contacting such as a fluidized bed, moving bed, slurry, etc. can be employed. Regeneration is accomplished by cutting off the flow of oil and passing a steam-air mixture from lines 4 and 5 through the catalyst bed under combustion conditions to burn carbonaceous deposits from the catalyst surfaces. Air contents of 10 to 30 (volume percent air on steam) and bed temperatures in the range of 500° to 2000° F. are used for regeneration.

A plurality of reactors can be operated in stages so that one or more reactors are in the regeneration stage while the others are in steam desulfurization operation. It is within the scope of the process to pass oil and catalyst through the reactor in slurry form and to regenerate catalyst in a separate regenerator. It is also within the scope of the process to omit regeneration and to periodically remove a portion of spent catalyst from the reactor and replace it with fresh catalyst. Steam desulfurization effluent is removed from reactor 3 by line 6. The effluent unreacted air, steam, hydrocarbon gases, H₂S, hydrocarbon vapors, desulfurized liquid oil and partially desulfurized oil. The effluent is passed into separator 7. In the case of a gas oil process feed a major amount, e.g., 50 to 100% of the desulfurized material will be in the vapor phase. If the feed is a vacuum residuum a major amount, i.e., 50 to 100 volume percent will be in the liquid phase. Gases and vapors are passed overhead from the separator by line 8. Liquid material is recovered from the separator by line 10. If desired, part of the oil, e.g., from 5 to 75 volume percent is recycled via lines 10 and 1. Desulfurized product is recovered from line 9.

The gas and vapors product are passed by line 8 through heat exchanger 2 and pressure reduction valve 11 into separator 12. Separator 12 is operated at a temperature in the range of 100° to 300°F. and a pressure in the range of 15 to 100 psia. Uncondensed gases are recovered by line 13. Desulfurized light oil is recovered by line 14 and water is recovered by line 14 and water is recovered by line 15. The process provides a means for the desulfurization of heavy oils without the use of expensive hydrogen. Instead low cost steam is the desulfurization agent and air can be used to oxidize, e.g., activate the feed. The catalysts employed in the process are active promoters of the hydrolysis reaction and are durable in continuous operations including regeneration.

Nicklin-Farrington

A process of T. Nicklin and F. Farrington; U.S. Patent 3,475,328; October 28, 1969 involves removing organic sulfur compounds from hydrocarbons in the gas phase. One or more hydrocarbons are mixed with steam and passed in the vapor phase over a catalyst containing uranoso-uranic oxide and/or uranium trioxide supported on an inert carrier such as aluminum oxide. The reaction temperature is between 350° and 500°C., the pressure being atmospheric or higher.

The uranoso-uranic oxide and/or uranium trioxide content of the catalyst may vary from 1 to 80%, preferably from 5 to 20%, and advantageously from 5 to 10%. Certain supports have been found to be more useful than others. Some supports, notably those of acidic character such as silica, although providing effective catalyst for short runs, suffer from the disadvantage of tending to cause carbon to be deposited on them. The effectiveness or otherwise of a catalyst made using a particular support may be readily ascertained by simple trial and experiment.

The process is preferably concerned with the use of catalyst supports for the uranoso-uranic oxide which are stable, i.e., which are of suitable mechanical strength, do not melt or sinter or undergo chemical change (for example reduction) under the conditions of the desulfiding reaction, and which have a surface area that is not materially affected under such conditions. According to one method of preparing the catalyst compositions, a compound of uranium that decomposes on heating to give uranoso-uranic oxide and/or uranium trioxide, for example, uranium nitrate, or uranium acetate, is dissolved in water and the catalyst support is added in powder form. The carrier may then be removed and heated to a temperature desirably below 500°C., and pelleted.

In another method, shaped pieces of the stable carrier are impregnated with a solution of a uranium compound as mentioned above and heated to a temperature desirably below 500°C. to form the uranoso-uranic oxide and/or uranium trioxide. The impregnation and heating may be repeated if necessary to ensure adequate coating of the carrier surface with uranoso-uranic oxide, and/or uranic trioxide. In yet another method, mixtures comprising a uranium compound as mentioned above, and a fusible salt which on strong heating gives an oxide which constitutes a stable carrier, are heated together until they melt and then heated to a temperature desirably below 500°C. so as to decompose the uranium compound and the salt. The product is ground and pelleted. Figure 46 shows a pilot plant apparatus which may be used for the conduct of this catalytic process.

FIGURE 46:   FLOW DIAGRAM OF DESULFURIZATION PROCESS USING URANIUM OXIDE CATALYST

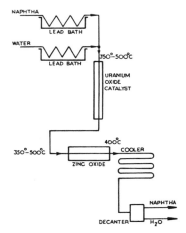

Source: T. Nicklin and F. Farrington; U.S. Patent 3,475,328; October 28, 1969

The following is a specific example of operation using such an apparatus. Esso Grade A naphtha and water after passing through meter pumps are vaporized and superheated to 350° to 500°C. in a gas fired lead bath before mixing together at the inlet to an upward flow reactor containing catalyst (3 ft.^{3}) made up in the form of 1/8 in. pellets. The desulfiding process proceeds at a temperature of from 350° to 500°C., and the emergent gases passed through a bed of zinc oxide granules (1 ft.^{3}) at a temperature of 400°C. to remove the H_2S. The desulfided product is cooled and the water is separated out by means of a decanter.

HYDROCRACKING

Mobil Oil

A process developed by T.-Y. Yan; U.S. Patent 3,546,100; December 8, 1970; assigned to Mobil Oil Corporation is one in which a rare earth exchanged crystalline aluminosilicate hydrocracking catalyst may be improved with respect to its cracking activity and selectivity by using water in controlled amounts to activate the catalyst cracking sites.

The hydrocracking operation is employed for many purposes including the cracking of high and low boiling hydrocarbon fractions such as distillate fractions boiling above about 300°F. and more usually those fractions boiling above about 400°F. and having an end boiling point as high as 900° to 1000°F. Hydrocarbon feeds including virgin gas oils, coker gas oils, cycle stocks and combinations thereof boiling in the range of 400° to 900°F. are often employed in hydrocracking operations for conversion to gasoline and jet fuel products.

The hydrocracking operation is generally carried out at a temperature selected from within the range of from about 400° to about 1000°F. but more usually the reaction temperature is selected from within the range of 500° to about 850°F. The hydrogen pressure in such an operation may be substantially any pressure selected from within the range of about 100 to about 3,000 psig but preferably pressures are selected at the lowest value within this range which will permit one to obtain the conversion and the product selectivity desired for a suitable catalyst on-stream life. The liquid hourly space velocity selected for hydrocracking is most usually selected from within the range of 0.1 to 10 and the molar ratio of hydrogen to hydrocarbon charge is maintained as low as possible in the range of 2 to about 80 and more usually in the range of from about 5 to about 50.

There is a net consumption of hydrogen during the hydrocracking operation and this consumption is dependent upon feed composition, severity of conversion and the olefin and aromatic materials encountered in the operation. Hydrogen consumption within the range of 500 to 3,000 scf/b. (standard cubic feet/barrel) are not unusual and are to be expected. Reaction temperature may also vary considerably for a given conversion level and may be expected by virtue of this process to be from about 5° to about 60°F. lower but more usually from about 10° to about 25°F. lower than would be normally experienced when hydrocracking with other types of hydrocracking catalysts.

Thus the activity of rare earth exchanged zeolites are significantly increased by operating in an atmosphere of limited moisture content. Such an activity increase permits a decrease in operating temperature to achieve a given conversion level. It has been found further that maintaining a moist atmosphere as described herein is particularly effective for improving the catalyst selectivity by reducing light ends production. That is, the production of C_1 to C_4 hydrocarbons was found to be significantly reduced.

The hydrogenation-dehydrogenation component or components which may be admixed with the rare earth exchanged crystalline aluminosilicates include metal oxides and sulfides of metals of the Periodic Table which fall in Group VI-A including chromium, molybdenum, tungsten and the like and Group VIII metals including cobalt, nickel, platinum, palladium, rhodium and the like and combinations of metal sulfides and oxides of the metals of Groups VI and VIII such as nickel, tungsten, sulfide, cobalt, molybdenum oxides and the like and hydrogenating components can be used in amounts ranging from 0.1 to about 10 and as high as 20 weight percent based upon the hydrocracking catalyst. Particularly suitable hydrocracking catalysts employed in the method of this process are the large pore crystalline aluminosilicates such as faujasite zeolites which are promoted with one or more platinum group metals and which have been prepared under conditions so as to contain after base exchange no more than about 5% sodium calculated as Na_2O and preferably no more than 1 or 2% of sodium.

OXIDATION

Commissariat à l'Energie Atomique

A process developed by H. Bottazi, B. Claudel, Y. Trambouze, A. de Calmes, H. Fould, and P.Y. Ledoray; U.S. Patent 3,444,098; May 13, 1969; assigned to The Commissariat à l'Energie Atomique, France relates to thoria catalysts for the combustion of gases and it is more especially concerned with catalysts to be used in afterburners for the exhaust gases of engines such as internal combustion engines or other thermal engines, of furnaces, ovens, etc. The process consists chiefly in preparing thoria by heating of a thermally decomposable thorium salt such as thorium nitrate or oxalate, possibly in the presence of a compound, capable of being decomposed by heat, of an element chosen among transition elements, such as uranium nitrate, or from the group of lanthanides, such as cerium nitrate, in proportions such that the percentage of oxide of the element with respect to the thorium oxide ranges from zero to some units percent, and preferably averages 1%, the rate of heating that is adopted ranging from 2° to about 10°C. per minute, and being preferably equal to 8°C. per minute, this heating being pursued up to a temperature ranging from 280° to 600°C., the heated products being kept at the final heating temperature for a long time, preferably at least ten hours.

The most common thorium salts, such for instance as the nitrate and the oxalate, are well adapted for carrying out the process. When these salts are heated, they decompose so as first to give intermediate products. The final decomposition products, to wit thoria is obtained only at temperature above 280°C. When these thorium salts, either taken alone or on a support, are heated in such manner that their temperature rises from 2° to 10°C. per minute, there is obtained, when heating is stopped at a temperature ranging from 280° to 600°C. a thoria which is catalytically active with respect to imperfectly burned gases.

However, the products of thermal decomposition must be subjected to the final heating temperature for a relatively long time, generally at least ten hours, in order to obtain a pure thoria free from anions. It will be noted that the catalytic power of the thoria that is obtained is maximum if the rate of heating is about 8°C. per minute and, when starting from thorium salts taken alone, when the final temperature remains between about 280° and 320°C.

The heating characteristics are narrowly connected with the passage of the various thorium salts through various hydration steps. Passage from the thorium nitrate through the step of pentahydrate is very advantageous because it permits of obtaining a thoria having a high catalytic power. This pentahydrate is in the form of small agglomerates, the grain size of which averages 0.5 mm., this grain size being besides variable with the hygrometric degree of atmospheric air, due to the hygroscopicity of this salt. The thoria that is obtained is in the form of small porous grains, the specific area of which averages 65 m.^{2}/g. or 30 m.^{2}/g. according as the starting material is nitrate or oxalate, respectively. The pure thoria that is thus obtained constitutes an excellent catalyst to finish the combustion of the nonburned portions of the exhaust gases of an engine. Advantageously this thoria will be used on a support.

However, although thoria already constitutes by itself an excellent catalyst for the after-burning of exhaust gases, its activity will be still enhanced by the addition of an activating or promoting agent. As promoting agent one may use elements chosen among transition elements or rare earths. Thus, it is advantageous to add to the thorium nitrate, previously to the thermal treatment thereof, a small amount of cerium nitrate. This last mentioned body is without influence on the course of the reaction, in view of the fact that it syncrystallizes with thorium nitrate, or on the final surface state of the thoria that is obtained. Although the presence of the promoting substance is not quite necessary and its concentration in thoria may be very variable and as high as some units percent, it was found that the best results were obtained when the concentration of the addition element averages 1% (calculated as oxide). When the addition element that is chosen consists of cerium oxide the maximum activation effect is obtained when its percentage in thoria reaches 0.96%.

The chief advantage of this catalytic thoria with respect to known catalysts for the after-burning of exhaust gases, such as iron, cobalt, nickel, platinum, vanadium compounds, lies in its catalytic activity which increases with the temperature. Increase of the temperature is unavoidable due to the exothermic character of the after-burning. It involves a quick deactivation of the known catalysts and their poisoning due to the presence in the exhaust gases of sulfur and lead tetraethyl and consequently a reduction of their activity which requires their frequent replacement. On the contrary, thoria prepared according to the process is not sensitive to poisoning agents even when it is overheated during its use to 600°C., at which temperature its catalytic activity is not altered.

A process developed by R. Bressat, A. de Calmes, B. Claudel and Y. Trambouze; U.S. Patent 3,459,682; August 5, 1969; assigned to Commissariat à l'Energie Atomique, France utilizes a highly reactive monophased oxidation catalyst consisting of a mixed oxide of thorium and uranium. The process for making this catalyst, includes heating a complex oxalate or mixed oxalate or uranium and thorium up to a decomposition temperature ranging from about 400° to about 500°C. Concerning the catalysts themselves, they consist of a monophased mixed oxides of thorium and uranium whose formula is:

$$U_xTh_{1-x}O_{2+y}$$

in which x is lower than 0.35 and y ranges between 0 and 0.3, such monophased oxides presenting, on the one hand, substantially the crystalline cubic structure with centered faces of thoria in which atoms of thoria have been substantially replaced by atoms of uranium and, on the other hand, a specific surface ranging from about 20 to about 30 square meters per gram. Concerning the methods for preparing these catalysts, they consist of, first, forming either a complex oxalate of general formula

$$(NH_4)_2[U_xTh_{1-x}(C_2O_4)_3]\cdot nH_2O$$

in which x has already the value it should have in the final mixed oxides desired and n is an integer, or a mixed oxalate whose formula is

$$U_xTh_{1-x}(C_2O_4)_2 \cdot nH_2O$$

in which x and n have the same meanings as above and, second, in decomposing under the ambient atmosphere and at a temperature ranging from about 400° to about 5000°C. the complex oxalate or mixed oxalate, this decomposition being preferably performed by raising the temperature gradually and linearly within a period of time of about three hours, up to the final decomposition temperature, for instance 500°C., and then by maintaining the product under this temperature for about 20 hours.

Standard Oil (Ohio)

A process described by R.K. Grasselli, and M.S. Friedrich; U.S. Patent 3,666,822; May 30, 1972; assigned to Standard Oil Company (Ohio) provides catalysts which are useful in the oxidation of olefins to aldehydes and conjugated dienes and in ammoxidation of olefins to nitriles. The catalysts comprise the combined oxides of uranium and molybdenum and the combined oxides of uranium and molybdenum in combination with arsenic, bismuth, tin, vanadium, iron, nickel and cobalt.

Another process described by R.K. Grasselli and M.S. Friedrich; U.S. Patent 3,666,823; May 30, 1972; assigned to Standard Oil Company (Ohio) provides catalysts of similar utility which comprise the combined oxides of uranium and arsenic on a catalyst support and the combined oxides of uranium and arsenic promoted by molybdenum, boron, vanadium, tin, nickel, bismuth, chromium, iron, manganese, zinc, tungsten, antimony, cerium, cobalt or rhenium.

LEAD- AND TIN-BASED CATALYSTS

DEHYDRODIMERIZATION

Allied Chemical

A process developed by W.P. Moore, Jr. and J.W. Mosier; U.S. Patent 3,435,089; March 25, 1969; assigned to Allied Chemical Corporation is one in which biallyl is produced by contacting propylene with PbO supported on an inert carrier having a surface area no greater than 1 m.^{2}/g. The reaction is conducted in the absence of molecular oxygen at a temperature of between 320° and 700°C. until the PbO is reduced to Pb$_2$O. The Pb$_2$O is subsequently converted back to PbO by heating in contact with a molecular oxygen-containing gas. The biallyl can be readily separated in high yield and purity from unreacted propylene which can be recycled. The arrangement of equipment which may be used in the conduct of this process is shown in Figure 47.

FIGURE 47:   SCHEMATIC VIEW OF LEAD OXIDE CATALYZED DEHYDRODIMERZATION OF PROPYLENE

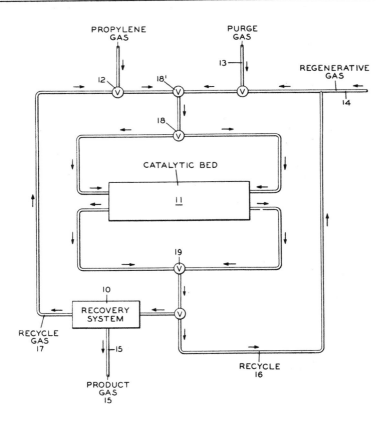

Source: W.P. Moore, Jr. and J.W. Mosier; U.S. Patent 3,435,089; March 25, 1969

The lead oxide catalyst bed 11 is preheated by a hot oxygen-containing gas to 600° to 700°C. after which propylene, flowing countercurrent to the heating gas, at 0.1 to 50 psia, preferably 14 to 18 psia, and space velocity of 2.5 to 40 min.$^{-1}$, preferably 5 to 30 min.$^{-1}$, is fed through valve 12 and feed control valve 18 to the reactor from either end. The entrance point alternates each cycle. Propylene enters the reactor cold and is heated over the lead oxide catalyst bed. When reaction temperature is reached, some lead oxide will be reduced and biallyl, CO_2, H_2O and small amounts of other products are formed. The predominant reactions proceed as follows:

$$2C_3H_6 + 0.5O_2 \xrightarrow{PbO} C_6H_{10} + H_2O$$

$$2C_3H_6 + 9O_2 \xrightarrow{PbO} 6CO_2 + 6H_2O$$

$$4PbO \longrightarrow 2Pb_2O + O_2$$

Product gases exit through valve 19 and are sent to a recovery system, designated 10, where biallyl may be recovered by any conventional means and discharged through line 15. Propylene is recycled for reuse to the synthetic stream through line 17. The synthetic cycle is followed by an inert separatory purge gas, such as steam, which enters through line 13 at 600°C. for 0.5 minute, then by hot air or a mixture of air and combustion gases which enter through line 14 and exit at line 16, the air being indirectly heated to a temperature of 500° to 800°C., preferably 625° to 725°C., at a space velocity of 1 to 8 min.$^{-1}$, preferably 6 to 8 min.$^{-1}$ for reoxidizing and reheating the catalyst bed, whereby the catalyst is restored according to the equation $2Pb_2O + O_2 \longrightarrow 4PbO$.

It appears that the reactions proceed according to the equations outlined above until about 60% of the PbO has been reduced to the suboxide, after which a competing consecutive reaction occurs that reduces the PbO still further to pure lead metal. When the oxide is reduced to lead metal, little biallyl is produced and cracking products and tars are primarily produced. Also, catalyst cannot be regenerated without special treatment requiring removal of the catalyst from the reactor. Thus, in accordance with the process, the synthesis must be closely controlled so that the oxide is not reduced to lead.

DEHYDROGENATION

Asahi Kasei Kogyo

A process developed by N. Kominami, K. Kawarazaki, M. Chono and H. Nakajima; U.S. Patent 3,520,915; July 21, 1970; assigned to Asahi Kasei Kogyo K.K., Japan for the preparation of acrylonitrile or methacrylonitrile by catalytic dehydrogenation comprises contacting propionitrile or isobutyronitrile in the gaseous phase at a temperature between 300° and 700°C. with a catalyst which is a stannous oxide-silica complex. The catalyst is formed by reacting a stannous halide with silica gel in an organic solvent at a temperature between 30° and 350°C., washing the stannous halide-silica reaction product with the organic solvent, hydrolyzing the resulting stannous halide-silica reaction product with an aqueous alkaline solution, removing the alkali substance and subjecting the resulting reaction product to heat treatment at a temperature between 300° and 700°C.

HALOGENATION

Kureha Kagaku Kogyo

A process developed by S. Seki and T. Watanabe; U.S. Patent 3,536,769; October 27, 1970; assigned to Kureha Kagaku Kogyo K.K., Japan is one in which ethylidene fluoride is prepared by reacting acetylene with anhydrous hydrogen fluoride using a cocatalyst comprising anhydrous tin tetrachloride and boron trifluoride.

In general, ethylidene fluoride is mixed with halofluoromethanes to be utilized as a refrigerant or a propellant and also is useful an an intermediate for manufacturing vinylidene fluoride and vinyl fluoride monomers. Ethylidene fluoride has been industrially manufactured by adding hydrogen fluoride to acetylene. However, this addition reaction does not proceed in the absence of a catalyst. A number of catalysts for such an addition reaction have been researched. Typical of these catalysts are chemical materials such as boron trifluoride (BF_3), tin tetrachloride anhydride ($SnCl_4$), fluorosulfonic acid and various other metal oxides and chlorides.

Typical processes for manufacturing ethylidene fluoride using BF_3 as a catalyst have been described in U.S. Patents 2,425,991; 2,762,849 and 3,190,930, etc. Though these processes have excellent characteristics such as high conversion rate and good recycling capability these processes have the following defects for use as industrial manufacturing processes. (1) as described in U.S. Patent 2,762,849, 1.57 mols of BF_3 is required per 2.04 mols of HF. Since it has a very low boiling point (-101°C.), it is thereby not retained in the reaction system, and therefore, in a continuous reaction, BF_3 has to be continually added in large amounts or separated and recycled by expensive equipment. This is economically disadvantageous. (2) In case the BF_3 concentration is decreased, or its catalytic action is reduced, vinyl fluoride is a by-product, which must be separated and purified. (3) While BF_3 is highly active as a catalyst for an addition reaction, BF_3 accelerates the carbonization of acetylene, vinyl fluoride, and the like, and therefore, not only does the reaction operation become difficult, but catalytic action is inhibited and catalytic life shortened.

An $SnCl_4$ catalyst, as described in U.S. Patent 2,830,099, has an excellent catalytic action in a liquid phase reaction, and does not produce vinyl fluoride. Further, it is comparatively long in catalyst life (in comparison with BF_3) and is easy to use under reaction conditions. However, heretofore the catalyst life and catalytic activity of $SnCl_4$, which are industrially important, have not been sufficiently studied. The following defects have been found.

(A) $SnCl_4$ catalyst is very low in activation velocity and requires an induction time of 2 to 4 hours on a laboratory scale, and 5 to 8 hours on a larger scale. It has been found that when water is introduced into a $SnCl_4$ catalyst system with HF or acetylene, the activation velocity is rapid, and when the water content is poor, the velocity is slow. However, the presence of water reduces catalyst life, and it is not desirable to introduce water into the reaction system. Further until the catalyst reaches an activated state, unreacted acetylene is present in comparatively large amounts and it is not easy to separate unreacted acetylene, ethylidene fluoride, and HF, and therefore, special expensive equipment is required.

(B) With an $SnCl_4$ catalyst, it is very difficult to separate a fresh catalyst and waste catalyst coexistent in the reaction phase and therefore, it is difficult to continuously carry out the reaction by continuously supplying fresh catalyst and removing waste catalyst. Also, fluorosulfonic catalyst alone, if it is not used in large excess amounts, has a low conversion rate and has a very short catalyst life.

It has been known that to compensate these defects, chlorides of tin, titanium and antimony can be added as a cocatalyst to extend catalyst life. However, even by such means, vinyl fluoride is produced and catalyst life is not sufficient. Other metal oxides and chlorides have been studied for use as catalysts, however, these have been too low in selectivity for ethylidene fluoride, and not only produce vinyl fluoride but require high temperatures accompanied by disadvantages such as carbonization.

However, it has been found desirable to prepare ethylidene fluoride by reacting anhydrous hydrogen fluoride with acetylene in the presence of a cocatalyst consisting of anhydrous tin tetrachloride and boron trifluoride. Reaction is preferably in the liquid phase, and the cocatalyst preferably comprises from 1/10,000 mol to 1 equivalent mol of BF_3 per mol of anhydrous tin tetrachloride. The process offers a very high conversion rate to ethylidene fluoride, and an activation velocity and cocatalyst life which are remarkably improved over the same parameters in previous processes.

POLYMERIZATION

Shepherd Chemical

A process developed by J.F. Shepherd and C.L. Bachler; U.S. Patent 3,207,696; September 21, 1965; assigned to The Shepherd Chemical Company relates to activators for use in polymerization of polysulfide resins and related resins. In polymerization or curing of polysulfide and related resins, an oxygen-donating material such as lead dioxide (lead peroxide) hereinafter designated as an activator, is added to a liquid polymer to induce curing of the polymer. The available types of activators tend to vary in activity. In general practice, more than a stoichiometric amount of an activator (7 1/2% or more by weight for lead dioxide) is normally added to the liquid polymer, thus this amount (approximately 50% excess) is used to insure the proper cure. Many theories are advanced as to the reasons for this necessary excess, such as limited surface area available, particle size distribution, amounts and types of impurities aiding reaction interference, and type of surface area. An object of this process is to standardize activators for use as curing agents for polysulfide and related resins, having a reproducible reaction rate and utilizing the total amount of the oxygen-donating material.

In the formation of these improved activators, a water slurry is prepared of an inert carrier of predetermined particle size and then the oxygen donating material is prepared in any of the known or accepted methods of manufacture and in which process the oxygen donating material is coated or deposited on the carrier thereby controlling the particle size of the resultant material. The essential steps in the process are shown in Figure 48 on the following page.

In a mixing tank 10, powdered litharge (PbO), powdered lime (Ca(OH)$_2$), and powdered barium sulfate (BaSO$_4$) are mixed with water to form a slurry. An agitator 11 mixes the powdered solids with the water to form the slurry. The slurry is transferred to a turbo-agitator 13, and chlorine gas is bubbled into the slurry to oxidoze litharge and deposit PbO$_2$ on the particles of a BaSO$_4$ carrier.

FIGURE 48:   BLOCK FLOW DIAGRAM OF PREPARATION OF LEAD DIOXIDE AS ACTIVATOR FOR POLYSULFIDE RESIN POLYMERIZATION

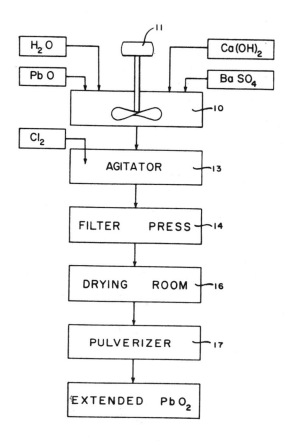

Source: J.F. Shepherd and C.L. Bachler; U.S. Patent 3,207,696; September 21, 1965

The oxidized slurry is then passed to a filter press 14 which separates the extended lead dioxide from the liquid of the slurry. The extended lead dioxide is then washed, dried in a drying room 16, and agglomerations of particles are broken up in a pulverizer 17 to form a powdered extended lead dioxide activator. The following is one specific example of the conduct of the process.

Example: To 1,270 parts of water at room temperature were added in the order given, 90 parts of finely divided commercial barium sulfate, 48 parts of lime (Ca(OH)$_2$), and 127 parts of commercial litharge (PbO). The water and solids were mixed to form a slurry. Chlorine gas was bubbled into the slurry while agitation was continued until 41 parts of chlorine gas had been added over a period of 1 1/2 hours. The slurry was allowed to stand overnight. The solids were filtered out, washed and dried at 230°F. to yield 243 parts of the improved lead dioxide activator. Analysis showed that the product included 52.6% lead dioxide and that 95.2% of the lead oxide had been converted to lead dioxide.

The agglomerates were broken up resulting in an average particle size of 0.31 micron. 6 parts of the improved activator were intimately mixed with 4 parts of dibutyl phthalate to make 10 parts of plasticized activator and the 10 parts of plasticized activator were intimately mixed with 90 parts of commercial liquid polysulfide resin, and the resin set to a tack-free mass in 11.8 minutes.

Solvay & Cie.

A process developed by P. Dassesse and R. Dechenne; U.S. Patent 3,356,668; December 5, 1967; assigned to Solvay & Cie., Belgium is a process for the polymerization of aliphatic 1-olefins in the presence of a catalyst obtained by commingling: (A) a material selected from the group consisting of tetraalkyl tin, tetraaryl tin, tetraalkyl lead, and tetraaryl lead, this material containing up to 40 carbon atoms per molecule; (B) a mineral acid salt of an element selected from the group consisting of titanium, vanadium, chromium, molybdenum, and tungsten; and (C) a halide of an element selected from the group consisting of aluminum, antimony and boron.

The essential feature of the process comprises adding to the mixture forming the catalyst at least one iodide of the class alkali metal iodides, including ammonium iodide, the mol ratio of catalyst component (C) to the iodide being in the range of 1.2:1 to 1.7:1 and the mol ratio of catalyst component (C) to catalyst component (A) is in the range of 0.3:1 to 2:1.

SILICA-ALUMINA CATALYSTS

CRACKING

Esso Research and Engineering

A process developed by D.W. Deed and T.K. Kett; U.S. Patent 3,630,886; December 28, 1971; assigned to Esso Research and Engineering Company is one in which high octane gasoline fractions are produced in a process comprising segregated cracking of virgin and recycle stocks coupled with distillation and extractive distillation to recover alkylation feedstock components and a high octane monocyclic aromatic petroleum fraction suitable for gasoline blending.

Since it may be necessary to reduce or eliminate the use of lead compositions in fuels, refiners are considering means of providing high octane fuel components to maintain octane levels adequate for existing high compression automotive engines. If lead is eliminated the octane loss would be about 6 to 10 numbers and additional processing of refinery gasoline components will be desired to maintain the octane number of the fuels at the required levels. The refinery processing units that are currently available to raise octanes are alkylation, reforming, isomerization, catalytic cracking and hydrocracking. Alkylation provides high octane paraffinic hydrocarbon components and reforming provides high octane aromatic components.

Since cat cracking is the major processing tool employed in modern refineries to reduce molecular weight, it would be desirable to use cat cracking as the basic step in providing the maximum quantity of materials that can be used directly in gasoline and/or upgraded by further inexpensive treating steps to provide high octane gasoline components.

A processing technique is described here in which segregated catalytic cracking steps and extractive distillation are integrated to form high octane naphtha and large yields of alkylation feedstocks. Briefly stated, it comprises the steps of cracking a virgin petroleum feedstock in a transferline type cracking zone, cracking cycle stock in a riser–dense bed cracking zone, distilling the combined cracked effluents, recovering alkylation feedstock and employing extractive distillation to recover a high octane monocyclic aromatic hydrocarbon fraction which is suitable for use in motor fuel.

Figure 49 shows the arrangement of apparatus which may be employed for the conduct of the process. Referring to the drawing, fresh feed is fed by line 1 to the lowermost part of transferline reactor 2. The feed is mixed with regenerated catalyst flowing in return line 3. Fresh makeup catalyst is added via line 4. Suitable fresh cracking feedstocks comprise hydrocarbon fractions boiling in the range of 450°F. to 1100°F., preferably 550°F. to 950°F. Preferred feedstocks include virgin atmospheric gas oils, virgin vacuum gas oils, hydrotreated gas oils, coker gas oils, fractions from solvent extraction, deasphalted oils and mixtures thereof. The preferred catalysts for the process are the crystalline aluminosilicate zeolite types.

In general, the chemical formula of the anhydrous crystalline zeolites employed, expressed in terms of mols, may be represented as:

$$0.9 \pm 0.2 Me_{2/n}O:Al_2O_3:XSiO_2$$

wherein Me is selected from the group consisting of metal cations, hydrogen and ammonia, n is its valence and X is a number in the range of 2 to 14, preferably 2.5 to 6.5. The crystalline aluminosilicate zeolites include synthetic crystalline aluminosilicates, naturally occurring crystalline aluminosilicates, and caustic treated aged clays in which a portion of the clay has been converted to crystalline zeolite. Synthetic materials include faujasites and mordenites. Natural materials are erionite, analcite, faujasite, phillipsite, clinoptilolite, chabazite, gmelinite, mordenite and mixtures thereof.

FIGURE 49:   FLOW SCHEME OF FLUIDIZED BED CATALYTIC CRACKING UNIT

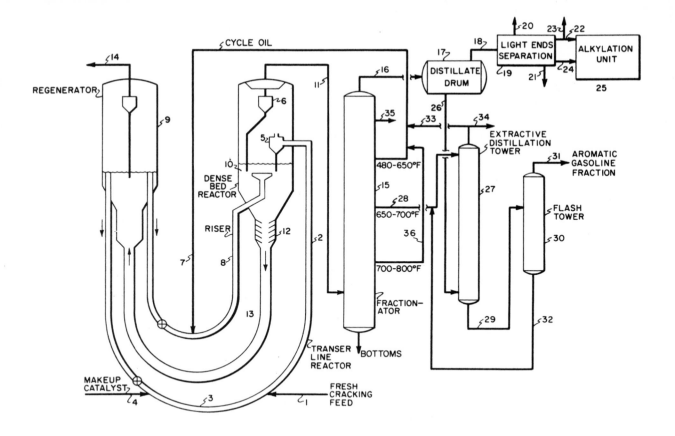

Source: D.W. Deed and T.K. Kett; U.S. Patent 3,630,886; December 28, 1971

Montmorillonite and kaolin clays can be treated to obtain crystalline aluminosilicates. All or a portion of the cations of the zeolites such as sodium cations can be replaced with hydrogen ions, ammonium ions, or metal cations such as rare earths, manganese, cobalt, zinc and other metals of Groups I to VIII of the Periodic Table. Matrix type fluid cracking catalysts in which the zeolite crystals are coated with or encapsulated in a siliceous gel are preferred zeolite type catalysts.

The mixture or dilute suspension of fluidized catalyst and feed in vapor or mixed vapor-liquid phase passes upwardly through transferline reactor 2 at a velocity in the range of from about 6 to about 50 ft. per second. The length-to-diameter ratio (L/D) of the reactor ranges from about 4 to about 50. The space velocity is in the range of 25 to 125 w./hr./w. Because the mixture of regenerated zeolite catalyst and fresh zeolite catalyst is very active, the fresh feed in transferline 2 is cracked in a few seconds; i.e., less than about 30 seconds, and more probably 0.5 to 10 seconds.

Effluent from the transferline reactor is initially separated in rough cut cyclone 5. Separated catalyst passes down the dipleg into the dense bed and cracked effluent passes up through the disengaging zone to cyclone 6. Segregated cycle oil, the source of which will be discussed hereinafter, is fed by line 7 into the bottom of riser 8. The cycle oil is mixed with regenerated catalyst from regenerator 9. The cycle oil is cracked in part in riser 8 and in part in fluidized dense bed 10. Cracked effluent passes through cyclone 6 into line 11 and catalyst is returned to the dense bed via the dipleg. Spent catalyst from the transferline cracking step and from the dense bed is stripped in stripper 12 and passed by line 13 to regenerator 9. Regeneration is conventional. Flue gas is recovered by line 14.

Cracking conditions in the transferline cracking zone and in the dense bed include temperatures in the range of 850°F. to 1050°F. and pressures in the range of 5 to 35 psig. The cycle oil is subjected to more severe cracking action because of the effect of the relatively long residence time; i.e., more than about 30 seconds.

The cracked effluent from the two cracking zones is passed by line 11 to fractionator 15. An overhead fraction having an end point in the range of about 390°F. to 430°F. is taken overhead from the fractionator by line 16. This fraction contains substantial quantities of C$_4$ hydrocarbons used in isoparaffin alkylation and substantial quantities of monocyclic aromatic hydrocarbons which boil in the gasoline boiling range. The overhead fraction is separated in distillate drum 17. An alkylation feed fraction having an end point in the range of from about 100°F. to about 130°F. passes by line 18 to light ends separation unit 19. The light ends unit is operated in the conventional manner to provide any desired type of separation. In this particular embodiment, for example, a gas fraction including C$_2$ minus hydrocarbons and other gases is recovered by line 20.

Propane and n-butane are recovered by line 21. A fraction containing C$_3$ and C$_5$ oleofins; i.e., propylene and pentenes, can be fed by line 22 to the alkylation unit, or alternatively any part of this fraction can be recovered by line 23. A C$_4$ fraction containing butenes and isobutane is passed via line 24 to isoparaffin alkylation unit 25. Alkylation is conventional operation with catalysts such as H$_2$SO$_4$ and HF at temperatures in the range of 20°F. to 100°F. and pressures in the range of 2 to 150 psig.

Returning to distillate drum 17, a hydrocarbon fraction containing substantial quantities of monocyclic aromatic hydrocarbons and typically boiling in the range of from about 115°F. to 410°F. is passed from the drum by line 26 to the lower section of extractive distillation tower 27. The predominantly monocyclic aromatic hydrocarbon fraction is contacted with a solvent under extractive distillation conditions. In extractive distillation the separation of different components of mixtures which have similar vapor pressures is effected by flowing a relatively high boiling solvent, which is selective for one of the components in the feed, down a distillation column as the distillation proceeds.

The relatively less soluble component passes overhead, while the selective solvent scrubs the soluble component from the vapor. The solvent containing the dissolved component is withdrawn from the bottom of the column and the dissolved component and solvent may be separated in an auxiliary unit. Tower 27 can be operated at temperatures in the range of 250°F. to 500°F. and pressures in the range of 0 to 25 psig. Conventional features of extractive distillation such as reboiler elements, reflux systems, bleed streams and pump-arounds have not been shown.

Any suitable solvent for monocyclic aromatic hydrocarbons can be used; however, it is a feature of the process that the solvent can be obtained from the process itself rather than from an external source. Specifically, the solvent can be obtained by recovering a particular fraction from cat fractionator 15. In a preferred embodiment, the solvent is an aromatic hydrocarbon fraction containing a major amount of three ring aromatic hydrocarbons. Thus, line 28 passes an aromatic fraction boiling in the range of from about 650°F. to about 700°F. to the upper portion of tower 27. The multicyclic aromatic fraction passes downwardly through the tower extracting monocyclic aromatics from the extractive distillation feedstock. The extract fraction is passed via line 29 to flash tower 30.

The monocyclic aromatic gasoline fraction is flashed overhead from tower 30 for recovery by line 31. The solvent is recycled via line 32. The nonaromatic raffinate from tower 27 can be recycled to the dense bed cracking step by line 33 and/or any portion of it can be recovered by line 34 as a product of the process. Since two ring aromatic hydrocarbons with fewer than three carbon side chains are not desirable recycle materials, an aromatic fraction of this type is removed from the process by line 35. For operations with insufficient coke make, part or all of this stream may be recycled. The monocyclic aromatic hydrocarbon fraction recovered from the process will have an unleaded octane number of from about 96 to about 102.

In this embodiment the cycle oil comprises two components from fractionator 15. A fraction boiling in the range of 700° to 800°F. is passed by line 36 to line 7 for admixture with a fraction boiling in the range of 480°F. to 650°F. and the mixed fractions are recycled to the dense bed reactor. The composition of the cycle oil is optional and any fraction or mixture of fractions amenable to dense bed cracking can be recycled for this purpose.

One of the major benefits of this process is that it provides an efficient, inexpensive means of recovering high octane aromatic motor fuel components as well as large quantities of alkylation feed. Recracking of desirable aromatic naphtha components is substantially reduced. Segregated cracking of selected fractions at the conditions that are most suitable for each fraction provides more efficient conversion to materials that can be used to raise gasoline octane without depending on lead compositions for octane boost.

Mobil Oil

A process developed by T. Dill; U.S. Patent 3,291,719; December 13, 1966; assigned to Mobil Oil Corporation is a cracking process conducted substantially at feed liquid phase temperature and pressure conditions in the presence of a highly active cracking catalyst.

Since the commercial development of catalytic cracking over 25 years ago, substantially all cracking operations in the industry have been of the type wherein petroleum oils boiling in the range above 400°F. have been fed in vapor form or mixed liquid-vapor form into catalyst-containing reactor at rather high temperatures, generally in excess of 800° F. in order to crack the oils and secure petroleum oil fractions boiling in the motor fuel oil range. In those cases wherein mixed liquid-vapor feed is fed to the reactor, rapid vaporization of the liquid portion of the feed occurs at the catalyst surface or just before reaching the same so that in effect the feed at the moment of cracking is substantially all in vapor form.

The ability for conducting catalytic cracking under relatively mild cracking conditions, i.e., under conditions of pressure and temperature such that the cracking operation is effected at much lower temperatures, which permits maintaining the feed either entirely or predominantly in the liquid phase, has not been completely satisfactory for a number of reasons. Liquid phase cracking at low temperatures has in the past resulted in undesirably low reaction rates and conversion.

This process relates to conducting catalytic cracking at low temperatures, in a predominantly liquid phase, with a suitable catalyst under conditions to limit the single pass conversion of feed to a low level in the range of 5% to 30% by weight and preferably in the range of 10% to 20% by weight of the feed being converted to material boiling the gasoline range. The reaction products having the desired boiling range are separated from the heavier material containing unconverted feed hydrocarbon, with the heavy material being returned to the reactor inlet or alternately, to a separate reactor for further conversion at similar low single pass conversion conditions. These steps are repeated until all or virtually all of the feed has been converted into the desired reaction products or until a desired level of conversion has been obtained.

The instant process has a number of advantages. The low temperature and low single pass conversion exposure in the reaction zone permit conversion of the hydrocarbon at conditions that minimize polymer formation on the catalyst. The liquid phase serves to retard formation of heavy polymers on the catalyst. The heat input to the process is minimized because of the lower temperatures utilized and because no heavy oils are vaporized and recondensed unnecessarily. The heat of the reaction is provided for the low conversion by the sensible heat of the total feed liquid, thus permitting operation at a desired catalyst/oil ratio independent of the reactor heat input requirement.

The separation of the desirable light reaction products can be easily accomplished by simple flash distillation. The equipment investment and operation costs are much lower because of the inherent simplicity of operation and because catalyst regeneration equipment is minimized or eliminated completely. The operation at controlled low conversion permits the removal of reaction products before their concentration in the reaction zone is high enough to result in appreciable further cracking and degradation to dry gas.

The process is based on the fact that the low conversion rate per pass of preferably 10 to 20% does not permit concentrations of converted materials in the reactor to reach the level at which undesirable recontacting and overcracking to coke and/or gas becomes appreciable. Further, any polymers that do form on the catalyst during the reaction can be removed in several ways before condensation thereof to coke-like residues. For example, any polymers formed can be removed from the catalyst by hydrogenating the polymer formed. The polymers formed under the conditions of operation specified herein are far easier to remove from the catalyst than the dense polymers or coke-like polymer product formed under more severe vapor phase conditions at higher temperatures.

According to the process, a charge stock is brought into contact with a superactive catalyst, to be described hereafter, at temperatures under 750° F. and preferably in the range of 300° to 650° F. and pressure sufficiently high so that a liquid phase is maintained in the reaction zone under conditions of low conversion, preferably in the range of 10 to 20% per pass with space velocity ranging upward from 5.0 v./v./hr. The gasoline formed in each pass is removed and the unconverted feed and other heavy materials are re-exposed to the low conversion conditions to obtain the desired ultimate conversion of fresh feed.

It is recognized that operation of a catalytic cracking process in the liquid phase and at low temperatures requires a catalyst having very high cracking activity. Suitable catalysts are aluminosilicates of ordered internal structure. The catalytic materials used in this process are superactive crystalline aluminosilicates, of either natural or synthetic origin having an ordered internal structure. These materials are possessed of very high surface per gram and are microporous. The ordered structure gives rise to a definite pore size, related to the structural nature of the ordered internal structure.

Several forms are commercially available. A 5A material indicates a material of "A" structure and a pore size of about 5 A. diameter. A 13X material is one of "X" faujasite type and 10 to 13 A. pore diameter and so on. There are also known materials of "Y" faujasite type, and others. Many of these materials may be converted to the "H" or acid form, wherein a hydrogen occupies the cation site. For example, such a conversion may be had by ion-exchange with an ammonium ion, followed by heating to drive off NH_3 or by controlled acid leaching. In general, the "H" form is more stable in materials having higher Si/Al ratios, such as 2.5/1 and above.

One material of high activity is H mordenite. Mordenite is a material occurring naturally as the hydrated sodium salt corresponding to:

$$Na_8(AlO_2)_8(SiO_2)_{40} \cdot 24H_2O$$

This mordenite material may be leached with dilute hydrochloric acid to arrive at an H or acid form. In a specific example, the mordenite material may be so treated as to have more than 50 percent in the acid form. Another type of high activity catalyst may be prepared by using Linde 13X molecular sieve, which is described in U.S. Patent 2,882,244. This material may be base exchanged with a solution of rare earth chlorides (containing 4% of $RECl_3 \cdot 6H_2O$) at 180° to 200°F. to remove sodium ions from the aluminosilicate complex and replace at least some of them with the chemical equivalent of rare-earth ions. After washing free of soluble material and drying, there is produced an REX crystalline aluminosilicate containing 1.0 to 1.5 percent (wt.) of sodium and about 25% (wt.) of rare earth ions calculated as Re_2O_3.

Similar preparations of high activity may be made by suitable preparation of a variety of crystalline aluminosilicates such as "Y" type faujasite, gmelinite, chabazite, and the like. For a fuller discussion of the nature of crystalline aluminosilicates and their method of preparation attention is also directed to U.S. Patent 3,033,778 to Frilette and U.S. Patent 3,013,989 to Freeman.

The crystalline aluminosilicate catalysts may be varied within wide limits as to aluminosilicate employed, cation character and concentration, and added components in the pores thereof incorporated by precipitation, adsorption and the like. Particularly important variables are silica to alumina ratio, pore diameter and spatial arrangement of cations. The cations may be protons (acid) derived by base exchange with solutions of acids or ammonium salts, the ammonium ion decomposing on heating to leave a proton. Polyvalent metals may be supplied as cations, as such or as spacing agents in acid aluminosilicates for stabilization. In addition to the rare earth metals mentioned above, other suitable cations for exchange in the aluminosilicates include, for example, magnesium, calcium, manganese, cobalt, zinc, silver, and nickel.

The above discussed catalysts possess activities too great to be measured by the "Cat. A" test which is a standard evaluation test, widely established and used for the evaluation of hydrocarbon cracking catalysts, both for preliminary evaluation and for control during commercial use by examination of activity. In this test, a specified Light East Texas gas oil is cracked by passage over the catalyst in a fixed bed, at a liquid hourly space velocity (LHSV) of 1.0, using a catalyst-to-oil ratio (C/O) of 4/1, at an average reactor temperature of 875°F., and atmospheric pressure (LHSV and C/O are expressed in volumes). The volume percentage of gasoline produced is the activity index (AI).

The method of this test is described more fully in National Petroleum News, 36, page R-537 (August 2, 1944). The control silica-alumina catalyst employed in the alpha rating test hereinafter more specifically described has an AI value of 46. To measure the activity of the instant superactive catalyst there has been developed a micro test method in which these catalysts are compared for relative cracking activity in the cracking of hexane with a conventional catalyst. Alpha is the measure of the comparative conversion ability of a particular superactive catalyst of the type above discussed when compared in the cracking of hexane with a conventional silica-alumina cracking catalyst (90% SiO_2-10%AlO_2) having an activity index as measured by the "Cat. A" test of 46.

Many such superactive catalysts have been found to have an α value of the order of about 1,000 where α is the comparative activity of the catalyst based upon conventional amorphous silica-alumina cracking catalyst as $\alpha = 1$. In order to use such catalyst with conventional equipment and processes available, particularly in the cracking of hydrocarbons, it is first necessary to modify the activity of such superactive catalysts.

One method for the adjustment of activity may be referred to as steam treating, or more shortly, steaming. It has been found that steaming can effect major decreases in the activity of the superactive catalysts utilized herein, and that controlled steaming can be utilized to acquire any desired degree of activity reduction. For example, a crystalline aluminosilicate of the 13X type which has been base-exchanged with a mixture of rare earth chlorides has a relative activity α, when freshly prepared of about 10,000. By controlled steaming in an atmosphere of steam for 5 to 40 hours, at 1300°F., its relative activity can be reduced to an α of about 10.

Another method of modifying such catalysts to reduce their activity is by dilution in a matrix of controlled activity or of little or no activity. Thus, a catalyst, such as RE 13X of $\alpha = $ app. 10,000 may be reduced readily to an activity useful in today's technology by incorporating in a matrix of amorphous silica-alumina, for example, of an activity of $\alpha = $ app. 0.5 to 1.0.

Thus, through combinations of the various methods of adjusting activity of the superactive catalytic materials any desired relative activity can be obtained. For example, the freshly prepared RE 13X of relative activity, $\alpha = $ app. 10,000 which was reduced by steaming to a material of $\alpha = $ app. 10, can be further reduced by compounding with an equal amount of catalytically inert material to an activity, $\alpha = $ app. 5.

The dispersing of the superactive aluminosilicates in a matrix, e.g., clay or inorganic oxides, may often be carried out as just indicated to dilute the very high activity. Moreover, the formation of pellets or beads is very desirable from the point of view of resistance to attrition in the cracking process. Generally spherical beads may be prepared by dispersing the aluminosilicate in an inorganic oxide sol according to the method described in U.S. Patent 2,900,399 and converted to a gelled bead according to the method described in U.S. Patent 2,384,946.

Figure 50 shows the layout of the equipment within which the operations may be conducted using these catalyst materials. In the drawing, a liquid phase multistep catalytic cracking schematic flow sheet is shown which utilizes three reactors. The crude oil feed enters a fractionator 12 through line 10 for separating the feed into two fractions. The lighter fraction leaves the fractionator 12 through line 14 passing into the reactor 16 containing the superactive catalyst. The effluent from the reactor 16 passes through a heater 18 and thence into a fractionator 20 which separates the effluent into a lighter gasoline containing fraction flowing through line 22 and a heavier fraction which is recycled via line 24 to the reactor 16.

FIGURE 50: FLOW DIAGRAM OF LIQUID PHASE CATALYTIC CRACKING PROCESS

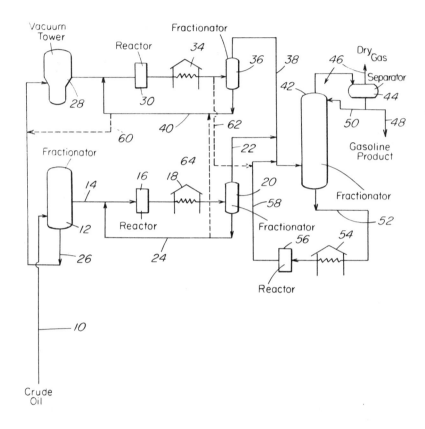

Source: T. Dill; U.S. Patent 3,291,719; December 13, 1966

The heavier fraction leaving the fractionator 12 through line 26 flows to a vacuum tower 28 and thence through line 32 to a reactor 30 containing superactive catalyst. The effluent from the reactor 30 passes through a heater 34 and thence into a fractionator 36 (flash drum) which separates the effluent into a lighter gasoline containing fraction flowing through line 38 and a heavier fraction which is recycled via line 40 to the reactor 30.

The gasoline containing fractions flowing in line 22 and in line 38 are merged and passed into a fractionator 42. The effluent from the upper portion of the fractionator 42 passes through a separator 44. Dry gas flows from the separator 44 through line 46 and the gasoline product flows through line 48. A portion of the gasoline product may be recycled to the fractionator 42 through line 50. The heavier fraction leaving the fractionator 42 through line 52 passes through heater 54 and thence into reactor 56 containing the superactive catalyst. The effluent is recycled into the fractionator 42 with the gasoline product containing streams through line 58.

Any polymers which remain in the liquid oil phase leaving the reactors can be reduced or removed by return of a portion or all of the uncracked oil to the feed preparation system, as shown in broken lines 60, 62 and 64. The conditions in the reactors are maintained so that the catalytic reactions occur in liquid phase at relatively low conversion rates per pass, preferably in the range of 10% to 20%. The velocity should be at least 5.0 v./v./hr. and the pressure can range up to 500 psig and higher, as necessary. The heat absorbed in this system is the sum of the heat required to vaporize the gasoline produced and the heat of reaction to produce the gasoline.

A process developed by J.P. Shambaugh; U.S. Patent 3,328,292; June 27, 1967; assigned to Mobil Oil Corporation also involves the treatment of hydrocarbons in the presence of superactive catalysts. According to this process catalytic cracking occurs in contact with a superactive catalyst in the bottom section of a fractionating column under liquid phase cracking conditions. The catalyst may be continuously drawn off, regenerated and returned to the column, and the products formed in the reaction pass upwardly into the top section of the fractionator for fractionating into the desired number of product fractions, thus eliminating need for separate fractionation and reaction vessels. Figure 51 shows the form of apparatus in which such an operation could be carried out.

FIGURE 51: FLOW DIAGRAM OF LIQUID PHASE CATALYTIC CRACKING PROCESS USING SUPERACTIVE CATALYST

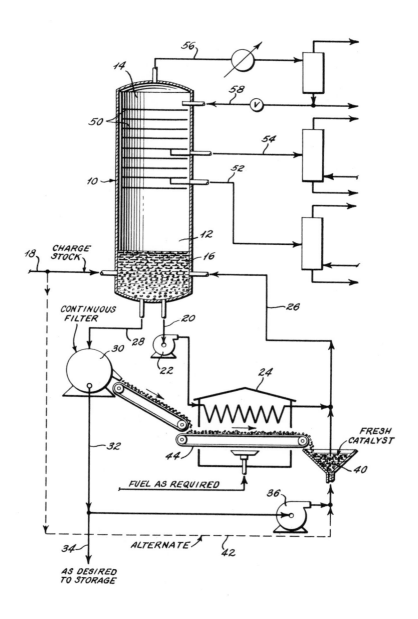

Source: J.P. Shambaugh; U.S. Patent 3,328,292; June 27, 1967

Referring to the drawing, a fractionating column 10 is shown which is divided into a bottom section 12 and a top section 14. A superactive catalyst 16, to be described in detail hereafter, is charged into the bottom section 12 of the fractionator. An inlet conduit 18 conducts the charge stock into the bottom section 12. As necessary to supply heat to the reaction catalyst slurry may be withdrawn from the bottom section 12 through outlet line 20, by a slurry pump 22, to a heater 24 and thence is returned to the bottom section 12 through line 26.

Another stream of catalyst slurry is drawn off through line 28 and is fed to a continuous filter 30 or other means of liquid-solid separation. The liquid product is removed from the separator through line 32 and may be removed to storage through line 34 or when desired to minimize the quantity of bottoms product, repumped into the bottom section 12 by pump 36 and line 26. The catalyst from the separator 30 in the form of filter cake, for example, may then be regenerated as for example in the heater 24, or in any other convenient form of regenerator, supplying all or a portion of the heat necessary for reaction, with supplemental fuel 35 fired into the heater 24 or the regenerator as necessary.

The catalyst may be discarded, or, after regeneration, the catalyst is reslurried at 40 by the clarified oil from pump 36 or by fresh charge stock from line 42. The reslurried catalyst is returned to the bottom section 12 through line 26 after merging with the recirculated catalyst slurry from the heater 24. The regenerated catalyst passing from the filter 30 through the heater 24 may be conveyed by a traveling grate shown schematically at 44.

The charge stock may be any hydrocarbon stream which is to be converted to a more valuable (usually lighter) product. It is contacted with the superactive catalyst 16 in the reaction zone in a predominantly liquid state. This is accomplished by operating under conditions that maintain the uncracked charge material in a liquid phase. Such conditions include temperatures in the range of 500° to 900°F., pressure in the range of 10 to 1,000 psia, space velocities in the range of 10 to 1,000 equivalent weights of oil per weight of catalyst per hour (w./hr./w.), and 0.01 to 10 equivalent weights of catalyst per weight of oil (C/O).

The pressure in the reaction section of the fractionator will be controlled by conventional overhead pressure controls, whereas the space velocity (w./hr./w.) catalyst to oil ratio (C/O), and recycle ratio will be controlled by the rate at which the catalyst-oil slurry is withdrawn from the tower bottom through line 20.

When only low conversion is desired, the clarified oil from the filter 30 may be rejected as a bottom product through line 34. Other means of regeneration other than combustion such as in the heater 24 may be utilized, such as removal of the heavy hydrocarbons by a solvent or other methods.

The top section 14 of the tower is provided with conventional fractionating plates 50 with a plurality of side streams drawn off at 52, 54 and 56. The side stream 52 and 54 are conducted to normal stripping, cooling and product blending or further processing. The stream 56 containing gas and gasoline is subjected to the normal recovery and separation steps with the reflux being returned via line 58.

The catalysts useful in this process are superactive catalysts which have a relative activity of as high as 10,000 times that of conventionally used catalysts in the cracking of hydrocarbons. Although technology is not available for achieving full use of these catalysts, it has been found that these materials exhibit product selectivity which is extremely attractive, since the ratio of gasoline yield to coke make in gas oil cracking has been found to be markedly greater than that of conventional catalysts.

Crystalline aluminosilicates are materials of ordered internal structure in which atoms of alkali metal, alkaline earth metal or metals in replacement thereof, silicon, aluminum and oxygen are arranged in a definite and consistent crystalline or ordered pattern. Such structure contains a large number of small cavities, interconnected by a number of still smaller channels. These cavities and channels are precisely uniform in size. The interstitial dimensions of openings in the crystal lattice limit the size and shape of the molecules that can enter the interior of the aluminosilicate and it is such characteristic of many crystalline zeolites that has led to their designation as "molecular sieves."

Zeolites having the above characteristics include both natural and synthetic materials, for example, chabazite, gmelinite, mesolite, ptilolite, mordenite, natrolite, nepheline, sodalite, scapolite, lazurite, leucrite, and cancrinite. Synthetic zeolites may be of the A type, X faujasite type, Y faujasite type, T type or other well known form of molecular sieve, including ZK zeolites.

Preparation of various examples of such zeolites is known, having been described in the literature, for example A type zeolite in U.S. Patent 2,882,243; X faujasite type zeolite in U.S. Patent 2,882,244; other types of materials in Belgium Patent 577,642 and in U.S. Patent 2,950,952. As initially prepared, the metal of the aluminosilicate is an alkali metal and usually sodium. Such alkali metal is subject to base-exchange with a wide variety of other metal ions. The molecular sieve materials so obtained are unusually porous, the pores having highly uniform molecular dimensions, generally between about 3 and possibly about 15 Angstrom units in diameter. Each crystal of molecular sieve material contains literally billions of tiny cavities or cages interconnected by channels of unvarying diameter. The size, valence and amount of the metal ions in the crystal can control the effective diameter of the interconnecting channels. There are commercially available from various sources materials of the A series and of the X faujasite series. A synthetic zeolite known as "Molecular Sieve 4A" is a crystalline sodium aluminosilicate having channels of about 4 Angstroms in diameter.

In the hydrated form, this material is chemically characterized by the formula:

$$Na_{12}(AlO_2)_{12}(SiO_2)_{12} \cdot 27H_2O$$

The synthetic zeolite known as "Molecular Sieve 5A" is a crystalline aluminosilicate salt having channels about 5 Angstroms in diameter and in which substantially all of the 12 ions of sodium in the immediately above formula are replaced by calcium, it being understood that calcium replaces sodium in the ratio of one calcium for two sodium ions. A crystalline sodium aluminosilicate having pores approximately 10 Angstroms in diameter is also available commercially under the name of "Molecular Sieve 13X." The letter X is used to distinguish the interatomic structure of this zeolite from that of the A crystals mentioned above. As prepared, the 13X material contains water and has the unit cell formula:

$$Na_{86}[(AlO_2)_{86}(SiO_2)_{106}] \cdot 267H_2O$$

The 13X crystal is structurally identical with faujasite, a naturally occurring zeolite. The synthetic zeolite known as "Molecular Sieve 10X" is a crystalline aluminosilicate salt having channels about 10 Angstroms in diameter and in which a substantial proportion of the sodium ions of the 13X material have been replaced by calcium. Molecular sieves of the X faujasite series are characterized by the formula:

$$M_{86/n}[(AlO_2)_{86}(SiO_2)_{106}] \cdot 267H_2O$$

where M is Na^+, Ca^{++} or other metal ions introduced by replacement thereof and n is the valence of the cation M. The structure consists of a complex assembly of 192 tetrahedra in a large cubic unit cell 24.95 A. on an edge. Both the so-called X and the so-called Y type crystalline aluminosilicates are faujasites and have essentially identical crystal structures. They differ from each other only in that type Y aluminosilicate has a higher SiO_2/Al_2O_3 ratio than the X type aluminosilicate.

The alkali metal generally contained in the naturally occurring or synthetically prepared zeolites prescribed above may be replaced by other metal ions. Replacement is suitably accomplished by contacting the initially formed crystalline aluminosilicate with a solution of an ionizable compound of the metal ion which is to be zeolitically introduced into the molecular sieve structure for a sufficient time to bring about the extent of desired introduction of such ion. After such treatment, the ion exchanged product is water washed, dried and calcined. The extent to which exchange takes place can be controlled.

Naturally occurring or synthetic crystalline aluminosilicates may be treated to provide the superactive aluminosilicates employed in this process by several means, such as base exchange to replace the sodium with rare earth metal compounds, by base exchange with ammonium compounds followed by heating to drive off NH_3 ions, having an H or acid form of aluminosilicates by treatment with mineral acid solutions to arrive at a hydrogen or acid form, and by other means. These treatments may be followed by activity adjusting treatments, such as steaming, calcining, dilution in a matrix and other means.

It should be noted that the catalysts used in this process may be a composite of the superactive aluminosilicate and a relatively inert matrix material, or it may consist only of the superactive catalyst. If the catalyst consists of a composite, it may be produced in the form of pellets, beads, or particles such as may be produced by spray drying. The matrix material may be any hydrous oxide gel, clay or the like. The matrix material used should have a high porosity in order that the reactants may obtain access to the active component in the catalyst composite. A high porosity matrix of the hydrous oxide type may be used in these composite catalysts, such as silica-alumina complexes, silica-magnesia, silica gel, high porosity clay, alumina, and the like.

The pellets or beads of the composite catalysts may be prepared by dispersing the aluminosilicate in an inorganic oxide sol according to the method described in U.S. Patent No. 2,900,399 and converted to a gelled bead according to the method described in U.S. Patent No. 2,384,946.

The crystalline aluminosilicate material must have a pore size or intracrystalline aperture or channel size sufficiently great to admit desired reactants. 5 A. is approximately the minimum pore size so acceptable. The composite may contain from 5 to 95 percent of the matrix material. Utilizing the conditions in the reaction of:

Pressure, psia	10 - 1,000
Temperature, °F.	500 - 900
Space velocity (w./h./w.)	10 - 1,000
C/O	0.01 - 10
Recycle ratio	0 - 10

a product distribution can be obtained as follows, expressed in percent weight of charge stock:

Gas	1 - 15
C4	3 - 25
Gasoline	20 - 75
Side streams	0 - 40
Bottoms	0 - 35
Coke	1/4 - 5

Standard Oil

A process developed by R.E. Evans and H.D. Zacher; U.S. Patent 3,644,199; February 22, 1972; assigned to Standard Oil Company is one in which conversion of a petroleum stock is effected catalytically in a vertical, elongated transport reactor having an internal elongated open-ended tube closable at its upper end. Passage of the fluid catalyst-oil dispersion through the internal tube is controlled by the degree of closure effected by the positioning of the closure means. Intensity of conversion is regulated by selection of catalyst-oil ratio and control of dispersion velocity in response to the degree of closure.

Catalytic petroleum conversion, such as catalytic cracking, has been effected conventionally in a dense fluid-bed reactor system with a fluidizable catalyst, typically silica-alumina containing 10 to 30 wt. percent alumina. In an improved apparatus catalyst and oil have been contacted in a pipe and transported cocurrently, usually vertically, into a dense fluidized bed of catalyst. The amount of conversion of heavy petroleum, such as a gas oil, into lighter fractions, such as gasoline or kerosene, is a function of space velocity, a measure of the amount of catalyst seen by the oil and the length of time during which catalyst and oil are in contact.

Historically, uncertainty in the space velocity requirement has been accommodated by designing for adequate catalyst inventory in the reactor. Seasonal variations in conversion requirements were met by adjusting the catalyst level in the dense-bed reactor. Development of molecular sieve catalysts, usually combining a crystalline aluminosilicate with the conventional silica-alumina catalysts, has led to the utilization of previously recognized advantages of the dilute-phase transport reactor. This process is customarily known as transfer-line cracking.

In this process, control of catalyst-oil contact time in a dilute-phase transport or "transfer-line" cracking reactor is achieved by providing a means for effectively altering the cross section area (and hence the volume) of the cracking reactor. This is done by providing an internal elongated concentric pipe which may be closed at its upper end in a controllable manner responsive to an external adjustment.

Figure 52 shows the form of apparatus which may be utilized to conduct such an operation. As shown there, petroleum gas oil, together with recycle gas oil, is introduced into the coaxial entry tube 13 of vertical transport cracking reactor 10 through line 11 which extends into entry tube 13. Regenerated and/or fresh catalyst is introduced through standpipe 12 into the annular space between the walls of line 11 and tube 13. Appropriate metering, valving, and dispersion or fluidizing steam inlets are not shown.

The fluidized mixture of catalyst and gas passes upwardly through reactor 10 while catalytic cracking of the gas oil to valuable hydrocarbon fractions of greater volatility occurs. In the course of the cracking reaction coke deposits on the catalyst. At the top of reactor 10, the mixture of coked catalyst, petroleum conversion products and unconverted gas oil passes into exit line 14 and then into stripping, product recovery and catalyst regeneration facilities, not shown.

Centrally of reactor 10 there is suspended riser tube 20 which extends vertically from a point near the bottom of reactor 10 upwardly to a point below the exit line 14. The riser tube 20 terminates at its upper end in flared section 21, adapted to receive, as a valve, tapered plug 22. The position of tapered plug 22 relative to flared section 21 can be varied by vertical adjustment of rod 23 from snug contact, whereby upward flow through riser tube 20 is completely shut off, to effectively complete separation, whereby upward flow through riser tube 20 is in no way impeded. Steam purge means, not shown, is provided near the top of riser tube 20 to prevent catalyst and oil accumulation when the tube is closed. Additional purge means, not shown, may be provided near the bottom of the tube for use when the riser tube is partially open. By appropriate adjustment of rod 23, plug 22 can be positioned relative to flared section 21 so that a predetermined flow through 20 can be achieved, corresponding to the flow attainable with a smaller effective diameter thereof.

Rod 23 is movable vertically by control means, not shown, through guide sleeve 24, situated in the top wall of reactor 10 and coaxial therewith. Sleeve 24 terminates at its upper end in packing gland 25. The lower end of sleeve 24 is open to the reactor and any accumulation of catalyst or oil in the sleeve may be removed by bleeding steam through bleed line 26 either continuously or intermittently.

Riser tube 20 is supported from reactor 10 by four support lugs 17 equally spaced about each vessel and welded thereto,

FIGURE 52: TRANSFER-LINE REACTOR DESIGN WITH MECHANICAL MEANS FOR CONTROL OF CONTACT TIME

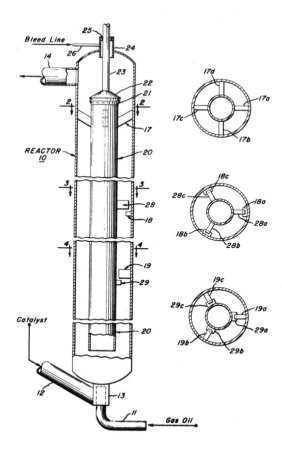

Source: R.E. Evans and H.D. Zacher; U.S. Patent 3,644,199; February 22, 1972

creating a downward angle of about 30° from the cross-sectional plane, normal to the reactor axis, designated as 2—2. The upper detail at the right of the figure presents the cross-sectional view downwardly at plane 2—2, showing the placement of welded support lugs 17a, 17b, 17c, 17d at 90° intervals.

Along the length of riser tube 20 guides are installed at intervals together with emergency stops, attached alternately to riser tube and reactor wall. Three guide bars 28, welded at equal spacings about the circumference of riser tube 20 extend laterally toward but not into contact with reactor wall 10. The guide bars overlap three emergency stops 18 welded with equal spacing to the wall of reactor 10. The center detail at the right of the figure presents a downward cross-sectional view at plane 3—3, showing the respective positions of guide bars 28a, 28b, 28c and emergency stops 18a, 18b, 18c at 120° intervals. At a spaced lower position, three guide bars 19 welded to the wall of reactor 10 overlap three emergency stops 29 situated therebelow and attached to the wall of the riser tube 20. The detail at the lower right of the figure presents a downward cross-sectional view at plane 4—4, showing the respective positions of guide bars 19a, 19b, 19c and emergency stops 29a, 29b, 29c at 120° intervals.

TRANSALKYLATION

Universal Oil Products

A process developed by E.L. Pollitzer, G.R. Donaldson and R.C. Hawkins; U.S. Patent 3,551,510; December 29, 1970; assigned to Universal Oil Products Company is a process for the production of an alkyl aromatic compound including the steps of alkylation, transalkylation and separation.

More specifically, the overall process comprises passing to an alkylation zone containing a solid phosphoric acid alkylation catalyst ethylene and a mixture of diethylbenzene and triethylbenzene, reacting the ethylene with such diethylbenzene and triethylebenzene at alkylation conditions, passing the effluent from the alkylation zone along with benzene, hydrogen and a chloride-containing component to a transalkylation reaction zone containing a hot, concentrated hydrochloric-acid extracted mordenite transalkylation catalyst, reacting the alkylation zone effluent with the benzene, hydrogen and chloride-containing component at transalkylation conditions, passing the transalkylation reaction zone effluent to a separation zone, separating from the separation zone benzene, hydrogen, hydrogen chloride, desired ethylbenzene and higher molecular weight polyethylbenzenes, recycling at least a portion of the unreacted benzene, hydrogen and hydrogen halide to the transalkylation reaction zone, recycling at least a portion of the polyethylbenzenes to the alkylation zone, and removing the desired ethylbenzene as product from the process. Figure 53 is a flow diagram showing the essentials of the process.

The first step of the process comprises passing to an alkylation reaction zone containing an alkylation catalyst, an olefin-acting compound, and polyalkyl aromatic compound. The olefin-acting compound, particularly olefin hydrocarbon, which may be charged to reaction zone 2 via line 1 may be selected from diverse materials including monoolefins, diolefins, polyolefins, acetylenic hydrocarbons, and also alkyl halides, alcohols, ethers, and esters, the latter including the alkyl sulfates, alkyl phosphates, and various esters of carboxylic acids. The preferred olefin-acting compounds are olefinic hydrocarbons which comprise monoolefins containing one double bond per molecule. Cycloolefins such as cyclopentene, methylcyclopentene, cyclohexene, methylcyclohexene, etc., may also be utilized, although not necessarily with equivalent results.

FIGURE 53: FLOW DIAGRAM OF ALKYLATION-TRANSALKYLATION PROCESS FOR ETHYLBENZENE MANUFACTURE

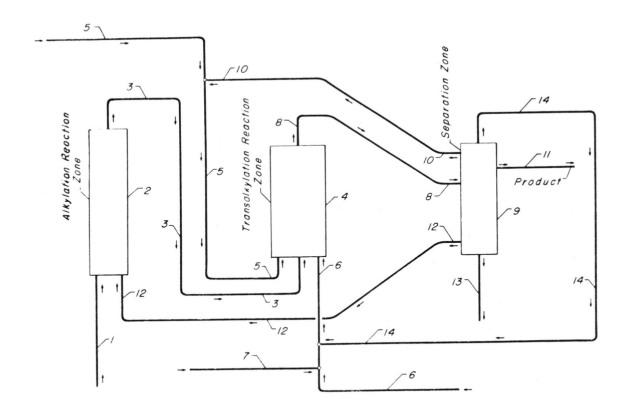

Source: E.L. Pollitzer, G.R. Donaldson and R.C. Hawkins; U.S. Patent 3,551,510; December 29, 1970

As stated above, the olefin-acting compound is passed to alkylation reaction zone 2 via line 1 and in the alkylation zone, the olefin-acting compound and polyalkyl aromatic compound via line 12, as hereinafter described, are reacted at alkylation conditions with an alkylation catalyst.

Alkylation reaction zone 2 is of the conventional type with an alkylation catalyst disposed therein in the reaction zone. The alkylation zone may be equipped with heat transfer means, baffles, trays, heating means, etc. The alkylation reaction zone is preferably of the adiabatic type and thus feed to the alkylation zone will preferably be provided with the requisite amount of heat prior to passage thereof to the alkylation zone.

The alkylation reaction zone contains an alkylation catalyst. A particularly preferred alkylation catalyst comprises a solid phosphoric acid-containing composite. The phosphoric acid-containing composite may be made by combining an acid of phosphorus such as pyro-, or tetra-phosphoric acid with the solid support. It is not intended to infer that the different acids of phosphorus which may be employed will produce catalysts which have identical effects upon any given alkylation reaction as each of the catalysts produced from different acids and by slightly varying procedures will exert its own characteristic action.

Triphosphoric acid, which may be represented by the formula $H_5P_3O_{10}$, may also be used as one of the starting materials for the preparation of the phosphoric acid-containing composite utilized in the alkylation reaction zone. A phosphoric acid mixture which is generally referred to as polyphosphoric acid may also be employed in manufacturing the composite. Polyphosphoric acid is formed by heating orthophosphoric acid or pyrophosphoric acid or mixtures thereof in suitable equipment such as carbon lined trays heated by flue acids or other suitable means to produce a phosphoric acid mixture generally analyzing from about 79% to about 85% by weight P_2O_5.

Tetraphosphoric acid, having the general formula $H_6P_4O_{13}$ which corresponds to the double oxide formula $3H_2O \cdot 2P_2O_5$ may be considered as the acid resulting when three molecules of water are lost by four molecules of orthophosphoric acid, H_3PO_4. The tetraphosphoric acid may be manufactured by gradual or controlled dehydration or heating of orthophosphoric acid and pyrophosphoric acid or by adding phosphorus pentoxide to those acids in proper amounts.

The phosphoric acid-containing composite may comprise a high surface area solid support. By the term high surface area is meant a surface area measured by surface adsorption techniques within a range of about 25 to about 500 or more square meters per gram and preferably a support having a surface area of approximately 100 to 300 square meters per gram. Therefore, satisfactory supports for the preparation of catalysts for use in the process include high surface area crystalline alumina modifications such as gamma-, eta-, and theta-alumina, although these are not necessarily of equivalent suitability. In addition to the aforementioned, gamma-, eta-, and theta-aluminas which may be utilized as solid supports, it is also contemplated that other refractory oxides and mixtures thereof, such as silica, zirconia, magnesia, thoria, etc. silica-alumina, silica-magnesia, alumina-silica-magnesia, alumina-thoria, alumina-zirconia, etc. may also be utilized as solid supports for the alkylation catalyst.

The phosphoric acid-containing composite utilized in alkylation reaction zone 2 may contain from about 8% or lower to about 80% or higher, by weight, of phosphoric acid, and preferably from about 10% to about 75% by weight of phosphoric acid.

In addition to the solid phosphoric acid alkylation catalyst which may be utilized in alkylation reaction zone 2, it is also contemplated that other alkylation catalysts such as a boron halide-modified refractory inorganic oxide catalyst as well as the various crystalline aluminosilicate alkylation catalysts may also be utilized.

The conditions utilized in alkylation reaction zone 2 may be varied over a relatively wide range. Thus, the desired alkylation reaction in the presence of the solid phosphoric acid catalysts, for example, may be effected at a temperature of from about 0° to 350°C. or higher, and preferably from about 0° to about 325°C. The alkylation reaction is usually carried out at a pressure from about atmospheric to about 200 atmospheres or more. The pressure utilized is usually selected to maintain the polyalkyl aromatic compound in substantially liquid phase. The liquid hourly space velocity will be maintained in the range of from about 0.1 to about 20 and preferably in the range of from about 1.0 to about 15.

When the alkylation reaction has proceeded to the desired extent, preferably with 100% conversion of the olefin-acting compound, the product (which may comprise both the reacted and unreacted materials) from the alkylation zone, which may be termed alkylation zone effluent, is withdrawn from alkylation reaction zone 2 via line 3 and passed to transalkylation reaction zone 4. Transalkylation reaction zone 4 is of the conventional type with a transalkylation catalyst disposed therein in the reaction zone. This reaction zone may also be equipped with heat transfer means, baffles, trays, heating means, etc. The transalkylation reaction zone is preferably of the adiabatic type and thus feed to the transalkylation zone will also be provided with the requisite amount of heat prior to passage thereof to the reaction zone.

As set forth hereinabove, the transalkylation reaction zone contains a transalkylation catalyst. A particularly preferred transalkylation catalyst comprises an acid-extracted crystalline aluminosilicate. It is especially preferred that the crystalline aluminosilicate be of the mordenite type. It has been found that, as contrasted with the prior art, the acid-extracted crystalline aluminosilicate catalyst, and especially the acid-extracted mordenite catalyst gave results definitely superior to what the art refers to as "acid-treated" mordenite catalysts, wherein the mordenite is only exchanged with a dilute mineral acid in order to replace cations with protons.

In the alkylation-transalkylation process, the mordenite, for example, is extracted with a hot, fairly concentrated acid, usually hydrochloric acid with the result that some alumina is actually removed from the lattice structure of the crystalline aluminosilicate. This acid-extracted mordenite catalyst has a higher silica alumina ratio than "acid-treated" mordenite and is a different catalytic structure than the so-called "acid-treated" mordenite alone. Further, by chemical analysis, the acid-extracted mordenite is a different composition of matter than the mordenite. In the preparation of these acid-extracted crystalline aluminosilicates, it is preferred to use mineral acids such as hydrochloric acid, phosphoric acid, aluminum chloride, sulfuric acid, and the like. However, the acid must be capable of removing alumina from the lattice structure of the crystalline aluminosilicate.

For example, hydrogen form mordenite having a silica to alumina ratio of about 15.8 can be acid-extracted with a hydrochloric acid solution maintained at a concentration of six mols of hydrochloric acid per mol of alumina present in the crystalline aluminosilicate. The mordenite can be contacted at temperatures in excess of 100°C. and preferably, in the range of from about 100°C. to about 110°C. Higher temperatures may, of course, be utilized at elevated pressures during this contacting procedure.

The alumina extraction may be continued for a period of several hours, and preferably, for at least a six hour period after which time the acid-extracted mordenite is washed to remove excess chloride and then calcined at temperatures in the range of from about 200°C. to about 900°C. and preferably, in the range of from about 500°C. to about 750°C. A reduction in the alumina content of this acid-extracted mordenite of from 9.5 to at least 7.3 weight percent is achieved and the silica to alumina mol ratio is increased from 15.8 to at least 21.5 after the acid-extraction showing that the alumina was actually removed from the lattice of the crystalline aluminosilicate. This acid-extracted mordenite catalyst has a higher silica to alumina ratio than the mordenite alone and is a particularly preferred transalkylation catalyst.

As set forth hereinbefore, effluent from alkylation reaction zone 2 along with aromatic compound from line 5, as hereinafter described, as well as hydrogen and halogen-containing component via lines 6 and 7, as hereinafter described, are passed to transalkylation reaction zone 4 wherein the alkylation zone effluent is reacted with the aromatic compound, hydrogen, and halogen-containing component at transalkylation conditions including a temperature in the range from about 100°C. to about 500°C. or higher, and preferably, in the range of from about 125°C. to about 475°C. The transalkylation reaction is usually carried out at a pressure in the range from about atmospheric to about 200 atmospheres or more. The liquid hourly space velocity will be maintained in the range from about 0.1 to about 20 and preferably in the range of from about 1.0 to about 15.

As set forth hereinbefore, an aromatic compound is passed to transalkylation reaction zone 4 via line 5. Particularly preferred aromatic compounds are the benzene hydrocarbons and of the benzene hydrocarbons, benzene itself is particularly preferred for passage to the transalkylation reaction zone. However, it is also contemplated that higher molecular weight aromatic compounds including toluene, ortho-xylene, meta-xylene, para-xylene, ethylbenzene, ortho-ethyltoluene, meta-ethyltoluene, para-ethyltoluene, 1,2,3-trimethylbenzene, 1,2,4-trimethylbenzene, 1,3,5-trimethylbenzene, normal propylbenzene, isopropylbenzene or cumene, normal butylbenzene, etc. as well as higher molecular weight aromatic hydrocarbons, including the hexylbenzenes, nonylbenzenes, dodecylbenzenes, pentadecylbenzenes, hexyltoluenes, nonyltoluenes, dodecyltoluenes, pentadecyltoluenes, etc. and mixtures thereof may be used.

Other suitable aromatic hydrocarbons which, at specified reaction conditions, depending upon the melting point of the aromatic chosen, would be in liquid form, would include those aromatic hydrocarbons with two or more aryl groups, such as diphenyl, diphenylmethane, and other polycyclic aromatics. Those aromatic hydrocarbons within the scope of this process containing condensed aromatic rings would include naphthalene, the alkylnaphthalenes, anthracene, phenanthrene, naphthacene, rubrene, etc. As stated hereinbefore, of the aromatic compounds and preferably aromatic hydrocarbons for use via line 5 to transalkylation reaction zone 4, the benzene hydrocarbons are preferred, and of the benzene hydrocarbons, benzene itself is particularly preferred.

We have found, that in contradistinction to other alkylation-transalkylation process flow schemes, that when hydrogen in an amount of from about 2:1 to about 20:1 hydrogen to hydrocarbon mol ratio is present during, for example, the transalkylation of ethylbenzene, superior results are unexpectedly obtained with the acid-extracted crystalline aluminosilicate described hereinabove. These results are unexpected inasmuch as hydrogen, in comparison to the prior art, is not needed to keep, for example, a noble metal active catalytic site from carbon deposition inasmuch as there is no noble metal present on the catalyst utilized in the transalkylation portion of this combination process.

In the flow scheme, hydrogen is represented as being passed to transalkylation reaction zone 4 via line 6. In addition, via line 7 and line 6, halogen-containing component in an amount of from about 0.001 to about 2.0 wt. percent may be commingled with the hydrogen passing via line 6 for reaction in transalkylation reaction zone 4. It is also contemplated that the halogen–containing component may be added simultaneously with, but independently of, the hydrogen stream passing to transalkylation zone 4.

Of the preferred halogen–containing components are the halogens and the halogen halides, and of the halogens, a preferred halogen is chlorine, and of the halogen halides, it is particularly preferred to utilize hydrogen chloride. However, the particular catalyst utilized, as well as the particular alkylaromatic hydrocarbon to be transalkylated, will dictate the choice of the halogen–containing component utilized in the alkylation–transalkylation process.

When the transalkylation reaction has proceeded to the desired extent so that a sufficient quantity of polyalkylated compounds are converted to monoalkylated compounds by reaction with the aromatic compounds furnished to the reaction zone, the products from transalkylation reaction zone 4 are withdrawn through line 8 and passed to separation zone 9 for recovery of the desired components therefrom.

In separation zone 9, unreacted aromatic compound, hydrogen, hydrogen halide, desired monoalkylated aromatic compound, and higher molecular weight polyalkylated aromatic compounds are separated by means such as, for example, fractional distillation. The unreacted aromatic compound is passed from separation zone 9 via line 10 for recycle to transalkylation reaction zone 4 via line 5, thereby obtaining economy of operation.

Hydrogen and hydrogen halide are recycled from separation zone 9 via line 14 to transalkylation reaction zone 4 via line 6 so that the net consumption of hydrogen and halogen–containing component is kept at a minimum. The product of the process, namely, the alkylaromatic compound, is withdrawn from separation zone 9 via line 11. Polyalkylated aromatic compound is withdrawn from separation zone 9 via line 12 where at least a portion of the polyalkylated aromatic compound is recycled to alkylation reaction zone 2. The heavier molecular weight polyalkylaromatic compounds are removed from separation zone 9 via line 13 along with any high–boiling condensation products that may have formed and accumulated during the processing operation.

CATALYST PREPARATION

American Cyanamid

A process developed by J.D. Colgan and W.E. Sanborn; U.S. Patent 3,403,109; September 24, 1968; assigned to American Cyanamid Company is a process for preparing an extruded silica–containing gel characterized in that a xerogel derived therefrom has a pore volume of from about 1.3 to about 2.0 cc./g. when measured after calcining the xerogel and which product has improved crushing strength. The process comprises (1) reacting (a) an alkali metal silicate with (b) a mineral acid, (c) in a water heel of a pH of between about pH 1.0 and pH 4.5, the silicate and the acid being added in an amount sufficiently to raise pH to at least about pH 9, and sufficiently to form a gel reaction product, (2) adding to the reaction product a mineral acid in an amount sufficient to lower the pH within a range of at least about pH 7.75 and about pH 8.75, and (3) thereafter aging the reaction product for a period of at least about 15 minutes, at a temperature of at least about 100°F.

In preparing extruded catalyst, different problems are encountered than are encountered in the preparation of bead catalysts. Generally speaking, in the preparation of extruded catalysts, the difficulties which are encountered on the small scale are magnified and intensified when attempts are made to produce them on a larger commercial scale. This is particularly true if the catalyst produced (1) is to be uniform and consistently prepared having a high degree of activity as calculated on a weight and volume basis as these terms are understood in the art, and (2) is to have a good (high) crush strength, a necessary prerequisite for fixed bed catalysts.

Thus, for example, if the ratio of base and activating materials, the proper use of lubricants, plasticizers and other variables, are not controlled within relatively narrow limits as to amounts, time of addition, other conditions of addition, and the like, the extrudates will have a poor (low) crush strength and will have a ready tendency to crumble, or the extrudates may alternatively be gummy and difficult to extrude. Further, if the extrusion mix is not of proper composition and content, it has been found in many instances that the formation of the extrudate is extremely difficult if not impossible in that the extrusion mix tends to ride on the extruder auger rather than being extruded.

If the plasticity or the consistency of the catalyst mass for extrusion is insufficiently fluid, it may be impossible to produce extruded pellets, in view of the fact that the extruder is unable to move the material through the die head. Catalyst mixtures having water contents in excess of those suited for the provision of extrudable mass are usually too free-flowing or too fluid to be properly extruded; in addition, this physical condition increases the difficulty of other processing problems, as for example, the drying of the extruded catalyst mixture. In some instances, this physical condition adversely affects the activity of the final catalytic product, as well as detrimentally affecting the crush strength. If the catalyst mixture is too plastic, after extrusion, extruded pellets have a strong tendency to adhere to one another, which characteristic may (a) destroy the saleability of a given production run, as well as (b) detrimentally affect the crush strength of the final extruded catalyst product. While it would seem that many of these difficulties could be readily overcome simply by adding more or less dry material, or more or less moisture to any given extrusion mix, practical information dictates to the contrary, either (1) because the extrusion mix after such modification simply is incapable of forming a good extrudate or (2) because such seemingly minor variations in a given procedure result in a mix which, while it can be extruded and produces an extrudate having a good appearance, has undesirably and possibly unacceptably low crush strength.

While the need for a satisfactory extrusion process for this type extrudate catalyst has been recognized by the prior art, no fully satisfactory process has previously been proposed for the extrusion of silica and silica-alumina catalyst supports with a high degree of crush strength associated with the extrudate. These silica compositions typically are very hard and abrasive when dried and have proved to be extremely difficult to extrude. While there has been some success with extrusion of soft, newly formed silica and silica-alumina hydrogels containing a high water content, no fully satisfactory procedure has previously been proposed for extrusion of older, previously dried or powdered silica and silica-alumina feed materials.

In the development work on the process, it was discovered that an extrudate having an unexpectedly high degree of crush strength may be prepared by extruding an extrusion feed which includes as a principal component, a partially dried hydrogel and/or a xerogel, and which is characterized by the fact that calcined xerogels derived from the hydrogel or the xerogel have a high pore volume.

Xerogel is a term typically used to describe the solid formed when a hydrogel is dried at approximately 250°F. A spray dried solid in the form of microspheres. Drying hydrogel at 250°F. removes essentially all of the free water and produces a material which is approximately 87% solids, 13% bound water.

Pore volume is determined by titrating the sample (usually calcined xerogel or calcined extrudates) with water at ambient temperature to saturation. Pore volume is then reported as cc (of water) per gram (of sample). It is a measure of the particle density of the sample. Note: See Cyanamid Manual, "Test Methods for Synthetic Fluid Cracking Catalysts," p. 21.

Macroporosity and microporosity are described as follows. Pores in high surface area materials such as activated silica gels are frequently divided into two or more distinct classes with respect to size. The smaller, the micropore system, contributes substantially all the surface area and is generated by activation. The larger capillaries, the macropores, are dependent on the method of manufacture, the type of agglomeration or degree of grinding for instance. Macropores contribute very little to surface area.

Micropores are arbitrarily broadly defined as pores whose diameters are less than 650 angstroms. These are pores within the range where capillary condensation of nitrogen occurs. Volume of pores in this range can therefore be measured by nitrogen absorption at high relative pressure.

Macropores are arbitrarily broadly defined as pores with diameters greater than 650 angstroms. The volume of macropores is generally calculated as the difference between the total pore volume and the volume of the micropores. Total pore volume may be measured by water titration to saturation. Particle density may also be measured (by mercury displacement, or, with extrudates, by measuring particle weights and dimensions), and used with known values of skeletal density to calculate pore volume. Macropore volume and pore size distribution in the macropore range (similar to the pore size distribution in the micropore range which can be obtained with nitrogen) can also be obtained by mercury penetration.

Bulk density is the weight per unit volume of bulk sample under prescribed conditions. Apparent bulk density (ABD) is the weight of sample which occupies a unit of volume when the unit of volume is rapidly filled by sample flowing by gravity from a low elevation (several inches). Compacted bulk density (CBD) is the weight of sample which will occupy a unit of volume when the sample is vibrated in a container to minimize container voids and maximize the weight of sample. Note: See Cyanamid Manual "Test Methods for Synthetic Fluid Cracking Catalyst," p. 9.

Extrudate crush strength is determined by selecting a sample of extrudates with length equal to from about 1.5 to about 3.0 times the extrudate diameter. The length of each extrudate is measured, and an average length is calculated. Each extrudate is then, in turn, placed between two flat plates and the force required to crush the extrudate is measured. The average crushing force for the sample is then calculated. Division of the average crushing force in pounds by the average extrudate length in inches yields a measure of extrudate crush strength which is termed "CSL." Units are pounds (of force) per inch (of length).

Prior to this process, a high degree of crush strength of catalyst support or of a catalyst formed therefrom, was normally thought by the skilled artisan to be associated with a high apparent bulk density, and low pore volume. Low apparent bulk density and high porosity previously were normally associated with a support or catalyst derived therefrom which could be easily crumbled such as by merely applying pressure between the fingers, whereby the catalyst or catalyst support would become finely divided, pulverized material. It has been discovered that xerogels or hydrogels when reduced to xerogels, characterized by low apparent bulk density and by high porosity impart unexpectedly a high order of crush strength to extrudates formed therefrom.

Hydrogels from which xerogels derive are normally prepared by a method known as a strike. Typically, in carrying out the conventional strike method, (1) an alkali metal silicate such as sodium silicate solution is reacted with a mineral acid such as sulfuric acid, hydrochloric acid, nitric acid, and the like, after which (2) the sol is gelled by appropriate means such as pH adjustment, (3) the final density of the gel is regulated by allowing the hydrogel to age, (4) the alkali ions are removed by base exchange with cations such as ammonium, and (5) the product is washed and dried.

According to this process three distinct methods have been found to obtain a hydrogel or a xerogel derived therefrom, which when employed as an extrusion feed obtains an extrudate having unexpectedly high crush strength, as well as desirable extrusion characteristics.

The first method is hereafter referred to as a "double-precipitation-strike." By this method, (1) approximately one-half of the total acid such as sulfuric acid and (2) alkali metal silicate such as sodium silicate, are added to a water heel while maintaining a pH of about 1.0 to about 4.5, the preferred pH being from about pH 2.5 to about pH 3.0. The silicate addition is then continued until there has been an increase in pH up to at least about pH 9, preferably about pH 10. The second half of the total acid is then added to bring the batch pH down to not lower than about pH 7.8 up to not more than about pH 8.7, preferably about pH 8.0 to about pH 8.5, and the gel is thereafter aged for at least about 15 minutes, preferably about 30 minutes, at a strike temperature of at least about 100°F., preferably from about 115°F to 120°F., with batch solids between about 4% and 9% and preferably at about 6%. In the initial above reaction of one-half the acid with the alkali metal silicate, it is not critical that the pH be acid, although the optimum results are obtained at the pH of about pH 1.0 to about pH 4.5, preferably about 2.5 to about 3.0 as stated above.

The second method is hereafter referred to as the "high salt-heel" strike. By this method, the standard strike procedure is employed (as discussed above) except that instead of reacting an alkali metal silicate with a mineral acid in a water heel, by the high salt-heel process an aqueous solution of a salt (typically prepared by reaction of an alkali metal compound with a mineral acid) is employed as a heel in substitution for a water heel normally employed in the ordinary standard strike method. The mols of salt (such as Na_2SO_4) added in ratio to the mols of silica present in the final batch after the strike may range from a low ratio of about 1:72 to a high ratio of about 1:4.5, preferably 1:36 to 1:9, of salt:SiO_2. The salt concentration in the water heel before beginning the strike may range from about 0.33% to about 5.3%, preferably about 0.66% to about 2.65%, based on the weight of water employed.

Thereby, by the high salt-heel method in intimate contact with a heel which includes the salt of an alkali metal in the concentration and ratio discussed above, alkali metal silicate is reacted with a mineral acid to form a reaction product. Additionally, however, by the high salt-heel method, it has been found necessary to employ an aging period of at least about 15 minutes at a temperature of at least about 100°F. at the standard pH of above about pH 6.5 up to about pH 8.0. Only in the extreme situation, where the maximum high salt concentration and the high ratio are employed, could a temperature possibly as low as about 80°F. be employed for purposes of this second method.

The sodium silicate normally employed is about 28.25 ±0.25% SiO_2, and although the acid concentration is not normally critical, the sulfuric acid employed is normally about 25 ±0.5%, on a weight basis. The silicate glass (solid) or water-glass (solution) preferably has a high $SiO_2:Na_2O$ molar ratio, normally about 3.25 $SiO_2:Na_2O$ molar ratio. Low SiO_2 content lowers yield, increases the required amount of acid reactant, and thus increases cost.

By the third process, hereafter referred to as the high-temperature process, either a conventional procedure or the above procedures may be followed for the strike method, except that (1) after the alkali metal silicate is added to a water heel thereby raising the pH typically to at least about 8.0 to about pH 11 — as dependent upon the solids content, mineral acid is added in an amount sufficient to lower or adjust the pH to form at least about pH 7 to about pH 8.5, and (2) the reaction-product gel is thereafter aged for a period of at least 15 minutes at a temperature of at least about 135°F. After the aging, in accordance with conventional strike methods, the pH is lowered to normally about pH 2.5 to about pH 4, preferably to about pH 3.0 to pH 3.5. The solids content of the aged product is preferably at about 7% by weight, but as stated above may vary. As noted above, if the alkali metal silicate is added in an amount to obtain a pH of at least pH 8.0, the pH need not necessarily be "lowered" but may be merely "adjusted" to the desired pH in the above stated range of about pH 7.0 to about pH 8.5.

If pH is already within this range, obviously further lowering or adjustment is purely optional. By each of the three methods discussed above, after the aging of the gel, the gel may be coated with alumina by any conventional procedure, such as by reacting alum and sodium aluminate, as is known to those skilled in the art. Also, it should be noted that any two or more of the above processes may be combined to produce a fourth method.

Additionally, it has been found that when employing any of the above three processes, an additional unexpected increase in the crush strength is obtained when the aged reaction product is filtered and washed to remove the alkali ions such as sodium sulfate prior to forming the spray-dried xerogel base.

It was also discovered that the extrudates having increased crush strength, extruded from a feed employing the gel reaction product of any one of the above three methods, have a low apparent bulk density and a high porosity when measured after calcining. It was found that an extrusion feed which obtains an extrudate having unexpectedly high crush strength is characterized in that a xerogel which prior to extrusion is calcined and measured for pore volume, exhibits a pore volume from about 1.3 to about 2.0 cubic centimeters per gram, preferably about 1.5 to about 2.0 cc. per gram.

Thus, by employment of any one of the methods, any hydrogel or any xerogel which is characterized in that the xerogels derived therefrom have low apparent bulk density and high porosity within the range discussed above, as measured by calcining the xerogel prior to measuring the pore volume, imparts an unexpectedly high order of crush strength to an extrudate derived by the extrusion of an extrusion feed which includes a major proportion of that (1) partially dried hydrogel, or (2) xerogel, or (3) mixtures of hydrogen and xerogel. The extrusion feed xerogels, which may optionally be milled, are mulled either with water or with undried or partially dried hydrogel slurry to produce a feed paste satisfactory for extrusion to conventional equipment.

The extrudates are normally dried and calcined, preferably at a temperature of approximately 1400°F. It has been further found that by employing a high temperature of calcination, an additional moderate increase in strength is obtained without a decrease in surface area. However, it is possible to obtain a high degree of crush strength without high temperature calcination. Calcination may be normally carried out at temperatures of about 1050°F. or more. In addition to the use of the hydrogels and xerogels, to produce extrudates having increased crush strength, it has been found that the xerogels also unexpectedly impart a high order of crush strength when formed into tablets by conventional tabletting operations.

Chevron Research

A process developed by G. Constabaris, B.F. Mulaskey, R.H. Lindquist and T.G. Chin; U.S. Patent 3,255,122; June 7, 1966; assigned to Chevron Research Company involves an improved method for the rapid determination of surface areas of high surface area catalysts.

The geometric surface of heterogeneous catalytic materials is one of their fundamental extensive properties. For a specific catalyst, an increase in activity per unit weight generally follows an increase in unit area. Sometimes this activity is desirable, in others it is not. Also process conditions cause loss of area on catalyst. The determination of surface areas is thus of great importance in both production and use of catalysts. Thus, in the production of catalysts, any increased speed in the measurement of the surface area is thus of great economic advantage since the sooner the area of the product is known, the sooner the operating conditions can be adjusted to minimize the amount of poor catalyst produced. Likewise, in the use of the catalyst where process conditions are causing the loss of desired catalyst activity by reduction of surface area, the quicker this is known, the sooner the process conditions can be changed to correct and offset the lowered catalyst activity.

Most prior methods of determining surface areas require elaborate equipment, skilled operators, and usually take a considerable amount of time to determine the surface area of one sample. Thus, the method of Brunauer, Emmett and Teller, the so-called BET method (J. Am. Chem. Soc., 60, 309 (1938)) takes one to several hours per sample analyzed and requires an elaborate glass apparatus with manometers, heaters and vacuum pump.

Therefore, it is an object of the process to provide a method for determining surface area of porous solids. A further object is to provide a method of obtaining indications as a function of the surface area, and which indications can be used to control catalyst manufacture or to adjust operating conditions in a catalytic conversion process in response to changes in surface area of the catalyst. This surface area indicating method is based on an entirely different phenomena than that applied in the BET method. It has the particular advantage of being able to yield surface area in a relatively short period such as 5 minutes for each dried sample. The equipment for carrying out the method can be relatively simple. Also, the method does not require complex calculations or electronic data processing apparatus for determining the surface area.

Preferably, the method is carried out by applying a pressure gas displacement technique to determine apparent and true skeletal density of the sample of high area porous solid. The method involves the steps of measuring the change in variable volume of a chamber containing the sample when gas is adsorbed on the catalyst surface and then measuring the same sample when adsorption is prevented from taking place. Gas adsorption can be prevented by covering the sample with the known volume of water or by using as a second gas a nonadsorbing gas such as helium. Where a nonadsorbing gas is used, the two measurements can be carried out in any order.

In this method, the basic principle is different, although superficially similar to that used in the BET method. While the BET method depends upon the use of a gas at a temperature at which it will condense onto the surface until the surface is covered, thereby giving an inflection point on the adsorption isotherm, the method uses gas at a temperature above the critical condensation temperature of the gas. The effect of the solid in contact with the gas is to introduce an imperfection into the gas. The effect of the force field contributed by the solid is an apparent shrinkage, i.e., the imperfection, of the gas in the presence of the high area solid. Hence, gas molecules are concentrated near the solid surface by the force field, and the apparent gas shrinkage is directly proportional to the area of the solid.

Therefore, the process is based upon the apparent shrinkage of the gas in contact with the surface area of the sample being tested. However, since a process is more readily described by use of the more familiar terms of adsorption, such terms are used to describe the process. The amount of gas that "disappears" which can be called the volume per unit weight adsorbed, "V_{STP}," is given by:

$$V_{STP} = Area \times \left[\begin{array}{l} \text{Constant X} \\ \text{Pressure X} \\ \text{Function of temperature and} \\ \text{\quad nature of the surface} \end{array} \right.$$

at constant temperature, pressure and with the same type of surface, this expression reduces to:

$$V_{STP} = Constant \times Area$$

In other words, the measured V_{STP} (i.e., the volume of gas which disappears) is a function of the surface area. For a particular type of surface, the constant can be obtained under fixed conditions of temperature and pressure. This is done by measuring the true surface area for the selected type of material, such as by use of the BET method, and obtaining the proportionality factor, which is the constant for that material. Thus, for other samples of the same type, this constant multipled into the measured V_{STP} gives the area.

Esso Research and Engineering

A process developed by H.B. Jonassen, G.P. Hamner, J.A. Rigney, R.B. Mason and S.N. Laurent; U.S. Patent 3,493,518; February 3, 1970; assigned to Esso Research and Engineering Company is one in which crystalline alumino-silicate zeolites which exhibit a unique X-ray diffraction pattern are prepared by first exchanging an alkali metal containing zeolite with a divalent metal which is less electropositive than the alkali metal and then contacting the divalent metal containing zeolite with a compound capable of forming an imido linkage between the divalent metal cations. These crystalline aluminosilicate zeolites are useful in hydrocarbon conversion processes.

The development of crystalline aluminosilicate zeolites of molecular sieve type has been tremendous in the recent past and such rapid development shows every sign of continuing into the future. The molecular sieves have exhibited outstanding catalytic activity in a wide variety of organic reactions, including most especially those involving conversions of petroleum hydrocarbons. Furthermore, molecular sieves have also been found to be extremely useful in separation processes wherein the singular ability of particular molecular sieves to adsorb one chemical compound species to the virtually complete exclusion of other chemical compounds which might be present is used.

In order to enhance a particular property of the molecular sieve as either a catalyst or adsorbent, it has been suggested by the art to replace the metal cations in the exchangeable positions in the zeolite crystal lattice. Such changes can affect the pore diameters in the zeolite thereby changing the zeolite's ability to adsorb different size molecules. Also, these introduced metals will change the electrostatic forces in the zeolite structure thus influencing the reactivity and relative acidity of the zeolite.

Other methods of influencing zeolite properties include deposition of metal compounds on the zeolite surface such as by impregnation; activation or deactivation of the zeolite surface with chemical agents such as steam, chlorine, hydrogen sulfide or similar compounds; and additionally deformation of the zeolite structure by thermal or acid treatment.

This process offers one more degree of flexibility in modifying zeolitic structures for particular catalytic applications. The imido superstructure is formed by contacting a crystalline aluminosilicate zeolite, which has been base exchanged with a suitable divalent metal cation as discussed above, with monosubstituted alkyl or aryl amines (which may be characterized as $R—NH_2$), ammonia or aqueous solutions of such compounds, e.g., ammonium hydroxide. The amines or ammonia may be introduced either as liquids, gases, or as solutions in selected solvents to effectuate the desired reaction forming the imido bridge. This operation leaves the electronegative sites on the zeolite previously bonded to the divalent cation occupied with either ammonium or alkyl or aryl ammonium ions. Careful heating, e.g. a temperature in the range between about 200° and 700°F., is sufficient to decompose the ammonium ions leaving hydrogen ions or substituted amines in these positions. Under selected conditions and with zeolites capable of admitting substituted amines, certain alkyl or aryl groupings may be incorporated in the zeolite crystal structure by this technique.

A process developed by L.V. Robbins, Jr., J.S. Anderson and C.E. Adams; U.S. Patent 3,558,476; January 26, 1971; assigned to Esso Research and Engineering Company yields a hydrocarbon conversion catalyst composition comprising a physical admixture of a conventional, amorphous, silica-containing cracking catalyst and a crystalline alumino-silicate zeolite suspended in an inorganic oxide gel matrix. The catalyst is particularly useful in the catalytic cracking of hydrocarbon feeds. When so used, the catalyst composition is more active and more selective than would be predicted from a consideration of its relative composition.

The most widely used catalytic cracking catalyst in the past was an amorphous silica-alumina gel catalyst containing, for example, 13% alumina and 87% silica. In recent catalysts, the alumina content has been raised to about 25 wt. percent. These catalysts are generally prepared from silica hydrogel or hydrosol, which is mixed with alumina to secure the desired silica-alumina composition; and, if desired, oxides of other metals such as magnesium, zirconium, or other Group II, III, or IV metals. However prepared, the final catalyst is amorphous in nature, and has pore openings of varying sizes ranging from less than about 5 A. in diameter to as much as 200 A. in diameter and higher.

This nonuniformity is a result of the amorphous character of these siliceous conventional cracking catalysts, and is responsible for certain undesirable characteristics. For example, in the very fine pores, a feed molecule encounters diffusion difficulties, and does not have free access over the entire catalyst surface. Also, certain product molecules cannot readily escape from the pore structure before being converted to undesirable lower boiling materials, e.g., dry gas, and coke. Difficulties and disadvantages such as these have contributed to the ever-present need for improved hydrocarbon conversion catalysts and catalyst supports.

Considerable interest within the petroleum industry has been directed to the use of crystalline alumino-silicate zeolite materials in hydrocarbon conversion catalyst systems. These well known materials, sometimes referred to as "molecular sieves," are characterized by a highly ordered crystalline structure with uniformly dimensioned pore openings, and are distinguishable from each other on the basis of composition, crystal structure, adsorption properties and the like. They are gaining wide acceptance as hydrocarbon conversion catalysts and catalyst supports due to substantially greater catalytic activity and selectivity to desired product.

However, the use of these crystalline materials for catalytic purposes does suffer from various drawbacks. For example, one of the problems encountered has been the difficulty of handling the extremely fine zeolite crystals, which can be less than 5 microns in size, in fluidized or moving bed processes. Further, the crystalline zeolite may often be unsuitable for direct use as a catalyst because of too high an activity which can lead to overconversion and runaway reactions. Also, the stability of certain of these alumino-silicate zeolite materials at high temperatures or upon steam treatment, is often too low for commercial acceptance. Steam stability refers to the ability of a catalyst to resist rapid deactivation in the presence of steam, and is used, for example, to assist in the regeneration of catalysts which have become deactivated as a result of coke deposition.

The catalyst is usually stripped of entrained oil by contact with steam and then treated with oxygen-containing gases at high temperatures to combust carbonaceous deposits. Still another disadvantage associated with the use of crystalline alumino-silicate zeolite catalysts resides in the fragility of the zeolite crystals which are commonly subject to considerable abrasion, breakage, and attrition loss when used in the form of a continuously moving stream such as in a fluidized operation.

The above disadvantages led to the development of combining the alumino-silicate zeolite crystals with a siliceous matrix, such as silica-alumina, so that the zeolite crystals become suspended in and distributed throughout the matrix. This composite catalyst will hereinafter be referred to as the "encapsulated" version, due to the coating of the zeolite crystals with siliceous gel. The encapsulated version of the catalyst consisting of crystalline alumino-silicate zeolite embedded in conventional siliceous materials such as silica-alumina, is characterized by a high resistance to attrition, high activity, exceptional selectivity and steam stability. It can be prepared, for example, by dispersing the zeolite crystals in a suitable siliceous sol, and gelling the sol by various means. Certain procedures for preparing this encapsulated catalyst are described in U.S. Patent 3,140,249. Other procedures involve the addition of zeolite crystals to a gelatinous precipitate of silica-alumina or silica-alumina hydrogel, and spray drying of the admixture to form spheroidal composite particles consisting of zeolite crystals encapsulated in silica-alumina gel matrix.

The encapsulated catalyst (i.e., crystalline zeolite distributed throughout and embedded in siliceous matrix) is, of course, more costly to manufacture than conventional cracking catalyst, such as silica-alumina. The added cost is however, overshadowed by substantially improved product yield and product distribution attributed to the presence of the crystalline zeolite component. It will be realized, however, that any reduction in the amount of encapsulated catalyst and its replacement with conventional cracking catalyst will markedly reduce total catalyst cost and will be highly desirable assuming that acceptable product yield, quality and distribution can be achieved.

Heretofore, the art has generally taught the catalytic use of either conventional amorphous siliceous gel type catalyst, or crystalline alumino-silicate zeolite catalyst, or the aforementioned encapsulated version of the zeolite catalyst which consists of the zeolite crystals embedded in a siliceous matrix. As mentioned, the encapsulated version has proven to be highly effective and will often be preferred to either the conventional amorphous gel catalyst or the crystalline zeolite catalyst per se. Substitution of either the amorphous or the crystalline zeolite catalysts with encapsulated catalyst is usually contemplated in terms of total replacement.

It has been surprisingly discovered, however, that the encapsulated version need not be the sole catalytic component in all cases, and that a physical admixture of the encapsulated version and conventional amorphous gel catalyst is a highly effective alternate which will be preferred to the encapsulated version per se in certain instances due to the economic gain to be realized by substitution of the more expensive encapsulated version with less expensive conventional catalyst.

Furthermore, although the encapsulation technique results in a "dilution" of the zeolite catalyst activity, in many instances catalytic activity may still be too high for existing plant facilities. In these instances, use of the aforesaid physical mixture of conventional amorphous catalyst and encapsulated catalyst will be desired. Other advantages associated with the use of this physical admixture accrue from the ease of adjustability of overall catalyst composition and characteristics in accordance with feed requirements or changing seasonal demands. Thus, product yields, distribution and quality can be readily adjusted for a given operation by simply varying the ratio of the components of the overall catalyst mixture.

This is particularly useful and convenient in fluidized operations wherein a portion of catalyst is continuously being withdrawn, regenerated and returned. Still another advantage derives from the use of such catalyst mixture with low quality, catalyst-contaminating feeds, owing to the distribution of contaminant over both components, thereby diluting its effect on the more expensive encapsulated component.

It has further been surprisingly discovered that, in addition to all of the above advantages, the admixture of encapsulated zeolite catalyst and conventional amorphous gel catalyst does not display the expected linear relationship resulting from the mere additive effect of the two components. Thus, when varying proportions of the encapsulated zeolite component and the conventional amorphous component are physically admixed, the catalytic cracking activity of the admixture is higher than would be predicted from the linear relationship between the two. This unusual effect is demonstrated over the entire range of proportions and has been observed with various other criteria, such as dry gas yield, coke make, gasoline production, etc. In all cases, the result obtained with the physical admixture of the encapsulated zeolite component and the conventional amorphous component is substantially better than the expected additive result.

Farbenfabriken Bayer

A process developed by G. Heinze and E. Podschus; U.S. Patent 3,296,151; January 3, 1967; assigned to Farbenfabriken Bayer AG is a process for the production of molecular sieve granules in spherical form. With the molecular sieve zeolites, the production of solid granules is difficult on account of the uniformly finely crystalline nature of these substances. They have so far usually been formed into granules by adding clay-like binders, such as kaolin, bentonite and attapulgite. However, these binders have the disadvantage that they do not impart a satisfactory resistance to abrasion to the granules and that, for hardening purposes, temperatures above 500°C. are necessary. These temperatures cannot be used with various zeolites on account of their thermal sensitivity. It has already been proposed to use silicic acid esters for binding purposes as these are hydrolyzed into a silica gel. The granules bonded therewith are not sufficiently hard and, in addition, the high cost of silicic acid esters opposes the use thereof.

This process for the production of molecular sieve granules in spherical form with a predominant content of molecular sieve, the granules being bonded with silicic acid, comprises stirring the powdery molecular sieve zeolites to be bonded with aqueous silica sol, to form a flowable suspension of pH 8 to 10, advantageously pH 8.2 to 9.0, mixing this suspension with comparatively small quantities of a second suspension of finely divided magnesium oxide in water in amount of 0.1 to 3.0% MgO calculated on dried activated granules and distributing the still-liquid gellable mixture of the two suspensions in drop form in a manner known per se in a liquid immiscible with water until the sol-gel conversion takes place.

It is surprising that the silica sols which are extremely sensitive to addition of electrolyte can be mixed with large quantities of the electrolytically strongly dissociated molecular sieve zeolites in aqueous suspension to give a suspension which is stable for several hours, and that furthermore such a suspension can be caused to gel almost immediately by adding a comparatively minimum quantity of an insoluble compound. As is known, even molecular sieve zeolites freed by careful washing from adhering impurities produce pH values between 9 and 11 on suspension in distilled water, i.e. pH values which are just in the range of the maximum flocculation sensitivity of aqueous silica sols. Consequently, it was not in a any way to be expected that compatibility exists between the two components.

The zeolite granules obtained according to this process constitute an essential technical advance, since, in addition to the advantages of the spherical form already mentioned above, they also have a number of other properties which are important for practical applications, such as a smooth and exceptionally abrasion-resistant surface, a high degree of hardness, a high zeolite content and an unreduced adsorption power of the enclosed zeolite despite the solid bonding. An additional advantage of the granules produced by this process is that they are resistant to water and the zeolite contained therein can easily be transformed by ion exchange with dilute salt solutions into other forms. This fact is also surprising, since without addition of zeolite, pure silica gel beads produced from silica sol burst because of internal stresses on coming into contact with water.

In carrying out this process, the powdery zeolite to be granulated is mixed with aqueous silica sol (advantageously of 15 to 40% SiO_2 content) into a flowable suspension of pH 8 to 10. On the other hand, a likewise aqueous suspension of hydrated magnesium oxide is also prepared by suspension of a finely divided magnesium oxide. It is advisable to allow this suspension to stand for at least 1 hour before use, so that the oxide can be hydrated. The two suspensions are then homogeneously mixed in a suitable proportion in a throughflow vessel with a high-speed stirrer or another mixing arrangement, in order immediately thereafter to flow through a nozzle into an organic liquid, in which the stream is split up into drops. The residence time of the drops in the organic phase is so chosen that the gelling process is initiated during this time and the gel balls which are formed have achieved the stability necessary for the after-treatment on leaving the organic phase. The process will now be described in further detail with reference to the drawing in Figure 54.

FIGURE 54:   APPARATUS FOR PRODUCTION OF SILICA BONDED ZEOLITIC MOLECULAR SIEVE GRANULES

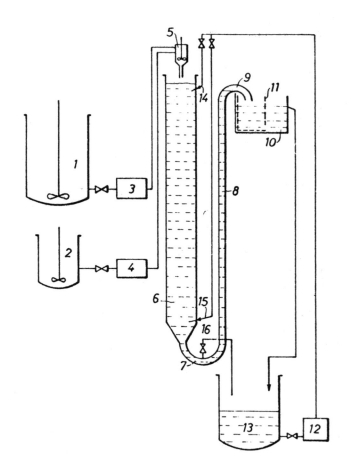

Source: G. Heinze and E. Podschus; U.S. Patent 3,296,151; January 3, 1967

The production of granules of a zeolite (Zeolite 4A) produced according to U.S. Patent 2,882,243 with a composition of

$$1.0 \pm 0.2Na_2O \cdot Al_2O_3 \cdot 1.85 \pm 0.5SiO_2 \cdot 0 - 6H_2O$$

will be described. In the figure, containers 1 and 2 are supply tanks for components A and B, which are supplied by means of proportioning pumps 3 and 4 in a stream, which is constant as a function of time, to a mixing nozzle 5 provided with a high-speed stirrer device. The combined components flow from the mixing nozzle 5 into the organic phase which is located in a pipe 6, the stream immediately being broken up into drops. The mixing nozzle can end above the liquid level or even dip thereinto.

The density of the organic phase is so chosen that the drops sink therein. The pipe 6 constitutes the actual granulating zone, along which the descending drops of the suspension solidify by gelling of the hydrosol content. The pipe 6 is connected through an elbow 7 to the pipe 8 of smaller diameter, which serves as the conveyor zone and which ends in an overflow bend 9. This overflow also simultaneously regulates the height in the granulation zone 6. The overflow 9 terminates above a collecting container 10 which is filled with the same organic liquid and from which the granules are discharged by means of a screening device 11. The reference 16 represents an emptying pipe.

The organic phase filling the system is not static but circulates in the direction of the arrows as indicated. A circulating pump 12 delivers from a supply container 13 to two inlet unions 14 and 15, which are arranged tangentially on the top and bottom ends of the granulation zone 6. If the diameters of the two vertical pipes 6 and 8 are in the ratio of for example 3:1, then provided the supply takes place through the inlet union 14, the upward flow in the conveyor zone 8 is nine times the downward flow in the granulating zone 6. However, if the supply quantity is distributed equally to the inlet unions 14 and 15, for the purpose of avoiding too strong eddy formation, then the upwardly flow important for the conveyance of the granules is 18 times the downward flow obtaining in the granulating zone.

Linear velocities of 5 to 30 cm./sec. are desirable for the upward flow. It is clear that the circulation quantity can be so regulated within a wide range that the downward flow in 6 is insignificant and is, at the most, of the order of the descending velocity of the particles and at the same time, an upward flow obtains in 8, which over-compensates the speed of descent of the particles and thus conveys them upwardly. This is further promoted by a constriction in cross-section caused by the here more densely packed granules themselves. The liquid stream discharges from the overflow 9 and the granules are rinsed in a collecting container filled with organic liquid. By means of inter-changeable wire strainers or other screening devices, for example a continuously revolving endless band, the granules are then gently lifted out of the liquid and exposed to a stream of hot air for drying purposes. As a result of being collected under liquid, the mechanical stressing during the assembly of the still fairly sensitive granulated elements is reduced, the strength of these granules being increased during the additional residence time until being lifted out of the liquid so that they are able to withstand the deformation due to the weight of the layers of granules disposed therabove.

The dimensions of the granulating device are as follows: the granulation zone 6 has a length of 2 m. and a diameter of 75 mm. and the conveyor zone 8 has a diameter of 25 mm. The organic liquid is trichloroethylene, which is at room temperature, and this is circulated by pumping at a velocity of 200 liters per hour. The supply at the inlet unions 14 and 15 is 100 liters per hour at each. A linear velocity of 0.6 cm./sec. is calculated for the downward flow in 6 and a velocity of 11.3 cm./sec. for the upward flow in 8.

For the component A, 10 kg. of crystalline sodium zeolite A with water content of 15% are suspended in 6.54 kg. of 14% silica sol with a specific surface according to BET of 200 m.^{2}/g. For this purpose, the finely powdered zeolite is consolidated into lumps by means of a roll press and these lumps are gradually incorporated into the silica sol by stirring. A thinly liquid homogeneous suspension is formed, and the pH of this suspension is adjusted to 9.0 by adding dilute hydrochloric acid. The suspension is stable for several hours.

The component B consists of a suspension of finely divided, hydrated magnesium oxide in water, which is prepared as follows: 60 g. of commercial "Magnesia usta extra leicht" with an MgO content of 83% are suspended in water by means of a high-speed stirrer device and the suspension is made up to 1 liter. The solution is left to stand for at least 1 hour before use.

The components A and B are supplied through proportioning devices to the mixing nozzle 5 in a ratio by volume of 5:1. The jet of the mixed suspensions leaving the nozzle is broken up in the trichloroethylene into drops, which slowly descend therein and solidify after 20 seconds before reaching the reversal loop. After drying at 110°C. and subsequent dehydration of the zeolite at 400°C., the granules which have a diameter of 3 to 5 mm. contain 90% by weight of anhydrous zeolite. The adsorption capacity of the hard granulates corresponds to the powder-like zeolite contained therein.

W.R. Grace

A process developed by W.A. Stover and W.S. Briggs; U.S. Patent 3,130,170; April 21, 1964; assigned to W.R. Grace and Company relates to the preparation of a clay cracking catalyst with improved cracking performance. Activated clays have been used for petroleum cracking catalysts for many years. During World War II, the demand for higher octane gasoline led to the development of synthetic catalysts. In the years following World War II, the synthetic manufacturers sold about 80 to 90% of the cracking catalysts used in the United States. However, the petroleum industry has revived interest in catalysts made from clays. This change has come about through the improvement of cheaper clay cracking catalysts.

It has been known that a number of factors determine the ability of a catalyst to perform. Among these are the impurity content, the pore volume of the catalyst, the pore diameter, and the overall surface area, all of which are quantities measurable by physical means. However, means have been established to reduce the impurity content and improve these catalytic properties.

One of the earliest, and still principally used methods of activating natural clays is to treat the clay with varying amounts of acid. When the clay, composed essentially of silica and alumina along with traces of other metallic compounds such as iron oxide, is exposed to acid, the acid serves to dissolve the alumina and other metallic compounds. It has been found that when the mixture of clay and acid solution is ammoniated, the alumina is reprecipitated on the surface of the silica, resulting in a suitable cracking catalyst. However, although the clay catalysts used are effective, there is still much room for improvement.

It has been found that a superior clay catalyst can be prepared by treating the clay with a sodium salt to form a sodium aluminum silicate complex, gelling the silicate and releasing the aluminum by treating with a strong mineral acid, adding a basic precipitant to precipitate the aluminum salts, filtering, drying, treating the filtrate with a material capable of base exchange with the sodium, washing, drying, calcining, and recovering the catalyst product.

The preferred procedure for carrying out the process is as follows: raw clay is mixed in an intensifier blender with sodium carbonate, $Na_2CO_3 \cdot H_2O$. The sodium carbonate can be present in from 35% to 80% by weight of the dry clay, preferably in the range of 40% to 50%. The mixture is mulled to an extrudable paste, calcined at a temperature of about 800°F. to 1400°F. for about 1.6 to 2 hours, preferably 1400°F. for about 1 hour, after which it is crushed, ground and slurried in water. Concentrated sulfuric acid (98% H_2SO_4) is added to give an acid concentration of about 3 to 15% by weight H_2SO_4, preferably about 5%, and an acid dosage of from 45 to 130% by weight, preferably about 70 to 100 weight percent. The mixture is heated to a temperature in the range of 180° to 210°F. for 1 to 2 hours, cooled below 100°F., and neutralized with ammonia.

Ammonia should be added to bring the pH to 5.0 to 10.0, preferably to pH 8.5. The slurry is filtered to remove the major part of soluble salts prior to drying, reslurried in water and refiltered and dried at about 160° to 230°F. The dried filter cake is base exchanged with a suitable aqueous solution, preferably containing 3% by weight $(NH_4)_2SO_4$ and 1% by weight NH_4OH. The dried solids are base exchanged, washed, redried, and pilled.

Kidde Process

A process developed by G.E. Kidde; U.S. Patent 3,039,974; June 19, 1962; assigned to Kidde Process Corporation involves the production of silica-alumina cracking catalyst using waste silicon tetrafluoride from phosphate rock processing operations in the form of fluosilicic acid as a major raw material.

Figure 55 illustrates the process as applied to a continuously-operated commercial plant designed for a daily production of 68.5 tons of anhydrous hydrogen fluoride, 10 tons of 35% hydrofluoric acid, 42 tons of synthetic fluid cracking catalyst and 25 tons of ammonium sulfate.

As indicated, 320 tons per day of fluosilicic acid is preheated with steam and reacted with 684 tons per day of aluminum sulfate. As will be explained below, the aluminum sulfate is produced in the hydrogen fluoride generator 9 and in dissolver 10, and comprises a 50% aqueous solution which must be maintained at a temperature above 260°F. in order to flow. The temperature in the reactor 11 is maintained at about 205°F. and the silica is precipitated in accordance with the basic reaction:

$$2H_2SiF_6 + 3Al_2(SO_4)_3 + 4H_2O \longrightarrow 3(AlF_2)_2SO_4 + 2SiO_2 + 6H_2SO_4$$

The slurry thus produced is filtered in filter 12, from which 300 tons per day of silica filter cake is obtained. This filter cake is washed countercurrently with hot water in 2 five-stage tray thickeners 13 and 14, and the washed slurry from the tenth stage (thickener 14) is filtered on a vacuum drum string discharge filter 15. A flocculating agent to improve filtration may be used.

FIGURE 55: PROCESS FOR THE PRODUCTION OF SILICA-ALUMINA FLUID CRACKING CATALYST

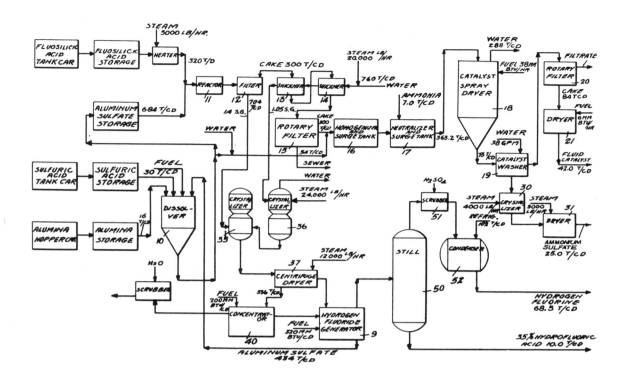

Source: G.E. Kidde; U.S. Patent 3,039,974; June 19, 1962

Aluminum sulfate (54 tons per day) from the dissolver 10 is mixed with the silica filter cake (300 tons per day) from the filter 15 in a Manton-Gaulin homogenizer 16. After this preliminary homogenization, gaseous ammonia (7 tons per day) is added with mixing at the tank 17 to precipitate the alumina at a pH of about 4.0. The pH for this precipitation may range from about 3.5 to about 5.5. The catalyst gel thus produced (363 tons per day) is dried in a 22' Swenson spray dryer 18, particular attention being paid to proper particle sizing with the ultimate object the production of a material in the 40 to 80 micron range. The equipment specified will produce a highly satisfactory material with about 85% there-of in the 40 to 80 micron range, about 5% larger than 80 microns and about 10% smaller than 40 microns.

After this drying, the fluid catalyst (73 tons per day) is washed at 19 with de-ionized water at a pH of about 5.5 for re-moval of all soluble salts. The washed slurry is filtered in a rotary filter 20 and the filter cake (84 tons per day) is dried at 21 to produce the final fluid cracking catalyst which is stored in bulk. The catalyst comprises 13% Al$_2$O$_3$ and 85% SiO$_2$. This composition may be readily varied to meet the requirements of the petroleum industry.

The overflow from the catalyst washer 19 contains 25 tons per day of ammonium sulfate and this is recovered by crystal-lization at 30 and drying in a centrifuge dryer 31. The dried product is stored in bulk. For production of hydrogen fluo-ride and hydrofluoric acid, the filtrate from filter 12, which has a specific gravity of 1.4, is fed to a crystallizer 35. The overflow from the thickener 13 is fed to a crystallizer 36 operating in multieffect relationship with the crystallizer 35. The crystals thus produced are dried in a centrifuge 37 to produce dried crystals of a complex compound having the following analysis: 25% alumina (Al$_2$O$_3$); 15% fluorine; 35% sulfate (SO$_4$); and 25% water. The mother liquor from the dryer 37 consists of 75% sulfuric acid, 1% fluorine, 3.3% sulfate, 2.2% alumina and 18.5% water. This mother liquor is concentrated at 40 to about 95% sulfuric acid, the vapors from the concentrator 40 being scrubbed at 41 to re-cover the hydrogen fluoride therein.

The concentrated mother liquor and dried crystals are reacted in the generator 9 at a temperature of about 850°F. The hydrogen fluoride vapors evolved from the generator are rectified in a carbon-lined rectifying tower or still 50 packed with Raschig rings. Anhydrous hydrogen fluoride is taken off overhead and the bottoms consist of 35% hydrofluoric acid. The overhead product is scrubbed in sulfuric acid at 51 prior to condensation in a shell and tube condenser 52.

The liquid product from the generator 9 is dry hydrated aluminum sulfate which is fed to the dissolver 10. The 50% aluminum sulfate for the process is produced in this dissolver, make-up sulfuric acid (30 tons per day) and alumina trihydrate (16 tons per day) being provided here.

Mobil Oil

A process developed by C.J. Plank and E.J. Rosinski; U.S. Patent 3,431,218; March 4, 1969; assigned to Mobil Oil Corporation relates to a method of converting natural clays to crystalline aluminosilicate zeolites, to the preparation of improved hydrocarbon conversion catalysts from such zeolites to the resulting catalysts, and to conversion of hydrocarbons in the presence of such catalysts. The process involves synthesizing crystalline aluminosilicate zeolite by treatment of natural clay with a caustic-containing solution under such conditions of time and temperature as to effect conversion of at least a portion of the reaction mixture to a crystalline aluminosilicate zeolite. The digestion may be carried out in the presence of an added source of silica.

A further aspect of the process involves compositing such crystalline zeolite with an inorganic oxide matrix and forming the composite into discrete catalyst particles, the crystalline zeolite having been subjected to ion-exchange with cations selected from the group consisting of hydrogen, ammonium, and polyvalent metals under such conditions as to reduce the alkali metal content, e.g., sodium content thereof, either prior or subsequent to the compositing step.

A process developed by G.T. Kerr, C.J. Plank and E.J. Rosinski; U.S. Patent 3,442,795; May 6, 1969; assigned to Mobil Oil Corporation involves a technique for increasing the silica to alumina ratio of zeolites. The stability of the exchanged crystalline aluminosilicates in the presence of heat, steam and acid, as well as their catalytic properties in general, are to a great extent dependent upon the silica/alumina ratio in the crystal lattice of the aluminosilicate. Generally speaking, the higher the silica/alumina ratio in the aluminosilicate, the greater the stability to heat, steam and acid.

In synthetic aluminosilicates, the silica/alumina ratio is essentially determined by the specific materials and the relative quantities of such materials used in the preparation of the zeolite. Naturally occurring zeolites are available, of course, with a fixed silica/alumina ratio. Hithertofore, no truly effective technique is known to alter drastically the silica/alumina ratio in natural crystalline aluminosilicates or in synthetic crystalline aluminosilicates after the latter have been formed. Since in many respects it would be advantageous to be able to convert an existing crystalline aluminosilicate to a crystalline material having a greater silica/alumina ratio, the desirability of a process which would make such conversion feasible from an economical and chemical standpoint would, quite obviously, be greatly desirable.

In accordance with this process, it has been found that it is possible to change drastically the silica/alumina ratio in crystalline aluminosilicates, modify their crystalline character to result in a shift to shorter metal-oxygen interatomic distances (measured as lattice cell constant a0) yet, at the same time, to obtain an improved zeolite or zeolite-like material having one or more enhanced catalytic properties.

In accordance with this process, the silica/alumina ratio of crystalline aluminosilicates may be significantly increased to form more useful catalytic composites by treating such aluminosilicates to remove part of the aluminum atoms in their anionic structure. Such removal is effected by a combined solvolysis-chelation technique.

In order to remove aluminum effectively from the crystalline aluminosilicate, it is essential that the aluminum first be removed from the tetrahedral sites in the anionic crystal lattice of the aluminosilicate. This step is effectuated through the formation of a hydrogen or acid zeolite and the subsequent solvolysis of the acid zeolite. More specifically, a portion of the cation of the zeolite must at some point be hydrogen. It is the solvolysis of those aluminum sites associated with such hydrogen ions which causes the zeolite to lose aluminum from the zeolitic framework.

Having removed aluminum from the tetrahedral sites of the aluminosilicate, the next problem is to separate this aluminum physically from the aluminosilicate (assuming such separation is desired). This may truly effectively be accomplished by means of a suitable complexing agent to complex the aluminum into a form which facilitates its separation from the aluminosilicate.

An extremely effective method of removing aluminum from the aluminosilicate with a minimum of effort and procedural steps involves the use of complexing agents which are effective not only to separate the aluminum physically from the aluminosilicate but to convert the aluminosilicate to its hydrogen or acid form prior to such separation, thus obviating the requirement for a preliminary acidification step.

Ethylenediaminetetraacetic acid (EDTA) may specifically be used in a process which takes place in a series of four stages. In the first stage, the zeolite undergoes attack by the hydrogen ions of the chelating agent with such hydrogen ions being substituted for at least a portion of the cations of the zeolite. In the second stage, the acid zeolite formed in the first stage solvolyzes to form a compound deficient in alumina. At this point in the process, the aluminum which has been removed from the tetrahedral sites of the aluminosilicate is still physically present on the aluminosilicate. In the third stage of the process, the aluminum hydroxide formed in the second stage reacts with excess of the chelating agent to form a quaternary ammonium hydroxide and an aluminum chelate. Finally, in the fourth and last stage of the process, the quaternary ammonium hydroxide reacts with the hydrolyzed acid zeolite formed in the second stage of the process to form the quaternary ammonium form of the alumina-deficient zeolite.

A process developed by G.T. Kerr, J.N. Miale and R.J. Mikovsky; U.S. Patent 3,493,519; February 3, 1970; assigned to Mobil Oil Corporation is a process for producing a hydrothermally stable catalyst composition of high hydrocarbon conversion activity which comprises calcining an ammonium-Y crystalline aluminosilicate in the presence of rapidly-flowing steam, base-exchanging the resultant steam product with an ammonium salt, treating the resultant exchanged product with a chelating agent capable of combining with aluminum at a pH between about 7 and 9, and recovering the final product. The resultant catalysts yield fantastically high α-cracking activities of about 500,000 to 3,000,000. Indeed, these are the highest activities ever observed for a hydrogen faujasite.

The ammonium-Y aluminosilicate is first calcined in the presence of rapidly flowing steam resulting, presumably, in the formation of lattice aluminum defects, aluminum containing cations, and other nonframework aluminum which exists as amorphous hydrated alumina. Base exchange with an ammonium salt, preferably ammonium chloride, transforms the product back into the ammonium form, and chelation preferably with the ammonium salts of ethylenediaminetetraacetic acid, and more preferably, with diammonium dihydrogen ethylenediaminetetraacetate removes the amorphous aluminum-containing material.

Chelation pH should be about 7 to 9, preferably, 7 to 8, to prevent further destruction of the aluminosilicate structure. As mentioned, while chelation may be with any agent capable of combining with aluminum, care should be taken that the reaction mixture pH is within the prescribed ranges. A finishing calcination in dry air, by conventional means, produces the superactive catalysts. In a preferred embodiment of this process, as a pretreating step, the ammonium-Y starting material is contacted with a solution of diammonium dihydrogen ethylenediaminetetraacetate in order to remove any amorphous agglomerates that may be present in the channels.

A process developed by E.J. Rosinski; U.S. Patent 3,520,828; July 21, 1970; assigned to Mobil Oil Corporation involves the preparation of composite particles of an inorganic oxide gel, e.g. of silica-alumina, having fines dispersed therein, e.g. of crystalline aluminosilicate zeolite. Process comprises forming a rapidly gelling hydrosol of the inorganic oxide and spraying the hydrosol into a gaseous medium to form a stream of particles which are suspended therein for a time sufficient to effect gelation. Fines are incorporated in the hydrosol prior to gelation by dispersing the fines in the hydrosol prior to spraying or by spraying the hydrosol into a gaseous medium containing fines suspended therein. The product, in which the fines constitute greater than 40% by volume, is useful as a catalyst, as for hydrocarbon cracking.

A process developed by N.Y. Chen and F.A. Smith; U.S. Patent 3,551,353; December 29, 1970; assigned to Mobil Oil Corporation involves increasing the silica/alumina mol ratio of crystalline aluminosilicates having a silica/alumina mol ratio greater than 10, e.g., mordenite, by a process comprising alternate steam and acid treatment.

A process developed by W.A. Stover and H.A. McVeigh; U.S. Patent 3,553,104; January 5, 1971; assigned to Mobil Oil Corporation yields a composite catalyst characterized by excellent selectivity and unusual heat stability. Thus when such catalyst is subjected to thermal treatment over a wide variety of temperatures, e.g., from about 1200° to 1750°F., there is virtually no change in its physical properties, e.g., pore volume, surface area, density, etc.

Such composite catalyst is made up of crystalline aluminosilicate particles contained in a porous silica or silica-zirconia gel matrix the gel having a pore volume of at least about 0.6 cc. per gram. The matrix further includes a weighting agent, preferably clay, in such quantity that the overall composite has a packed density of at least 0.3 gram/cc.

The foregoing composites are made by admixing with an aqueous alkali metal silicate a particulate weighting agent such, as, e.g., a kaolin clay, desirably as a dispersion in water, so as to coat the clay particles with alkali metal silicate. This admixing is conveniently done at room temperature although of course higher or lower temperatures may be employed if desired. The mixture is then heated, generally to a temperature of from about 100° to 160°F. and acid is added to adjust the pH to from about 8 to 10. This temperature is maintained for a time of about 1 to 6 hours or longer. At this point, if a silica-zirconia-weighting agent (e.g., clay) matrix is desired, a zirconium salt is added, desirably as an aqueous solution thereof. Acid is then added to reduce the pH to about 4 to 7 and form a silica gel-weighting agent or silica gel-zirconia gel-weighting slurry, which is then admixed with a slurry of crystalline aluminosilicate. The resulting composite is separated and dried in the form of particles suitable for fluid catalytic conversion.

The composite catalysts produced show unusual selectivity and are particularly desirable in their ability to crack hydrocarbons to relatively high yields of gasoline while having low coking tendencies. This is of great value when dealing with "dirty" feed stocks, e.g., heavy gas oils and "recycle" stocks, which ordinarily give off appreciable coke yields when subjected to cracking.

REGENERATION

Gulf Research and Development

A process developed by J. Senyk and R.S. Toohey; U.S. Patent 3,232,711; February 1, 1966; assigned to Gulf Research and Development Company relates to the quantitative determination of catalyst carbon content, using a photometric technique, and more particularly, to on-line analysis of catalyst for carbon content and to the use of such analysis as a basis for the control of catalytic regeneration operations.

The carbon or coke content of the catalyst employed in catalytic hydrocarbon conversion processes is one of the most important parameters by which the efficiency of such conversion can be controlled, as even a relatively slight increase in the carbon content of a catalyst can result in a significant reduction in the activity and/or selectivity of such catalyst. By way of illustration, published data indicate that an increase in the carbon content of a synthetic silica-alumina cracking catalyst of as little as from one percent to 1.5 percent can result in a reduction in conversion from 65 to 63.5 percent and an increase in coke yield from about 6.66 percent to 6.74 percent, based on the feed. For a catalytic cracking unit charging 40,000 barrels per day of fresh feed, a change in the activity and selectivity of the cracking catalyst of this magnitude would mean a reduction of about 600 barrels per day of converted products, principally gasoline, as well as an outright loss of about 32 barrels per day of total products to additional coke deposition on the catalyst.

Notwithstanding the importance of the carbon content of a catalyst to the operating efficiency of catalytic hydrocarbon conversion processes, such as catalytic cracking of gas oils to gasoline, close control of such processes with respect to carbon on catalyst has been difficult to achieve. Thus, when catalyst carbon content is obtained by conventional laboratory methods, a considerable time may elapse under normal conditions between the withdrawal of a catalyst sample for analysis and the reporting of results. During night operation, even longer periods without analytical information as to the carbon content of the catalyst may occur, as laboratory analyses are ordinarily not obtainable at night. Consequently, a considerable loss of products may have already been suffered by the time the discovery is made that the process has been operated at less than optimum efficiency with respect to the carbon content of the catalyst.

Although it is possible during intervals in which laboratory analyses are not available to control process conditions on the basis of visual estimates of the carbon content of the catalyst, this method of operation has not proved entirely satisfactory, and in fact is inherently unreliable since even small differences in carbon content that are not normally distinguishable by visual observation can make a significant difference in results. For example, one study has indicated that an increase in carbon content of a cracking catalyst of as little as 0.05 percent could mean a loss of approximately $100 per day in a cracking unit of the size indicated above.

Automatic, rapid, on-line analysis of catalyst for carbon content has been proposed previously to alleviate the problems indicated above, but has not been found entirely satisfactory, as the analytical methods hitherto available for on-line analysis have been merely variations of relatively cumbersome laboratory methods that are not well suited for in-plant operation.

It has been found that dependable, accurate automatic on-line analysis for carbon on catalyst can be carried out by a photometric technique, whereby close control of regenerated catalyst carbon content can be achieved, together with proportionately improved product yields from the reaction in which the regenerated catalyst is employed.

Briefly, this process involves providing a sample of catalyst that is representative of the catalyst mass whose carbon content is to be determined and disposing the catalyst sample on a support. A uniform test surface is then formed in the catalyst sample, and this uniform surface is subjected to light radiation of known intensity. The intensity of the radiation reflected from the test surface is detected and converted to an output whose intensity is related to the carbon content of the catalyst.

The essential pieces of apparatus involved in the conduct of this process are shown in Figure 56. The numeral 1 denotes a catalyst standpipe, the upper portion of which is connected to a fluidized catalyst regeneration vessel 56, shown in fragment, and the bottom portion of which is connected to the feed inlet line 58 of a catalytic cracking reactor 66, shown in fragment. Standpipe 1 permits freshly regenerated catalyst 60 to flow in fluidized form from a fluidized bed 62 of catalyst that is being regenerated in the regenerating vessel 56, at a rate controlled by slide valves 64, into the feed line 58, where it is premixed with hot cracking charge stock in the desired proportion.

Catalyst bed 62 is regenerated by combustion of carbon previously deposited thereon during the catalysts of hydrocarbon cracking or like reactions in vessel 66. Combustion of such carbon is effected by continuous introduction of air from blower 68, control valve 70, heater 72, and standpipe 74 into regenerator 56.

Numeral 2 refers to an upper or catalyst withdrawal section of a catalyst sample line, or bypass conduit, for withdrawing catalyst from catalyst standpipe 1 and for transporting the same to analysis means 5. Numeral 4 refers to a lower or

FIGURE 56: APPARATUS FOR QUANTITATIVE DETERMINATIONS OF CARBON ON CATALYSTS

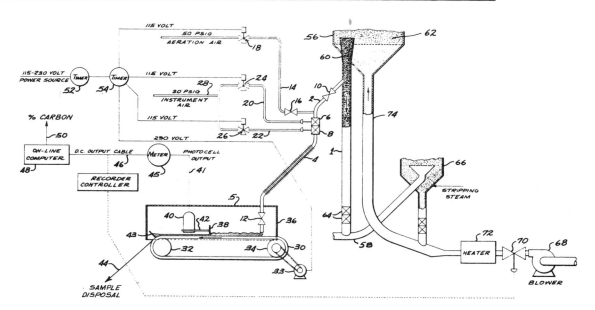

Source: J. Senyk and R.S. Toohey; U.S. Patent 3,232,711; February 1, 1966

catalyst cooling section of the catalyst sample line. Line 4 is provided with spaced fins for rapid disspiation of heat from the hot, freshly regenerated catalyst into the atmosphere, prior to passage into analysis means 5. Flow through catalyst sample lines 2 and 4 is controlled by pneumatically operated slide valves 6 and 8 and also by manually operated valves 10 and 12. Lines 2 and 4, together with control valves 6 and 8 form sampling means interconnecting a catalyst analysis means 5 and a source of freshly regenerated catalyst whose carbon content is to be analyzed.

Line 14 indicates a catalyst aeration air line for aerating catalyst in lines 2 and 4 and for clearing the catalyst sample line, whereby plugging of the catalyst sample line by a compacted mass of catalyst is avoided. Flow of aeration air for the catalyst sample line is through line 14 and is controlled by solenoid operated valve 18 and by manually operated valve 16. Slide valves 6 and 8 are operated by an auxiliary supply of instrument air delivered to the system through line 28 and to the diaphragm motors of valves 6 and 8 by lines 20 and 22 respectively. The flow of instrument air through lines 20 and 22 is regulated, respectively, by solenoid operated valves 24 and 26.

Numeral 30 refers to a conveyor belt, which is desirably formed from some heat resistant material such as steel, although heat resistant material need not be used when the cooling capacity of the catalyst cooling line is sufficiently large to prevent damage to the belt by the catalyst samples. Belt 30 forms a support on which the sample of catalyst to be ana- lyzed can be disposed for photoelectric scanning. Numeral 34 indicates the drive roller for conveyor belt 30, and numeral 32 indicates the idler roller. Drive roller 34 is driven by electric motor 33. Roller 34 and motor 33 can be operatively associated by a drive belt, as shown, or alternatively, by a suitable gear train. Numeral 36 indicates a hood, formed from sheet metal or the like, enclosing the upper surface of the conveyor belt 30 to prevent the entrance of stray light and air circulation.

Numeral 38 denotes a leveling bar attached at each end to the inside surfaces of hood 36. Bar 38 is mounted so as to provide a small clearance, say one quarter inch, between its lower surface and the upper surface of the conveyor belt 30. This bar, together with the conveyor belt 30, acts as a means for forming a uniform, smooth, planar test surface, normal to the direction of the light radiation from a source in search unit 40, in the catalyst samples disposed on the up- per surface of conveyor belt 30. The action of the leveling bar on the catalyst sample prior to photoelectric scanning is very important, as the photoelectric search unit of analyzer means 5 is sensitive to changes in density of the catalyst sample and in the angle of repose of the catalyst sample.

Numeral 40 denotes the photoelectric search unit component of analysis means 5. Search unit 40 is supported by a bracket 42 mounted on the left-hand side of leveling bar 38. Numeral 43 indicates a scraper adapted to remove catalyst samples from the upper surface of belt 30 after they have been scanned by search unit 40. Catalyst so removed drops to inclined plane 44 and thence to a catalyst disposal vessel, or hopper, not shown.

Numeral 41 represents an electrical connection for conducting the electrical photocell output of the analyzer search unit 40 to the meter 45, which amplifies and visually indicates the magnitude of the intensity of such output. This output intensity is a function of the quantity of light radiation reflected from the test surface of the catalyst sample and of the amount of carbon on the catalyst sample. In the illustrated embodiment, the photocell output is electrically conducted from the meter 45 by way of DC output cable 46 to an on-line computer 48. Computer 48 is previously programmed with calibration information relating photocell output intensity to catalyst carbon content. The computer functions to apply the necessary scaling factor to the photocell output and converts the information thus obtained to a second, digitized output that is indicative of percent carbon on catalyst. This second output is passed through electrical conductors 50 to a logging typewriter, not shown, which in turn prints out the percent carbon on the catalyst sample.

The information obtained in the above-indicated manner can be used manually to control a variable in the catalytic regenerating process, such as the rate at which regeneration air is introduced into the regenerator, or alternatively, a variable in the catalytic conversion reaction, such as the rate at which charge stock is introduced to the reactor, so as to permit a greater or lesser quantity of carbon to remain on the regenerated catalyst, if desired. In a more complex operation, the percent carbon on catalyst as determined by the computer also can be stored by the computer for use in the overall correlation of the process to obtain an optimum or near optimum balance of operating conditions. In an alternative operation, there can be employed instead of or in addition to the computer a conventional instrument adapted continuously to indicate and/or record the strength of the photocell output with increasing time, or a conventional controller — which can be indicating and/or recording — as shown by dashed lines, or neither, that will not only continuously monitor the strength of the photocell output signal but that also will control a predetermined process variable, such as regeneration air rate, in a predetermined manner.

In operation of the system illustrated, the over-all sampling and analytical system is actuated by schedule timer 52, which is set to operate at desired sampling intervals. Schedule timer 52 in turn actuates sequence timer 54, which in turn actuates, in the desired sequence, conveyor drive motor 33, solenoid operated valves 24 and 26 and control valves 6 and 8, and solenoid operated valve 18. With manually controlled valves 16, 10 and 12 open, freshly regenerated catalyst is withdrawn from standpipe 1, caused to pass through catalyst sample line 2, control valves 6 and 8 and through air-cooled catalyst sample line 4.

Catalyst flows in the direction indicated because of a slight superatmospheric pressure in standpipe 1. The catalyst so withdrawn is then disposed on the upper surface of moving conveyor belt 30. As the catalyst sample moves beneath leveling bar 38, a smooth, planar test surface is formed in the catalyst sample. As the test surface moves beneath the scanning element of analyzer search unit 40, the light diffusely reflected from the test surface is detected and converted to an electrical output by the measuring photocell component 84 of search unit 40. This output is amplified and indicated by meter 45 and converted to percent carbon on catalyst by computer 48.

In a specific embodiment there is employed a catalyst sample line comprising an upper or sample withdrawal section approximately 8.5 feet in length and a lower portion, or catalyst sample cooling section, fabricated from a 14.5 foot length of finned steel tube having an outside diameter of one inch. The upper section of the catalyst sample line is provided with a pair of pneumatic slide valves and an aeration air line formed of one-half inch pipe, as shown. The analyzing means comprises a steel conveyor belt six inches in width mounted on a pair of rollers whose axes are spaced five feet apart.

One of the conveyor rollers is driven by an explosion-proof electric motor. The search unit and reflection meter employed in this embodiment is a Photovolt photoelectric reflection meter, Model 610, containing a Model 610-Y search unit. The schedule timer is a Flexopulse continuous cycle timer having a cycle time of one hour. The cycle timer is a Polyflex reset cycle timer having a cycle time of one minute. The leveling bar is spaced approximately one-quarter inch above the upper surface of the conveyor belt and the aperture in the protective light shield of the photoelectric search unit is spaced approximately one-half inch above the upper surface of the conveyor belt.

The upper end of the catalyst sample line is connected to the regenerator standpipe that connects the lower section of the regenerator vessel with the feed inlet line of the cracking reactor. In the present embodiment the regeneration vessel contains an inventory of approximately 500 tons of fluidized semisynthetic silica–alumina cracking catalyst which is undergoing regeneration by combustion of carbon deposits thereon at a temperature of about 1150°F. Combustion is maintained in the regenerator vessel by introduction therein of 650,000 pounds per hour of undiluted air.

At a preset sampling time, the schedule timer activates the cycle timer. The cycle timer in turn activates the following elements in the order named: the conveyor motor, the solenoid operated valve that controls the first pneumatic slide valve, the solenoid operated valve that controls the second pneumatic slide valve, and the solenoid operated valve that controls the aeration air. With the above-indicated, automatically operated valves in an open position, except valve 18 which is closed when energized, and with all manually controlled valves in an open position, catalyst passes from the standpipe through the withdrawal section of the catalyst sample line, through the catalyst cooling section of the catalyst sample line and onto the upper surface of the conveyor, where it is deposited in the form of mounds.

A uniform test surface is formed in the nonuniformly mounded catalyst sample by the leveling bar as the catalyst is conveyed toward the search unit. As the catalyst passes beneath the search unit, diffuse light reflectance is measured and converted into percent carbon on catalyst, as described. At the conclusion of a preset sampling period, the schedule timer switches off the sequence timer, which then resets to zero, whereby the timer-controlled valves 24 and 26 are caused to close, aeration valve 18 is opened, and the conveyer belt is switched off.

Assuming a satisfactory operating level of 0.35 percent by weight carbon on catalyst, and assuming, by way of example, an observed reading of 0.40 percent by weight carbon on catalyst at a given sampling time, a slight increase in the regeneration air flow rate to the regenerator vessel, say 20,000 pounds per hour, is effected by manual adjustment of air blower control valve 70. At the next sampling time, the carbon content of the catalyst is again observed. If it has returned to the desired operating level of 0.35 percent by weight, no further adjustment will be made. On the other hand, if the percent coke on catalyst remains above the desired level, a further increase in the rate of introduction of regeneration air to the regenerator vessel will be made.

Mobil Oil

A process developed by F.A. Smith; U.S. Patent 3,394,075; July 23, 1968; assigned to Mobil Oil Corporation involves operating a cyclic process comprising catalytic cracking and regeneration so that relatively high coke burning rates are achievable in the process. This is made possible by the use of superactive crystalline aluminosilicate containing catalysts which retain catalyst activity and selectivity at relatively high residual coke levels, higher than that permissible with amorphous silica-alumina catalysts without destroying their activity and high selectivity for converting hydrocarbon feeds to desired product.

Therefore, the operation described includes and is particularly directed to maintaining relatively high uniform residual coke levels on regenerated catalyst particles in cooperation with effecting a relatively uniform deposition of coke on the catalyst particles during the conversion step. This control on coke level and deposition permits maintaining a closer temperature control on the individual catalyst particles selected in the system.

Phillips Petroleum

A process developed by J. Van Pool; U.S. Patent 3,223,650; December 14, 1965; assigned to Phillips Petroleum Company involves an improved method and apparatus for regenerating solid particle-form contact mass material. It provides better control of catalyst outlet temperatures and eliminates or at least substantially reduces the occurrence of afterburning in the regenerator. It provides for increased conversion of carbon to carbon monoxide rather than carbon dioxide so that less heat is evolved, or so that — at the same heat load — more carbon can be burned. Finally, it provides better separation of catalyst particles from flue gas and more efficient cooling of the catalyst particles.

The operation of the process will now be described in further detail with reference to Figure 57. Spent catalyst having carbonaceous deposit thereon is introduced into regenerator 10 through a plurality of feed conduits 11. Combustion air is passed through one or more inlet ducts 12 provided at spaced intervals around regenerator 10. A portion of the flue gas resulting from the burning of the deposits on the particle-form contact material passes upwardly through the bed of contact material and is removed through duct 13.

Since the particle-form contact material entering through conduits 11 is relatively cool, e.g. about 900°F., the flue gas removed through duct 13 does not have the requisite temperature to produce afterburning. The remainder of the flue gas passes downwardly through the portions of the bed which have already been subjected to the heat of combustion, resulting in additional burning and even higher temperature e.g. 1050° to 1150°F.

One or more indirect heat exchanging coils 14 are located in a lower section of the bed to withdraw heat from the descending particle-form contact material and flue gas. Below cooling coils 14 are located a plurality of radially-directed inverted channels or flue gas collectors 15. These downwardly-directed troughs provide suitable shielded chambers into which the flue-gas can be withdrawn from the gravitating particle-form contact material. The channels can be progressively wider from inner end to outer end and are blocked off on the outer ends. Perforated pipes 16 enter through the wall 17 under each channel and connect in the central core space of regenerator 10 with a ring-type header 18.

Below the flue gas collectors 15 is located a horizontally disposed plate 19. A plurality of downwardly-directed conduits 21 feed the contact material from the bottom of the bed to outlet pipe 22. Plate 19 and conduits 21 define a plenum chamber 23 in the bottom of the regenerator 10. The particle-form contact material is passed from outlet pipes through conduits 24 into a lift tank 25. The flue gas is removed from the enclosed region under channels 15 through the downwardly-directed pipes 26 into the plenum chamber 23. Contact material carried over by the gas through the pipes 26 is separated in the plenum chamber 23 and returned to the contact material withdrawal stream through outlet pipe 22. Particle-free gas is withdrawn through outlet conduit 27 to a flue gas stack (not shown).

FIGURE 57: DIAGRAM OF MOVING BED REGENERATOR KILN

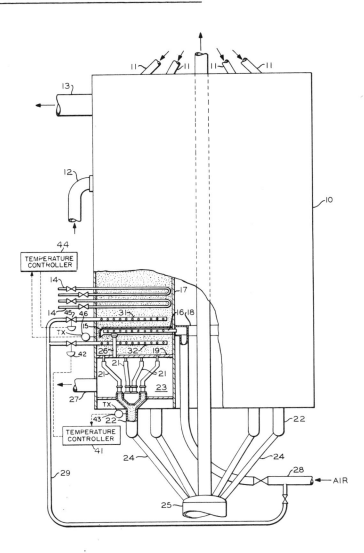

Source: J. Van Pool; U.S. Patent 3,223,650; December 14, 1965

Relatively cool air, e.g. atmospheric temperature, is passed through line 28 to ring header 18, and then through perforated pipes 16 into the region under channels 15. The cool air mixes rapidly with the hot flue gas withdrawn from the bed,
cooling it substantially. This cool air also cools channels 15, pipes 26, plate 19, and conduits 21, thus cooling the contact material by indirect heat exchange.

In accordance with this process, cool air is passed through line 29 and introduced directly into the bed or contact material
through perforated pipes 31 and 32. Pipe 31 is located at a distance above channels 15 and below cooling coils 14. Air
passing through perforated pipe 31 mixes directly with the flue gas and contact material and substantially reduces the temperature thereof, thus completely eliminating or substantially reducing the possibility of afterburning in the area. The
flow of air through pipe 31 can be controlled to increase the back pressure on the burning zone, thus forcing a larger portion of the flue gas upwardly through the bed and reducing the oxygen concentration in the burning zone.

As the contact material in the upper portion of the bed is relatively cool, the flue gas will be cooled below the temperature at which afterburning can occur, thus increasing the percentage of the carbon converted to carbon monoxide rather
than carbon dioxide. As approximately 10,140 Btu are released per pound of carbon converted to carbon monoxide in

comparison to the release of approximately 14,500 Btu per pound of carbon converted to carbon dioxide, this results in a decrease in the over-all heat release and thus increases the capacity of the regenerator.

Pipe 32 is located in the bed of contact material below channels 15 and above plate 19. Air passing through perforated pipe 32 mixes directly with the contact material, substantially cooling the contact material, and aids in separating any remaining flue gas from the contact material below channels 15. As most of the flue gas has already been withdrawn into channels 15, the air from pipe 32 has to cool only the contact material itself thus providing more efficient cooling. Accurate control of the outlet temperature of the contact material can be obtained through manipulation of the temperature and/or rate of flow of the air in pipe 32.

While only one pipe 31 and one pipe 32 have been illustrated for purposes of simplicity, a plurality of pipes uniformly distributed over the cross-section of the contact material bed can be utilized at both locations. The perforated pipes 31 and 32 can be connected to the air supply either from the exterior of the regenerator 10 or by ring headers similar to ring header 18. The air flow and air temperature for pipes 31 and 32 can be controlled by common equipment or by separate means. Suitable means, such as a temperature controller, can be utilized to control the temperature and/or rate of flow of air for pipes 31 and 32. For example temperature controller 41 can regulate valve 42 in pipe 32 responsive to the temperature of the contact material in outlet pipe 22 as indicated by temperature transmitter 43. Similarly, temperature controller 44 can regulate valve 45 in pipe 31 responsive to the temperature of the contact material in the region adjacent or above channels 15 as indicated by temperature transmitter 46.

A process developed by J.B. Rush; U.S. Patent 3,494,856; February 10, 1970; assigned to Phillips Petroleum Company relates to the removal and disposal of catalyst fines contained in effluent gaseous streams. In order to prevent atmospheric contamination by the catalyst fines, it is necessary to subject the flue gases to rather expensive treating steps.

One system which has been used involves passing these gases through electrostatic precipitators. However, large precipitators are required and these are quite expensive. Another procedure which has been proposed for the removal of catalyst fines comprises scrubbing the flue gas with an oil to remove the fines. However, the scrubbing oil must then be treated to remove the fines. Filtering operations to perform this function are quite expensive, and there still remains the problem of disposing of substantial volumes of the fines, which are then present in the form of a paste with the scrubbing oil.

In accordance with this process, a method is provided for removing catalyst fines from gaseous streams in a relatively inexpensive manner and for disposing of the removed catalyst fines. This is accomplished by first scrubbing the flue gas with an oil to remove the catalyst fines. The resulting oil containing the catalyst fines is then introduced into a distillation unit wherein asphalt is being produced. It has been discovered that substantial volumes of catalyst fines can be incorporated into the resulting asphalt in this manner without materially affecting the desired properties of the asphalt. This permits removal of substantial quantities of fines from a refining operation without extensive disposal procedures. A flow diagram showing the essentials of this process is given in Figure 58.

Referring now to the drawing in detail, there is shown a distillation column 10. A crude oil to be refined is introduced into column 10 through a conduit 11. The distillation column, which actually can comprise one or more columns, produces several product streams. As illustrated, a gaseous product is removed through an overhead conduit 12. Side stream withdrawal conduits 13, 14 and 15 remove gasoline, distillates and virgin gas oil, respectively. A topped crude kettle product is removed through a conduit 16 which communicates with a vacuum distillation column 17. Column 17 is operated to produce an asphalt kettle product which is removed through a conduit 18 and an overhead vacuum gas oil product which is removed through a conduit 19, the latter communicating with conduit 15.

The two gas oil streams passed through conduits 15 and 19 are introduced into a catalytic cracking reactor 22. This reactor, in combination with a regenerator 23, constitutes a fluidized catalytic cracking unit, which can be of any known type. Spent catalyst from reactor 22 is conveyed to regenerator 23 through a conduit 24, and regenerated catalyst is returned to reactor 22 through a conduit 25. The cracking products from reactor 22 are removed through a conduit 26 and delivered to suitable fractionation or other processing units, not shown.

An effluent flue gas stream, which contains appreciable quantities of catalyst fines, is removed from regenerator 23 through a conduit 28. In accordance with this process, conduit 28 communicates with a scrubbing column 29. A cooler 28' is disposed in conduit 28 to reduce the temperature of the flue gas stream from approximately 1100°F. to approximately 500°F. Scrubbing oil is introduced into the upper region of column 29 by a conduit 30. Makeup scrubbing oil is supplied to conduit 30 through a conduit 31 which extends from conduit 19. Conduits 19 and 31 have respective valves 32 and 33 therein to control the flows of vacuum gas oil to reactor 22 and scrubber 29 respectively. An effluent flue gas stream substantially free of catalyst fines is removed from the top of scrubber 29 through a conduit 34. An oil stream containing entrained catalyst fines is removed from the bottom of scrubber 29 through a conduit 35.

All or a substantial portion of the oil system removed through conduit 35 is introduced into the feed to vacuum distillation unit 17. To this end a conduit 36, which has a valve 37 and a heater 38 therein, communicates between conduit 35 and

FIGURE 58:   FLOW SCHEME FOR CATALYST FINES RECOVERY IN CATALYST REGENERATION

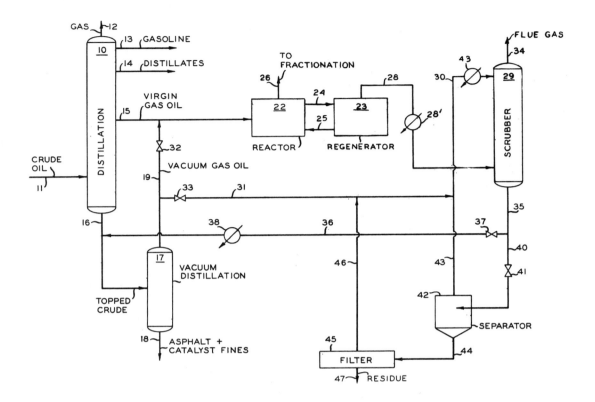

Source: J.B. Rush; U.S. Patent 3,494,856; February 10, 1970

feed conduit 16 to distillation column 17. The catalyst fines entrained in the scrubbing oil are thus contained in the asphalt kettle product from column 17.

In certain operations of this type, it may not be possible or practical to incorporate all of the catalyst fines into the asphalt product. For example, excessive amounts of such fines may result in an asphalt product which does not meet certain specifications. Under these circumstances, the remainder of the catalyst fines can be removed by a filtering operation. A conduit 40, which has a valve 41 therein, communicates between conduit 35 and the inlet of a separator 42. This separator can be a cyclone separator or a centrifuge, for example. An effluent oil stream which has substantial amounts of catalyst fines removed therefrom is recovered from the overhead of separator 42 and passed by a conduit 43 to conduit 30.

A conduit 44 extends from the bottom of separator 42 to a filter 45 to pass a residue catalyst containing oil stream to the filter. The resulting filtrate is removed through a conduit 46 which communicates with conduit 30. The residue, which is in the form of a paste of catalyst fines and oil, is removed through line 47 for disposal. A cooler 48 is provided in conduit 30 to cool the scrubbing oil to a desired scrubbing temperature. The amount of cooling required depends to a large extent on the amount of makeup oil which is introduced through conduit 31. In normal operation, the maximum permissible amount of catalyst fines is delivered by conduit 36 to distillation column 17.

A process developed by J.B. Godin; U.S. Patent 3,546,132; December 8, 1970; assigned to Phillips Petroleum Company is one in which fixed catalyst beds of silica promoted by incorporation of oxides of a metal of Group III-A or IV-B of the Periodic Table are regenerated by the steps of (1) heating and flushing the bed with an inert material to remove the more volatile components therefrom; (2) passing a mixture of steam and air through the bed to effect oxidative stripping of contaminants from the bed; (3) burning the residue of contaminants in the bed; and (4) removing free oxygen and water from the bed.

In the operation of fixed-bed catalytic polymerization and cracking reactions wherein catalyst regeneration by combustion of carbonaceous deposits is carried out, it is important to remain below a substantially fixed predetermined maximum regeneration temperature. This is because the catalyst generally has a given maximum temperature, which, if exceeded, would adversely affect the activity or physical properties of the catalyst or both. Thus, temperature control, during regeneration, is extremely important for this reason alone, since, otherwise, cost of replacement of catalyst would be prohibitive.

When utilizing catalysts of silica promoted by an oxide or oxides of a metal of Group III-A or IV-B, the predetermined maximum regeneration temperature is in the range from 1700°F. to 1800°F. Generally, if such catalyst bed reaches 1800°F. or higher during regeneration, the catalyst life will be substantially cut in half. When the catalyst bed is raised to about 1800°F. during two regenerations, the activity of the catalyst is entirely gone. However, the higher temperatures which are below the maximum regeneration temperature are generally the most efficient for the removal of the combustible deposits in the catalyst bed. Thus, heretofore, it has been necessary to operate at rather high temperatures (approaching 1800°F.) when regenerating the silica catalyst. This regeneration procedure is undesirable in that it requires very close temperature control in order to prevent the regeneration temperatures from exceeding the maximum predetermined temperature and resulting in undesired destruction of catalyst activity.

The conduct of regeneration according to this process will now be described in further detail with reference to the flow diagram in Figure 59. The feed stream is supplied to reactor 10 via conduit 11 and removed from the reactor 10 via conduit 12. The feed stream is normally heated (by means not shown in the drawing) to a desired initial reaction temperature before it is introduced into reactor 10. Conduits 13, 14, 15, and 16 supply regeneration fluids to the silica catalyst bed within reactor 10.

Thus, when valve 17 in conduit 11 is open, the heated hydrocarbon feed will pass from conduit 11 into the upper region of reactor 10. As the feed stream passes through reactor 10, the desired reaction such as polymerization, or cracking will occur, and the product stream is removed from the lower region of reactor 10 via conduit 12. The catalytic reaction will result in various unwanted carbonaceous deposits forming on the surface of the silica catalyst bed within reactor 10 which reduce the catalyst bed activity. When the maximum allowable amount of carbonaceous deposit has formed within the catalyst bed in reactor 10, valve 17 is closed, reactor 10 is drained, and the regeneration process is begun.

FIGURE 59: FLOW SCHEME FOR STEAM-OXYGEN REGENERATION OF FIXED BED CRACKING CATALYSTS

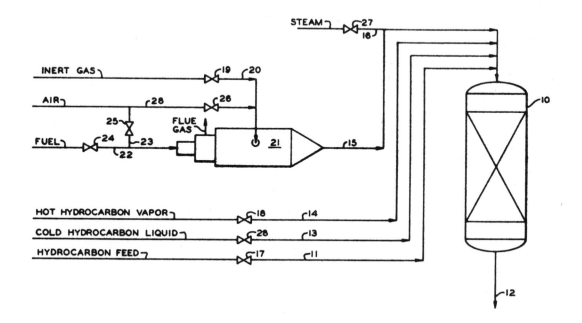

Source: J.B. Godin; U.S. Patent 3,546,132; December 8, 1970

According to the regeneration process as illustrated in the drawing, valve 18 is opened to allow a hot hydrocarbon vapor to pass into the upper portion of reactor 10 via conduit 14. This hydrocarbon vapor is an inert, nonpolymerizable hydrocarbon (i.e., substantially completely nonreactive in the catalytic environment) selected from alkanes, cycloalkanes, and aromatic hydrocarbons. Examples of suitable hydrocarbons include ethane, butane, pentane, hexane, cyclopentane, cyclohexane, benzene, naphthalene, and the like. The vapor is heated to a temperature sufficiently to heat the silica catalyst within reactor 10 to a temperature in the range of from about 200° to about 400°F. This will result in the more volatile contaminants vaporizing and being purged from the catalyst bed. This heating and purging action will continue for a time sufficient to heat the entire silica bed to the desired temperature and remove the majority of the more volatile contaminants therefrom.

For most operations, the hot hydrocarbon vapor is passed through the silica bed from 0.1 to 6 hours. Next, valve 19 is opened to allow an inert gas (i.e., substantially completely nonreactive in the catalytic environment) such as nitrogen or helium and the like to pass from conduit 20 through heater 21, conduit 15 and into the upper region of reactor 10. Heater 21 can be any indirect heat exchange type heating device such as illustrated in the drawing. As illustrated, the heater 21 is a double shell type heater having a fuel burner disposed with another shell surrounded by an indirect fluid heating outer shell chamber.

Fuel and air are passed to the burner within the inner heating shell via conduits 22 and 23, respectively, the reactive proportions of each are controlled by the action of valves 24 and 25, respectively. Thus, heater 21 will heat the inert gas to the temperature of the silica catalyst bed (to a temperature in the range of about 200° to about 400°F.). The inert gas will thereby purge reactor 10 and remove any remaining hydrocarbon vapors therefrom.

Next, valve 10 is closed and valves 26 and 27 are opened to thereby allow air to pass from conduit 28 through heater 21 and into conduit 15 wherein it is admixed with steam from conduit 16. The mixture of steam and air enters the upper portion of reactor 10 and passes through the catalyst bed and out conduit 12. The air is heated by heater 21 and the steam is normally superheated steam maintained at a temperature from about 500° to about 700°F. Valves 26 and 27 are manipulated to yield a steam-air ratio which heats the silica bed within reactor 10 to a temperature within the range of from about 500° to about 700°F. to thereby effect oxidative stripping of contaminants therefrom. The ratio of the air to the steam is low enough to prevent combustion of the contaminants within the bed. This oxidative stripping process will remove volatile contaminants from the bed which were not removed by the hot inert hydrocarbon vapor. This process also heats the bed in preparation for the combustion step. This step will continue until the entire catalyst bed is heated to the desired temperature and until the desired amount of volatile contaminants are stripped therefrom (usually from 0.1 to 8 hours).

Next, the combustion process is begun by increasing the flow of heated air through conduit 15 passing to the upper region of reactor 10. Thus, the flow of heated air through conduit 15 is increased until combustion is initiated in the upper region of the silica bed within reactor 10. The combustion front will thereafter move from the top to the bottom of reactor 10, and will be maintained at a temperature in the range of from about 700° to about 1200°F. by adjusting the flow of air through conduit 15, for example.

After the combustion zone or front has passed through the entire length of the catalyst bed, the bed can be purged and cooled with inert material and put back on stream. However, it is preferred that a secondary combustion process be carried out by closing valve 27 and allowing heated air to pass through column 10 via conduit 15. The heated air will maintain the column at a temperature within the range of about 700° to about 1100°F., preferably about 1100°F. for a time sufficient to combust any remaining carbonaceous residue that was not burned by the steam-air combustion step described above. The heated air will generally pass through the column from a period of about 0.1 to about 4 hours.

If the secondary combustion step is carried out, the column is then dried and purged with inert gas, allowed to cool, and then put back on stream. A preferred method of drying and cooling the column comprises closing valves 24 and 25 while leaving valve 26 open to thereby allow cool air to pass through reactor 10 until the reactor is cooled to a temperature that is preferably lower than about 400°F., preferably about 300°F.

Next, valve 26 is closed and valve 19 opened to thereby allow cool inert gas to sweep through column 10 and remove any free oxygen therefrom. Next, valve 19 is closed and valve 18 is opened to allow hot hydrocarbon vapor (maintained generally at a temperature within the range of about 200° to about 400°F.) to pass through column 10 and remove any remaining water vapor therefrom. If it is desired to place reactor 10 immediately back on stream after regeneration, then valve 18 is closed and valve 28 is opened to allow cold liquid hydrocarbon to flush through the silica catalyst bed. The cold liquid hydrocarbon is vaporized as it is passed through the catalyst bed until the bed is cooled to a temperature of 100°F. or lower. The catalyst bed has now been completely regenerated and cooled, and is ready to be placed back on stream.

Standard Oil

A process developed by S.F. Kapff and L.T. Wright; U.S. Patent 3,414,382; December 3, 1968; assigned to Standard Oil Company involves determining the amount of carbon deposited on catalyst utilizing a chamber having a sample inlet and outlet means, a heater means, a heater control means, a gas flow regulator means, and a recorder means adapted to receive an output signal for producing a record of the differential temperature versus time to measure the coke content of the sample. The rate of progress of a combustion zone in a solid catalyst bed provides the measure of the carbon content on the catalyst sample. Figure 60 shows the essential apparatus elements involved in the conduct of this process.

FIGURE 60:   ALTERNATIVE APPARATUS FOR DETERMINATION OF CARBON DEPOSITS ON CRACKING CATALYSTS

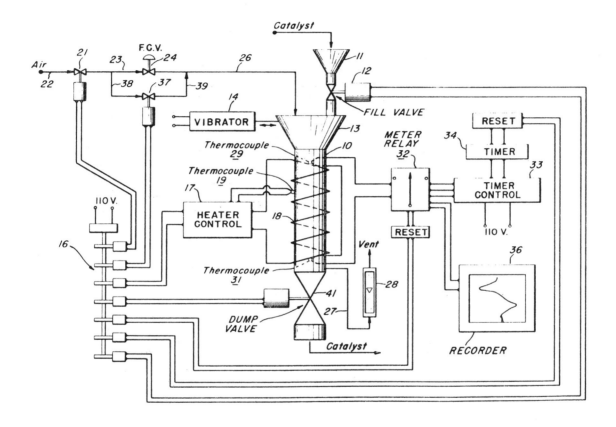

Source: S.F. Kapff and L.T. Wright; U.S. Patent 3,414,382; December 3, 1968

A sample of fluid catalytic cracking catalyst having coke deposited thereon is introduced into chamber 10 through hopper 11, solenoid operated fill valve 12 and funnel 13. An electrically operated vibrator 14 facilitates the flow of the catalyst sample into the chamber and serves to improve reproducibility of catalyst density in the chamber from sample to sample. The sample of catalyst to be tested is placed in hopper 11 and programmer 16 is then started. Operation is automatic from this point on.

Programmer 16 closes the circuit to solenoid fill valve 12 permitting the sample to flow through funnel 13 into chamber 10, and then closes fill valve 12. The programmer 16 then closes the circuit to heater control 17 which heats chamber 10 by passing an electrical current through resistance wire winding 18. The heater control 17 regulates the current flowing through the winding 18 so that the temperature of the chamber as sensed by thermocouple 19 is maintained at a temperature of about 1500°F. The programmer 16 is set so that sufficient time is provided for heating of the catalyst sample to a constant high temperature, after which time the circuit to solenoid-operated air supply valve 21 is energized permitting combustion air to flow through line 22, valve 21, line 23, flow control valve 24 and line 26 into closed funnel 13, thence into the catalyst sample initiating a combustion zone in the top portion of the sample. Products of combustion are withdrawn from the bottom of chamber 10 through a fritted metal filter via line 27 and through flow meter 28 to vent.

Thermocouple 29 located in the catalyst sample near the top of the catalyst bed in chamber 10 is connected differentially with thermocouple 31 which is also located in the catalyst bed but at a location near the bottom of the bed. The output signal from the differential thermocouple circuit is received by motor relay 32 which closes the circuit of timer control 33 when differential thermocouple output signal reaches a predetermined value. Timer control 33 in turn energizes the circuit to timer 34 to initiate operation of the timer when the combustion zone reaches thermocouple 29. As carbon is burned from the catalyst the combustion front, or zone, progresses down the catalyst bed at a rate which is inversely proportional to the amount of carbon contained in the sample.

When the combustion front reaches thermocouple 31 causing the differential thermocouple output signal received by meter relay 32 to reach a predetermined value, the meter relay 32 energizes a shut-off signal to the timer control 33 which in turn stops the timer 34. Timer 34 then registers the elapsed time required for the combustion zone to progress through the catalyst bed from thermocouple 29 to thermocouple 31. The timer 34 may be calibrated in terms of elapsed time or, more conveniently, it may be calibrated in terms of coke content of the catalyst. A strip-chart recorder 36 is connected to the meter relay 32 to receive and record the output from differentially-connected thermocouples 29 and 31 to produce a record of the differential temperature, or temperature difference, versus time. The recorder is started and stopped at the times described for starting and stopping heater 18. In this embodiment, the programmer 16 is preset so that sufficient time is permitted for the combustion zone to pass through the sample when the sample contains the maximum anticipated amount of coke.

Upon elapse of this preset time interval the programmer 16 turns off the heater control 17 which in turn opens the heater circuit stopping the current flow through winding 18. Simultaneously the programmer 16 energizes the circuit to solenoid-operated cooling air valve 37 which permits an increased flow of air, from air supply line 22 and valve 21 through lines 23, 38, 39 via line 26 through the funnel 13 and into the chamber 10, to cool the catalyst sample. The programmer 16 is preset to provide sufficient time for cooling of the catalyst. Upon elapse of this preset cooling time, the programmer 16 energizes solenoid-operated dump valve 41 which releases the spent catalyst sample from the chamber 10, and then closes the dump valve 41. The programmer 16 then closes cooling air valve 37, air supply valve 21, and energizes the reset circuits of the meter relay 32 and the timer to place same in their initial condition for the next analysis. The instrument is then in readiness to receive a new sample and make another analysis.

Texaco

A process developed by D.P. Bunn, Jr. and H.B. Jones; U.S. Patent 3,394,076; July 23, 1968; assigned to Texaco, Incorporated involves the regeneration of catalyst in a fluid catalytic cracking process wherein a hydrocarbon is contacted with a fluidized solid catalyst in a reaction zone effecting conversion of at least a portion of the hydrocarbon to desired conversion products with the concomitant deposition of coke on the catalyst. This process especially relates to a process and apparatus for regenerating fluidized solids wherein the solids are transported through an elongated reaction flow path by inducing a swirling motion in the dense phase bed of a regeneration zone thereby extending the path of particles traversing the regeneration zone and increasing their residence time therein. The apparatus employed in the conduct of such a process is shown in Figure 61.

Virgin gas-oil in line 2 and regenerated catalyst from line 3 are introduced through a fresh feed riser 10 into reactor 11. A second feed stream in line 5, advantageously a recycle gas-oil, and regenerated catalyst from line 6 are introduced into reactor 11 through recycle feed riser 12. In reaching reactor 11, recycle feed riser 12 passes internally through stripper 15 without being in open communication therewith and extends into the dense phase bed which has a level 16 in the reactor 11. Products of cracking and a small amount of entrained catalyst leave the bed in reactor 11 at level 16 and pass cyclone separator 60 wherein entrained catalyst is separated and returned to the dense phase bed through dip-leg 61. Separated gaseous products are discharged from cyclone 60 through line 62 to plenum 63 which may also collect gaseous products from other cyclone separators not shown. Product vapors from plenum 63 are discharged through product line 64 to fractionation and recovery equipment, not shown.

In the course of the catalytic cracking process, coke is deposited on the fluidized solids catalyst. Fluidized solids catalyst is removed from the reactor 11 through standpipe 13, and passed to stripper 15 wherein entrained and occluded hydrocarbons are displaced by stripping steam introduced through steam ring 17. Stripped catalyst from the bottom of stripper 15 is passed through return pipe 14, slide valve 18 and return inlet or inlet conduit 21 into regenerator 20. Spent catalyst return inlet 21 communicates with regenerator 20 through tangential inlet 22 in the lower cylindrical wall of regenerator 20.

A dense phase bed having an upper level 25 is maintained in regenerator 20 and return pipe 21 introduces the used catalyst below level 25. Return pipe 21 introduces the used catalyst as a dense phase directing into the dense phase of regenerator 20 thereby avoiding the localized high temperature which results when oxygen rich air meets high carbon catalyst such as

FIGURE 61:   FLOW DIAGRAM OF FLUID CATALYTIC CRACKING UNIT SHOWING REACTOR AND REGENERATOR

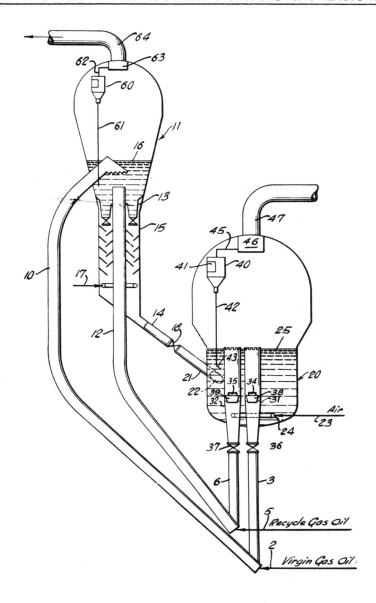

Source: D.P. Bunn, Jr. and H.B. Jones; U.S. Patent 3,394,076; July 23, 1968

may be encountered in transporting catalyst from a reactor to a regenerator as a suspension in the burning air. Tangential inlet 22 introduces the catalyst particles into the regenerator with a horizontal component of velocity. Since the dense phase bed is confined by the cylindrical wall of vessel 20, a swirling motion is imparted to the particles. Catalyst is withdrawn from regenerator 20 through drawoff standpipes 31 and 32 which are located circumferentially remote from inlet 22. In this way, catalyst introduced through inlet 22 follows a peripheral path from inlet to outlet. Such a peripheral path is substantially greater than the straight line distance between the inlet and outlet. Thus the path that the catalyst is forced to take is elongated and the catalyst consequently has a longer residence time for the removal of the coke thereon.

Oxygen containing gas, for example, air, is introduced into regenerator 20 through air line 23, air ring 24 and nozzles 30. Air ring 24 is concentrically aligned in the base of regenerator 20 below dense phase level 25. Nozzles 30 are advantageously spaced about the bottom portion of the air ring 24 or may be inclined in the direction of swirl.

The catalyst particles coming through the tangential inlet 22 are suspended in the gas in the regenerator and the resulting suspension exhibits many of the characteristics of a fluid. The coke or carbonaceous material on the catalyst particles is burned away by the oxygen in the regeneration gases.

Catalyst is withdrawn from regenerator 20 through draw-off standpipes 31 and 32. Standpipe 31 communicates through slide valve 36 and standpipe 3 with fresh feed riser 10. Standpipe 32 communicates through slide valve 37 and standpipe 6 with recycle feed riser 12. Draw-off standpipes 31 and 32 are in open communication at their tops with the interior of the regeneration chamber at the top of dense phase 25 of the fluidized bed therein. Preferably, the tops of the draw-off standpipes 31 and 32 are separated whereby the affects of fluctuations in bed level are minimized.

Apertures or windows 34 and 35 are provided in the lower portions of draw-off standpipes 31 and 32 respectively. Windows 34 and 35 are located in their respective standpipes so that they occupy less than 180° of the circumference of the standpipe and are preferably located so that they face away from inlet 22, the point where the spent catalyst return pipe 21 enters regeneration chamber 20. In this way, the portion of the respective standpipes behind the windows acts as a shroud tending to prevent bypassing of catalyst directly from the inlet to the outlet standpipe.

In order to funnel a greater amount of catalyst through the windows 34 and 35 as described above baffles or scoops 38 and 39 are affixed respectively to the draw-off standpipes 31 and 32 at the bottom of the windows 34 and 35. These baffles flare upwardly and outwardly from the base of each window 34 and 35 to an elevation somewhat below the top of the windows. These baffles flaring out about the face of the windows serve a dual purpose. In addition to the swirl established in a horizontal plane, a toroidal flow path is induced by the flow of gases into the center of the contacting zone in regenerator 20 which provides a vertical component to the particle motion. Solids separating in the disperse phase of the contacting zone tend to fall downwardly at the walls of the contacting zone. Scoops 38 and 39 are provided at the bottom of the draw-off hopper windows to receive the downwardly circulating catalyst and direct the descending catalyst into the draw-off hoppers 31 and 32.

The scoops 38 and 39, at the same time, deflect rising gases away from the draw-off windows. They serve as a funnel for downwardly traveling catalyst particles, thereby directing a larger and more continuous amount of catalyst particles through the windows. These baffles also deflect the upwardly traveling regeneration gases away from the windows so that they are prevented from entering therein.

Combustion gases leaving the dense phase bed at level 25 and entrained catalyst pass through the disengaging space in the upper portion of regenerator 20 to gas-solids separating cyclone 40 having inlet 41. Cyclone inlet 41 is oriented to receive gases rotating in the same direction as the catalyst particles introduced into regenerator 20 through inlet 22 without reversal of direction. Although, only a single cyclone is shown, it will be understood that a plurality of cyclones may be assembled to provide two or more stages of separation and a plurality of single or plural stage assemblies may be employed depending upon the gas handling capacity of the particular cyclone system employed and the total amount of gas to be handled.

Solids separated in cyclone 40 are returned to the bed in regenerator 20 by cyclone dip-leg 42. Dip-leg outlet 43 is oriented to direct such returned solids in the direction of swirl of catalyst introduced through tangential inlet 21. Surprisingly, the amount of solids separated from the effluent gases by cyclone 40 may exceed the amount of catalyst circulated from reactor 11 to regenerator 20 through inlet line 21. Typically the catalyst returned through the dip-leg 42 may be about 120 percent of the catalyst circulated through the dense phase inlet line 21, and so the return of this catalyst in the direction of the swirl substantially augments the swirling flow. Gases from cyclone 40 are passed through line 45 to plenum 46 which may also receive effluent gas from other cyclones not shown.

In an example of this process, catalyst from a reactor vessel containing a high percentage of carbonaceous material is introduced into the regenerator vessel through a tangential inlet. The tangential inlet induces a clockwise motion to the catalyst. Air is introduced into the vessel at a point below the catalyst inlets. The air passes through the catalyst bed, reacting with the carbonaceous deposits on the catalyst to form carbon dioxide, carbon monoxide, and water vapor. The reaction products leave the dense phase catalyst bed and flow through cyclones, which have the inlets oriented to enhance the clockwise flow. The regenerated catalyst leaves the regenerator vessel through a catalyst withdrawal hopper which is shrouded on the side adjacent the inlet to prevent bypassing of catalyst directly from inlet to outlet.

Exemplary conditions in a fluid catalytic cracking unit regenerator operating as described above are as follows:

Regenerator bed temperature, °F.	1150
Regenerator top pressure, psig	20
Spent catalyst to regenerator, tons per minute	16.5
Entrained catalyst returned by cyclones to regenerator bed, tons per minute	20

Specific coke burning rate, lb. coke/hr./lb. catalyst	0.08
Carbon on spent catalyst, wt. percent	1.0
Carbon on regenerated catalyst, wt. percent	0.4
Regenerator vessel i.d., feet	20
Direct distance from inlet to outlet, feet	8
Peripheral distance from inlet to outlet, feet	40

It is obvious from the above data, that the peripheral patch dictated by the tangential inlet and shrouded withdrawal hopper provides a path twice that which would result if the inlet and outlet were merely on opposite sides of the vessel. Also, the peripheral path is as much as five times the length of the path typically provided in designs which do not provide features to induce a peripheral path. The longer path is desirable to prevent bypassing of partially regenerated (high carbon level) particles to the reactor, where desirable cracking reactions would be suppressed due to the incomplete regeneration.

DEMETALLIZATION

Of the various metals which are to be found in representative hydrocarbon feedstocks some, like the alkali metals, only deactivate the catalyst without changing the product distribution; therefore they might be considered true poisons. Others such as iron, nickel, vanadium, and copper markedly alter the character and pattern of cracking reactions, generally producing a higher yield of coke and hydrogen at the expense of desired products, such as gasoline and butanes. For instance, it has been shown that the yield of gasoline, based on cracking feed disappearance to lighter materials dropped from 93 to 82% when the laboratory-measured coke factor of a catalyst rose from 1.0 to 3.0 in commercial cracking of a feedstock containing some highly contaminated marginal stocks.

This decreased gasoline yield was matched by an increase in gas as well as coke. If a poison is broadly defined as anything that deactivates or alters the reactions promoted by a catalyst then all of the four metals mentioned above can be considered poisons. It is hypothesized that when deposited on the surface of a catalysts, Fe, Ni, V and Cu superimpose their dehydrogenation activity on the desired reactions and convert into carbonaceous residue and gas some of the material that would ordinarily go into more valuable products. The relatively high content of hydrogen in the gases formed by metals-contaminated catalysts is evidence that dehydrogenation is being favored. This unwanted activity is especially great when nickel and vanadium are present in the feedstocks.

Metal poisoning of cracking catalysts is a major cost item in present day refining and is a bottleneck in upgrading residual stocks. Current methods of combatting metal poisoning are careful preparation of feedstocks to keep the metals content low and catalyst replacement to control metals levels on the catalyst. An alternate solution, demetallizing the catalyst, which would avoid discarding of expensive catalyst, and enable much lower grade, highly metals-contaminated feedstocks to be used, is possible.

Du Pont

A process developed by L.G. McLaughlin; U.S. Patent 3,224,979; December 21, 1965; assigned to E.I. du Pont de Nemours and Company involves the removal of traces of heavy metals from a silica-alumina cracking catalyst by contacting it with aqueous solutions of mercapto compounds.

In this process, the regenerated catalyst, which is now substantially free from carbon deposits, is introduced into an aqueous solution of a mercapto compound at a temperature ranging from about 75°F. to about 250°F. and preferably ranging from about 100°F. to 175°F. for a period ranging from 1 to 8 hours and preferably from 2 to 6 hours.

The treated catalyst and the solution containing the mercapto compound are then passed through a filter. The mercaptan containing filtrate is returned to the mercaptan solution bath and the catalyst residue is passed to an oven where it is dried at a temperature ranging from 125° to 175°C. The silica-alumina catalyst thus treated is substantially free of all metallic contaminates and is reintroduced along with hydrocarbon raw materials into the cracking zone of the cracking process.

For practical reasons the removal of metallic impurities from the catalyst by the aqueous solution of mercapto compounds is preferably performed upon only a portion of the total catalyst being used in the process. However, it is a continuous operation such that a portion of the catalyst being used in the cracking process is always being treated. In this manner a constant supply of silica-alumina catalyst from which the metallic impurities have been removed is reintroduced into the cracking process.

The exact portion of the total amount of catalyst used in the process which is withdrawn for removal of the metallic impurities during any given period is dependant, for the most part, on the particular hydrocarbon raw material being cracked.

For example, when cracking low molecular weight gas oils which are relatively low in metallic impurity content, as little as 5 percent or less of the total catalyst need be treated during a twenty-four hour continuous cracking cycle.

On the other hand, when cracking high molecular weight hydrocarbon, such as residual materials which are high in metallic contamination, as much as 40 percent or more of the total catalyst could be treated during an equal twenty-four hour operating cycle. In general, however, the proportion of catalyst that is treated in the process during a twenty-four hour operating cycle usually averages in the range from 5 to 40% of the total catalyst used and more likely from about 10 to 25%.

In an alternative and preferred method of practicing this process, a cation exchange resin is also used to remove the metallic impurities from the mercaptan solution such that the mercaptan solution can be continuously reused. The cation exchange resin can be introduced into the metallic removal process at a stage subsequent to the above described filtration operation where the catalyst and the aqueous mercaptan solution are separated. In this instance, instead of returning the recovered mercaptan solution directly back to the cleaning tank it is first treated with the cation exchange resin, passed through a screening unit to remove the cation exchange resin, and then is returned to the cleaning tank.

As another alternative the cation exchange resin can be present in the mercaptan solution such that the metallic impurities are removed from the silica-alumina catalyst by the mercaptan solution and from the mercaptan solution by the cation exchange resin substantially simultaneously. Following this method, the cation exchange resin is separated from the catalyst by screening. This is possible because the particle size of the commercially employed catalyst is such that it will pass through a 100 mesh U.S. standard sieve and the particle size of the cation exchange resin as produced commercially is in the range of 10 to 50 mesh U.S. standard sieve. Subsequent to separation of the catalyst from the cation exchange resin the catalyst is oven dried as described above and reintroduced along with the hydrocarbon raw materials into the cracking zone of the catalyst cracking process.

Esso Research and Engineering

A process developed by B.L. Schulman; U.S. Patent 3,373,102; March 12, 1968; assigned to Esso Research and Engineering Company involves demetallization, not by treatment of regenerated catalyst, but by feed pretreatment to remove metals.

The metals are postulated to exist as large high boiling ring complex aromatic compounds called porphyrins and characterized as "volatile" or "nonvolatile." The volatile porphyrins have an average vapor pressure equivalent to a hydrocarbon boiling at about 1150°F., while the nonvolatile porphyrins are possibly polymeric in character and have a boiling point more than about 1300°F. These complex contaminating compounds are adsorbed on the catalyst during the cracking operation and are thought to split open, leaving behind the metal as a contaminating deposit. As nickel is the worst offender as a contaminant, the metal or concentration content of an oil is here expressed in terms of "equivalent nickel" by the relation:

$$\text{Equivalent Ni} = \text{Ni} + 1/5 \text{ (Vanadium)} + 1/10 \text{ (Iron)}$$

Concentrations of metal are given in parts per million by weight (wppm). Various methods have been tried heretofore for removal of metals from the oil feeds but they have not been entirely successful or have been too expensive or too cumbersome.

According to this process, a high boiling oil feed which contains high boiling polynuclear aromatic hydrocarbons and metals and includes those aromatic hydrocarbons having metal atoms in large ring complexes is removed from a pipe still and/or vacuum still and is mixed with finely divided fresh or regenerated cracking catalyst particles at an elevated temperature but below a temperature for cracking. The catalyst particles are then separated as a slurry from the oil in a liquid cyclone separator or hydroclone or "Dorrclone." The liquid hydroclone is exceedingly efficient and gives 90 to 95% yield of purified catalytic cracking oil feed.

Preferably more than one stage of treatment of the oil feed with catalyst particles is used. The treated oil is taken overhead and the spent catalyst is withdrawn from the bottom of the hydroclone. The treated oil is then catalytically cracked in accordance with conventional practice and the catalytically cracked products are fractionated into a gas fraction, a gasoline fraction, a light aromatic heating oil, an aromatic heavy cycle gas oil or recycle oil fraction and a bottoms fraction.

The aromatic heavy cycle oil is a good solvent for polynuclear or heavy aromatic hydrocarbons. These heavy aromatic compounds with which the contaminating metals are associated are preferentially and easily adsorbed from gas oils by fresh or regenerated catalyst particles. Light aromatic heating oil or even naphtha may be used as a solvent but the heavy cycle oil is preferred because of its better solvent properties.

The spent cracking catalyst particles are mixed with a portion of the heavy cycle gas oil fraction separated from the catalytically cracked products and the mixture passed to a hydroclone to separate regenerated catalyst particles from the cycle oil fraction which acts to wash off and dissolve the adsorbed contaminants or impurities from the catalyst particles. Preferably more than one stage of washing or regenerating the catalyst is used. In the hydroclone, the washed or regenerated catalyst particles are separated from the heavy cycle gas oil wash liquid and recycled to the oil treating step to remove heavy or high boiling aromatic hydrocarbon compounds and the like from the oil feed.

The cycle oil wash liquid separated in the hydroclone contains impurities dissolved in the oil wash or removed from the catalyst particles and a portion of this wash liquid may be recycled as part of the wash liquid used to regenerate the catalyst particles. Part of the contaminated cycle oil wash liquid is passed to the pipe still or vacuum still where the high boiling aromatic hydrocarbon compounds are separated from the heavy cycle gas oil and removed as bottoms and discarded or part of the contaminated wash liquid may be passed to the product fractionator of the catalytic cracking unit where high boiling aromatic hydrocarbon compounds are removed as bottoms and discarded.

Heavy cycle gas oil separated from the catalytic cracking products fractionator is the aromatic wash liquid or solvent used to regenerate the spent adsorbent catalyst particles. This is one of the advantages of this process as make-up wash liquid or solvent is readily available from the catalytic cracking products fractionator as the heavy aromatic cycle gas oil stream. The use of the cycle stock makes the solvent loss zero, since any of the cycle oil solvent which remains in the catalyst or oil feed will go eventually to the catalytic cracking unit which is its ultimate goal anyway. The cycle oil passes to the catalytic cracking step with treated feed. Any fine catalyst particles containing cycle oil solvent which are carried over into the catalytic cracking unit go through the cracking step to crack the occluded solvent and then the spent fine catalyst particles are discarded from the final stages of regeneration and reactor fractionator bottoms. The functioning of this process will now be described in further detail with reference to Figure 62.

The reference number 10 designates a line for feeding oil to the furnace 11. The oil feed may be a crude petroleum oil, topped crude, residual oil from an atmospheric or vacuum still, a heavy gas oil, extraneous stocks, etc., which contains metal contaminants, high Conradson carbon and polynuclear aromatic hydrocarbons including those having four or more rings. The oil feed to be treated is heated to a temperature between about 600°F. and 800°F., and the heated oil is passed through furnace 11 and line 11' to atmospheric still 12 to separate vapors from a bottoms fraction withdrawn through line 13. The vapors pass overhead through line 14 and may be subsequently separated into gas, naphtha and heating oil fractions. The bottoms from line 13 are passed to the lower portion of a vacuum pipe still 16 to separate a distillate oil from higher boiling hydrocarbons or vacuum pipe still bottoms. The still 16 is under a pressure of between 40 and 150 mm. of mercury. It is not essential to have both an atmospheric still and a vacuum still as either one may be omitted if desired.

The bottoms fraction from vacuum still 16 and boiling above 1100° to 1200°F. is withdrawn through line 18 and may be discarded or passed to a visbreaking step to produce additional catalytic cracking feed stock. The distillate hydrocarbons boiling below about 1200°F. pass overhead through line 22, are cooled to between 400°F. and 500°F. in cooler or condenser 24 and the liquid hydrocarbons are mixed with an oil slurry of finely divided catalytic cracking catalyst from line 26 which is partially spent coming from a second separation stage later to be described. The distillate hydrocarbon feed has a boiling range of about 600°F. to 1200°F. The oil-catalyst mixture is passed through pump 28 where additional mixing takes place and the resulting mixture is introduced into hydroclone 32 to separate catalyst particles from purified oil. The catalyst-oil mixture is maintained at a temperature between about 400°F. and 500°F. for between 0.25 and 1 minute. The distillate oil feed in line 22 has about 1 to 3 wppm equivalent nickel, being composed of 0.2 to 2.0 wppm of vanadium, about 0.1 to 2.0 wppm of nickel and about 1 to 10 wppm of iron.

If desired, a side stream may be withdrawn from the vacuum still 16 through valved line 33 provided with a cooler 33' and the cooled side stream passed into line 22 ahead of pump 28. This side stream may be 50% or less of the total distillate stream in line 22 and would have a higher metals level than the overhead distillate in line 22 and could also be used as a wash oil stream. Other oil feeds may be used as catalytic cracking oil feeds. For example, all or a portion of the heating oil fraction, boiling above 450°F., from line 14 may be passed directly to hydroclone 32.

The finely divided cracking catalyst used for treating oil feeds may be any conventional cracking catalyst, such as silica-alumina catalyst, containing 10 to 35% alumina or it may be an acid treated bentonite clay or synthetic silica-alumina magnesia catalyst or silica-magnesia catalyst, etc. The catalyst is finely divided and has a large surface area as the adsorption of the impurities or contaminants in the oil is a surface effect and the more surface that is presented by the catalyst particles, the better adsorption and removal of the aromatic metal-containing compound or compounds which is obtained. The heavy or high boiling aromatic hydrocarbon compounds are preferentially adsorbed by the catalyst particles. Cracking catalyst which is off-specification because it is too fine can be used. The catalyst particles in hydroclone 32 absorb the impurities from the oil feed being treated.

The catalyst is preferably fresh catalyst having a surface area of 450 to 600 m.^{2}/g., but may be regenerated catalyst having a surface area of 75 to 150 m.^{2}/g. The catalyst particles are of a size below 150 microns, preferably between about 5 and 100 microns. The amount of fresh catalyst particles used is between 0.05 and 0.50 part by weight to 1 part

FIGURE 62: FLOW SCHEME OF PROCESS FOR DEMETALLIZATION OF CATALYTIC CRACKING FEED STOCKS

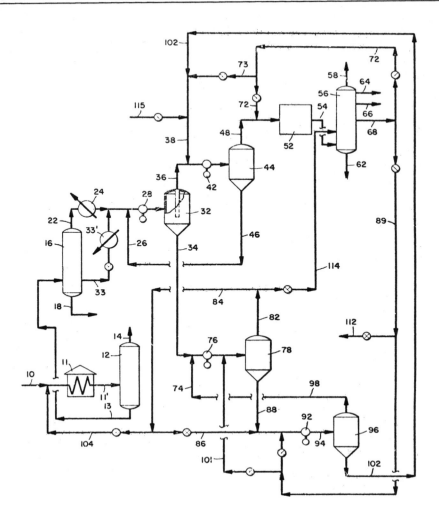

Source: B.L. Schulman; U.S. Patent 3,373,102; March 12, 1968

by weight of oil to be treated. Surface area of the catalyst particles controls the amount of catalyst used. In the examples hereinafter described, the surface area of the catalyst is between 400 and 500 m.^{2}/g. The amount of catalyst used is inversely proportional to the surface area of the catalyst.

The catalyst is separated from the oil in the hydroclone 32 which is similar in its action to a cyclone separator. The hydroclone 32 is exceedingly efficient and separates 95 to 99.9% of the catalyst particles from the oil. The separated spent catalyst particles with some occluded oil are withdrawn as a slurry from the hydroclone 32 through bottom line 34 and are regenerated by a washing or solvent treatment, as will be hereinafter described in greater detail. The oil slurry in line 34 is about 50 weight percent catalyst and 50 weight percent oil.

The separated oil is withdrawn overhead from the hydroclone 32 through line 36 and is mixed with a clean washed or solvent regenerated finely divided cracking catalyst of the same type as above described and is introduced through line 38 and is further mixed with the oil in pump 42. The temperature during mixing in pump 42 is between about 400°F. and 600°F. and the time of treatment is between about 0.25 and 1 minute. The slurry of regenerated catalyst particles from line 38 comprises 50 weight percent catalytically cracked heavy cycle oil wash liquid and the rest catalyst.

The mixture from pump 42 is introduced into a second hydroclone 44, which is of the same construction as hydroclone 32, to separate partially spent catalyst particles from purified oil feed. The separated catalyst is withdrawn from the bottom of the hydroclone 44 through line 46 as a slurry. The catalyst particles in hydroclone 44 adsorb further amounts of impurities from the oil feed being treated. The slurry of partially spent catalyst particles in line 46 and cycle stock is the one which is passed through line 26 as above described and introduced into the first hydroclone 32 as the adsorbent catalyst particles. The slurry in line 46 contains 50 weight percent oil and 50 weight percent catalyst. The efficiency of hydroclone 44 is of the same high order as hydroclone 32. While two stages of hydroclones are preferred and shown on the drawing, more than two stages may be used. In some cases only one stage may be necessary. The amount of catalyst particles introduced via line 38 is between about 0.05 and 0.50 part by weight to 1 part by weight of oil.

The purified oil with a lower metal content and lower aromatic high boiling liquid content is withdrawn overhead from the second hydroclone 44 through line 48 and is passed as oil feed to the catalytic cracking unit 52. The purified oil has between about 10 to 40% of the metals of the untreated oil. The cracking unit 52 is diagrammatically shown and includes a reactor and a catalyst regenerator (not shown). The cracking unit may be a fixed bed, a moving bed or fluid bed unit but is preferably a fluid unit. The temperature during the cracking in the reactor is between about 850°F. and 1000°F. and the temperature during regeneration is between about 1000°F. and 1250°F. The w./hr./w. is between about 2 and 40. The catalyst is a conventional silica-alumina catalyst containing 10 to 35% alumina and having a particle size mostly below 60 mesh and mostly between about 5 and 150 microns.

The hot catalytically cracked products leave the cracking unit through line 54 and are introduced into bottom portion of the product fractionator 56 for fractionating products into a gaseous hydrocarbon fraction which is taken overhead through line 58, a gasoline fraction which is taken off as a side stream 64 and a heating oil fraction which is taken off as a side stream through line 66 below the level of line 64. An aromatic heavy cycle gas oil or recycle fraction is withdrawn from a lower portion of the product fractionator 56 through line 68. The heavy cycle gas oil has a boiling range of 550°F. to 600°F. to 800° to 900°F. A portion of this heavy cycle gas oil is preferably recycled through line 72 to the catalytic cracking unit 52 for further cracking. Another portion of the heavy aromatic cycle oil from line 72 may be passed through line 73 for supplying solvent liquid to line 36 leading to hydroclone 44. The heavy cycle gas oil has about 40 to 50 weight percent aromatic hydrocarbons predominantly 2 and 3 ring aromatic hydrocarbons (50 to 60%) and alkyl benzenes and some 4 ring aromatic hydrocarbons.

Returning now to the spent catalyst particles withdrawn from the first hydroclone 32 as a slurry through line 34, the spent catalyst slurry is mixed with a partially spent wash oil from line 74 and the mixture is further mixed in pump 76. The wash or solvent oil dissolves and washes off the contaminating material on the catalyst particles. The wash oil comprises heavy aromatic cycle gas oil which is partially spent as it comes from a second washing step to be presently described. The temperature of the resulting mixture in pump 76 is between about 400°F. and 600°F. and the time of treating is between about 0.25 and 1 minute. The amount of wash oil used is between 0.5 and 2 parts by weight to 1 part by weight of partially spent catalyst withdrawn through line 34.

The oil-catalyst mixture or slurry is introduced into a third hydroclone 78 for separating partially regenerated finely divided catalyst from the spent wash or solvent oil. The spent solvent oil is withdrawn overhead through line 82 and a portion thereof may be passed through lines 84 and 86 for recycling to hydroclone 78 for the washing or regenerating of the finely divided catalyst. The partially regenerated finely divided catalyst slurry withdrawn from the bottom of the hydroclone 78 through line 88 is preferably mixed with fresh heavy aromatic cycle gas oil from line 89 which takes cycle oil from line 68. The resulting mixture is further mixed in pump 92. The slurry in line 88 comprises 50% by weight catalyst and 50% by weight of oil.

The temperature of the mixture in pump 92 is between about 400°F. and 600°F., and the time of treating is between about 0.25 and 1 minute. The amount of heavy cycle gas oil used is between about 1 and 10 parts by weight to 1 part by weight of the partially regenerated catalyst particles.

The slurry of mixture is passed from pump 92 through line 94 to a fourth hydroclone 96 to separate a slurry of regenerated or washed catalyst particles from partially spent heavy cycle wash or solvent oil. The separated partially spent wash or solvent oil is withdrawn overhead through line 98 and comprises the oil passed through line 74 above referred to, for admixture with the spent catalyst particles withdrawn from the first hydroclone 32 through line 34 above mentioned and described. The catalyst particles in hydroclone 96 are further washed to dissolve and remove adsorbed metal organic or contaminating material from the catalyst particles.

Regenerated catalyst or washed catalyst particles are withdrawn as a slurry from the bottom of the fourth hydroclone 96 through line 102 and these regenerated catalyst particles comprise the fresh or regenerated catalyst particles introduced into line 36 from line 38 above mentioned. The slurry in line 102 comprises 50 weight percent solids and 50 weight percent oil.

The aromatic wash or solvent liquid used in washing the spent and partially spent catalyst particles comprises heavy cycle oil above described and is supplied by line 68 withdrawn as a side stream from the product fractionator 56. A portion of the withdrawn heavy aromatic cycle gas oil is passed through line 89 and introduced into line 88 for washing the partially regenerated catalyst particles in the last hydroclone 96. Or a portion of the cycle gas oil in line 89 may be passed through line 101 as wash liquid and introduced into line 34 to provide some fresh wash oil for washing spent catalyst from hydroclone 32. If desired, some of the cycle oil from line 68 may be withdrawn from the system through line 112.

Instead of sending all or a portion of spent solvent or wash liquid from lines 82 and 84 to the pipe still 12 and/or vacuum still 16 through line 104, all or a portion of the wash liquid from line 82 may be passed through line 114 to a region below withdrawal line 68 and above inlet line 54 of the product fractionator 56 where the high boiling contaminants are rejected as high boiling compounds and withdrawn as fractionator bottoms through line 62. This is particularly suitable if the light catalytic cycle oil from line 66 or naphtha from line 64 is used as the wash liquid. This minimizes the wash liquid ending up in the feed to the catalytic cracking unit 32. If the spent wash liquid is introduced into the atmospheric and/or vacuum still 16 with the catalytic cracking oil feed, the high boiling contaminants in the spent wash liquid are rejected with the bottoms fraction withdrawn through line 13 from the atmospheric still 12 and/or line 18 from the vacuum still 16. If desired, more than two catalyst regenerator hydroclones (78 and 96) may be used or only one may be used in some cases.

Make-up catalyst particles for treating the oil feed are preferably introduced into line 38 from inlet line 115. As above pointed out, there is substantially no loss of solvent or wash liquid as it is supplied by the catalytic cracking unit and any solvent and/or oil feed which remains in the finely divided adsorbent cracking catalyst used for treating oil feed, will eventually go to the fluid catalytic cracking unit. The extremely fine catalyst or fines comprising the adsorbent cracking catalyst which find their way to the fluid catalytic cracking unit, will be lost or discarded from the system. Make-up solvent for line 89 is readily available from the product fractionator 56 as a heavy cycle gas oil withdrawn as a side stream through line 68.

Where only a single stage catalyst purification stage is used, the purified oil in line 36 is passed directly to the catalytic cracking unit 52. Fresh or regenerated catalyst particles are introduced into line 22 through line 26. Spent catalyst particles from line 34 are treated with fresh heavy cycle gas oil from line 68 via line 101 and the mixture passed to hydroclone 78. Spent wash cycle gas oil passes overhead from line 82 to pipe still 12 and vacuum still 16. Regenerated or washed catalyst particles from line 88 in this modification are recycled as fresh catalyst via line 26. While a single stage may be used, the use of multistages improves metals removal.

Gulf Research and Development

A process developed by W.H. Humes and M.M. Stewart; U.S. Patent 3,226,335; December 28, 1965; assigned to Gulf Research and Development Company involves the reactivation of deactivated catalysts by contacting the catalysts with unstabilized silica sol solution having a tendency toward transition to the gel form but prior to appreciable transition to a gel.

After a silica-alumina cracking catalyst has been on stream for an extended duration, its activity becomes reduced. The activity can be partially restored by burning carbonaceous deposits from its surface. However, the burning operation does not remove metal contaminants. The treatment with unstabilized silica sol is highly effective for the reactivation of a cracking catalyst which has been deactivated by metals, such as nickel and vanadium, contained in a hydrocarbon passing over the catalyst. On the other hand, it was found that when silica sol is stabilized against gel formation prior to catalyst treatment, its effectiveness for reactivating the catalyst is greatly diminished. The diminished effectiveness of the stabilized silica sol is probably due to a growth in the size of colloidal silica particles upon stabilization to about 200 A. resulting in the inability of these particles to penetrate very small catalyst pores. Figure 63 shows one version of such a silica sol catalyst reactivation procedure.

The numeral 10 represents a reactor for performing a hydrocarbon conversion process, such as cracking, wherein the hydrocarbon charge enters through line 12 and vapor product is discharged through line 14. A fluid cracking material such as alumina, silica-alumina, silica-magnesia, silica-zirconia, etc. essentially free of other active metals, is charged to reactor 10 through line 16. Deactivated catalyst is removed from reactor 10 and is passed to regenerator 20 through line 22. Air is charged to regenerator 20 through line 24 for burning carbonaceous material from the catalyst surface and excess air and combustion products are discharged from regenerator 20 through line 26. While carbonaceous material is removed by the burning operation, metal contaminants cannot be removed from the catalyst in regenerator 20.

Catalyst is removed from regenerator 20 at a temperature of about 1000°F. to 1100°F. and a portion is passed through line 28, containing valve 30, to catalyst cooler 32. A coolant enters coils in catalyst cooler 32 through line 34 and is removed therefrom through line 36. The coolant cools the catalyst by indirect heat exchange at least to a temperature of 500°F. and preferably to a temperature in the range 300°F. to 400°F. It is important for several reasons that the catalyst be cooled prior to contact with silica sol. If the catalyst is too hot, very high temperature steam will form upon contact of the sol solution with the catalyst.

It is well known that high temperature steam induces considerable deactivation in a silica-alumina cracking catalyst. Furthermore, excessive expansion of water into steam upon the catalyst will fracture the catalyst subdividing it into exceedingly small particles. Also, if catalyst surfaces are at an excessively high temperature a spray of silica sol liquid

FIGURE 63:   PROCESS FOR DEMETALLIZING CRACKING CATALYST BY SILICA SOL TREATMENT

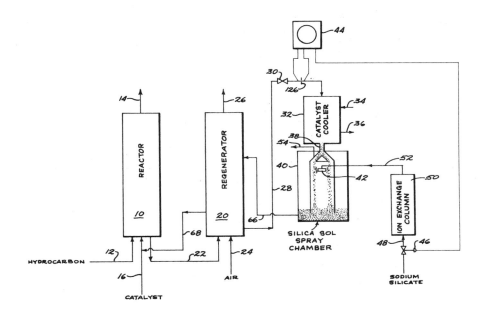

Source: W.H. Humes and M.M. Stewart; U.S. Patent 3,226,335; December 28, 1965

will vaporize immediately upon contact with the catalyst, depositing silica only at the most exposed regions of the catalyst surface.   On the other hand, at lower catalyst surface temperatures the sol remains in the liquid state long enough to permit flow into small catalyst surface pores, carrying silicic acid thereinto, and resulting in silica deposition in these hidden regions of the catalyst surface as well as in the more exposed surface regions.

Cooled catalyst falls through line 38 into spray chamber 40.   The rate at which catalyst is charged to spray chamber 40 is adjusted at valve 30.   This rate is ascertained by measuring pressure drop across orifice 126.   This rate can be used to control the sodium silicate rate to an ion exchange column by means of recorder-controller 44 and pneumatic control valve 46 which is responsive thereto.   Not only is the silica sol flow rate to the spray chamber maintained directly proportional to catalyst flow rate but, more importantly, the rate at which silica sol solution is prepared is continuously maintained directly proportional to catalyst flow rate.

Since the silica sol is unstabilized when used and immediately upon preparation starts to become transformed into a gel whereby its efficiency for catalyst reactivation is impaired, it is important that only as much silica sol is prepared as is required for current use so that silica sol is never subjected to storage prior to being sprayed upon the catalyst.   This control is advantageously effected by means of controller 44 which continuously pneumatically regulates valve 46 in aqueous sodium silicate line 48 leading to ion exchange column 50.   In this manner only sufficient aqueous silica sol is prepared and discharged from ion exchange column 50 through line 52 to deposit between 0.5 and 10 percent by weight, and preferably about 2 to 4 percent by weight, of SiO_2 upon the catalyst currently flowing past spray nozzle 42.

Water vapor is removed from the spray chamber through line 54 and relatively dry SiO_2 impregnated catalyst falls to the bottom of spray chamber 40.   If required, additional drying can be performed in chamber 40.   The dry, impregnated catalyst is returned to regenerator 20 through line 66 where it becomes calcined at regenerator temperature prior to returning in a highly active condition to reactor 10 through lines 68 and 16.

Figure 64 shows an alternative version of the silica sol catalyst reactivation procedure.   The apparatus designated by 70 is a reactor for catalytically cracking a hydrocarbon stream which enters the reactor through line 72.   Vaporous conversion products discharge from reactor 70 through line 74.   A fluidized silica-alumina cracking catalyst enters reactor 70 through line 76.   Deactivated catalyst is removed from reactor 70 and is passed to regenerator 80 through line 82.   Air flows into regenerator 80 through line 84 for burning carbonaceous material from the catalyst surface.   Excess air and combustion products are discharged through line 86.

FIGURE 64: ALTERNATIVE PROCESS FOR DEMETALLIZING CRACKING CATALYST BY SILICA SOL TREATMENT

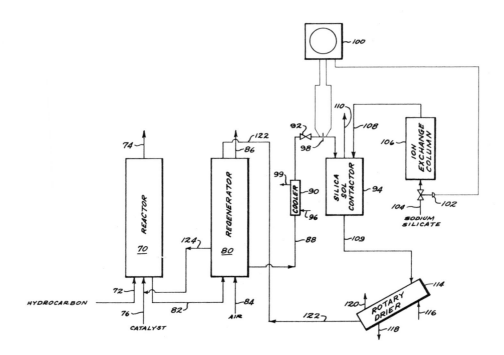

Source: W.H. Humes and M.M. Stewart; U.S. Patent 3,226,335; December 28, 1965

Catalyst is removed from the regenerator 80 at an elevated temperature of about 1000°F. to 1100°F. and is passed through line 88, containing cooler 90 and valve 92, to silica sol contacting chamber 94. Coolant enters cooling coils within cooler 90 through line 96 and is discharged therefrom through line 99. Cooler 90 indirectly cools the catalyst to a temperature within the range 300°F. to 400°F. As explained above, there are several important reasons for cooling the silica-alumina catalyst prior to its contact with silica sol.

The rate at which catalyst enters silica sol contactor chamber 94 is adjusted by means of valve 92 and this rate is ascertained by measurement of pressure drop across orifice 98. By means of recorder-controller 100 and pneumatic control valve 102, which is responsive to the controller, the catalyst flow rate determines not merely the rate of silica sol flow to chamber 94 but, more importantly, also determines the rate at which silica sol solution is prepared. Controller 100 pneumatically regulates valve 102 in aqueous sodium silicate line 104 leading to ion exchange column 106. In this manner only sufficient aqueous silica sol is prepared in ion exchange column 106 and discharged through line 108 to deposit a predetermined percent of SiO$_2$ upon the catalyst currently flowing into contactor 94.

Following contact of silica sol and catalyst in chamber 94, water vapor is removed through line 110 and wet catalyst impregnated with SiO$_2$ is removed through line 109 to rotary drier 114. Wet catalyst is dried in rotary drier 114 having internal heating coils to which heating fluid is charged through line 116 and from which heating fluid is removed through line 118 while water vapor is removed overhead through line 120. Dry catalyst is returned to regenerator 80 through line 122 wherein it is calcined at the elevated temperatures of the regenerator. The calcined SiO$_2$ impregnated catalyst returns to reactor 70 through lines 124 and 76.

Nalco Chemical

A process developed by K. Odland; U.S. Patent 3,258,430; June 28, 1966; assigned to Nalco Chemical Company involves improvement of the operational efficiency of metal contaminated silica-alumina cracking catalysts by treating such catalyst with a metal ion-free water, which has a pH within the range of 2.5 to 5.5, for a period of time sufficient to remove at least a portion of the metal contaminants from the catalysts. After such treatment, the water is removed from the catalyst and the catalyst returned to the catalytic cracking unit from whence it was taken, or to other units, where it is ready for the further conversion of various types of petroleum hydrocarbons. In a preferred mode of operation, the pH of the metal ion-free water is within the range of 3.4 to 4.0, and most preferably the pH is maintained within the range of 3.4 to 3.7. Figure 65 illustrates the essentials of the operation of the demetallization process.

FIGURE 65:   PROCESS FOR DEMETALLIZING CRACKING CATALYSTS BY WATER WASHING

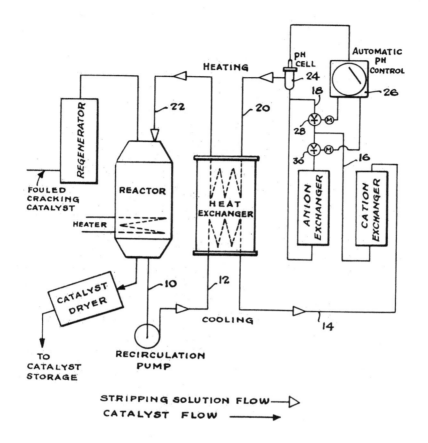

Source: K. Odland; U.S. Patent 3,258,430; June 28, 1966

After the catalyst has been treated by a conventional regenerator, it is then passed into a reactor where it is contacted by the metal ion-free water which has been heated to temperatures usually in excess of 150° F. After this washing treatment, the catalyst is dried and returned to the catalytic cracking unit for further treatment of petroleum hydrocarbons. The water flows from the reactor through drain line 10, and is pumped through line 12 into the heat exchanger where it is cooled to below about 140° F. After cooling, the water then flows through line 14 to the cation exchange resin where the metals are removed. After the cation exchange treatment, the water flows through lines 16, 18 and 20 to a hot leg of a heat exchanger where it is heated for reuse in the treating of additional metal-fouled catalyst. It enters the catalyst treating reactor through line 22.

When the water flows through line 18 after contact with the cation exchange resin, it contacts pH cell 24 connected to pH controls 26. When the pH recorder indicates excess acidity, the system shown in the drawing operates to bypass at least a portion of the cation exchange effluent to an anion exchange resin which is utilized as an acidity reducing unit. In operation the pH controller actuates pneumatically controlled motorized throttle valves 28 and 30; closing valve 28 while opening valve 30. Under these conditions, the cation exchange treated water passes to the anion exchange resin for a period of time sufficient to raise the pH to the desired level, at which point the pH controller closes valve 30 and opens valve 28.

It is contemplated that for minor pH control, valves 28 and 30 would always be partially open to allow a small portion of the water to contact the anion exchange resin. From the above it may be gathered that several advantages are afforded by the practices of this process. An important gain realized is that only metal contaminants are removed from silica-alumina catalyst. There is no loss of catalyst alumina. This may be attributed to the mild treating conditions. Another advantage is that there is little degradation of the ion exchange resins due to the temperature regulated water. A most important concept is the use of an anion exchange resin in conjunction with a cation exchange resin to maintain careful pH control of the water used to treat the catalyst.

It is within the scope of this process that the cation metal free water may be used for one catalytic treatment and then discharged to suitable waste lines, or returned to other areas of petroleum refinery for purposes of utilization in the petroleum refining processes such as for example, in cooling various lines for purposes of heat exchange.

Sinclair Research

A process developed by E.H. Burk, Jr., H. Erickson and A.D. Anderson; U.S. Patent 3,122,510; February 25, 1964; assigned to Sinclair Research, Incorporated is a process for the removal from solid oxide hydrocarbon conversion catalysts of metals, e.g. Ni, V and Fe, which poison the catalytic activity of the catalysts. The method includes vapor phase chlorination of the catalyst to provide or remove the poisoning metals in the chloride form, and a wash with a liquid aqueous medium to remove poisoning metal constituents. Figure 66 shows the arrangement of apparatus used in the conduct of this process.

FIGURE 66:   PROCESS FOR REMOVING NICKEL AND VANADIUM FROM CRACKING CATALYST BY CHLORINATION

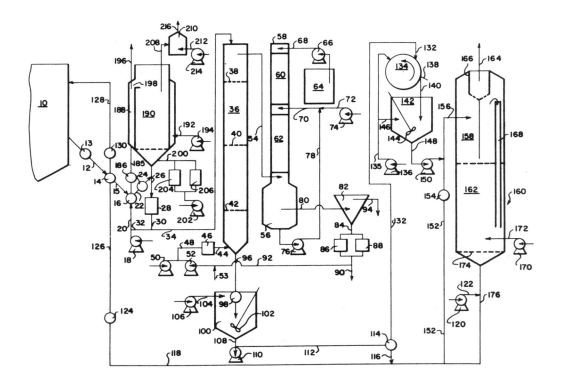

Source: E.H. Burk, Jr., H. Erickson and A.D. Anderson; U.S. Patent 3,122,510; February 25, 1964

In the drawing 10 represents the catalyst regenerator associated with a conventional hydrocarbon catalytic conversion system. The regenerator 10 is provided with a line 12, having a valve 13, for drawing off continuously or intermittently a fraction of poisoned catalyst from the regenerator. The catalyst is permitted to flow to the junction 14 and through the line 15 to the junction 16 where it may be conducted by a fluid, preferably air from the compressor 18 and pipe 20, through the pipe 22, having the valve 24, to the conduit 26. Conduit 26 allows the catalyst to fall by gravity into the cooler 28 where the temperature of the catalyst is allowed to fall from the 900° to 1175°F. temperature of regeneration to the chlorination temperature of about 350° to 1000°F. After the catalyst has cooled sufficiently it is withdrawn from the cooler 28 by the pipe 30. A conveying fluid, once more preferably air from the compressor 18 and the line 32 conveys the catalyst through line 34 to the chlorinator 36.

The chlorinator is generally an elongated chamber made of Monel or other chlorine resistant material and may be provided with one or a plurality of internal grids 38, 40, 42 for gas distribution and break up of catalyst particle agglomerates. The chlorinating agent is brought to the chlorinator 36 from the conduit 44 and heater 46. The heater is provided to give the agent the required temperature of chlorination. The chlorinating agent enters the heater 46 from the mixing conduit 48, having been pumped into this conduit by one or both of the pumps 50 and 52 which lead from suitable sources of the components which make up the chlorinating agent; for example, pump 50 may be connected with a source of chlorine gas while pump 52 is connected to a source of carbon tetrachloride by line 53.

The drawing shows apparatus for recovering excess chlorinating agent from the chlorinator 36. In the recovery apparatus shown, which is exemplary of various systems which can be adopted for recovery of the various chlorinating agents mentioned above, the excess chlorination vapor is withdrawn from the chlorinator by the line 54 to the tank bottom 56 of recovery vessel 58. This vessel, which may be supplied with beds 60 and 62 of inert solid contact material, may be supplied at the top with a reagent such as a strong caustic soda solution from the storage vessel 64, pump 66 and line 68. The vessel may also have a line 70 for water from the line 72 and pump 74, and for recycle fluid from the tank bottom 56 through pump 76 and line 78.

This particular system, of course, is designed to neutralize the waste chlorine and flush out the metallic chlorides resulting from the chlorination. The aqueous mixture of chlorides is drawn from the tank bottom through the line 80 to the settling tank 82. Carbon tetrachloride settles to the bottom of this tank from which it may be drawn by the line 84 and adsorptive dryers 86 and 88 to storage by the line 90 or to reuse in the chlorination by line 92 and pump 52. The aqueous part of the mixture from tank bottom 56 goes to the top of the settling tank 82 where it may be withdrawn by line 94 to waste.

The chlorinated catalyst leaves the chlorinator 36 by conduit 96. The catalyst particles, freed of accessible iron and vanadium, pass through valve 98 to slurry tank 100 which is advantageously provided with stirrer 102. In the slurry tank the chlorinated catalyst is quickly stirred into a large volume of distilled or deionized water from the line 104 and pump 106. This very dilute slurry is quickly removed from the slurry tank 100 by the line 108 and pump 110 and conveyed through line 112. Valve 114 may be adjusted to direct the catalyst slurry through line 116 whence it may be conveyed through the line 118, by a fluid, e.g. air, from the compressor 120 and line 122. The conduit 118 has the valve 124. The catalyst slurry passes to the conduit 126, junction 14 and conduit 128, which has the valve 130, back to the regenerator 10.

When demetallizing is to be conducted on a continuous basis, the junction 14 is arranged to conduct the streams of poisoned and treated catalyst free from contact with each other. Alternatively the valve 130 may be closed, allowing the catalyst to flow from junction 14 through line 15 and back into any or all parts of the demetallizing system.

Alternatively, valve 114 may be adjusted to direct the catalyst slurry through line 132 to the filter 134 which may advantageously be a rotary vacuum drum filter. The cake of catalyst particles on the filter may be rinsed by distilled or deionized water from line 135 and pump 136, scraped from the filter by doctor blade 138 and fall through the path or conduit 140 into the reslurry tank 142 which advantageously is provided with a stirrer 144 and a line for distilled or deionized water 146. This catalyst slurry is drawn by line 148 and slurry pump 150 to the conduit 152 provided with the valve 154 from which the catalyst slurry may be returned to the regenerator by the line 118 as previously described. Alternatively, valve 154 may be closed, directing flow from slurry pump 150 into the conduit 156 and the dryer section 158 of the dryer-calciner 160.

In the dryer, the free water contained in the catalyst slurry is evaporated preferably by contact with flue gases leaving the calcination section 162, and the water vapors are exhausted through line 164. The dryer may also be equipped with the cyclone separator arrangement 166 for removal of catalyst fines from the exhaust vapors. The catalyst particles flow out of dryer section 158 through the pipe 168 to the calcination section 162. In the calcination section the catalyst may be raised to calcination temperature of about 900° to 1200°F. or more, advantageously by burning a fuel in the catalyst bed. Pump 170 and line 172 are provided to convey fuel or a fuel-combustion supporting gas mixture into the bed. Calcined catalyst from the bed is allowed to fall through the screen 174 to the line 176 whence it may be conducted to other parts of the apparatus previously described.

A sulfiding treatment of the catalyst after removal from the cracking operation and before chlorination is quite advantageous especially to enhance nickel removal. For example, in many cases it has been found that only 5 to 10% of the nickel oxide in the poisoned catalyst is removed without sulfiding. It is theorized that metals, especially Fe and Ni, present in poisoned catalysts may be largely in solid solution in the catalyst matrix. The metal ions being mobile in solution at elevated temperatures, it has been found possible to concentrate the metals at the catalyst surface by treatment with a sulfiding agent, such as hydrogen sulfide at elevated temperatures. The sulfiding agent converts the metal ions at the surface to metal sulfides which seem less soluble in the matrix. Diffusion of metal ions transports additional metal to the surface where it is in turn converted to the sulfide. Thus, a continuing process concentrates the metals as sulfides on the catalyst surface whence they are more readily susceptible to chlorination.

The sulfiding step can be performed by contacting the poisoned catalyst with elemental sulfur vapors, or more conveniently by contacting the poisoned catalyst with a volatile sulfide, such as H_2S, CS_2 or a mercaptan. The contact with the sulfur-containing vapor can be performed at an elevated temperature generally in the range of about 500° to 1500°F., preferably about 800° to 1300°F. Other treating conditions can include a sulfur-containing vapor partial pressure of about 1.0 to 30 atmospheres or more, preferably about 0.5 to 25 atmospheres. Hydrogen sulfide is the preferred sulfiding agent. Pressures below atmospheric can be obtained either by using a partial vacuum or by diluting the vapor with gas such as nitrogen or hydrogen.

The time of contact may vary on the basis of the temperature and pressure chosen and other factors such as the amount of metal to be removed. The sulfiding may run for instance, at least about 5 or 10 minutes up to about 20 hours or more depending on these conditions and the severity of the poisoning. Temperatures of about 900° to 1200°F. and pressures approximating 1 atmosphere or less seem near optimum for sulfiding and this treatment often continues for at least 1 or 2 hours but the time, of course, can depend upon the manner of contacting the catalyst and sulfiding agent and the nature of the treating system, e.g. batch or continuous, as well as the rate of diffusion within the catalyst matrix.

The illustration also shows a suitable arrangement for sulfiding the catalyst when this is to be performed before chlorination. When sulfiding is to be performed, valve 24 is closed so that the catalyst particles flowing to junction 16 are conveyed by the fluid from line 20 through the line 185 having the valve 186 to an outer chamber 188 of sulfider 190.

The catalyst particles and the sulfider itself are raised to the sulfiding temperature of say about 1150°F. advantageously by burning a fuel in the bed of particles in this chamber. The fuel, with or without the addition of combustion supporting gas, may be supplied by line 192 and pump 194, and exhaust gas may be vented through line 196. The heated catalyst particles may flow into the sulfiding chamber as through the opening 198. A sulfiding gas, e.g. H_2S or CS_2, is passed to the bottom of the sulfider 190 by the line 200 from the pump 202. When the sulfiding gas is wet, passage to the line 200 may be through the adsorptive dryers 204 and 206. Exhaust sulfiding gas is passed out of the sulfider 190 through the line 208, advantageously to a burner 210. This burner may be provided with a line 212 and pump 214 for supplying combustion-supporting gas, and with a vent 216 to the atmosphere.

A process developed by G.B. Hoekstra; U.S. Patent 3,140,923; July 14, 1964; assigned to Sinclair Research, Inc. involves removing the metal contaminants from a fluidized bed of cracking catalyst using apparatus comprising in combination a catalyst regenerating chamber, a heated oxygen-treating chamber, a sulfiding chamber, a chlorinating chamber and a quench chamber, the chambers being connected by valved lines and arranged for sequential catalyst movement by gravitational flow from the oxygen-treating chamber through the sulfiding and chlorinating chambers in sequence to the quench chamber.

The apparatus also includes means for contacting the gas-treated catalyst with an aqueous liquid in the quench chamber and automatic temperature sensing means in the quench chamber to control the entrance of the catalyst into the quench chamber. The catalyst flow rate from the chlorinating chamber to the quench chamber controls the catalyst bed levels in and total catalyst flow through the catalyst demetallization chambers.

A process developed by H.G. Russell and W.B. Watson; U.S. Patent 3,150,075; September 22, 1964; assigned to Sinclair Research, Incorporated involves an improved procedure for the utilization of hydrocarbon feedstocks which contain large amounts of poisoning metals. Large quantities of mineral oil petroleum crudes, fractions thereof, and hydrocarbons derived therefrom, contain harmful amounts of metal impurities, such as nickel, vanadium and iron. These impurities are frequently present in such large amounts that utilization of the hydrocarbon as a catalytic cracking feedstock is a real problem since the metals accumulate on catalysts in cracking, adversely affecting the product distribution of cracking yields by increasing coke and gas make and decreasing gasoline make. Thus catalytic cracking of the hydrocarbons is uneconomical because the metal impurities harmfully affect selectivity of the catalyst.

For this reason, such stocks have not heretofore been utilized to the fullest possible extent as cracking feedstocks, but rather have been diverted to use as a fuel to supply the power needs of the refinery, or have been sold for use as low-value fuels outside the refinery. In this process, heavy residual hydrocarbon oils containing more than about 1.5 parts per million of vanadium and/or more than about 0.6 part per million of nickel are converted by catalytic cracking in the presence of steam to lower-boiling products such as gasoline while depositing metal contaminants and about 20 to 60 lbs. of coke per barrel of feed on the catalyst. Cracked products are fractionated to remove gasoline and other desirable low-boiling cuts. Catalyst is continually circulated between the reactor and a large capacity regenerator.

The heat given off in regeneration is used to make steam, some of which can be passed to the reactor as a diluent for cracking, some of which can be used to strip hydrocarbon values from the catalyst before it is regenerated, and some of which can be used for electric power generation by passage through a turbine. Heat in the regenerator may also be employed to produce a circulating heat medium. Catalyst is continually removed from the cracking system (from the reactor loop in this case rather than from the regenerator) for demetallization. Thus by treating the heavy hydrocarbon feed, preferably a residual petroleum oil feed containing a high proportion of metal contaminants and coke-forming materials, to catalytic cracking, valuable gasoline and other low-boiling materials are produced while still supplying the power needs of the refinery. The flue gas from the regenerator can be employed in the catalyst demetallization process for drying catalyst. Figure 67 is a block flow diagram of this process.

FIGURE 67: FLOW DIAGRAM OF INTEGRATED CATALYTIC CRACKING PROCESS SHOWING DEMETALLIZATION STEP

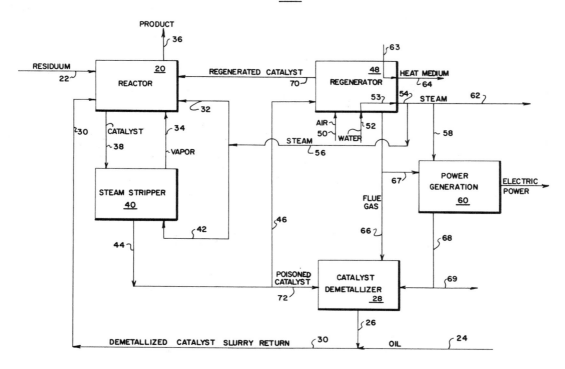

Source: H.G. Russell and W.B. Watson; U.S. Patent 3,150,075; September 22, 1964

The petroleum residual oil is fed to the catalytic cracking reactor 20 by way of line 22, preferably directly from the crude oil stills (not shown). Some oil also is fed by line 24 to the exit line 26 from the catalyst demetallizer 28 to convey demetallized catalyst by line 30 to the reactor 20. Steam is fed to the reactor by line 32 and stripping steam, carrying vaporized hydrocarbons is conveyed to the reactor by line 34. Cracked products are withdrawn from the reactor by line 36.

Catalyst, containing relatively large amounts of coke and some poisoning metal, is continually withdrawn from the reactor 20 by line 38 which conducts the catalyst to stripping chamber 40. Steam enters this chamber from line 42 and passes countercurrently to the catalyst, picking up vaporizable hydrocarbons entrained in the catalyst. Catalyst, still containing coke and metal poisons, is withdrawn from the stripper by line 44 and the greater proportion is sent by line 46 to the regenerator 48. Air for burning off coke is conducted to the regenerator by line 50 and water or low temperature steam is introduced by line 52 to the heat exchanger 53 in the bed of catalyst.

The water or low-pressure steam passes through the coil or other suitable apparatus in indirect heat exchange with the burning coked catalyst and is withdrawn as high temperature steam by the line 54, whence part of it may be conducted by line 56 to use in the reactor and/or stripper, part by line 58 to a turbine 60 for generation of electricity and part by line 62 to other uses in the refinery.

Liquid heat exchange medium may be passed in indirect heat exchange with the burning coked catalyst through the line 63 and may be withdrawn by line 64 to any suitable use. Flue gases from the regenerator may be conducted by line 66 for use in the catalyst demetallizer as will be explained below, or by line 67 to the generator turbine. Also, exhaust steam from electricity generation may be brought by line 68 to the demetallizer for use as a diluent as will also be explained below. Alternately the exhaust steam may be taken by line 69 for general refinery uses. Regenerated catalyst is returned to the cracking reactor by line 70. A slip stream of catalyst is continually removed from the line 44 for conveying to the demetallizer 28 by line 72. For severely poisoned catalysts, the treatment may be repeated.

A process developed by H. Erickson and H.G. Russell; U.S. Patent 3,216,951; November 9, 1965; assigned to Sinclair Research, Incorporated involves carrying out an aqueous oxidizing wash procedure for the removal from solid oxide hydrocarbon conversion catalystis of metals, e.g. Ni and V, which poison the catalytic activity of the catalysts. Such a procedure usually includes contacting the contaminated catalyst with a sulfiding agent before the wash to put the metal contaminant in the sulfide form.

The metal poisons may then be converted to a dispersible form by a liquid aqueous oxidation medium. Such an oxidizing medium may consist of water, with or without the addition of small amounts of, preferably mineral, acids, to which chlorine, air, ozone or other molecular oxygen-containing gases may be added. The inclusion in the liquid aqueous oxidizing solution of nitric acid provides for increased vanadium removal, especially when the sulfiding is preceded by a treatment of the catalyst at elevated temperature with a molecular oxygen-containing gas. A suitable form of apparatus for the conduct of this washing process is shown in Figure 68.

FIGURE 68: APPARATUS FOR DEMETALLIZATION BY AQUEOUS OXIDIZING WASH PROCEDURE

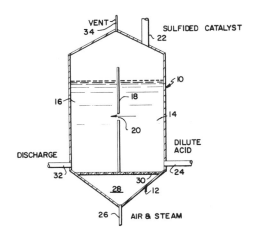

Source: H. Erickson and H.G. Russell; U.S. Patent 3,216,951; November 9, 1965

The reaction vessel comprises an outer upper cylindrical shell 10 with a lower conical-shaped bottom 12. The vessel and its component parts may be made of stainless steel or any other suitable material chemically resistant under reaction conditions. The cylindrical shell 10 is divided into two compartments 14 and 16 by a vertical weir or weirs 18. This weir has an aperture 20 positioned below the level of the catalyst slurry and such that a vertical plane normal to the weir would bisect the weir and intersect the opening. The vessel may advantageously be provided with a plurality of weirs to form any number of compartments desired.

A solids intake conduit 22 leads to a first compartment in the upper chamber and is in communication with a sulfiding chamber (not shown). The liquid intake conduit 24 is positioned near the bottom of the upper chamber and is in communication with a source (not shown) of a dilute nitric acid solution or any other suitable reagent which when aerated will form the liquid oxidizing agent. An air and steam inlet conduit 26 is provided to the conical-shaped plenum chamber 28 which is defined by the porous reactor bottom 30. An outlet conduit 32 is provided for removing the treated catalyst from the last compartment of the reactor. The cylindrically-shaped chamber is also equipped with vent means 34 to facilitate removal of effluent products from the reactor and conveniently may lead to the atmosphere or to a scrub tower or any other means to minimize the discharge of polluting gases.

In the preferred mode of operation, a previously sulfided metal-contaminated catalyst falls down through pipe 22. The addition of the catalyst, having the high temperature of the sulfidation is frequently sufficient to impart to the resulting slurry the almost boiling temperature at which the oxidation is effectively conducted. This temperature may conveniently be maintained by introduction of steam into the system. Alternatively, the sulfided catalyst may be slightly cooled before introduction into the reaction vessel to avoid undue vaporization of the liquid component of the oxidizing medium.

The catalyst flows through conduit 22 into compartment 14. The liquid component of the oxidizing agent, flowing through conduit 24, enters the bottom of the vessel to form a slurry with the catalyst. Effective contact between the catalyst and the liquid component of the oxidizing agent is achieved by agitation due to the air and steam flowing from conduit 26 through the plenum chamber 28, and through the porous bottom 30. This arrangement obviates the need for mechanical mixers in the reaction vessel. Aeration of the liquid forms the oxidizing medium.

The weir 18, the top of which is above the level of the catalyst slurry, is in fluid tight engagement with the porous bottom 30 and the only means of egress of the slurry from compartment 14 is through the aperture 20 in the weir. Motion and heat are again imparted to the catalyst particles passing to compartment 16 where oxidation is continued in the same manner as in compartment 14. When, as shown, only two stages are provided, the oxidized catalyst slurry is led from the final stage through conduit 32 back to the hydrocarbon conversion reactor perhaps by way of a filter (not shown) and a subsequent washing operation. The following is a more specific example of the operation of this process.

The cracking catalyst had a size range of about 20 to 150 microns and comprised a synthetic gel silica-alumina composite containing about 13% alumina. This catalyst was introduced to a cracking reactor along with a vacuum residuum derived from a West Texas crude oil and having an API gravity of 15.1, a Conradson carbon content of about 8.8 weight percent, a viscosity of about 400 seconds Saybolt at 210°F., and an initial boiling point above about 1000°F. and containing 24.7 ppm of nickel, 39.9 ppm of vanadium and 21.3 ppm of iron. In the reactor the hydrocarbon and solid catalyst are heated to about 850°F. at a pressure of about 8 psig and a WHSV of about 15. Under these conditions, a 30 to 40% conversion of the feed to lighter materials is effected with the effluent being substantially free of the metal contaminants along with the associated coke formers. Catalyst is taken from the reactor and its carbon content is reduced from about 2 to 0.5 weight percent through contact with air in the regenerator.

A slip stream of regenerated catalyst analyzing 0.5% coke, 370 ppm nickel, 1,300 ppm vanadium and 840 ppm iron is continuously withdrawn from the regenerator at a daily inventory rate of 75% and sent to demetallization where it is held for about 2 hours in a zone where it is contacted with air at about 1300°F. and then sent to a sulfiding zone where it is fluidized with H_2S gas at a temperature of about 1050°F. for about 2 hours.

The sulfided catalyst is discharged to a cooler where the temperature of the catalyst is reduced to about 900°F. and introduced through the top of a two compartment reactor. A dilute aqueous solution of nitric acid enters first compartment of the reactor at a velocity of about 0.1 ft. per minute to mix with catalyst and form a slurry. Air and steam are admitted to the reactor through the plenum chamber and the porous plate at a rate of about 0.01 ft. per minute. The heated, agitated and partially oxidized slurry passes from the first to last compartments through a 3 inch aperture in the partitioning weir and remains in the reactor for a total residence time of about 30 minutes. The catalyst is then discharged from the reactor and directed to a wash treatment effective to remove nickel and the available vanadium. The metals level of the cracking catalyst after demetallization analyzes 151 ppm nickel, 950 ppm vanadium and 650 ppm iron.

A process developed by H. Erickson, J.P. Gallagher and H.G. Russell; U.S. Patent 3,222,293; December 7, 1965; assigned to Sinclair Research, Incorporated is a demetallization process involving a simplified mode of contacting the catalyst with an aqueous oxidizing medium, wash water and a dilute ammonium hydroxide solution. The process uses a modified pan filter for the contact procedures.

Pan filters are widely used in industry. In general the filter consists of several horizontal wedge-shaped pans which rotate around a central core through filling, filtering and dumping stages. A filter cake is formed on a cloth or screen near the bottom of these pans. Vacuum is applied to the underside of this filter cloth or screen to accelerate filtrate flow through the cake. After filtrate removal the pan proceeds to a dump position where a mechanism rotates the pan on a horizontal axis to turn the pan upside down for cake discharge. After discharge the pan is rotated back to its normal position ready to receive more slurry.

A conventional pan filter is modified to provide for the catalyst contact procedures of this process. A provision is made for air blow-back through the filter cloth during about the first third or half of the cycle and deeper pans are provided to hold the volume of a 20 to 40% solids catalyst slurry.

A process developed by H. Erickson and H.G. Russell; U.S. Patent 3,295,897; January 3, 1967; assigned to Sinclair Research, Incorporated involves a coordinated system for demetallization of regenerated catalyst and return of demetallized catalyst to the reactor in the form of an aqueous slurry. Figure 69 shows the arrangement of equipment for the conduct of this process.

The figure shows a fluidized catalytic cracking reactor 10 which is provided with a feed line 12 for introduction of synthetic gel, silica-alumina cracking catalyst and a hydrocarbon gas oil feedstock. The reactor is also provided with the exit line 14 for the removal of cracked product hydrocarbons, including gasoline, which are generally in a vapor state. The reactor is also provided with the standpipe 16 for the removal of catalyst which is in need of regeneration. This catalyst is conveyed from standpipe 16 through pipe 18 conveniently by air from the source 20 to the regenerator 22. In the regenerator this air, with, if desired, additionally-introduced air, burns coke from the catalyst. The regenerator is also provided with the vent 24 for removal of exhaust gases. Regenerator standpipe 26 is provided for removal of

FIGURE 69: FLOW DIAGRAM OF INTEGRATED PROCESS FOR SULFIDING THEN CHLORINATING REGENERATED CATALYST TO EFFECT DEMETALLIZATION

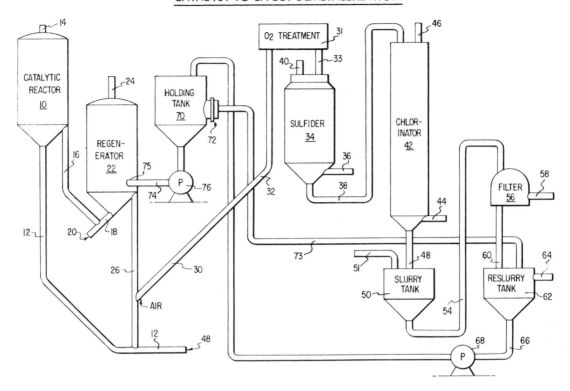

Source: H. Erickson and H.G. Russell; U.S. Patent 3,295,897; January 3, 1967

regenerated catalyst from the combustion zone maintained in the regenerator. This catalyst may be conveyed by feed hydrocarbon from the source 48 through the pipe 12 to the reactor. The regenerator standpipe 26 is provided also with the tap-line 30 whereby a selected amount of catalyst may be conveyed through pipe 32 to the demetallization system.

The demetallization system shown includes means for oxygen-containing gas treatment, sulfiding, chlorinating, washing, filtering, and rinsing the poisoned catalyst. The pipe 32 conveys catalyst in a stream of air to the calciner 31 where the catalyst is treated at an elevated temperature to enhance subsequent vanadium removal. Next the catalyst passes by line 33 to the sulfider 34 wherein the catalyst, at an elevated temperature, is subjected to the action of a sulfiding gas, for example, hydrogen sulfide, from the source 36, to enhance later nickel removal. Catalyst is withdrawn from the sulfider by pipe 38, and exhaust gases are removed by line 40.

Catalyst is then conveyed to the chlorinator 42 wherein it is contacted with a chlorinating vapor, for example, a mixture of chlorine gas and carbon tetrachloride, from the source 44. Exhaust gases are withdrawn from the chlorinator by line 46, and catalyst gravitates out of the chlorinator by line 48 to the slurry tank 50 where it is contacted with a liquid aqueous medium, for example, water, from the source 51. In other procedural schemes the compartment may provide for contact of the catalyst with other aqueous media, such as a dilute ammonia solution, weak acids, chelating solutions, or aqueous reducing agents. The slurry may be conducted by line 54 to the filter 56 from which filtrate may be removed by line 58, and catalyst filter cake by line 60. The cake may be brought to the reslurry tank 62 for contact with additional water from the source 64.

A plurality of such slurrying, filtering, washing, etc., stages may be provided, but eventually a slurry demetallized catalyst having about 5 parts of water per part of catalyst is conducted through line 66 by centrifugal pump 68 to the elevated holding tank 70 positioned about 2 to 5 feet above the point of entry of demetallized catalyst into the regenerator to provide at least a small gravitational force aiding pump 76. The difference in height between pump 68 and the point of entry into the regenerator 75 is usually about 40 to 50 feet or higher, depending upon the design of the system. In the holding tank the catalyst slurry is settled and water through filter 72 is returned through line 73 to the reslurry tank to increase the catalyst concentration in the slurry to about 1.5 parts of water per part of catalyst before pumping it into the regenerator 22 through line 74 with progressing cavity pump 76.

TITANIUM-CONTAINING CATALYSTS

POLYMERIZATION

Titanium halides in conjunction with aluminum alkyls or alkoxides provide the basis for a wide variety of catalysts used in the manufacture of high-density polyolefins obtained in low pressure polymerization processes. The reader is referred to "Polyolefin Processes" by Marshall Sittig published by Noyes Development Corp., Park Ridge, N.J. (1967) for more details of the various permutations and combinations involved in such catalyst compositions. Only a few representative compositions will be cited here as examples.

Asahi Kasei Kogyo

A process developed by Y. Takashi, I. Aishima, Y. Kobayashi and Y. Tsunoda; U.S. Patent 3,494,910; February 10, 1970; assigned to Asahi Kasei Kogyo K.K., Japan is a process for producing a polyolefin having high isotacticity in which an α-olefin is polymerized by using a ternary catalyst consisting of $TiCl_3$, an organoaluminum compound and R_3SbX_2 where R is an alkyl group of 2 to 5 carbon atoms and X is a halogen.

Dart Industries

A process developed by A.P. Haag and M. Weiner; U.S. Patent 3,623,846; November 30, 1971; assigned to Dart Industries Inc. involves controlling particle size during condensation or desublimation of a material such as titanium trichloride. The technique involves injecting a vapor stream into an enlarged chamber maintained at a temperature below the condensation temperature of a material in the vapor stream. An inert gas is intermittently puffed through apertures in the enlarged chamber for buoying particles of material condensed from the vapor into the entering vapor stream so that additional material condenses thereon, thereby enlarging the particle size.

Titanium trichloride has been found to be a valuable catalyst material; particularly for the polymerization of polypropylene. The titanium trichloride apparently occurs in more than one allotropic form and it is found that the effectiveness of the catalyst depends on the form in which it occurs. This form is at least in part determined by the technique employed for forming the titanium trichloride.

A technique that has been employed in the past involves the reaction of titanium tetrachloride vapor with sponge titanium at elevated temperature to produce titanium trichloride by the reaction $3TiCl_4 + Ti = 4TiCl_3$. The product of the reaction is a hot gas or vapor stream, including excess titanium tetrachloride, titanium trichloride vapor, and usually an inert carrier gas. The titanium trichloride is condensed into a powder by quenching the hot gas and the powder is recovered.

One of the parameters of interest in the production of polypropylene is the particle size of the polymer produced, and it appears that, in at least some of the polymerization techniques employed, the particle size of the catalyst employed in the polymerization reaction can affect the crystal size of the polypropylene. It is therefore desirable to have a technique for controlling the particle size of titanium trichloride in the original manufacturing process.

Figure 70 shows the apparatus which may be used. There is proveded a storage tank 10 for containing liquid titanium tetrachloride. Liquid is passed from the storage tank 10 to a vaporizer 11 which heats and evaporates the titanium tetrachloride to produce a vapor thereof. An inert gas supply 12 is provided for mixing inert gas such as nitrogen, helium, hydrogen, or argon with titanium tetrachloride vapor coming from the vaporizer 11. The inert gas provides additional gas volume for sweeping the titanium tetrachloride vapor through the balance of the described system. If desired, the gas may be added in the vaporizer to assist in its operation, and it may be preheated if the possibility exists of cooling to the dew point of the titanium tetrachloride.

FIGURE 70: APPARATUS FOR MANUFACTURE OF TITANIUM TRICHLORIDE OF CONTROLLED PARTICLE SIZE

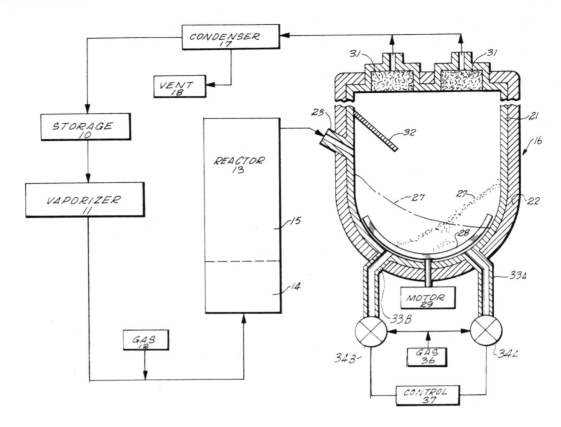

Source: A.P. Haag and M. Weiner; U.S. Patent 3,623,846; November 30, 1971

The mixture of titanium tetrachloride vapor and sweep gas is injected into the bottom of a conventional reactor 13 which in one embodiment comprises a steel cylinder about one foot in diameter and about ten feet long, heated to about 900°C. A lower zone 14 in the reactor is filled with pieces of graphite so that the gas or vapor stream passing thereover is pre-heated. The upper portion 15 of the reactor is filled with sponge titanium with which the titanium tetrachloride reacts to produce titanium trichloride according to the reaction set forth above.

Effluent from the top of the reactor is a vapor stream containing unreacted titanium tetrachloride vapor, the sweep gas from the gas supply 12, and titanium trichloride vapor since the reactor is maintained at a temperature substantially above the sublimation temperature of titanium trichloride. The vapor stream is then injected into a desublimer or settler 16 described in greater detail later. The desublimer is maintained at a temperature below the sublimation temperature of titanium trichloride and above the boiling point of titanium tetrachloride so that solid titanium trichloride is condensed in the desublimer and vaporous titanium tetrachloride and the sweep gas pass through.

The vapor stream from the desublimer 16 then passes to a conventional condenser 17 where the titanium tetrachloride is condensed to a liquid and recycled to the storage tank 10. The sweep gas and other injected gases, described later, are passed from the condenser to a vent 18. The desublimer or settler 16 comprises a metal shell 21 forming an enlarged chamber having a sufficiently large cross section that the flow of gases upwardly therethrough is insufficient to buoy a substantial portion of the particles accumulating in the chamber and carry them out with exiting gases. The shell 21 is surrounded by a temperature control jacket 22 for maintaining the temperature of the shell below the condensation temperature of titanium trichloride and above the dew point of titanium tetrachloride. The temperature control jacket 22 is a conventional element such as electric heaters or a surrounding jacket of heat transfer fluid so that either heating or cooling of the shell can be provided. Any other conventional means for maintaining the temperature of the shell 21 in the aforementioned range may be employed.

A pipe 23 is provided on a side of the shell 21 for conducting vapor from the reactor 13 to the interior of the shell. The pipe 23 is fabricated of a heat resistant metal since it is maintained at a temperature of from 750° to 900°C. to prevent titanium trichloride condensation on the walls. Since the desublimer is below the sublimation temperature, solid titanium trichloride condenses inside the desublimer and settles to the bottom thereof as a loose, finely-divided powder 27. A rotating paddle 28 driven by a motor 29 moves about the bottom of the desublimer to stir the powder 27 and prevent accumulations of powder on the wall of the chamber. If desired, the paddle 28 may be extended any distance up the sides of the shell 21 to remove accretions of solid titanium trichloride.

The vapor stream entering the chamber through the pipe 23 includes unreacted titanium tetrachloride and sweep gas as mentioned above. These gases exit from the settling chamber or desublimer through filters 31 which serve to prevent any particles of titanium trichloride buoyed by the gases from leaving the settling chamber. After passing through the filters, the gases pass to the condenser 17 as described above. If desired, means may be provided for back flushing gas through the filters 31 for removing accumulations of particles therefrom.

A baffle plate 32 is provided between the gas inlet pipe 23 and the filters 31 to inhibit direct flow of gases therebetween and minimize pass of uncondensed titanium trichloride vapor from the settler. Near the bottom of the settler, a pair of gas inlet pipes 33 are provided through the shell 21. Each of the gas inlet pipes 33 is connected to a separate conventional solenoid controlled valve 34. A gas source 36 supplies pressurized inert gas to both of the solenoid valves 34. A controller 37 is also connected to both of the solenoid valves 34. The controller 37 comprises a conventional multiple contact timer, or the like, for selectively opening and closing the valves 34 according to a selected time schedule.

During operation of the system described, titanium trichloride particles are formed by condensation in the cooler de-sublimer and settle to the bottom of the settler as a loose powder 27 which accumulates over the gas inlet pipes 33. The particles initially formed upon desublimation of the titanium trichloride are in the order of about one micron diameter, which for some catalyst uses is considered too fine. Therefore, in order to cause particle growth, an inert gas is injected in the bottom of the settler through one of the gas inlet pipes 33a by opening the solenoid valve 34a connected to the inert gas supply 36. The flow of gas into the chamber through the accumulated powder 27 buoys up the powder and temporarily suspends it in the vapor within the desublimer.

Some of the cool vapor having particles of titanium trichloride suspended therein is drawn into and mixed with the stream of hot vapors being injected into the chamber through the pipe 23. The consequent cooling of the injected gas stream causes condensation or desublimation of the titanium trichloride vapor which preferentially occurs on the existing particles which serve as condensation nuclei. The condensation of additional titanium trichloride on the particles causes an increase in their size. It is found with such a technique that instead of one micron particles accumulating in the settler, that particles in the range of from 2 to 8 microns are readily prepared, which represents a volume increase in the order of from 8 to 64 times the original volume of the particles.

If the flow of inert gas into the bottom of the settler is steady, two undesirable effects may occur. Continuous flow of gas through accumulated powder in the bottom of the settler may channel so that the gas flows through substantially the same path at all times and only a portion of the particles are buoyed up to a region where they may be drawn into the hot vapor stream. Likewise, in a settler having continuous flow, a degree of elutriation may occur whereby the finest particles are carried in the upwardly moving gas stream so as to accumulate on the filters 31 and block flow therethrough.

In order to substantially preclude these effects, inert gas is injected through accumulated powder in the bottom of the settler at spaced intervals so that intermittent buoying of the particles occurs. An intermittent flow of gas through the accumulated powder raises particles of powder off the bottom and mixes them with vapor in the desublimer, but does not carry any substantial portion to the filters 31 to cause clogging or blinding thereof. The flow of inert gas through the powder buoys up particles in the path of the gas stream, and also draws adjacent particles into the gas stream so that they are also buoyed up. The flow of gas is then terminated so that an insubstantial quantity of powder is lifted to the level of the filters.

In order to minimize the effect of channeling through the powder, and to provide a thorough mixing of powders in the desublimer, a plurality of gas inlets are provided at the bottom. In the illustrated embodiment, two gas inlet pipes 33a and 33b are employed, each with its separate solenoid controlled valve 34a and 34b, respectively. In operation, one of the valves, 34a for example, is opened for an interval so as to inject gas into the desublimer through accumulated powders 27 and buoy the powders up in the chamber. The valve 34a is then closed to minimize elutriation of particles and the powder slowly resettles to the bottom of the desublimer, possibly in an arrangement different from the original accumulation as illustrated in phantom in the drawing.

After a selected time interval to permit the settling of at least the coarser particles, another solenoid valve 34b is opened for an interval to buoy up powder accumulated over the gas inlet 33b. The solenoid valve 34b is left open for an interval and then shut and a time permitted to elapse before the first solenoid valve 34a is reopened. Thus, the intermittent gas puffing cycle in the desublimer proceeds: first valve 34a open, both valves closed; second valve 34b open, both valves closed; first valve 34a open, both valves closed, etc.

Since the growth of particles depends on their coming in contact with the hot vapor stream containing vaporous titanium trichloride, the degree of growth obtained will depend on the number of times the particle is mixed into the vapor stream. It will also be apparent that after a large number of particles has been buoyed up by a period of flow of inert gas into the desublimer, many particles will remain buoyed up for an interval and slowly settle out at a rate determined by the usual rate of flow of gas through the desublimer from the reactor, the buoying gas viscosity and density, the particle size, and also the density and turbulence of the vapor in the desublimer.

In any event, the larger and heavier particles will settle out more rapidly and the finer particles will remain buoyed up in temporary suspension in the vapor for a longer period of time. In this manner, smaller particles have a higher degree of probability of being mixed with the hot vapor stream for particle growth. Since the finer particles encounter the hot gas stream more often there is a preferential growth on finer particles as compared with larger particles so that a good degree of particle size uniformity is obtained in the titanium trichloride powder in the desublimer.

In a specific example, titanium trichloride was made by reduction of titanium tetrachloride with sponge titanium. In this process about 50 pounds per hour of titanium tetrachloride vapor was passed through a bed of sponge titanium at a temperature of about 900°C. In addition, about 10 cubic feet per hour of argon was mixed with the vaporized titanium tetrachloride in order to keep the vapor moving through the system as desired. Under the selected reaction conditions, about 5 to 10 pounds per hour of titanium trichloride vapor was produced and was passed to a desublimer and settler through a 3/4" i.d. graphite tube maintained at about 900°C. The pressure in the desublimer was maintained in the range of from about 5 to 15 psig.

The desublimer comprised a steel cylinder about 4 feet i.d. and 16 feet tall, with fine stainless steel filters at the top. A pair of standard 1/4" pipes are connected to the bottom of the desublimer on opposite sides thereof, and conventional solenoid operated valves are provided on the two lines so that argon gas can be injected into the bottom of the desublimer. The walls of the desublimer were maintained in the range of from about 300° to 350°C. so that the gases and vapors reaching the filters at the top of the desublimer were at a temperature of less than about 400°C. In this way, titanium tetrachloride is maintained as a vapor and titanium trichloride condenses.

After about 12 to 18 hours of operation of the reactor and desublimer, titanium trichloride powder accumulated in the bottom of the settler to a depth of about two feet. Throughout the period of operation, argon, nitrogen, or helium gas is admitted at intermittent intervals alternately through the two gas inlets by means of solenoid valves operated by an automatic timer. In one mode of operation, argon gas at about 20 psig was left on for about 5 seconds on one valve, turned off for 15 seconds, turned on for 5 seconds on the other valve, turned off for 15 seconds, and so on.

In another mode of operation, nitrogen gas was injected alternately through the two gas inlets at a rate of about 7.2 cubic feet per minute in a cycle with 1 minute one valve on, 1 minute both valves off, 1 minute the other valve on, 1 minute both valves off, and so on. In each of these modes of operation, titanium trichloride particles principally in the size range of from about 2 to 8 microns accumulated in the desublimer. If desired, titanium tetrachloride vapor at elevated temperature can be employed in lieu of inert gas for reducing the volume of noncondensable material in the system.

Esso Research and Engineering

A process developed by E. Tornqvist; U.S. Patent 3,424,774; January 28, 1969; assigned to Esso Research and Engineering Company relates to a catalyst comprising cocrystalline beta-$TiCl_3 \cdot xAlCl_3$ in combination with an organometallic compound and in particular an aluminum trialkyl, an aluminum dialkyl halide or a mixture thereof. Such a catalyst system is particularly useful for the cyclotrimerization of conjugated diolefinic materials, for example, the preparation of trans, trans, trans-1,5,9-cyclododecatriene from 1,3-butadiene.

Tokuyama Soda

A process developed by K. Azuma, K. Shikata, and K. Yokokawa; U.S. Patent 3,399,184; August 27, 1968; assigned to Tokuyama Soda K.K., Japan yields a catalyst for producing alpha-olefin high polymers which is obtained by reacting a low valent titanium halide, a polyalkylhydrosiloxane, dicyclopentadienyl titanium, and if desirable, a dialkyl zinc. The low valent titanium halides and polyalkylhydrosiloxanes, the components making up the catalyst, are reagents that are readily obtainable commercially. On the other hand, dicyclopentadienyl titanium can be readily synthesized by reducing dicyclopentadienyl titanium dichloride in a hydrocarbon solvent using a sodium amalgam. The catalyst composition consisting of the reaction product of these three components has an activity whose duration is great as well as a fast polymerization speed. Further, the catalyst components used, except the metal halides, are all soluble in hydrocarbon solvents. Again, the fact that the procedure for synthesizing the catalyst is simple as well as that common equipment will suffice satisfies the requisites of commercial catalysts.

Ziegler

A process developed by K. Ziegler, H. Breil, H. Martin and E. Holzkamp; U.S. Patent 3,546,133; December 8, 1970;

154

yields polymerization catalysts for polymerizing olefins such as ethylene and its homologs comprising the product formed by mixing a heavy metal compound with an alkyl aluminum dihalide. The heavy metal compound is a compound of a metal of Groups IV-B, V-B, VI-B or VIII of the Periodic System or manganese, such as the salts, freshly precipitated oxide or hydroxide. The alkyl aluminum dihalide may be an admixture with a monohalide such as in the form of a sesquihalide.

A process developed by K. Ziegler, H. Breil, E. Holzkamp and H. Martin; U.S. Patent 3,574,138; April 6, 1971 involves a catalyst for polymerizing olefins such as ethylene and its homologues in the form of the product obtained upon mixing an aluminum triaryl or aralkyl with a salt, freshly precipitated oxide or hydroxide of a metal of Group IV-B, V-B or VI-B of the Periodic System, including thorium or uranium. Preferable salts are halides, such as titanium chloride or zirconium chloride.

VANADIUM-CONTAINING CATALYSTS

OXIDATION

American Cyanamid

A process developed by W.B. Innes; U.S. Patent 3,282,861; November 1, 1966; assigned to American Cyanamid Co. is one in which an exhaust gas catalytic composition, employing V_2O_5 and a copper compound such as copper oxide as necessary components, and preferably additionally employing at least one catalyst selected from the noble metals (1) overcomes the overheating problem specific to automobile exhaust gas converters, (2) is characterized by a low ignition temperature, thereby requiring only minor warming up periods, (3) obtains a high percentage conversion of both carbon monoxide and total hydrocarbons, particularly of ethylene and olefins of three or more carbon atoms, and (4) has a high stability to both lead poisoning and to steaming conditions while concurrently being highly active at a small cross-section of the catalyst particle.

According to this process, the oxidation catalyst comprises a major portion of alumina, and as catalytic agents, between about 2 and about 10% vanadia (V_2O_5) and between about 2 and about 10% of copper, or copper compound such as copper oxide, the percentage of copper being expressed in terms of the weight of elemental copper. Additionally, it is advantageous to include from between 0.01% (100 ppm) to 0.03% (300 ppm) of palladia, or an equivalent amount of some other noble metal. The noble metal normally is employed as a salt. The particular noble metal employed should be one which will more effectively catalyze the oxidation of hydrogen, carbon monoxide, ethylene, propylene, and higher olefinic components of the combustion exhaust gases at low temperatures, i.e., at temperatures less than 400° to 450°C., than the V_2O_5 catalytic component of the catalyst. Thus, according to this process, the exhaust gas converter catalytic composition includes vanadia and a copper compound, preferably copper oxide, as essential components, and advantageously, includes a minor percentage of one or more such as platinum, palladium, rhodium, iridium, ruthenium, and osmium. Other copper compounds typically include either or both cupric and cuprous salts of chloride, cyanide, nitrate, ferricyanide, ferrocyanide, fluoride, sulfide, sulfite, thiocyanate, acetate carbonate, oxides, and mixtures thereof.

The utilization of a catalyst in an automobile exhaust gas converter raises many problems not ordinarily encountered in the catalyst art. Accordingly, in order to have the required crush strength, stability to steaming and to lead poisoning, permeability to gas flow to prevent excessive back-pressure, catalytic activity for the conversion of hydrocarbons and carbon monoxide, low ignition temperature, and other requirements, it is critical that the catalytic composition have a pore volume of about 0.7 to about 0.9 cc/g., and a particle size having a minimum cross-section of about 1/20 inch (0.05 inch) to about 1/5 inch (0.20 inch).

The catalyst compositions of this process may be prepared by impregnating a suitable alumina base with a prescribed amount of V_2O_5, i.e., vanadium pentoxide, followed by impregnating with the second catalytically active materials described hereinabove. Alternatively, the impregnation procedure may be reversed, i.e., the second catalytic materials being the first impregnant, followed by impregnation with V_2O_5.

A second and important aspect of this catalyst composition relates to a composition which in essence may be described as a physical mixture of a V_2O_5 impregnated alumina with an alumina base material activated with the second catalytic material. This method of preparation and the catalyst compositions resulting therefrom result in substantial flexibility in catalyst compositions in that the relative amounts of catalytically active components may be varied quite readily by the simple addition or subtraction of one of the optional components.

Since units for different cars will require varying amounts of low temperature heat release, depending on location of the catalytic unit and exhaust composition, this flexibility is important. Additionally, where the catalytically active material includes a noble metal, by using physical mixtures of the two essential catalytic components, the noble metal is more easily recovered after use, which is an important economic advantage. As compared with V_2O_5-alumina catalyst heretofore used for the oxidation of hydrocarbon combustion exhaust gases, the catalyst composition has several marked advantages. It results in more effective oxidation of harmful ethane, propylene and carbon monoxide, in the environment of an automobile exhaust converter. The presence of at least one copper compound and preferably a third catalytic agent, as for example, platinum, results in a substantially faster warmup of the catalyst bed in comparison with that of V_2O_5-alumina catalyst. Faster warmups of the catalytic bed result principally from the oxidation of hydrogen and carbon monoxide by the second catalytic component which performs more effectively at low temperatures than the V_2O_5-alumina portion of the catalyst composition.

This oxidation at low temperatures of the above-identified constituents of combustion exhaust gases results in the rapid release of heat due to the oxidation of these constituents, which in turn causes the entire catalyst bed to warm up substantially immediately and thereby renders the V_2O_5-Al_2O_3 component active for hydrocarbon oxidation. Thereby one obtains the desirable low ignition temperature.

Random catalytic composition catalysts of solely a copper compound such as CuO, and/or solely a noble metal, although very active and highly effective oxidation catalysts for hydrogen and carbon monoxide, are in themselves not suited for catalytic muffler usage in the absence of the V_2O_5-alumina catalytic component. This is true because their use (without vanadia) requires the supply of sufficient additional oxygen to completely react with the carbon monoxide and hydrogen before very effective conversion or oxidation of hydrocarbon constituents of the exhaust gases can be realized. This means high air or oxygen requirements which in turn means large capacity air induction devices which are not always adaptable to an automobile exhaust gas converter system.

Additionally, where high oxygen requirements exist, the heat released from the rapid and near complete oxidation of carbon monoxide, hydrogen and to a lesser extent hydrocarbons, results in a rise of the temperature of the catalyst bed to very high temperatures which can damage the catalyst and the container thereof, as well as affect nearby components in an automobile or other apparatus employing an internal combustion engine if the container and exit pipe are not well insulated, as discussed above. Further, hot exhaust gases also present a safety problem, such as possibly causing a fire.

Moreover, catalysts containing noble metals (without V_2O_5) as an active catalytic component have a high susceptibility to lead poisoning from leaded fuels at high temperatures, and as is well known the noble metal catalysts are comparatively expensive.

The combination of vanadium pentoxide (V_2O_5) and a copper compound such as CuO, preferably with at least one catalytic noble metal such as palladium or platinum, with alumina base material in the amounts described has to a substantial extent overcome the disadvantages of either a copper compound or a noble metal catalyst employed alone. Thus, warmup times are substantially reduced, resulting in a catalyst composition that is effective in very short periods of time, and a catalyst is provided which is effective for ethene (ethylene), as well as higher olefinic hydrocarbons, and furthermore results in substantial carbon monoxide oxidation (up to about 40%) without requiring large capacity air induction devices. Additionally, such a catalyst maintains activity after long usage with leaded gasoline.

Chemical Process

A process developed by R.B. Egbert, F.F. Oricchio and T.J. Gluodenis; U.S. Patent 3,407,215; October 22, 1968; assigned to Chemical Process Corporation involves the production of phthalic anhydride (PAA) by fluid bed catalytic air oxidation of o-oxylene in which a bromine promoter is used in the vapor phrase together with a V_2O_5—SO_3—K_2O catalyst on a highly porous silica support having a surface area greater than 450 m.^{2}/g. and a pore diameter less than 100 A. and greater than 25 A. whereby yield, selectivity and catalyst loading are substantially increased.

Most PAA made commercially by vapor phase catalytic oxidation is made by oxidation of naphthalene with air. The trend during the last ten years has been towards large installations employing fluid catalytic beds. It has been recognized for some time that the production of PAA by the vapor phase oxidation of o-xylene rather than naphthalene would be economically advantageous due to the much lower cost of o-xylene in many cases.

However, the only commercial production of PAA by the vapor phase oxidation of o-xylene has been with a fixed bed catalyst, but this has the disadvantage of requiring individual reactors of small capacity and exceedingly high air ratios, e.g. 30 to 1 by weight and higher. High air to feed ratios present difficulties in recovery of the PAA and high installation and operating costs. Furthermore, individual reactors of small capacity further increase investment costs in large installations.

It has been long recognized that the use of a fluidized catalytic bed such as is used in the vapor phase oxidation of naphthalene would be highly desirable for the vapor phase oxidation of o-xylene to PAA in order to eliminate the aforesaid disadvantage of fixed catalytic beds. However, until recently no one has succeeded in finding a commercially feasible way of doing this, although a number of attempts have been made.

The basic problem has been that with the fluidized bed technique using the catalysts conventionally used for the vapor phase oxidation of naphthalene, the yields of PAA from o-xylene are too low for commercial feasibility. The reason for this is that the fluidized catalysts most suited for the vapor phase oxidation of naphthalene promote the oxidation and ultimate severance of the benzene ring when used for the vapor phase oxidation of o-xylene under reactor conditions required for reasonable conversion of the o-xylene. This results in the formation of CO, CO_2 and maleic anhydride at the expense of PAA. Essentially, then, the problem resides in poor selectivity in oxidation of both methyl carbons of the o-xylene as distinguished from the ring carbons under the conditions required for reasonable conversion of the o-xylene. Selectivity is defined as the ratio of oxidized o-xylene in which the methyl carbon or carbons of the o-xylene have been partly or wholly oxidized but neither of them removed but in which the benzene ring of the o-xylene remains intact, to the total o-xylene converted.

The process for the first time provides a commercially feasible method for the catalytic, vapor phase oxidation of o-xylene to PAA with an oxygen containing gas, particularly air, in a fluidized catalytic bed in which high PAA yield and high selectivities are achieved.

This is achieved by carrying out the reaction at relatively low temperatures and with relatively low air to o-xylene ratios in the presence of one or more of a particular class of reaction promoting agents in the vapor phase together with a particular class of porous silica-supported vanadium oxide catalysts. The promoting agent in the presence of such catalysts promotes the selective oxidation of the methyl carbons of the o-xylene to form phthalic anhydride at relatively low temperatures and at relatively low air ratios thereby providing high PAA yields and selectivities. Selectivities and yields are increased partially because the promoting agent permits the use of lower reaction temperatures with the particular catalyst, it being recognized that the higher the reaction temperature the greater the attack on the benzene ring.

The presence of such promoting agent with the present catalysts is not only selective in reducing oxidation of the benzene ring to a minimum, but it is also selective in suppressing oxidation of the o-xylene to benzoic acid and to incomplete oxidation products such as phthalide, o-toluic acid and o-tolualdehyde at the relatively low reaction temperatures used. The promoting agents of this process are selected from the group consisting of elemental bromine, hydrogen bromide and other bromine compounds which have the property of dissociating into bromine or HBr at reactor conditions.

The use of bromine and bromine compounds in the vapor phase catalytic oxidation of a large variety of aromatic and alkyl aromatic compounds, including benzene, naphthalene and o-xylene, to a large variety of oxidation products, including phthalic anhydride and maleic anhydride, has been suggested in U.S. Patent 2,954,385, filed October 2, 1957. This patent refers to the benefits of using such compounds in fixed as well as fluidized catalytic beds, and among a large variety of classes of catalysts it refers to unsupported vanadium oxide and also to vanadium oxide supported on a large variety of porous and nonporous supports, including silica. The use of large air/o-xylene ratios, relatively high temperatures, and very short contact times typical of fixed catalytic beds, are disclosed.

This disclosure has been of no help, but rather has been detrimental in the finding of a commercially feasible method of vapor phase, fluid bed catalytic oxidation of o-xylene to PAA because the use of bromine and bromine compounds with the particular vanadium oxide catalysts disclosed provide no discernible improvement over the PAA yield and selectivity achieved without the bromine, which are too low for commercial practicability.

Furthermore, the relatively high air ratios taught by that patent present the aforesaid problems referred to with respect to the fixed bed processes. Also it has been found that the presence of bromine has no effect, and in some cases a negative effect, on yield in the vapor phase, fluid bed oxidation of benzene and naphthalene. Furthermore, as the temperature of reaction and air/o-xylene ratio are increased to the temperatures and ratios disclosed in this patent, the benefits of bromine and bromine compounds in promoting selective oxidation of o-xylene to PAA decrease sharply, even with catalysts of the process. Although the earlier patent cited recommends porous catalyst supports with bromine compounds for ring rupturing oxidations and nonporous catalyst carriers for oxidation of alkyl side chains, it has been found that satisfactory results cannot be achieved in the vapor phase, fluid bed oxidation of o-xylene to PAA by the use of bromine promoters with nonporous supports or even porous supports lacking certain properties. In fact, it has been found that the use of bromine additives is effective only with catalysts having certain unique properties.

The particular class of fluid catalysts with which the bromine promoting agents are effective, consists essentially of vanadium oxides, particularly vanadium pentoxide and tetroxide, fluxed with a compound of SO_3 and alkali metal oxide. The catalyst is supported on a highly porous, hydrated, amorphous, particulate silica support, i.e., silica gel particles, having a surface area of not substantially less than 450 square meters per gram and a mean pore diameter which is not substantially less than 25 to 30 angstroms and which is preferably less than 100 angstroms, and more preferably less than about 80 to 90 angstroms. The catalyst surface area of the supported catalyst is preferably greater than about 100, more preferably greater than 160, square meters per gram for an equilibrium supported catalyst formed by applying the catalyst to solid particles of the silica and is preferably greater than about 40 or 45, more preferably greater than 60, square meters per gram for an equilibrium supported catalyst formed by adding the catalyst to the silica prior to its formation into solid particles.

Dainippon Ink Seizo

A process developed by H. Kakinoki, I. Kamata and Y. Aigami; U.S. Patent 3,215,644; November 2, 1965; assigned to Dainippon Ink Seizo K.K., Japan relates to the use as a catalyst of KHV_2O_6 or a composition principally of the same in preparing phthalic anhydride by oxidizing either ortho-xylene, naphthalene or a mixture thereof.

It has been known for a long time that the vanadium pentoxide type catalyst was useful as an oxidation catalyst of organic compounds, and various compositions thereof have been proposed. To obtain satisfactory results, it was however generally required that a specific catalyst be selected and used depending upon the type of material to be oxidized or the oxidation mechanism.

For example, as the catalyst for obtaining phthalic anhydride by air oxidation of naphthalene there have been known those in which vanadium pentoxide has been added to such as pumice, aluminum sponge, carborundum, silica gel, etc. That in which vanadium pentoxide has been caused to be absorbed in silica gel and referred to as the so-called German catalyst is known to be particularly excellent. It has been previously reported in a Publication Board Report by the Department of Commerce in Washington that a yield of 87% is possible when this catalyst is used.

However, even though this catalyst is used in the reaction for preparing phthalic anhydride by oxidizing ortho-xylene or a mixture of ortho-xylene and naphthalene, the yield is at most only less than 50%. In this case, unless an entirely different catalyst is used, good results are not obtainable.

It has been found that KHV_2O_6 or a composition containing the same as a principal component was used with very great advantage as a catalyst in preparing phthalic anhydride by oxidizing in a gaseous phase ortho-xylene, naphthalene or a mixture thereof.

According to this process, the KHV_2O_6 or the composition containing the same as a principal component is prepared generally by mixing a vanadium compound and a potassium compound in a ratio such that the molar ratio of vanadium to potassium (V:K) becomes 1:0.25 to 0.75 and reacting the two under acidic condition. In this connection, when the vanadium compound and the potassium compound are mixed such that the molar ratio of vanadium to potassium (V:K) becomes that in which the proportion of potassium is less than 1:0.25 or more than 1:0.75 and reacted under a condition similar as above, the proportion of those potassium-vanadium containing compounds other than KHV_2O_6 that are contained in the obtained catalyst composition becomes large. Consequently, it is undesirable in that either the yield of phthalic anhydride decreases, the life of the catalyst is shortened conspicuously or the coloration of phthalic anhydride obtained after a relatively short number of hours increases.

In preparing the KHV_2O_6 or a compound containing the same as its principal component, as the vanadium compound to be used as the starting material, any of the vanadium compounds such as vanadium pentoxide (V_2O_5), ammonium metavanadate (NH_4VO_3), etc. may be used. As regards the other starting material, the potassium compound, this also may be any of the potassium salts or potassium compounds such as potassium sulfate (K_2SO_4), potassium chloride (KCl), potassium carbonate (K_2CO_3), potassium nitrate (KNO_3), potassium acetate (CH_3COOK), potassium formate (HCOOK), etc. However, since the KHV_2O_6 that is used as the catalyst in the process is present only in an acidic and anhydrous state, it is necessary that the obtained KHV_2O_6 or the compound containing the same must be maintained under acidic conditions.

The KHV_2O_6 or a compound containing the same as its principal component can also be used in the gaseous phase oxidation reaction of ortho-xylene, naphthalene or a mixture thereof by causing their adherence to any of the known carriers such as activated alumina, fused alumina, titanium oxide, pumice, aluminum sponge, silica gel, gypsum, etc.

Therefore, in preparing the KHV_2O_6 or a composition containing the same as its principal component, the vanadium compound and the potassium compound are mixed together with a medium such as, for example, water such that the ratio V:K comes within the range of 1:0.25 to 0.75. Then this mixture is either coated to the surface of a carrier, a carrier impregnated with the mixture or the mixture and a carrier is mixed to a paste-like consistency. These are then dried under acidic conditions, preferably while feeding thereto an acidic gas, at above 40°C. before or after molding, and thereafter is burned at about 200° to 500°C. The catalyst may also be prepared by adding in excess to an acidic solution or suspension containing the aforementioned vanadium compound and potassium compound a reducing acid such as, for example, SO_2 gas or oxalic acid to make a vanadyl liquid, then coating or impregnating the carrier with the same followed by drying this at above 40°C. and molding, and thereafter burning at about 200° to 500°C.

Petro-Tex Chemical

A process developed by R.O. Kerr; U.S. Patent 3,288,721; November 29, 1966; assigned to Petro-Tex Chemical Corp. yields improved vanadium-phosphorus catalysts containing alkali metals which are particularly useful for the oxidation of hydrocarbons to dicarboxylic acid anhydrides. Although catalysts for the oxidation of butene to maleic anhydride have been suggested, these prior art catalysts have not proven to be satisfactory as commercial catalysts. Generally the catalysts have produced only low yields of product or have not had satisfactory catalyst life. It is one of the objects of this process to provide improved catalysts for the oxidation of olefins to dicarboxylic acid anhydrides at high yields. It is also an object to provide an improved catalyst which is effective to produce high yields of dicarboxylic acid anhydrides for prolonged periods of time. According to this process, it has been found that in the oxidation of hydrocarbons to dicarboxylic anhydrides the catalysts comprising particular ratios of vanadium, oxygen and phosphorus lose their catalytic activity due to the loss of phosphorus. It has further been found that the phosphorus may be stabilized by the addition of particular amounts of elements of Group Ia of the Periodic Table. The elements of Group Ia are the alkali metal elements. The catalysts which have the phosphorus stabilized are effective to produce maleic anhydride for very long periods of time at high yields.

The catalysts comprising vanadium, oxygen, phosphorus and Group Ia atoms are chemically combined in a complex. It is difficult to determine the exact chemical arrangement of the atoms. Oxides of vanadium and phosphorus are present when the catalyst is being used to oxidize the hydrocarbons to maleic anhydride. The atomic ratio of the phosphorus and vanadium should be about 1.0 to 2.0 atoms of phosphorus per atom of vanadium. Expressed in terms of the oxide, the ratio of P_2O_5 to V_2O_5 will be from about 1.0 to 2.0 mols of P_2O_5 per 1.0 mol of V_2O_5.

Preferably, the ratio of atoms of phosphorus to atoms of vanadium will be from about 1.1 to 1.6 atoms of phosphorus per atom of vanadium. The atomic ratio of the total atoms of Group Ia elements to phosphorus should be between about 0.003 and 0.125 atom of Group Ia elements per atom of phosphorus. The best results have been obtained when the ratio of Group Ia atoms to phosphorus atoms has been from about 0.01 to 0.060 atom of elements of Group Ia per atom of phosphorus. When the Group Ia atom is introduced into the catalyst preparation in the form of a compound, for example, as lithium hydroxide or potassium chloride, the weight of the Group Ia metal compound will be from about 0.05 to about 5.0 weight percent of the total weight of the vanadium, phosphorus and oxygen.

The atomic ratio of oxygen to the remaining components of the catalyst, when the catalyst is in the process of being used to catalyze the oxidation, is difficult to determine and is probably not constant due to the competing reactions of oxidation and reduction taking place during the reaction at the high temperatures. Perhaps at room temperature the ratio of oxygen to phosphorus may be about 2 to 5 atoms of oxygen per atom of phosphorus and the ratio of oxygen to vanadium may be from about 2 to 5 atoms of oxygen per atom of vanadium. The overall ratio of oxygen to the combined atoms of vanadium and phosphorus at room temperature then would be about 4 to 10 atoms of oxygen per combined atoms of vanadium and phosphorus.

The advantage of the addition of a Group Ia stabilizer may be demonstrated by making comparative runs with and without the Group Ia phosphorus stabilizer. A vanadium–phosphorus catalyst complex was prepared with an atomic ratio of phosphorus to vanadium of about 1.3 atoms of phosphorus per atom of vanadium. This catalyst was used to oxidize butene to maleic anhydride. The catalyst initially gave high yields, but after 300 hours of operation the yield fell off sharply. Another catalyst was prepared which contained as a phosphorus stabilizer 0.4 weight percent of lithium hydroxide based on the total weight of vanadium, oxygen and phosphorus. A marked increase in catalyst life was evident.

The catalyst may be prepared in a number of ways. The catalyst may be prepared by dissolving the vanadium, phosphorus and alkali metal compounds in a common solvent, such as hot hydrochloric acid and thereafter depositing the solution onto a carrier. The catalyst may also be prepared by precipitating the vanadium and phosphorus compounds, either with or without a carrier, from a colloidal dispersion of the ingredients in an inert liquid.

In some instances the catalyst may be deposited as molten metal compounds onto a carrier; however, care must be taken not to vaporize off any of the ingredients such as phosphorus. The catalyst may also be prepared by heating and mixing anhydrous forms of phosphorus acids with vanadium compounds and the alkali metal compound. The catalysts may be used as either fluid bed or fixed bed catalysts. In any of the methods of preparation heat may be applied to accelerate the formation of the complex. Although some methods of catalyst preparation are preferred, any method may be used which results in the formation of the catalyst complex containing the specified ratios of vanadium, phosphorus and alkali metal.

United Coke and Chemicals

A process developed by H.L. Riley; U.S. Patent 3,182,027; May 4, 1965; assigned to United Coke and Chemicals Company Limited relates to the manufacture of glasses which may be used in the production of catalysts in some reactions, for example the oxidation of polynuclear aromatic compounds, and especially the oxidation of naphthalene to phthalic anhydride.

In this process, vanadium pentoxide is dissolved in molten potassium pyrosulfate in a molecular ratio of $K_2O:V_2O_5$ between 1:1 and 6:1, and preferably about 4:1. The solution is allowed to cool and harden, and surprisingly a solid glass is formed. For the production of a catalyst this is crushed to form catalyst particles. The temperature at which the vanadium pentoxide is dissolved in the molten potassium pyrosulfate should not exceed 400°C. if excessive fuming is to be avoided. In fact solution will occur at temperatures below 300°C., but the rate of solution is much lower at such temperatures. In practice there is little fuming and the rate of solution is satisfactory if the temperature is about 350°C. Although solid potassium pyrosulfate may be melted and used in this process, it is advantageous to form it in situ by heating potassium sulfate and sulfuric acid. This may conveniently be done in a stainless steel vessel, and the vanadium pentoxide may be stirred into the melt while it is in this vessel. The vanadium pentoxide is preferably in the form of flakes, since these have been found to give rise to less fume than the powdered variety during production of the melt.

Solid potassium pyrosulfate may also be produced in situ from, for example, potassium chloride or potassium carbonate and sulfuric acid, but this method is generally less convenient. It is important to heat the molten glass to a temperature higher than that to which it will be subjected as a catalyst in order to drive off any excess of sulfur trioxide present. This ensures that little or no sulfur trioxide will be evolved when the catalyst is in use. Figure 71 shows in perspective and in vertical section a suitable form of apparatus for use in this process.

The apparatus shown comprises a stainless steel reaction vessel 1 housed in a casing 2, which is mounted in a frame 3 on trunnions 4. A handle 5 is provided for tilting the vessel. An electrical resistance element 6 surrounds the vessel 1 so that its contents may be heated externally, and the remaining space 7 between the vessel and the casing 2 is filled with slag wool or other thermal insulation. To prevent the highly corrosive molten glass from damaging the casing 2 or the resistance element 6, the reaction vessel 1 is provided with a long spout 8 extending beyond the edge of the casing, and with a tightly fitting stainless steel collar 9.

FIGURE 71:   APPARATUS FOR PRODUCING A VANADIUM-POTASH GLASS CATALYST BY A BATCH PROCESS

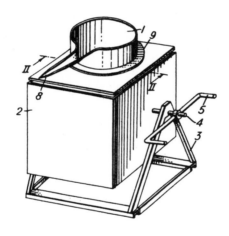

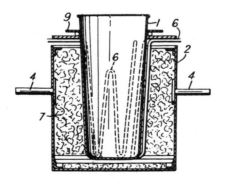

Source: H.L. Riley; U.S. Patent 3,182,027; May 4, 1965

The following example illustrates the conduct of this process and the following raw materials were used:

	Kg.
K_2SO_4 (powder, British Pharmacopoeia Codex quality 99%)	27.9
H_2SO_4 (98% chemically pure)	16.0
V_2O_5 (flake)	7.23

The potassium sulfate was slowly added to the sulfuric acid, which was gently heated in the stainless-steel vessel 1. The vessel was then heated to about 300°C., and the addition of vanadium pentoxide started. Starting with the vessel warm from a previous run, this preliminary operation (formation of pyrosulfate) occupied 2 1/2 hours, and the addition of the vanadium pentoxide required about 1/2 hour. The temperature of the melt was then raised to 400°C. This occupied a further 1 1/2 hours.

Pouring the melt into shallow stainless-steel trays was the next operation. If this is done at 400°C. excessive fuming occurs. This can be completely avoided by allowing the vessel to cool to 300°C. before pouring, which is time consuming. In fact the contents of the vessel were cooled to 300°C. by the addition of 13.5 kg. of coarsely crushed glass from a previous batch. When this had melted, the contents of the pot were poured out. This occupied about another 1/2 hr. Stirring during the addition of the vanadium pentoxide was carried out by means of a stainless-steel tube, closed at the bottom end, which also acted as the thermocouple sheath. The total time for making this batch, which weighed 47.6 kg. (excluding the glass added for cooling), was 5 hours.

The solid glass was then subjected to a preliminary crushing in a double-roller claw crusher. This reduced the glass to a maximum size of about 10 mm. The coarsely crushed material was then fed to a grinding mill consisting of two discs, containing pegs, one stationary and the other revolving at high speed. This reduced the material to less than 150 microns in size. It was found necessary to use an air purge on the grinding mill, as otherwise the heat developed caused the melt to soften and clog. The crushing and grinding operations occupied 1 1/2 hours.

As is stated above, it is important to heat the molten glass to a temperature higher than the temperature at which the catalyst will be used. The following table shows the effect on the composition of the melt of heating the glass of this example at 400°C.

	Percent SO_3	Percent V_2O_5	Percent V_2O_5/ Percent SO_3
Calculated for $4K_2S_2O_7:1V_2O_5$	53.4	15.2	0.28
Time of Heating, hrs.:			
0	51.4	14.8	0.29
1	52.0	15.3	0.29
2	50.8	15.2	0.30

It will be noted that there was little loss of SO_2 even after 2 hours heating at 400°C.

A process developed by H. Markham, P.H. Pinchbeck and P.P. Gaynor; U.S. Patent 3,591,525; July 6, 1971; assigned to United Coke & Chemicals Company Limited, England yields a catalyst consisting of a glass of vanadium pentoxide and potassium pyrosulfate absorbed by silica gel particles. Such a catalyst is produced by continuously introducing a preheated mixture of silica gel particles and particles of the glass into a hot fluidized bed of the particles. The catalyst, when used in the production of phthalic anhydride, gives a better yield than a similar catalyst produced by a batch process. Figure 72 shows the arrangement of equipment employed in the conduct of the process.

In this process a mixture of vanadium pentoxide and potassium pyrosulfate is melted, cooled to a glass, cast, ground to about 100 BS mesh, weighed and mixed with silica gel particles of 40 to 100 A. pore diameter and of particle size between 50 and 300 BSS mesh.

The mixture is withdrawn from a reservoir 1 by suction through a flexible pipe 2 in a closed hopper 3 in which the suction is set up by an air ejector 4 in a branch 5 from a compressed-air ring main 6. To prevent the discharge of dust particles into the atmosphere a filter 7 is interposed between the ejector 4 and the hopper 3. The hopper 3 discharges into a second closed hopper 8 through a pipe 9. The charging of the hopper 3 is intermittent, and while it is taking place a valve 10 in the pipe 9 is closed.

From the hopper 8 the mixture passes under the control of a continuously driven, rotary star valve 11 into an ejector 12. Compressed air from the ring main 6 flows through a meter 13, a preheater 14 and a pipe 15 to this ejector, and the mixture of particles is thus carried in a stream of hot air through a pipe 16 to the lower part of a column 17 containing a

FIGURE 72:   APPARATUS FOR PRODUCING A VANADIA-POTASSIUM PYROSULFATE GLASS CATALYST ON SILICA GEL CATALYST BY A CONTINUOUS PROCESS

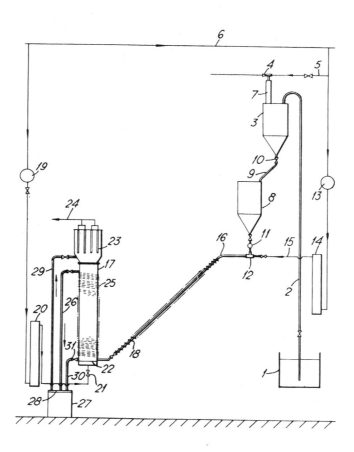

Source: H. Markham, P.H. Pinchbeck and P.P. Gaynor; U.S. Patent 3,591,525; July 6, 1971

fluidized bed of particles. The temperature in the pipe 16 is maintained at a little over 100°C. by a heating element 18 around the pipe. This is found to be advantageous in removing superficial or physically absorbed water from the particles of the silica gel before these enter the fluidized bed.

The bed is fluidized by compressed air which flows from the main 6 through a meter 19 and a preheater 20, which is externally heated electrically, to the base of the column 17 under the control of a valve 21 and emerges through a distribution ring or grid plate 22. This air flows upwards through the column to maintain the particles fluidized in the column. At the top of the column there are filters 23, through which the air passes to flow to a chimney through a pipe 24.

On entering the column 17, which is maintained at a high temperature by means of external electrical heating elements 25, the particles are rapidly heated to a temperature between 300° and 400°C., at which the glass of vanadium pentoxide and potassium pyrosulfate melts. Impregnation of the silica gel then occurs.

The product is removed from the top of the bed through an overflow pipe 26, which delivers it into a drum 27 through a cover 28, drums being successively brought beneath and closed by the cover. A pipe 29 runs from the cover 28 to the column 17 to allow the pressures to equalize. The reactor 17 may be completely emptied when required through a pipe 30 controlled by a valve 31.

As an example one column 17 was 11 feet high and 16 inches in diameter. The bed temperature was maintained at 370° to 380°C. and the air velocity was 1/2 foot per second in the reactor. A mixture of 4 parts of silica gel and 1 part of the powdered glass of vanadium pentoxide and potassium pyrosulfate was added at a rate of 200 lb./hr. through the pipe 16, which was 1 1/2 inches in diameter and maintained at 110°C., and in which air was flowing at a linear velocity of

10 feet per second. The product overflowing from the bed was suitable for use as a catalyst in the oxidation of naphthalene. The advantage obtained by means of the process is shown by the results obtained in a prolonged test. A silica gel was supplied by a manufacturer to the following specification:

pore volume	0.7 to 0.8 cc/g.
surface area	350 to 450 m.^{2}/g.
pore diameter	60 to 90 A.
size grading	
> 295 microns	10% maximum
295 to 152 microns	25% maximum
152 to 76 microns	remainder
76 to 53 microns	10% maximum
< 53 microns	5% maximum

Some of this silica gel was used with glass to make catalyst according to this process, and more of the same gel and the same glass were used to make catalyst by the batch process described in U.S. Patent 3,226,338. Both catalysts were then used to produce phthalic anhydride from naphthalene at optimum operating conditions for phthalic anhydride conversion.

With the catalyst produced by the batch process, the proportion of naphthoquinone initially produced was 0.8%, and after 6 months operation had increased to 2.0%. The reactor temperature then had to be increased in order to decrease the naphthoquinone proportion with a resultant reduction in phthalic anhydride yield. With the catalyst produced according to this process, the initial proportion of naphthoquinone was 0.4% and after 6 months this had only risen to 0.9%.

POLYMERIZATION

Farbwerke Hoechst

A process developed by H. Schaum and G. Hörlein; U.S. Patent 3,551,395; December 29, 1970; assigned to Farbwerke Hoechst AG, Germany relates to the manufacture of amorphous copolymers consisting of ethylene, higher α-olefins and, if desired, a small amount of a diolefin with modified organometal mixed catalysts in suspension. Alkoxy derivatives of perhalogenated open-chain or cyclic olefins are used as catalyst reactivators and the yields are so high that the complicated removal of catalyst from the polymer may be dispensed with. The products obtained are characterized by advantageous industrial properties.

The process provides for the copolymerization of ethylene with an α-olefin of the formula $R-CH=CH_2$, in which R represents an aliphatic linear or branched hydrocarbon radical containing fewer than 7 carbon atoms, if desired with such a quantity of a diolefin that the terpolymer contains thereof 2 to 8% by weight, preferably 2 to 4% by weight, in suspension, under a pressure from 0 to 30 atmospheres (gauge) at a temperature within the range from -30° to +50°C., with a coordination catalyst comprising a tri- to pentavalent vanadium compound and an organoaluminum compound, both of which are soluble in the dispersion medium used, while stirring or otherwise mechanically stirring the reactants about, which comprises carrying out the polymerization:

(1) in the presence of a halogenated hydrocarbon as dispersion medium in which the copolymer is insoluble under the reaction conditions,

(2) with the vanadium compound in a concentration from 0.001 to 0.1 mmol per liter of dispersant,

(3) at an Al:V-ratio from 20 to 200, preferably from 30 to 100, and

(4) in the presence of an alkoxy derivative of a perhalogenated, open-chain or cyclic olefin in which at least one double bond is in the α-position to the alkoxy group.

Monsanto

A process developed by M.R. Ort, E.H. Mottus, M.M. Baizer and D.E. Carter; U.S. Patent 3,546,083; December 8, 1970; assigned to Monsanto Company involves an electrolytic method for making olefin, e.g., ethylene, polymerization catalysts. These catalysts are made electrolytically from anodes of transition metals such as vanadium or manganese and metals such as aluminum or alloys of aluminum and vanadium or manganese, a methylene dihalide especially methylene dichloride and an electrolyte such as $HOAlCl_2$.

Also, the transition metal, e.g., V or Mn, catalyst component can be made separately as can the nontransition metal, e.g., Al component. Preferably, a small amount of an olefin substantially inert toward polymerization, i.e., a nonreactive olefin such as cyclohexene, is used during the electrolysis to promote conductivity. Also, preferably all materials are substantially pure and are dry during electrolysis and polymerization, except for small known amounts of water which may be added to promote conductivity in electrolysis and/or the polymerization. It is also preferred to blanket the electrolysis cell with an inert gas, such as nitrogen, to exclude oxygen which may tend to poison some of the catalysts. The catalysts are normally soluble in the methylene dihalide medium in which they are made at least in sufficient concentration to be useful as polymerization catalysts. Figure 73 shows the apparatus used in the integrated catalyst preparation-polymerization process.

Vessel 1 is an electrolysis cell, 2 is an anode made of an alloy of aluminum and vanadium having about 2.5% vanadium therein, 3 is an aluminum cathode and stirrer 11 is provided for agitation. Line 21 is for the purpose of introducing nitrogen to blanket the reaction in the electrolysis cell and line 22 is for the purpose of venting nitrogen. Instead of anode 2, two separate anodes can be used, one of aluminum and the other of vanadium with the ratio of the surface area on the vanadium anode to the aluminum anode being such as to give about the same catalyst composition as the 2.5% vanadium alloy. Into the electrolysis cell through line 20 is introduced a preelectrolysis solution consisting of methylene dichloride, aluminum trichloride and an equimolar amount of water based on the aluminum trichloride plus a small amount of cyclohexene to promote conductivity in the cell. The direct current flowing through the cell is adjusted to a sufficiently high level to avoid the formation of any substantial amounts of strictly chemical catalyst.

Catalyst is charged via line 23 to polymerization vessel 4 which is agitated by stirrer 5. Through line 24 ethylene, containing about 20 volume percent hydrogen, is continuously charged to vessel 4 and through line 25 makeup hexane is charged to vessel 4. From the top of the polymerization vessel through line 26 ethylene which has not been polymerized plus some vaporized dichloromethane and hexane flow to condenser 6. From condenser 6 gaseous ethylene goes through line 27 to compressor 7 which delivers the ethylene back to polymerization vessel 4 through line 29. Through line 28 condensed hexane and methylene dichloride from condenser 6 go to enclosed basket centrifuge 8 for the purpose of washing centrifuged polyethylene cake. Alternatively, if it is decided not necessary or desirable to wash the polyethylene cake in the centrifuge, the condensed hexane and methylene dichloride can be returned directly to the polymerization vessel.

From the bottom of polymerization vessel 4 a slurry of polyethylene in hexane and methylene dichloride is taken through line 30 and introduced to centrifuge 8. From centrifuge 8, via line 31, hexane and methylene dichloride containing catalyst which has been separated from the polyethylene is returned to the polymerization vessel and, under such conditions, only makeup catalyst need be added to the polymerization vessel via line 23. Alternatively, if it is not desired to reuse the catalyst recovered from the centrifuge, the hexane and methylene dichloride containing the catalyst can be distilled to remove the catalyst, the solvent condensed and returned to the polymerization vessel.

Polyethylene separated from the slurry in the centrifuge is withdrawn from the centrifuge via line 32 and goes to dryer 9. In dryer 9 the hexane and methylene dichloride remaining with the polyethylene are evaporated and taken via line 34 to condenser 10. The condensed methylene dichloride and hexane from condenser 10 are returned to the polymerization vessel via line 35. From dryer 9 the dried polyethylene product is removed via line 33. An alternative method of operating is to bypass centrifuge 8 with the slurry in line 30 and charge the slurry directly to dryer 9.

FIGURE 73:   FLOW DIAGRAM OF OLEFIN POLYMERIZATION PROCESS USING ELECTROLYTICALLY-PRODUCED VANADIUM-ALUMINUM CATALYST

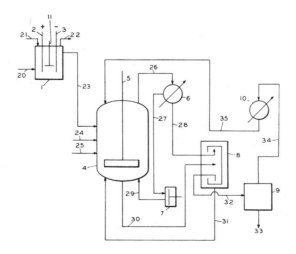

Source: M.R. Ort, E.H. Mottus, M.M. Baizer and D.E. Carter; U.S. Patent 3,546,083; December 8, 1970

Solvay & Cie

A process developed by A. Delbouille and H. Toussaint; U.S. Patent 3,479,326; November 18, 1969; assigned to Solvay & Cie, Belgium is one in which olefins are polymerized and copolymerized in the presence of a catalyst constituted of the reaction product of a compound of an element of Group IVb of the Periodic Table (such as TiCl4), a compound of an element of Group Vb of the Periodic Table (Vanadium halide or alkoxide) and a copolymer of vinyl alcohol containing 1 to 20 mol percent of polymerized vinyl alcohol and an activator which is a metal of Groups I to III of the Periodic Table or a hydride or organometallic compound of the Groups I to III metals.

CHROMIUM-CONTAINING CATALYSTS

AUTOMOTIVE EXHAUST TREATMENT

Universal Oil Products

A process developed by T.V. De Palma and M.W. Perga; U.S. Patent 3,594,131; July 20, 1971; assigned to Universal Oil Products Company utilizes a converter for catalytic conversion of fluid streams, such as internal combustion engine exhaust gases, which embodies a catalyst retaining section having a reservoir section therein. The reservoir section is established by a partition that prevents flow of fluid through one part of the catalyst section. A flanged end of a movable perforate partition spans the catalyst bed and keeps all the catalyst material in a compacted state. As catalyst material shrinks or is lost by mechanical attrition or chemical spallation, the flanged end of the movable perforate partition forces fresh material from the reservoir section to an area of the bed where fluid flow exists. In a modified embodiment, the movable perforate partition exposes additional catalyst material to fluid flow at the other end of the catalyst section as catalyst material is lost.

Suitable oxidation catalysts for use in reducing exhaust gases of an internal combustion engine includes metals of Groups I, V, VI, VII, and VIII of the Periodic Table, particularly chromium, copper, nickel, and platinum. These components may be used singly, or in combination of two or more, etc., and will generally be composited with an inorganic refractory oxide support material, such as alumina, silica-alumina, silica-alumina-zirconia, silica-thoria, silica-boria, or the like.

The physical shape of the catalysts may be such that they are in the form of spheres, cylinders, or pellets, typically having a dimension of one-sixteenth to one-quarter inch, although larger particles or smaller particles may be employed where desirable. Mixed sizes of catalyst material may also be well utilized especially as a means to provide for a low temperature catalytic oxidation process. Also, the catalytic material may be in the form of impregnated fibers which, in turn, may be placed in a mat-like bed arrangement. It is also contemplated that the catalytic material be formed into a rigid material corresponding to that of the catalyst section.

Figure 74 shows one form of catalytic device which may be used in the process. The converter is shown to include an outer housing 60 which has an elongated rectangularly shaped tubular section 61, to which are connected end closure sections 62 and 63, respectively. They have flanged ends to permit various types of connections. Within outer housing 60 is a fixed perforate partition 64 having perforations or apertures 65 therethrough. Partition 64 may be attached to the outer housing in various ways.

In other words, it may be welded along the entire interior of housing 60 or it may be welded at one end and supported at the other edges via grooves, thereby providing a rigid but slideable fit. The slideable fit is preferred, since it permits independent expansion of the partition. Also within the outer housing is the movable perforate partition 67 with the apertures or openings 68 therethrough. This movable partition 67 has a flanged end 70 which extends adjacent to the fixed perforate partition at 71. A fixed blocking partition 72 is disposed adjacent to the movable perforate partition 67 in proximity to the flanged end.

This blocking partition overlaps a portion 73 of the movable perforate partition to thereby block flow of fluids through that portion. Thus, in effect, a reservoir of catalyst material is formed at that end of the catalyst retaining section. The perforate partitions as well as the blocking partition may extend to the side of the housing 60. If that be the case, the housing itself will serve as part of the end closure means. To completely enclose the catalyst retaining section 75, a fixed perforate closure partition 76 is attached to the outer housing and disposed adjacent the movable perforate partition 67 to overlap the movable perforate partition over a portion 77. This perforate closure partition may be imperforate if desired. Of course, located within catalyst retaining section 75 is catalyst material 79. The movable perforate partition in this

FIGURE 74:   SECTION OF CATALYTIC CONVERTER FOR AUTOMOTIVE EXHAUST GASES WITH MECHANICAL MEANS FOR KEEPING CATALYST BED COMPACT

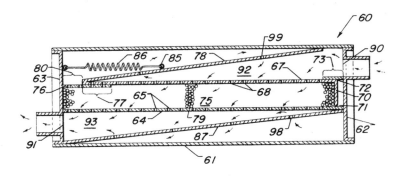

Source: T.V. De Palma and M.W. Perga; U.S. Patent 3,594,131; July 20, 1971

embodiment has a transverse portion 78 extending from the perforate face of the catalyst retaining section to prevent flow of fluids through portion 80 of the fixed closure perforate partition. Transverse portion 78 also serves as a means to vary the cross-sectional area of passageway 92. The portion 78 extends completely to housing 60 and is movable thereon. Attached to portion 78 at 85 is a tension spring 86 which in turn is attached to end section 63 of the housing. The tension spring serves as the pressuring means to maintain the flanged portion 70 in pressure contact with the catalyst material within catalyst retaining section 75. Tension spring 86 may be complemented with adjustment means (not shown) which could be used for varying the tension or the pressure on the catalyst material within the catalyst retaining section 75. On the opposite side of the catalyst retaining section is located a distribution plate 87 which varies the cross-sectional area of passageway 93. Ports 90 and 91 are located in end sections 62 and 63, respectively, to provide communication into the interior of housing 60.

As is, the converter may be used with the incoming fluid coming through port 90 or with the incoming fluid coming through port 91. For sake of simplicity, it is assumed that the fluid enters through port 90; then as the fluid passes through port 90 it enters a passageway or manifold 92 which is defined by the face of the perforate partition 75 and by the face of portion 78 of movable partition 67. Since passageway 92 decreases in cross-sectional area in the direction of flow, the velocity head of the incoming fluid will develop into a uniform pressure head across the face of the catalyst retaining section.

The uniform pressure head will also be effectuated by the fact that the passageway 93 defined by plate 87 and the opposite face of the retaining section is increasing in cross-sectional area in the direction of flow. Thus, the uniform pressure head will create uniform flow through the catalyst material in catalyst retaining section 75. As shrinkage, attrition, or chemical spallation reduces the volume of catalyst material within the retaining section 75, the flanged portion of perforate plate 67 will be moved by the action of spring 86 to compact the catalyst material and thus prevent any bypass.

Actually, new material or fresh catalyst particles within the reservoir section will actually be introduced to contact the flow of fluids. Also since the movable partition 67 is moving at the same time as flanged portion 70 the overlapped portion 77 of fixed closure partition 76 will become larger. Since in this embodiment the fixed closure partition 76 is perforated, the perforations in plate 67 will be exposed to more perforations in plate 76. Thus, the effective size of the catalyst retaining section 75 becomes greater as particles are lost. It is to be noted that portion 78 and plate 87 have spaced apart openings 99 and 98 provided therein.

These openings are utilized to create a muffling effect within the converter.   Thus, the converter will serve as a muffler when utilized on the exhaust stream of the internal combustion exhaust engine. From the foregoing description, it is seen that this particular converter is of such a design that little or no bypass of fluids will occur when catalyst material is lost. It is also seen that since the catalyst reservoir is not defined by a perforated plate, it in itself will not cause mechanical breakdown or crushing of catalyst materials. In addition, it is to be noted that the particular converter may be constructed in such a manner that the various components will be capable of expanding and contracting relative to each other as the temperature of the apparatus rises. Also, it is seen that as catalyst is lost the actual size of the catalyst retaining section becomes greater.

It is desirable that the components within the converter be made of lightweight relatively thin gauge material, whether of ordinary steel or an alloy, such that the assembly is relatively lightweight and such that the temperature effects may also be accommodated by some material flucture without causing breakage of seams and joints. The material used should also be of a character that is able to withstand the high temperatures that may result from the operation of a converter.

A process developed by T.V. De Palma and R.S. Carleton; U.S. Patent 3,644,098; February 22, 1972; assigned to Universal Oil Products Company relates to an improved form of a converter adapted to hold a bed of solid contact material to treat an exhaust gas stream. More specifically, the apparatus provides the means for the catalytic conversion and purification of the exhaust gases from an internal combustion engine, with the utilization of a design that provides for easy placement and removal of the catalytic bed and also that overcomes the problem of heat losses by being adapted for direct connection to the exhaust ports of the engine.

Figure 75 shows such a device in longitudinal and transverse section. Outer housing 1 in this particular arrangement has been formed by a casting operation known to those skilled in the art, and is of a suitable material adapted for casting, such as an iron alloy. The front face of outer housing 1 has flanged edges 2 provided around its entire periphery. Flanged edges 2 are provided with spaced-apart threaded holes 3, to be used for connection of the cover plate. The cover plate 4 is designed to completely enclose the space within housing 1 and has holes 5 thereon for inserting of bolts 6. A gasket or other suitable sealing means 7, made of a material capable of withstanding temperatures of 2000°F., is provided between cover plate 4 and the flanged surfaces or edges 2.

FIGURE 75:   SECTION VIEWS OF AUTOMOTIVE EXHAUST CONVERTER ATTACHED DIRECTLY TO ENGINE MANIFOLD

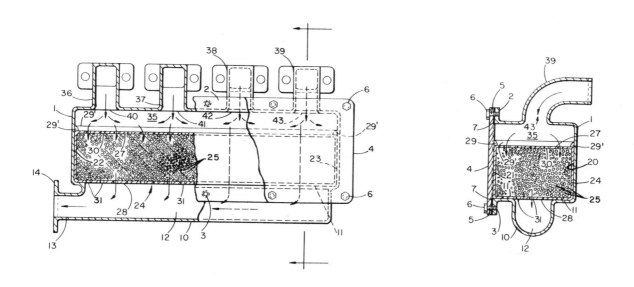

Source: T.V. De Palma and R.S. Carleton; U.S. Patent 3,644,098; February 22, 1972

In this particular embodiment the outer housing 1 has a lower longitudinal section 10, which is of a narrower cross-section than is the main rectangular portion of housing 1. Thus is provided an internal surface 11, which serves as the surface of placement for the catalyst cartridge. The space 12, established by lower longitudinal extension 10, provides the outlet manifold section for the treated exhaust gases. A narrower cross-section outlet pipe means 13, having flanged portions 14 for direct connection to the conventional exhaust system of an internal combustion engine, is located at the left end of manifold section 12, although it may be disposed at any location thereof.

Located within outer housing 1 and designed to abut walls 20, 21, 22, and 23, thus preventing any quantity of exhaust gases from passing thereby, is catalytic cartridge 24, shown containing subdivided catalyst particles 25. For most efficient converter operations, the cartridge should be filled to capacity. Suitable oxidation catalysts include the metals of Groups I, V, VI, VII, and VIII of the Periodic Table, particularly chromium, copper, iron, nickel, and platinum. These components may be used singularly, in combinations of two or more, etc., and will generally be composited with an inorganic refractory oxide support material, such as alumina, silica alumina, silica alumina zirconia, silica-thoria, silica-boria, and the like. The cartridge, as shown has imperforate sidewall portions abutting walls 20, 21, 22, and 23 of housing

1. The top and bottom walls 27 and 28, are perforate wall sections which have been attached permanently to the sidewalls, after the catalyst particles 25 were inserted. Although not shown, it is contemplated that the construction of cartridge 24 will be such that the construction will prevent buckling during the operation of the converter. This may be accomplished in various ways. For example, the walls of the cartridge may be accordioned along the length so as to absorb any expansion due to temperature differentials. Also contemplated as being within the scope of this improvement is a cartridge-type construction that embodies slotted wall sections to permit the various sections to be slidably supported within each other. In addition, it is contemplated that instead of having permanent enclosed particles, the cartridge is to be constructed in such a way as to permit opening and closing of at least one wall portion, thereby permitting access to remove contaminated catalyst particles and replace them with rejuvenated particles.

The cartridge is inserted into housing 1 via the opening in the front of the housing 1, and placed on surface 11 to abut housing 1 at 20, 22, and 23. Also noted are projections 29 located on cover plate 4 and projections 29' located on the inner walls of outer housing 1. These projections serve to hold the cartridge 24 onto surface 11 when outer housing 1 is sealed by cover plate 4 and serve to prevent exhaust gases from passing thereby. Gasket 7 is then inserted onto flanged pieces 2, cover 4 is inserted over this gasket, and both are bolted into place via bolts 6, to abut cartridge 24 at 21. It is to be noted that the cartridge should not be so tightly abutted against the walls of the housing as to foreclose any expansion.

Horizontal plate 28 has apertures 31 provided therein. It is also noted that, similarly, plate 27 has apertures 30 located along its length. Thus is established a cartridge having an inlet perforate wall section 27 and an outlet perforate wall section 28 to establish flow of the exhaust stream through the subdivided catalyst particles 25. The cartridge is disposed within housing 1 so that the top inlet perforate wall section 27 is spaced from the housing and particularly from the inlet means introducing the exhaust steam into the housing, thereby establishing an inlet manifold section 35.

The inlet means comprises in this particular embodiment four conduits 36, 37, 38, and 39, which are spaced apart so as to coincide with the exhaust ports of an internal combustion engine. Necessarily, flanged portions are provided on each conduit so as to establish a connecting means to the exhaust ports of the engine. It is noted that these conduits have been cast in this particular embodiment at the same time as the outer housing 1, thus establishing a one piece construction. These conduits communicate with inlet manifold section 35 via ports 40, 41, 42 and 43 through the top of housing 1.

DEHYDROGENATION

Chevron Research

A process developed by B.F. Mulaskey, H.F. Harnsberger and R.H. Lindquist; U.S. Patent 3,346,658; October 10, 1967; assigned to Chevron Research Company relates to chromia-alumina catalysts and their preparation for use in processes of the cyclically operated, so-called adiabatic, fixed bed type, for dehydrogenating butane and mixtures of butane and butenes and butadiene.

In the so-called adiabatic, fixed bed, butane dehydrogenation process, normal butane or a mixture of butane and butenes at elevated temperatures of 900° to 1200°F. are passed at subatmospheric pressure in the range of 1 to 10 psia through a bed of chromia-alumina catalyst particles preheated to the reaction temperature, at a space velocity of 0.5 to 3 volumes per volume of catalyst per hour. Frequently, heat retentive, catalytically inert, refractory solid particles are mixed with the catalyst particles, the heat required for the endothermic dehydrogenation reaction being abstracted from the preheated catalyst and inert solids.

The temperature of the catalyst bed decreases as it gives up heat absorbed by the reaction, and at the same time a carbonaceous deposit or coke is laid down on the catalyst. The feed is stopped after an on-stream period of 5 to 30 minutes. The bed is then contacted with a heated stream of oxygen-containing gas at 900° to 1200°F. for an equivalent length of time, which serves to burn off the carbonaceous deposit to regenerate the catalyst, and thereby to restore the bed to the initial elevated temperature, where the cycle is repeated. Thus, the catalyst is alternately and repeatedly exposed at frequent intervals to hydrocarbon vapors under conversion conditions and to oxygen-containing gases under regeneration conditions.

Usually several reactors are used in parallel, and while one reactor is in the reaction period or cycle, another is in the regeneration period. The requirements for a catalyst to be used in such a process are very stringent, for the catalyst must be extremely rugged to withstand the repeated oxidation and reduction and yet have good activity. It is desirable that the catalyst have a favorable coke:conversion ratio because the heat released when the coke is burned is retained in the catalyst bed to supply heat for the endothermic dehydrogenation reaction. The catalysts used commercially are composed essentially of chromia (Cr_2O_3) supported on alumina, and they may be promoted with a minor amount of an alkali metal oxide such as sodium or potassium oxide to improve selectivity. An unusual feature of the process is that, where in most catalytic processes it is desirable to use the most active catalyst available, in the cyclically operated fixed bed butane dehydrogenation process it is not desirable to use the most active catalyst. There is a great danger in the process that, if excess coke is produced, the temperature reached in the catalyst bed during regeneration will be so excessively high as to severely damage the catalyst and/or equipment. It is a characteristic of the process and catalysts that coke production increases with increasing temperature and increasing activity. Therefore, if excess coke is ever produced, causing a high temperature on regeneration, in the subsequent conversion cycle there will be an increased production of coke, leading to a still higher temperature on regeneration, and so on, in a runaway self-destructive manner.

When starting up the process with a fresh catalyst charge, it has frequently been found that the catalyst is overly active or has an unfavorable coke:conversion ratio such that it is necessary to use low feed and air temperatures and low per-pass conversions for a time ranging from a few days to a week or more until the catalyst has lost its initial high activity. As the catalyst loses activity, the temperature can be raised to increase conversion and/or to compensate for a lower coke:conversion ratio. The activity of the catalyst may then reach a stable desired activity, but usually with known prior art catalysts the activity continues to decline, and the temperature must be continually increased to maintain desired production rates.

In a particular process, therefore, the catalyst used should have a desired activity within a safe operating range of activities, and preferably not exceeding a maximum safe activity. The desired activity, safe operating range of activity, and maximum safe activity are not absolute intrinsic properties of the catalyst, but depend also on the manner in which the process is operated. Since in processes of this type the catalyst particles are often mixed with inert heat retentive solid particles in the catalyst bed, the activity desired of the catalyst per se depends on the relative amounts of catalyst and inert used. The more active the catalyst, the more it can be diluted with inert solids. When the catalyst activity declines, however, the apparent activity decline is magnified in proportion to the amount of inert solids used, and the observed general rule has been that the higher the catalyst activity, the more rapidly it will decline.

Thus, the most desired property for the catalyst itself is that it have stable activity, provided that this stable activity is adequate to maintain desired conversion at fixed dilution with inert solids, and provided that the coke:conversion ratio is suitable. The desired conversion may also depend on plant circumstances, and the desired catalyst activity may vary accordingly. Where butane is expensive, for example, lower per-pass conversion may be economically more attractive than high conversion operation because better selectivity may be obtained, producing more butadiene and less light gases per pound of butane.

Catalysts of highly unusual and desirable properties can be prepared by heat treating a high surface area alumina support at 1100° to 1600°F. for 2 to 24 hours to reduce its BET (nitrogen absorption) surface area, then impregnating with sufficient chromium compound decomposable to Cr_2O_3 to give 25 to 40 weight percent Cr_2O_3 on the finished catalyst, and then heating the chromium-impregnated alumina in an oxygen-free atmosphere at 1100° to 1700°F. for 2 to 48 hours to reduce the active chromia surface area, as measured by CO chemisorption, to less than 15 micromols CO per gram of catalyst.

In this way, catalysts were prepared which had initially lower activity than previously used catalyst, but the new catalysts were characterized by their activity increasing (instead of declining) during an initial period of use in the butane dehydrogenation process, by their activity thereafter remaining relatively stable, and by their activity ultimately declining more slowly than that of the previously used catalysts. It has been found that even in the case of these new improved catalysts the abovementioned problem of too high an activity can sometimes arise, by virtue of the catalyst activity increasing too much during the initial period of use. The process then provides methods for adjusting the initial activity of chromia-alumina catalysts, including the new catalysts described immediately above, such that too high an activity is prevented from being attained in the process.

In accordance with the method, a chromia-alumina catalyst which would normally be overly active for butane hydrogenation, when first used in a fixed bed cyclic dehydrogenation process, is prepared for use in the process by treating the catalyst with flowing hot vapor free of oxygen at between 1400° and 1800°F. until the butane dehydrogenation activity of the catalyst has been decreased to lower than desired for use in the process, and then treating the catalyst with flowing hot vapor containing oxygen at between 1000° and 1800°F., but preferably not hotter than treated in the preceding step, until the butane dehydrogenation activity of the catalyst has been increased to an activity desired for use in the method.

In one embodiment, the treating with hot vapor free of oxygen, and the treating with hot vapor containing oxygen, are controlled with respect to temperatures and times so as to adjust the butane dehydrogenation activity of the catalyst to substantially the maximum activity at which excessively high temperatures are not reached in the catalyst bed during regeneration in startup operation of the method at desired high per-pass conversion operation, i.e., the maximum safe activity.

Phillips Petroleum

A process developed by E.W. Pitzer and H.R. Sailors; U.S. Patent 3,219,587; November 23, 1965; assigned to Phillips Petroleum Company relates to an improved catalyst regeneration step in a butane dehydrogenation system in which n-butane is dehydrogenated in contact with a particulate catalyst consisting essentially of alumina and chromium oxide Cr_2O_3. In one plant of this type, operation involves a cycle including about one hour on dehydrogenation (on-stream) followed by an hour on regeneration. Thus, only about one half of the investment in catalyst and chamber capacity is actually in productive use at any given time. By changing the regeneration method so that only 5 minutes or less is required for each catalyst regeneration allows at least twelve times as much time on dehydrogenation as on regeneration when using a one hour dehydrogenation period. Thus, the use of the improved regeneration scheme of the method proportionately increases the production of butenes for a given amount of catalyst and reaction chamber capacity.

The advantage of the short regeneration time requirement may also be taken partially in the form of a shortened process cycle with the attendant increase in efficiency and yield of olefins and partially in the form of reduced catalyst and reactor requirement. In this modification, the dehydrogenation period can be 15 minutes and the regeneration period one to five minutes. Improvement in selectivity of conversion in the shortened conversion period results in a higher ultimate yield and the increased percentage of the process cycle spent on dehydrogenation reduces the catalyst and reactor requirement needed. The optimum ratio of time spent on dehydrogenation to time on regeneration is a matter of economics and can be determined by well known methods.

For the practice of this method, it is necessary that the reactor be of such shape that the regeneration gas path is short enough to enable regeneration gas to pass through the catalyst bed in the required quantities without exceeding the velocity of sound. In n-butane dehydrogenation, reaction conditions include a temperature in the range of 1000° to 1200°F., a pressure below 50 psig, a space velocity in the range of 500 to 10,000 v./v./hr., and a catalyst consisting essentially of alumina and Cr_2O_3 in which the Cr_2O_3 may vary from 10 to 60% by weight. The preferred catalyst composition is 80% Al_2O_3 and 20% Cr_2O_3 (by weight).

The operation of the process involves contacting coked catalyst with regeneration gas containing free-O_2 in the range of 5 to 35 volume percent, preferably air, at sufficiently high space velocity that the increased velocity decreases the regeneration temperature substantially over maximum regeneration temperatures which are produced by regeneration with air at low space velocities. The method is based on the fact that increasing the space velocity of regeneration air charged to coked catalyst beds results in increasing regeneration temperatures only up to a limit, and that further increases in space velocity result in decreasing regeneration temperatures. This decrease in temperature with increased space velocity is due to increasingly incomplete utilization of the oxygen in the air as the contact time of the air decreases below a critical value. Thus, the heat released for each unit of regeneration air can be controlled so that the exit temperature of the air does not exceed that at which the catalyst starts to be damaged.

HALOGENATION

Daikin Kogyo

A process developed by K. Kato and S. Ogawa; U.S. Patent 3,431,067; March 4, 1969; assigned to Daikin Kogyo Kabushiki Kaisha, Japan is one in which anhydrous chromium fluoride is heated at 300° to 750°C. in a stream of a gaseous mixture of molecular oxygen and water vapor to provide an improved activated amorphous fluorination catalyst. At least 6 weight percent of water vapor is used based on the weight of chromium fluoride being activated.

Previously, it has been known that some halohydrocarbons can be fluorinated by vapor-phase reaction with hydrogen fluoride in the presence of a metallic halide fluorination catalyst. In British Patent 799,335 it is disclosed that chromium oxyfluoride is effective in catalyzing such vapor-phase fluorination reactions. The chromium oxyfluoride has been prepared by heating hydrated chromium fluoride in the presence of molecular oxygen at an elevated temperature, 350° to 750°C. It is also disclosed that it is essential to employ hydrated chromium fluoride as the material to be activated, because an effective fluorination catalyst cannot be prepared from anhydrous chromium fluoride, CrF_3.

However, as described in British Patent 821,211, when certain halohydrocarbons are fluorinated with hydrogen fluoride over the chromium oxyfluoride catalyst obtained by the above described process, the catalytic activity thereof is markedly reduced in a relatively short time due to the formation of carbon deposits on the catalyst, resulting in decreased catalyst life. Although the contaminated catalyst may be regenerated by contacting it with oxygen at 500°C. this requires an interruption of the fluorination process and usually requires up to 2 hours to regenerate the catalyst satisfactorily, making its commercial use difficult. Moreover, the employment of the catalyst in fluorination processes requires a relatively high temperature if it is desired to obtain the fluorinated products in a high yield.

Investigations have revealed the possibility of obtaining a fluorination catalyst having a more extended life and higher catalytic activity from anhydrous chromium fluoride which has been considered incapable of providing an activated fluorination catalyst in the aforementioned prior art. In this method, the material to be activated is anhydrous chromium fluoride, which may be prepared by any of the conventional methods. For instance, it may be prepared by fluorinating chromium trioxide (CrO_3) with an excess amount of hydrofluoric acid in the presence of a reducing agent and heating the resultant hydrated chromium fluoride at a temperature of higher than 600°C. The anhydrous chromium fluoride may be mixed with graphite and/or silicon dioxide, shaped into desired forms such as tablets, pellets, etc., and then subjected to the activation process. The graphite, silicon dioxide, or mixture thereof, is employed in the range of from 1 to 10 weight percent, preferably from 2 to 4 weight percent, based on the weight of anhydrous chromium fluoride employed. The silicon dioxide employed alone, or in admixture with graphite, facilitates the prevention of corrosion of the activating reactor and helps to increase the porosity of the catalyst. Silicon dioxide may be further added to the CrF_3 tablets or pellets so as to be interspersed among them in the range of less than 40 weight percent, preferably from 5 to 20 weight percent, based on the weight of the CrF_3 tablets or pellets.

Another process developed by K. Kato and S. Ogawa; U.S. Patent 3,455,840; July 15, 1969; assigned to Daikin Kogyo Kabushiki Kaisha, Japan involves the use of an improved fluorination catalyst comprising (1) an activated chromium oxyfluoride prepared by heating chromium fluoride at from 300° to 750°C. while supplying thereto a continuous stream of a gas mixture containing molecular oxygen and water vapor, the water vapor being supplied in the range of at least 2 weight percent in total amount based on the weight reduced to anhydrous chromium fluoride and (2) 1 to 50 weight percent of at least one rare earth compound selected from the group consisting of a rare earth fluoride, rare earth chloride and rare earth oxide, based on the weight reduced to anhydrous chromium fluoride.

POLYMERIZATION

Montecatini Edison

A process developed by P. Saccardo, G. Trada, V. Fattore and J. Herzenberg; U.S. Patent 3,412,040; Nov. 19, 1968; assigned to Montecatini Edison, SpA, Italy involves a single-component catalyst for the polymerization of alpha-olefins, consisting essentially of a silica-containing catalyst carrier reacted with chromyl chloride (CrO_2Cl_2) and/or chromyl fluorochloride (CrO_2FCl), alone or in the presence of chromyl fluoride (CrO_2F_2). The catalyst is thereafter activated for subsequent direct use in the polymerization process by heating the single-component catalyst in oxygen or inert gas to a temperature of 300° to 600°C.

The treatment with chromyl chloride may be effected in the liquid phase where the liquid chromyl chloride contacts the catalyst carrier directly. In a modification of the liquid phase technique, the chromyl chloride can be dissolved in a solvent (e.g., 1,2-dibromoethane nitrobenzene, chloroform and carbon tetrachloride). In another advantageous method of treating the catalyst carrier with chromyl chloride, the latter is constituted as part of a gas stream which is employed to contact the carrier.

In this case, it is desirable to convey the chromyl chloride to the carrier in a gas stream containing an inert gas (e.g., nitrogen), i.e., a gas incapable of reacting with the chromyl compound. According to another feature of this method, the chromyl chloride (CrO_2Cl_2) is reacted with hydrogen fluoride to produce chromyl fluoride, which reacts with a siliceous-catalyst carrier. The reaction also yields chromyl fluorochloride which is capable of reacting with silica to yield silicon tetrafluoride and deposits a chromyl-containing compound upon the surface of a carrier.

It has been found that the technique of the process is surprisingly effecting in depositing relatively large quantities of chromium on that carrier. Catalyst systems of this type, however, are most effective when the chromium content is between substantially 0.1 and 10% by weight of the catalyst system and preferably between 0.1 and 4% by weight when the catalyst is employed in the polymerization of alpha-olefins. Activation of the catalyst system is effected by heating it to temperatures ranging between substantially 150° and 800°C. in the presence of oxygen, an oxygen-containing gas or an inert gas (e.g., nitrogen).

Activation temperatures between 300° and 600°C. are most desirable when the catalyst system is to be employed for the production of alpha-olefins. It should be further noted that catalyst systems prepared using chromyl chlorides and chromyl fluorides, in conjunction with or without activation have been found to evidence synergistic activity giving rise to catalyst efficiencies in excess of those obtained when the chromyl fluoride and the chromyl chloride are used alone. Still higher efficiences will result when activation is employed. The following is a specific example of the operation of the process, described with reference to Figure 76.

In such an operation, 2 g. of chromyl chloride were placed in a Monel bubbler 10 through which a gaseous mixture was passed containing 90% nitrogen and 10% hydrogen fluoride, with a throughput of 500 ml./minute via inlet tubes 11 and 12. The aqueous mixture emerging from tube 13 of the bubbler 10 and consisting of chromyl chloride, chromyl fluoride, chromyl fluorochloride, hydrogen chloride, hydrogen fluoride and nitrogen was passed at room temperature through a stainless-steel reactor 14 containing 100 g. of silica-alumina (Davison, 87% by weight silica and 13% by weight alumina).

When no more chromyl chloride remained in the bubbler 10, the flow of hydrogen fluoride was stopped, while the flow of nitrogen was maintained in order to purify the catalyst from occasional acid vapors physically entrapped on it. The

FIGURE 76: APPARATUS FOR PRODUCTION OF SUPPORTED CHROMIUM POLYMERIZATION CATALYST

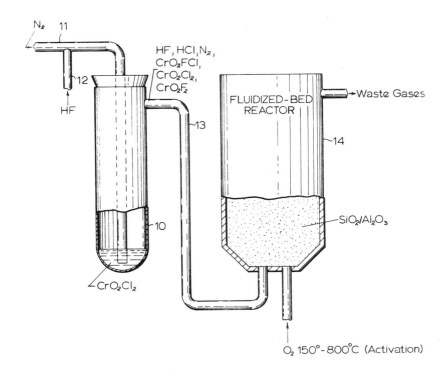

Source: P. Saccardo, G. Trada, V. Fattore and J. Herzenberg; U.S. Patent 3,412,040; November 19, 1968

olefin-polymerization catalyst contained 1.37% by weight chromium. Further quantities of nitrogen were passed through the system to act as the inert gas providing the activation atmosphere, while the temperature of the silica-alumina was raised up to 520°C. at which it was held for 2 hours. After cooling, the catalytic system thus obtained was utilized with good results for polymerizing ethylene to a solid product.

National Distillers and Chemical

A process developed by R.S. Solvik and W. Kirch; U.S. Patent 3,446,754; May 27, 1969; assigned to National Distillers and Chemical Corporation is one in which a chromium oxide catalyst containing a support such as silica, alumina, thoria, zirconia and the like, is activated by inductive heating to a temperature of at least 1000°F. for at least 60 minutes. Conventional catalyst processes, especially those used in the polymerization art, activate chromium oxide type catalysts in a nonreducing atmosphere by drying and heating air externally and then passing it through a catalyst bed at a constant rate until the catalyst reaches a desired temperature.

The catalyst is normally held at this temperature for a specified period of time. If external heat is supplied to the system it normally is supplied by either (1) heating the walls of the activation chamber by passing hot gases or fluids through a jacket around the activation chamber, or by (2) affixing electrical resistance heaters to the outside surface of the activation chamber. Such an activation process is described in U.S. Patent 2,825,721.

The process comprises activating by induction heating a particulate chromium oxide catalyst by contacting, at activation temperatures achieved solely through the use of electromagnetic induction heating, the catalyst with a nonreducing gas at a velocity sufficient to maintain the catalyst in a fluidized state, maintaining the catalyst at activation temperature for a specified length of time, and thereafter collecting the activated catalyst in a dry container. The basic improvement over the prior art is the discovery that controlling and maintaining the desired activation temperatures by the use of electromagnetic induction modifies the properties of a conventional catalyst, while it is in the fluidized state, in some unknown manner, where a catalyst is produced. Therefore, use of this catalyst under polymerization conditions produces polymers with properties unobtainable when catalysts activated in the conventional manner were used. It has been found that polymers of lower molecular weight (i.e., higher melt index) are synthesized when catalysts activated in accordance with the process are utilized under proper polymerization conditions, as compared with polymers produced with catalysts activated by conventional hot air jacketed heating. Accordingly, the products of this method are more useful for conversion to finished products in conventional plastics conversion equipment.

This phenomenon has extreme importance in the field of the slurry polymerization processes (where these types of catalysts have found wide use) where the major variable for controlling polymer molecular weight has been synthesis temperature. In such processes, synthesis temperatures have to be maintained relatively low to prevent polymer softening or solubilizing, hence, polymer molecular weights are necessarily typically high (low melt index) which limits the ability to convert the polymers to finished products. The use of induction heating to activate and produce catalysts which synthesize lower molecular weight polymers for a given synthesis temperature greatly expands the area of operation of slurry-type reactors.

The catalysts suitable for activation as herein disclosed are, in general, supported chromium oxides which are well known as to their composition and their use as polymerization catalysts. As set forth in U.S. Patent 2,825,721 of Hogan, et al., the catalyst may be prepared by depositing chromium oxide (e.g., Cr_2O_3) or a chromium compound calcinable to chromium oxide, on a suitable support and activating to leave part of the chromium on the support in hexavalent form. The support may be selected from one or more of the following members: silica, alumina, thoria, zirconia, silica-alumina, silica-thoria, silica-zirconia, acid treated clays, and other materials generally known as catalyst supports. In its preferred form, the catalyst is a silica gel supported chromium oxide, where at least a portion of the chromium is in the hexavalent state.

As previously noted, the catalyst is obtained by inductively heating a chromium oxide catalyst at activation temperatures for a specified period of time under fluidized conditions. After such activation, the catalyst is cooled, purged with an inert gas, and collected in a dry container. The fludizing gas is required to maintain the catalyst particles in a suspended and separated state where the inductive powers of the activation can easily reach and affect each catalyst particle. The fluidizing gas also helps to remove any water that may be contained in the catalyst and therefore assists in obtaining a dry, activated material.

FIGURE 77: APPARATUS FOR ACTIVATION OF PARTICULATE SUPPORTED CHROMIUM OXIDE POLYMERIZATION CATALYSTS

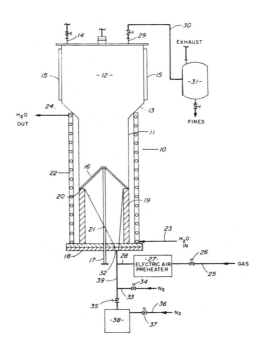

Source: R.S. Solvik and W. Kirch; U.S. Patent 3,446,754; May 27, 1969

The activation is accomplished solely by inductively heating the fluidized catalyst at a temperature of 1000° to 2000°F., preferably 1200° to 1800°F., for 60 minutes to 20 hours, preferably 4 to 12 hours. This activation at elevated temperatures may be preceded by gradually inductively heating the fluidized catalyst particles to an intermediate temperature of approximately 400 to 800°F., and holding the particles at that temperature for 30 minutes to 2 hours. An electromagnet induction coil operating at 1 to 20 kilocycles, preferably 8 to 12 kilocycles, is wrapped about the fluidization portion of the activation apparatus to provide heat to the system via the inductive forces.

Figure 77 on the previous page shows a suitable form of apparatus for the conduct of the process. The activator 10 is an elongated cylindrical chamber having a lower fluidization chamber 11 constructed of Inconel-X and an upper disengaging chamber 12 constructed of stainless steel. The upper disengaging chamber 12 is somewhat larger in cross-sectional area than the fludization chamber 11 and is connected thereto by a stainless steel frusto-conical connecting section 13. The catalyst to be activated is charged to the activator through port 14 in chamber 12. Electrical strip heaters 15 are affixed to the outside of disengagement zone 12 to minimize heat loss and prevent readsorption of water by the catalyst.

At the bottom of the fluidization zone 11 there is a conical 200 mesh screen 16 affixed to a rod 17 extending through the bottom flange 18 of the catalyst activator through a water-cooled packing gland (not shown). An insert 19 affixed to the bottom flange 18 of the catalyst activator extends up into the activator and is machined at the top to provide a seat 20 for the conical screen 16 to prevent leakage of catalyst around the screen. A conical section 21 is positioned below the screen to facilitate discharge of the catalyst.

An induction coil 22 is wrapped around fluidization zone 11 to provide heat to the system via inductive forces. In addition, the induction coil 22 extends somewhat below fluidization zone 11 to provide additional preheat to the fluidizing gas. Induction coil 22 is fabricated from copper tubing and is water cooled, the cooling water entering at 23 and leaving at 24. The fluidizing gas is admitted through conduit 25 controlled by valve 26. The gas is passed through an electric preheater 27 and then to fluidization chamber 11 via conduit 28. The velocity of the gas is maintained by any appropriate means (not shown) at a level sufficient to keep the catalyst particles in a fluidized state. The gas passes through the activator 10 exiting through port 29 in chamber 12. The gas and any entrained particles pass via conduit 30 to separator 31 where such particles are removed from the exhaust gases.

During activation, screen 16 is in the lowered position forming a seal to prevent catalyst from falling below the screen. After activation, the purge gas, such as nitrogen passed through conduit 33 and controlled by valve 34, is turned off, screen 16 is raised and catalyst discharges through the discharge port 32 to a dry container 38 via conduit 39, controlled by valve 35. The container is blanketed with an inert gas, such as nitrogen, which is admitted through conduit 36 controlled by valve 37. Appropriate thermocouples (not shown) and other control apparatus are advantageously placed to monitor and regulate the induction chamber temperatures, measure and control catalyst preheating temperatures, and measure and control fluidization gas temperatures, etc. With such a system, sensitive temperature control can be attained where accurate activation temperature programs can be easily programmed and reproduced.

Phillips Petroleum

A process developed by J.I. Stevens, J.E. Cottle and W.T. Wise; U.S. Patent 3,151,944; October 6, 1964; assigned to Phillips Petroleum Company involves the activation and regeneration of chromium oxide-containing catalyst masses which are utilized in the polymerization of olefins at relatively low temperatures. This is another method for this activation or regeneration which is more economical than processes previously employed for the activation or regeneration. The catalyst is heated to a temperature of approximately 950°F., held at that temperature for several hours, and then cooled in a particular manner. Activation in a fluidized bed has been found to be especially suitable. The activating gases are used in an amount sufficient to maintain the catalyst in a fluidized state but insufficient to supply the heat necessary for the activation. This additional heat is supplied by indirect heat exchange.

Figure 78 shows the overall arrangement of apparatus which may be employed in the conduct of the process. Fluidization air is supplied through conduit 10, this air being compressed in compressor 11, cooled in cooler 12, and condensate removed therefrom in vessel 13. In order to dry the air, a series of fluidization air dryers, illustrated schematically as 14, 16 and 17 is used. Suitable drying agents include calcium sulfate, calcium chloride and silica-alumina. Using a series of dryers such as illustrated, provides a method of supplying a continuous stream of dried air since one chamber can be on standby, one can be under regeneration, while the third is in use.

Regeneration air is supplied by means of conduit 18, this air being compressed in compressor 19, cooled in cooler 21, condensate removed therefrom in vessel 22 and heated in heater 23. This regeneration air is supplied by means of conduit 24 containing branches 26, 27, and 28 to dryers 14, 16 and 17, respectively. This regeneration gas is vented by means of conduits 29, 31, and 32. The dry air should have a dew point preferably below -60°F. at 40 psia. Now returning to the fluidization air, this air is passed from vessel 13 to the activator chambers 33 and 34 (provided with cyclone collectors 35 and 40, respectively) by means of different paths depending upon the particular stage of the activation process. The first stage of the treatment comprises heating the catalyst to a temperature of 500°F. over a period of 7 hours, undried

FIGURE 78: FLOW DIAGRAM OF PROCESS FOR ACTIVATION AND REGENERATION OF CHROMIUM OXIDE-CONTAINING POLYMERIZATION CATALYSTS

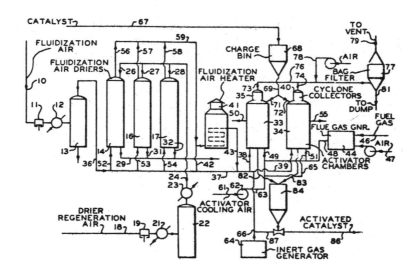

Source: J.I. Stevens, J.E. Cottle and W.T. Wise; U.S. Patent 3,151,944; October 6, 1964

air being suitable for this process, keeping in mind that a gas of reduced free oxygen content is necessary when the method is one of catalyst regeneration. This air can be supplied by conduits 36, 37 and 38 or 39, depending upon which activator chamber is on stream. The heating for the fluidization air is supplied by fluidization heater 41, the air passing to this heater through conduit 42 communicating with conduit 36.

Extending from heater 41 is conduit 43, this conduit communicating with conduit 37. The temperature can be controlled by passing all the gas through heater 41 or, and more preferably, the air heater 41 can be operated at a uniform temperature and an increasing amount of the air supplied to the activator chamber can be passed through. The flow rate of the fluidization air depends, of course, upon the particle size of the catalyst but a suitable range is from 0.15 to 1.5 feet per second.

Additional heat is supplied to the activator chamber, this being supplied by flue gas generator 44 provided with fuel supply conduit 46 and air supply conduit 47. The flue gas passing therefrom is introduced into the activator chambers by means of conduits 48 and 49 or 51. Conduits 50 and 55 are provided for removal of flue gas from chamber 33 and 34, respectively. As pointed out previously the heat supplied by this flue gas constitutes the major portion of the heat for the catalyst treatment.

Following this heating to 500°F., a further period of 3 hours is used to raise the temperature from 500° to 750°F. After reaching 750°F., further heat treatment is done in the presence of dry air. For this purpose, the air is passed through one of dryers 14, 16 or 17 from conduit 36 by means of conduits 52, 53 or 54. The dried air is passed by conduits 56, 57 or 58 to conduit 59 which communicates with air heater 41 and conduit 37 by means of conduit 42. In the same manner as temperature was controlled on the undried air, this temperature control is obtained for the dry air. The temperature is increased from 750° to 950°F., the preferred activation temperature, over a period of approximately 2 hours. The temperature is maintained at 950°F. for 5 hours, utilizing the dry air as the fluidization gas.

At the end of this phase of the treatment the catalyst is cooled in a stepwise manner. Over a time interval of 2 hours, dry air is passed over the catalyst, this reducing the temperature from 950°F. to approximately 750°F. As with the heating, the major portion of the heat exchange for cooling is indirect, activator cooling air being supplied by conduit 61 which extends to pump 62. Extending from pump 62 are conduits 63 and 65 which communicate with conduits 49 and 51, respectively. In this step the temperature of the fluidization air is gradually decreased.

The last stage of the cooling is done in the presence of inert gas or dry air and the catalyst is cooled from the 750° to 100°F. over a period of time of 10 hours. An inert gas generator 64 is shown having conduit 66 extending therefrom, this conduit communicating with conduit 37 to supply inert gas to chambers 33 and 34. An inert gas such as one having less than 5 ppm of carbon monoxide and oxygen and from 0.1 to 0.2% by volume of hydrogen is satisfactory. The dew point is preferably reduced to −60°F. at 40 psia. Finally, the catalyst flow path will be described. This catalyst, supplied through conduit 67, is introduced into charge bin 68 and, by means of conduits 69, 71, and 72 to chambers 33 and 34. Conduits 73 and 74 are provided extending from the top of the activator chambers, these conduits communicating with conduit 76 which extends to filter 77, conduit 76 being provided with auxiliary air conduit 78.

This auxiliary air is provided to cool the exhaust gases passing to the filter. From filter 77, conduit 79 extends to a vent and conduit 81 extends to a dump. Following activation or regeneration, the catalyst is passed from the bottom of the activator chamber 33 or 34 by means of conduit 82 or 83 to collection chamber 84. As needed, activated catalyst is removed from collector 84 and passed to the point of use, a pneumatic gas transfer line being shown as conduit 86, this being supplied with inert gas by means of conduit 87 which, in turn, is connected to conduit 66.

A process developed by D.W. Walker and E.L. Czenkusch; U.S. Patent 3,351,623; November 7, 1967; assigned to Phillips Petroleum Company relates to a catalyst system for ethylene polymerization. In the polymerization of ethylene, the melt index (indicative principally of the molecular weight of the polymer) is closely proportional to the reaction temperature, lower temperatures producing polymer with a low melt index (high molecular weight) and higher temperatures producing polymer with a high melt index (low molecular weight).

Thus, a polymer can in general be produced having a specific melt index which is suitable for its intended application, be it for pipe, film, bottles, or the like. Operationally speaking, however, it is desirable to operate at the highest reaction temperature which will produce polymer having the required melt index. This high reaction temperature is desirable because it provides the largest temperature differential across the reactor wall, and therefore the best heat transfer, between the inner reactor wall in contact with the reaction mixture and the outer reactor wall in contact with the coolant. Olefin polymerization reactions are exothermic and heat must be carried away, but the coolant temperature must not be so low that it will cause polymer to deposit upon the inner walls of the reactor. Consequently, polymerization processes utilizing higher reaction temperatures can be cooled more effectively without danger of losing control of the process.

It has been found that such a reaction temperature-melt index relationship can be substantially improved when polymerizing ethylene over a supported chromium oxide catalyst system by the utilization of a vanadium beta-diketone, exemplified by vanadium acetylacetonate, and an organometal, exemplified by triethylaluminum. That such a beneficial result would be obtained from such a combination of catalytic materials was entirely unexpected. When an organometal, such as triethylaluminum, is added to the supported chromium oxide catalyst system, a mild improvement in the reaction temperature-melt index relationship is realized.

However, when vanadium acetylacetonate is added to the supported chromium oxide, there is a poisoning effect and essentially no polymer of any description is formed. It is readily seen, therefore, that the substantially beneficial results obtained when both triethylaluminum and vanadium acetylacetonate were added to the supported chromium oxide system were completely surprising and unobvious. Such a combination results in a very large improvement in the temperature-melt index relationship. In some cases the process permits an increase in operating temperature of 20°F. or more.

MOLYBDENUM-CONTAINING CATALYSTS

ALDOLIZATION

Esso Research and Engineering

A process developed by W.J. Porter, Jr., J.A. Wingate and J.A. Hanan; U.S. Patent 3,248,428; April 26, 1966; assigned to Esso Research and Engineering Company is a process where aldehydes or ketones or mixtures thereof are aldolized and dehydrated in the presence of a solid, insoluble catalyst comprising the reaction product of molybdenum oxide and magnesium oxide.

The liquid phase aldolization of aldehydes or ketones in the presence of strong bases is well known in the art. The hydroxy carbonyl or aldol compounds formed in the presence of these catalysts are generally subjected to a subsequent dehydration step in order to produce the corresponding alpha,beta-unsaturated aldehydes or ketones. Difficulties are experienced in such process, however, because the strong bases tend to promote further reaction of the products which initially result from the simple condensation of two carbonyl molecules. Consequently, the yields of alpha,beta-unsaturated carbonyl compounds from such processes are generally unsatisfactory, and the isolation of pure products is difficult.

The production of alpha,beta-unsaturated carbonyl compounds by passing carbonyl reactants in the vapor phase at elevated temperatures over a fixed catalyst bed is also known. However, the yields of alpha,beta-unsaturated products are generally low per pass and, in addition, one must contend with the disadvantages inherent in vapor phase operations.

More recently, it has been discovered that alpha,beta-unsaturated carbonyl compounds are produced by heating, for example, aldehydes in the liquid phase with metal soap catalysts which are soluble therein. To a greater or lesser degree, the soaps of a wide variety of metals, for example, cobalt, manganese, magnesium, zinc, lead and the like, have been found to be catalytically active. Generally, the conversions and selectivities to alpha,beta-unsaturated aldehydes are high in the soap-catalyzed processes; however, use of the soluble soap catalysts presents an additional problem. Before the alpha,beta-unsaturated product can be processed further, for example, by hydrogenation to the corresponding alcohols, the soluble catalyst content must be reduced to below about 10 to 20 ppm. While suitable demetalling processes are known to the art, this additional processing step necessarily adds to the ultimate cost of the product.

These and other disadvantages of the prior art processes for producing alpha,beta-unsaturated carbonyl compounds of increased molecular weight from carbonyl compounds of relatively lower molecular weight are overcome by this process. It has now been surprisingly found that alpha,beta-unsaturated carbonyl compounds can readily and conveniently be produced by contacting two molecules of the same or different monomer carbonyl compounds in liquid phase and at elevated temperatures with a catalyst comprising the reaction product of molybdenum oxide and magnesium oxide. High conversions of the reactants and high selectivities to the desired alpha,beta-unsaturated carbonyl products are realized by this process. Furthermore, since the oxide reaction product catalyst is substantially insoluble in the reactants and in the products, its use in liquid phase aldolization processes as described here effectively eliminates the necessity for subjecting the products to a demetalling step.

As noted above, the catalysts of the process comprise molybdenum oxide and magnesium oxide in chemical combination. It appears that at elevated temperatures the molybdenum and magnesium oxides react to form a hard, durable, amorphous, spinel-type material. The ratio of oxides used to produce the MoO_3-MgO reaction product may be varied considerably. For example, from about 0.5 to 15 parts by weight of molybdenum oxide per part of magnesium oxide can be used. However, the reaction product obtained upon calcining a mixture containing from about 0.8 to 5 parts by weight of MoO_3 per part of MgO is preferred and especially that product derived from equimolar amounts of the two oxides, that is,

3.6 parts of MoO$_3$ per part of MgO. It is, of course, not required that MoO$_3$ and MgO per se be used to obtain the desired reaction product. Any inorganic or organic compounds of the metals which upon calcining will form the respective oxides, e.g. nitrates, carbonates, hydroxides, acetates, oxalates, and the like, may also be used. Calcining temperatures will vary somewhat depending upon the nature of the compounds employed, but in general will be between about 300° to 1500°F., e.g. 1100°F.

The oxide reaction product may be utilized in various forms, for example, the product per se may be crushed to suitable size and packed to form a solid catalyst bed, or it may be powdered and pelletized, pilled or compacted into various shapes. It may also be employed as the active ingredient in combination with a catalyst support or carrier such as alumina, bauxite, coke, kieselguhr, limestone, silica gel and similar materials well known in the catalyst art. Those materials characterized as high surface area supports, i.e. those providing surface areas in excess of about 100 square meters per gram, are to be preferred, especially alumina. Any of the means commonly employed to produce supported oxide catalysts may be used (Watts U.S. Patent 2,888,396). Generally, the supported catalyst will contain from about 1 to 20 weight percent, preferably 5 to 15 weight percent, of the reaction product of the molybdenum and magnesium oxides. Although compositions outside of these ranges can be used, they are effective to a lesser degree.

The aldolization process is carried out in the liquid phase at elevated temperatures. That is to say, the carbonyl reactant or reactants, either in the presence or absence of an inert solvent, e.g. aliphatic or aromatic hydrocarbons and the like, is contacted as a liquid with the molybdenum oxide–magnesium oxide reaction product at a temperature between about 50° and 600°F. The exact temperature to be used is not critical and is somewhat dependent on the particular carbonyl reactant or combination thereof in the feed. Whether or not a diluting solvent is employed, as well as the time of contact with the catalyst which may vary from about 5 minutes to about 4 hours, will also to some extent determine the reaction temperature. With most carbonyl reactants, and especially with the C$_1$ to C$_{10}$ aliphatic carbonyl compounds, a temperature of between about 100° and 400°F., with contact times between about 15 minutes to 2 hours is preferred.

The pressure at which the process is carried out must exceed the vapor pressure of the reaction mixture at the operating temperature, that is, the pressure should be sufficient to maintain a liquid reactant phase in contact with the catalyst. Otherwise, the pressure is not critical and atmospheric, subatmospheric or superatmospheric pressures may be employed. With the lower molecular weight reactants, e.g. C$_1$ to C$_{10}$ carbonyl compounds, pressures slightly above atmospheric are required to maintain the reactants in the liquid phase. On the other hand, with the higher molecular weight reactants, e.g. C$_{11}$ to C$_{20}$, it may be desirable to use slightly reduced pressures. Where the reaction is carried out under superatmospheric pressures, the autogeneous pressure of the reactants is generally sufficient. Where higher pressures are desired, the reaction may be carried out under pressures produced by the addition of gases such as nitrogen or hydrogen.

The process may be carried out in a batch, semicontinuous, or continuous manner. Any apparatus may be used where the necessary temperature and liquid phase conditions can be maintained. It is preferred, however, to employ a system from which the water of reaction can be removed as it forms.

Figure 79 shows a suitable form of apparatus for the conduct of this process. n-Butyraldehyde is passed in liquid phase at

FIGURE 79: APPARATUS FOR LIQUID PHASE CATALYTIC ALDOLIZATION PROCESS

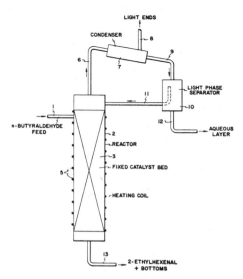

Source: W.J. Porter, Jr., J.A. Wingate and J.A. Hanan; U.S. Patent 3,248,428; April 26, 1966

a rate of 0.98 volume of feed/volume of catalyst/hour via line 1 into reactor 2. Reactor 2 is packed with a fixed bed 3 of 0.25" diameter by 0.25" length pills of the molybdenum oxide-magnesium oxide reaction product. The catalyst bed is maintained at a temperature between about 250° and 450°F., e.g. 426°F., with electrical resistance heaters 5 or other suitable means. Pressure within the tower is the autogenous pressure of the reaction mixture, e.g. about 195 psig. The liquid in reactor 2 is maintained at a level sufficient to cover the fixed catalyst bed. Aldolization of the butyr-aldehyde and dehydration of the condensation product occur in the packed sections of the reactor to produce a mixture of unreacted n-butyraldehyde, water, and 2-ethylhexenal.

Under the process conditions, an azeotropic mixture of n-butyraldehyde and water refluxes into the uppermost portion of the column. The azeotrope is removed through line 6 and passed into condenser 7. Light ends in the feed or formed during the reaction are vented from condenser 7 through line 8, while the liquid condensate is passed through line 9 into phase separator 10. The upper layer of condensate which forms in separator 10 is predominantly n-butyraldehyde. This phase is recycled through line 11 into reaction tower 2. The water phase, which contains a minor proportion of n-butyraldehyde dissolved therein, is removed via line 12. The n-butyraldehyde in the water phase may be recovered and recycled if so desired. The aldolization and dehydration reaction product, 2-ethylhexenal, is removed from the catalyst bed through line 13 for further purification and processing.

AMMONOXIDATION

Knapsack-Griesheim

A process developed by K. Sennewald, W. Vogt, J. Kandler, R. Sommerfeld and G. Sorbe; U.S. Patent 3,254,110; May 31, 1966; assigned to Knapsack-Griesheim AG, Germany employs improved catalysts in the preparation of acrylo-nitrile from propylene and of methacrylonitrile from isobutylene.

The reaction of olefins with ammonia in the presence of oxygen which results in the formation of nitriles has been known for some time already. Catalysts which are suitable for use on an industrial scale and which enable good yields, high degrees of conversion and a great efficiency of the catalyst to be obtained are also known. For example, a catalyst which comprises iron oxide, bismuth oxide and molybdenum oxide can be used with particular advantage in a flowing bed reactor or in a fluidized bed reactor.

The catalysts, however, have the drawback of being very expensive which, in particular, is due to their content of bismuth oxide. When a catalyst of the aforesaid kind is prepared, the cost of the bismuth oxide amounts to two-thirds of the cost of the active oxidic components. Besides, when using the bismuth salts their physiological properties have to be taken into consideration.

For these reasons high requirements have to be met with respect to the working up and recovery of such catalysts. When the reaction is carried out in a fluidized bed which is the advantageous mode of operation there is, for example, always rubbed off a certain portion of the catalyst. These pulverulent portions have to be separated and the bismuth contained therein has to be recovered by decomposition, if possible in a quantitative yield, and this requires additional apparatus and acid.

Now it has been found that catalysts of appropriate composition which do not contain bismuth can also have a great efficiency and enable high degress of conversion and high yields to be obtained. These efficient catalysts are combina-tions of oxides of the transition elements with molybdenum oxide. In addition to these oxides the catalysts contain a carrier, which preferably is silicic acid. As transition elements are preferably used, the elements of the first transitional period, for example, Ti, V, Cr, Mn, Fe, Co, Ni, Cu, and Zn, the elements of the group of iron, viz. Fe, Co, and Ni, as well as V, Cr and Mn having proved particularly suitable. The catalysts contain either one or more of the metals in the form of its or their oxides in combination with molybdenum oxide. The catalysts used advantageously com-prise iron oxide and/or nickel oxide and molybdenum oxide and, if desired, another oxide of a transition metal deposited on a carrier. The catalysts may contain phosphorus pentoxide as a further component.

These catalysts may then be used in a form of reaction apparatus as shown in Figure 80. The starting gas mixture is intro-duced via conduit 17 through perforated bottom 2 into reaction tube 3, catalyst substance 4 charged into the reaction tube attaining approximately height limit 5. The gas whirls up the catalyst and entrains it into quiescent vessel 6 whose inside cross-sectional area is approximately five times as large as that of reaction tube 3 and in which the speed of flow of the gas is reduced and drops to below the speed of discharge of substantially all grain sizes, so that the catalyst falls back into reaction tube 3. Small quantities of the catalyst, in particular of the pulverulent portion of the catalyst, which do not fall back are separated in cyclone 7 and fall back through fall pipe 8 into reaction tube 3. The lower end of fall pipe 8 is screened by means of cup 9 in order to prevent the gas introduced at 17 from penetrating from the bottom into fall pipe 8. The gas leaves cyclone 7 via exhaust pipe 10. Reaction tube 3 is maintained at the desired reaction temperature while quiescent vessel 6 and exhaust gas pipe 10 are maintained at about 300°C. Along the whole length

FIGURE 80:   AMMONOXIDATION REACTOR FOR ACRYLONITRILE PRODUCTION FROM PROPYLENE

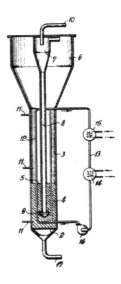

Source: K. Sennewald, W. Vogt, J. Kandler, R. Sommerfeld and G. Sorbe; U.S. Patent 3,254,110; May 31, 1966

of reaction tube 3 are distributed short pipes 11 which serve for the arrangement of thermoelements, for measuring the pressure and for introducing further quantities of gas.   For dissipating the heat of the exothermic reaction, heat exchangers 12 are provided which may be charged with a heat transferring agent via recycle conduit 13 by means of pump 16.   By means of the same recycle system the reaction may be started by heating.   The heat is removed from the cycle via cooler 14 and supplied via heater 15, which cooler 14 and heater 15 may be used optionally.

In an alternative form of reactor design, as shown in Figure 81, the catalyst is returned outside the apparatus.   The starting mixture is introduced via conduit 17 at necked–down portion 18 into reaction tube 3 in which it whirls up

FIGURE 81:   ALTERNATIVE AMMONOXIDATION REACTOR DESIGN FOR ACRYLONITRILE PRODUCTION FROM PROPYLENE

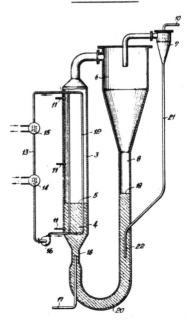

Source: K. Sennewald, W. Vogt, J. Kandler, R. Sommerfeld and G. Sorbe; U.S. Patent 3,254,110; May 31, 1966

catalyst 4 which, in the state in which it is charged into the apparatus, approximately attains height limit 5 and it partly discharges it into the cyclone-shaped separator 6 from where the catalyst falls down through down-pipe 8. In down-pipe 8 the catalyst comes approximately to height limit 19. The catalyst is returned from down-pipe 8 through connecting bend 20 into reaction tube 3 by means of the current of gas entering at 18 and passing in bubbles through the catalyst. On leaving separator 6, the gas enters cyclone 7 in which the last constituents of the catalyst are retained and returned to the cycle via down-pipe 21 which is immersed within down-pipe 8 at 22 into the catalyst. The exhaust gas escapes via pipe 10.

Reaction tube 3 is provided with means 11 serving for the arrangement of thermoelements, for measuring the pressure and for the introduction of further quantities of gas. Heat transferring cycle 13, which includes pump 16, enables heat to be eliminated from reaction tube 3 via heat exchange elements 12 or the reaction to be initiated by heating. The heat transferrer may be cooled by cooler 14 and heated by heater 15.

Nitto Chemical Industry

A process developed by K. Hiroki, T. Shizume, Y. Nakamura, T. Yoshino, H. Kamio, S. Irie and Y. Kawamura; U.S. Patent 3,346,617; October 10, 1967; assigned to Nitto Chemical Industry Co., Ltd., Japan is one in which improvement in both the percent conversion of isobutylene and percent yield of methacrylonitrile in the gas-phase ammoxidation of isobutylene have been found when the composition of a specified bismuth phosphomolybdate catalyst employed for the ammoxidation is made "more alkaline," either by the addition to the bismuth phosphomolybdate catalyst of an alkali metal or alkaline earth metal, or by the substitution of arsenic and/or antimony for a part or all of phosphorus in the phosphomolybdate composition, or further by the addition to the substituted molybdate of an oxide or hydroxide of an alkali or alkaline earth metal. The bismuth phosphomolybdate catalysts referred to include compositions expressed by the empirical formula: $Bi_{6-12}P_{0.5-5}Mo_{12}O_{46-61}$. The atomic ratio of the molybdenum to the alkali and/or alkaline earth metal to be added may range preferably 1-5 to 12 and most preferably 1-2 to 12.

Methacrylonitrile is produced in good yield, according to one aspect of the process, by contacting a mixture containing isobutylene, molecular oxygen and ammonia in the gas phase with a catalyst having the empirical formula: $P_nMo_{12}Bi_dX_eY_fO_g$, where X is selected from the group consisting of arsenic and antimony, and Y is selected from the group consisting of alkali metals and alkaline earth metals, and $n = 0-5$, $d = 6-12$, $e = 0.5-5$, $f = 0.5$, and at least 1, when $e = 0$, $g = 46-61$ and $n+e = 0.5-5$.

Standard Oil (Ohio)

A process developed by J.L. Callahan, B. Gertisser, and J.J. Szabo; U.S. Patent 3,354,197; November 21, 1967; assigned to The Standard Oil Co. (Ohio) relates to the oxidation of olefin-ammonia mixtures to unsaturated nitriles, such as propylene-ammonia to acrylonitrile, using an improved oxidation catalyst consisting essentially of oxides of the elements bismuth and molybdenum, and optionally, phosphorus, promoted by oxides of boron and bismuth.

The Callahan, Foreman and Veatch U.S. Patent 2,941,007 describes the oxidation of an olefin such as propylene and the various butenes with oxygen and a solid catalyst composed of the oxides of bismuth, molybdenum and silicon, and optionally, phosphorus. This catalyst selectively converts propylene to acrolein, isobutylene to methacrolein, α- and β-butylene to methyl vinyl ketone and to butadiene, etc. High yields are obtainable, although in the case of the butenes, careful control of reaction conditions may be required in order to direct the reaction in favor of either methyl vinyl ketone or butadiene, depending upon which of these alternative products is desired.

The Idol, Jr., U.S. Patent 2,904,580 employs the same catalyst to convert propylene, ammonia and oxygen to acrylonitrile, at approximately atmospheric pressures and elevated temperatures. Excellent conversions, usually in the range of 40 to 80%, nitrogen basis, of useful products are obtainable.

In accordance with this process, the catalytic activity of such bismuth oxide-molybdenum oxide catalysts is greatly enhanced or promoted by the combination therewith of a mixture of boron and additional bismuth in the form of their oxides, referred to as promoters. The promoters are best applied by impregnation or surface coating of the catalyst, after its formation. Further, it has been determined that a portion of the supplemental bismuth oxide promoter can be replaced with manganese oxide, and that phosphorus oxide can also be present as a supplemental oxide.

The proportions of boron oxide and bismuth oxide, with or without phosphorus oxide and/or manganese oxide, are important in obtaining the optimum enhanced activity. The boron oxide concentration, calculated as boron, should be within the range from about 0.5 to 1% by weight; and the amount of bismuth oxide, calculated as bismuth, should be within the range from about 5 to 10% by weight, although more than 10% can be used if desired. If manganese oxide is employed, it can be used on a bismuth oxide equivalent weight basis, but not more than about one-third of the promoter bismuth oxide, calculated as bismuth, can be replaced by manganese oxide. While this catalyst may be employed without any support, it is desirable to combine it with a support. A preferred support is silica because the silica improves the catalytic activity of the catalyst. The silica may be present in any amount, but it is preferred that the catalyst contain between 25 to 75% by weight of silica. Many other materials such as Alundum, silicon carbide, alumina-silica, alumina, titania and other chemically inert materials may be employed as a support which will withstand the conditions of the process.

The catalyst may comprise phosphorus, also present in the form of the oxide. Phosphorus will affect, to some extent, the catalytic properties of the composition, but the presence or absence of phosphorus has no appreciable effect on the physical properties of the catalyst. Thus, the composition can include from 0%, and preferably from at least 0.1%, up to about 5% by weight of phosphorus oxide, calculated as phosphorus.

The promoter is incorporated with the catalyst base by impregnation thereof, using an aqueous solution, dispersion, or suspension of a boron compound and of a bismuth compound, with or without a manganese compound, either the oxide, or a compound thermally decomposable in situ to the corresponding boron oxide, bismuth oxide, and manganese oxide, respectively, without formation of other deleterious metal oxide residue, for instance, ammonium phosphate, ammonium tetraborate, ammonium permanganate, manganese nitrate, bismuth nitrate, boric acid, bismuth hydroxide, manganese hydroxide, bismuth phosphate, and bismuth borate. The phosphorus-containing compounds also add phosphorus to the catalyst. After impregnation with such solution, employed in a concentration and amount to provide the desired amount of bismuth and boron, and optionally, manganese, the catalyst base is dried, and then calcined at a temperature above that at which the compounds applied are decomposed to the oxides. Temperatures in excess of 800°F., but below that at which the catalyst is deleteriously affected, usually not in excess of about 1050°F., can be used.

The basic catalyst composition comprises bismuth oxide and molybdenum oxide, the bismuth-to-molybdenum ratio Bi:Mo being controlled so that it is at all times above 1:3. There is no critical upper limit on the amount of bismuth, but in view of the relatively high cost of bismuth and the lack of an improved catalytic effect when large amounts are used, generally the atomic ratio bismuth to molybdenum of about 3:1 is not exceeded. The nature of the chemical compounds which compose the basic catalyst is not known. The catalyst may be a mere mixture of bismuth and molybdenum oxides, with or without phosphorus oxide, but it seems more likely that the catalyst is a homogeneous micro mixture of loose chemical combinations of oxides of bismuth and molybdenum, with, optionally, phosphorus, and it is these combinations which appear to impart the desirable catalytic properties to this catalytic composition. The catalyst can be referred to as bismuth molybdate or, when phosphorus is present, as bismuth phosphomolybdate, but this term is not to be construed as meaning that the catalyst is composed of these compounds.

The bismuth and boron, and optionally, manganese, compounds added thereto as promoters may or may not enter into the chemical composition of the catalyst. Bismuth added later with boron produces a different result from boron added to a catalyst composition containing more than the usual amount of bismuth, i.e., that stoichiometrically equivalent to the weight of added boron, and has a different function, since the enhanced catalytic effect is not obtained when boron oxide is combined with a composition previously containing the same excess of bismuth. Therefore, the promoted catalytic effect may be due to some complex boron oxide-bismuth oxide combination formed on the surface of the catalyst. In any event, the boron and bismuth are present in the form of their oxides.

HYDROCRACKING

Esso Research and Engineering

A process developed by F.B. Sprow and G.W. Harris; U.S. Patent 3,575,847; April 20, 1971; assigned to Esso Research and Engineering Company is one in which coal extracts containing suspended solids are hydrocracked in a fixed bed down-flow reactor. Bed plugging is minimized by using substantially spherical catalyst granules having a minimum diameter at least ten times as great as the maximum dimensions of the suspended solids and maintaining a flow rate above the minimum at which occlusion of the bed results, preferably at least 1,000 pounds per hour per square foot.

Figure 82 shows the overall coal treatment process including the hydrocracking step using the preferred spherical catalyst. The detail at the lower left of the figure shows the large number of relatively horizontal areas which exist in the extruded cylindrical catalyst which promote the deposition and collection of solid material which ultimately leads to a blocking of certain flow paths, so that the catalyst area exposed in the blocked flow paths becomes useless insofar as promoting the reaction is concerned. Both the flat ends of the cylinders and the relatively horizontal sides of the cylinders which are horizontally disposed would tend to encourage the deposition and collection of the suspended solids.

In the detail at the lower right of the figure, by contrast, it is seen that only the upper portion of the sphere would allow the fines to collect, and as to the sloping surfaces other than the upper portion, the fines would be washed off by the flowing liquid before bridging could occur. Further, by reason of the spherical shape of the catalyst, the points of contact between the catalyst granules would be limited to a very small area in the case of spherical catalysts as opposed to the possibility of long lines of contact where cylindrical catalysts may be involved, for example, where two granules are

FIGURE 82: FLOW DIAGRAM OF PROCESS FOR COOL EXTRACT HYDROCRACKING USING SPHERICAL CATALYST

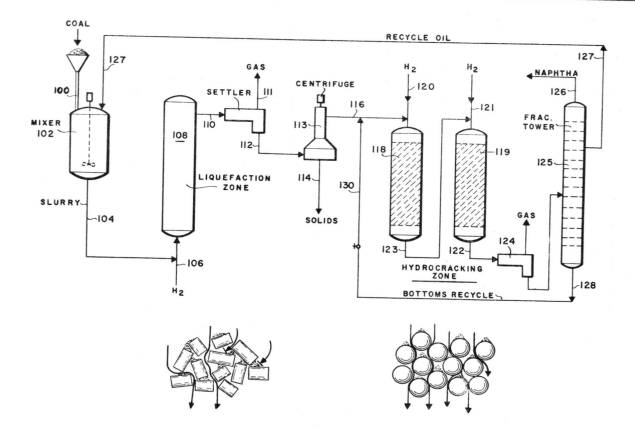

Source: F.B. Sprow and G.W. Harris; U.S. Patent 3,575,847; April 20, 1971

parallel and side by side. Referring back to a consideration of the overall process, raw coal is seen to be fed by way of line 100 into a mixer 102 where a slurry is created and withdrawn by way of line 104. Any suitable coal-like material can be used, for example, subbituminous coal, bituminous coal, lignite and asphalt. The coal is generally ground to a particle size of about 8 to 300 mesh, and may be dried before it is fed into the mixer 102.

The slurry may be mixed with hydrogen (introduced by way of line 106) and introduced into a liquefaction reactor 108. Within the liquefaction reactor, the coal is allowed to dissolve under conditions of high temperature and pressure, such as a temperature within the range of 650° to 850°F. and a pressure from 350 to 2,500 psig. The hydrogen treat rate (if hydrogen is used) may be fairly low, and may suitably range from 100 to 1,000 scf/bbl. of total slurry charge. In the lique-faction reactor, the coal is depolymerized and partially thermally cracked, and a product is withdrawn by way of line 110 which comprises the coal extract, depleted hydrogen-donor solvent and undissolved solids. The hydrogen and non-condensable gases are separated from the liquid and solid components and are removed by way of line 111 while the slurry is carried by line 112 to a solids-liquid separation unit such as the centrifuge 113.

Solids are removed from the centrifuge by way of line 114, and the clarified liquid is passed by way of line 116 into a hydrocracking zone, suitably comprising two reactors, 118 and 119. In the hydrocracking zone, the clarified oil is contacted with hydrogen introduced by way of lines 120 and 121 and is passed sequentially (via line 123) in downflow across stationary beds of spherical catalyst granules in the reactors 118 and 119. The clarified oil is preferably in the liquid phase, but may be in the mixed liquid-and-vapor phase, while the hydrogen obviously will be maintained in the gas phase and dissolved in the liquid phase. The catalyst within the hydrocracking reactor is substantially spherical, and is at least ten times larger in minimum diameter than the largest dimension of the particles being entrained in the clarified liquid.

The products of the hydrocracking reactor are removed by way of line 122, the hydrogen separated therefrom by means 124, and the liquid is fractionated in tower 125 to obtain a naphtha stream which is removed by way of line 126 for

185

further treatment, a recycle oil which is removed by way of line 127 and a bottoms product which is removed by way of line 128. The bottoms stream is preferably recycled to extinction by way of line 130. The recycle oil in line 127 has received hydrogen by reaction in the hydrocracking reactors 118 and 119 and is therefore suitable for use as a hydrogen donor solvent. This material, boiling within the range of 350° to 750°F. is recycled and admitted into the mixer 102 as a slurrying oil for the coal 100, as well as providing the donor hydrogen for the liquefaction reaction.

The catalyst to be used is substantially spherical, with a minimum diameter at least ten times as great as the average maximum dimension of the largest 5% of the solid particles carried in the clarified oil. The catalyst may have only a hydrogenation activity or it may have both hydrogenation and cracking activity. Where the catalyst has only a hydrogenating activity, thermal cracking will cause the reduction in average molecular weight, and the catalyst will assist in hydrogenating the fragments. Suitable catalysts are cobalt molybdate, nickel molybdate, nickel tungsten, and palladium. Various substrates can be used such as kieselguhr, alumina, silica, faujasites, etc. The cobalt molybdate catalyst is preferred, and may have 3.4 weight percent cobalt oxide, 12.8 weight percent molybdenum oxide, and 83 weight percent alumina. The catalyst may range from 2 to 5 weight percent cobalt oxide and from 10 to 15 molybdenum oxide.

Phillips Petroleum

A process developed by K.L. Mills; U.S. Patent 3,235,508; February 15, 1966; assigned to Phillips Petroleum Company relates to a method for preparing catalysts comprising finely divided, nonporous solid oxide bases.

It is known that various reactants, especially hydrocarbons, can be converted either thermally or catalytically to some desired product. Many catalysts, both supported and unsupported, are known for hydrocarbon conversion reactions. However, many of the known catalysts are difficult to prepare in a desirable form for use in a particular hydrocarbon conversion process. For example, catalysts prepared from finely divided, nonporous solid oxides by impregnation with a metal catalyst element, and subsequently dried, do not retain the low density and fine characteristic of the original oxide carrier. Instead, the wet cake dries to a hard, relatively densified cake which can be broken into hard granules.

Although the cake or granules can be ground to fairly fine particles, e.g., 100-mesh and finer, the grinding does not effect subdivision even approaching the fineness of the original solid oxide carrier, which ordinarily has an average particle size of not over 100 millimicrons and more often less than 50 millimicrons.

This process is directed to an improved method for preparing catalysts from such finely divided, nonporous solid oxide carriers where the discrete particle sizes of the carrier are retained for use in the subsequent hydrocarbon conversion process.

Broadly, according to the process, a catalyst composition is provided comprising a finely divided, nonporous pyrogenic oxide having average particle sizes of less than about 100 millimicrons impregnated with a metal catalytic promoting agent dispersed in a liquid nondeleterious dispensing medium such as a liquid hydrocarbon. Further, according to the process, a method is provided for preparing slurries of supported catalysts suitable for direct charging to the reaction zone which comprises the steps of slurrying a finely divided, nonporous solid oxide carrier (also known as "finely divided pyrogenic oxides") with an aqueous solution of a metal catalytic element to impregnate the support with the catalytic element, mixing the slurry thus formed with a liquid hydrocarbon that is nondeleterious to the catalytic element, milling the mixture thus formed until a uniform dispersion is obtained, subjecting the dispersion thus obtained to flashing conditions to free the dispersion of water, and recovering a slurry of the catalyst in the hydrocarbon substantially free of water as a product of the method.

The catalyst slurry obtained according to the process described above can be charged directly to a conversion process, e.g., catalytic hydrogenation such as hydrocracking, especially where the catalytic element does not need any special activation treatment prior to contacting with the reactant material in the conversion zone. In such an operation, it is ordinarily preferred that the hydrocarbon employed as the finely divided catalyst dispersing medium be the same hydrocarbon material as the reactant hydrocarbon employed in the conversion process.

Alternatively, when a preliminary activation step or treatment is required for the catalytic element prior to use, the slurry of finely divided catalyst in hydrocarbon recovered after flashing to remove water can be heated or otherwise treated to remove the hydrocarbon dispersing medium to recover a dry catalyst consisting of the finely divided support impregnated with the desired metal catalytic element. The dry catalyst recovered is then subjected to the desired activation treatment prior to charging the activated catalyst to a conversion process.

Figure 83 illustrates the preparation of such a catalyst for a process for hydrocracking topped crude. A finely divided pyrogenic alumina is introduced into impregnation zone 10 by way of line 11 and is contacted therein with an aqueous solution of cobalt nitrate and ammonium molybdate introduced through line 12. The alumina base is impregnated in zone 10 by thoroughly mixing same with the aqueous solution of catalytic element until a uniform slurry is obtained.

FIGURE 83: BLOCK FLOW DIAGRAM OF OVERALL PROCESS FOR PREPARING AND UTILIZING HYDROCRACKING CATALYSTS CONTAINING MOLYBDENUM

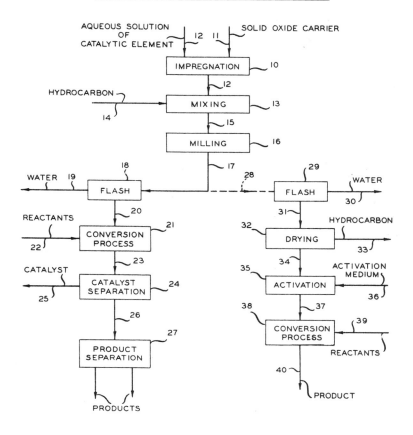

Source: K.L. Mills; U.S. Patent 3,235,508; February 15, 1966

The mixing in zone 10 can be continued until it is certain that the alumina carrier is impregnated with the desired amount of the catalytic element, which is cobalt-molybdate in this embodiment.

An aqueous slurry of the impregnated finely divided pyrogenic alumina carrier is removed from zone 10 and passed to mixing zone 13 where the aqueous slurry is intimately mixed with a hydrocarbon dispersing medium introduced by way of line 14. The hydrocarbon dispersing medium employed is topped crude, which is also the reactant for the subsequent hydrocracking process. The amount of hydrocarbon mixed with the aqueous slurry in zone 13 will be sufficient to provide a final mixture containing hydrocarbon as a major proportion of the mixture. The mixture obtained in zone 13 is then passed by way of line 15 to milling zone 16 where the mixture obtained in zone 13 is subjected to high-shear milling. Milling zone 16 can be any known milling or grinding device suitable for subjecting the mixture to high-shear milling so as to form a uniform dispersion of the catalyst in the hydrocarbon. Suitable devices include three-roll paint mills, pigment mills, ink mills, ball mills, colloid mills, homogenizers or similar devices.

The high-shear milling effected in zone 16 generally causes heating of the dispersion being passed through the mill so that the effluent removed from the milling zone ordinarily can be subjected to flash conditions without further heating to remove water from the dispersion. However, if the high-shear milling does not cause a sufficient temperature rise of the mixture, the mixture in line 15 passed to milling zone 16 can be heated or, if desired, the effluent removed by line 17 from milling zone 16 can be heated to the desired flashing temperature for removing water. Generally, the effluent removed from milling zone 16 is at a temperature ranging from about 225° to 400°F., preferably from about 250° to 350°F. to facilitate substantially complete water removal by flashing from the milling zone effluent. The mill effluent in line 17 at an elevated temperature such as set forth above is passed to flash zone 18 where the water is flashed off from the dispersion and removed by way of line 19. A liquid colloidal suspension of finely divided catalyst in topped crude, which is substantially water-free, is removed from flash zone 18 by way of line 20. The liquid colloidal suspension of catalyst in topped crude removed from the water flash zone is passed directly to a hydrocracking reaction zone 21. Topped crude reactant and hydrogen are charged to hydrocracking zone 21 along with the colloidal suspension which is operated under conventional conditions of temperature, pressure and reaction time to convert the topped crude to the desired product. The reactant and hydrogen can be charged to zone 21 by way of line 22. The hydrocracking reaction improves the gravity of the reactant hydrocarbon, which is ordinarily a material of poor properties, as well as effecting the removal of undesirable contaminants such as sulfur compounds, nitrogen compounds, organometallic compounds, and the like. The condition obtaining in zone 21 ordinarily ranges from a temperature of about 600° to 900°F., a pressure ranging from about 100 to 5,000 psig, a liquid hourly space velocity ranging from about 0.2 to 10, and a hydrogen flow rate of about 500 to 10,000 cubic feet per barrel of oil feed. The catalyst usually amounts to about 0.2 to 5 weight percent, preferably about 1 to 5 weight percent of the oil charged. When the oil-catalyst suspension is prepared with a higher concentration of catalyst than desired in the conversion zone, catalyst-free oil can be charged in sufficient amount to reduce the catalyst concentration to the desired level.

The hydrocracking zone effluent removed by line 23 is processed for catalyst and product separation. The catalyst can be separated, for example, by fractionation of the hydrocracker effluent to produce a slurry oil containing the catalyst, and the catalyst recovered therefrom by filtration or centrifuging. The recovered catalyst can be regenerated by burning off the carbon in a suspended solid type of operation. The above procedure is generally referred to in the drawing as zone 24. The catalyst removed from zone 24 by line 25 can be recycled to the hydrocracking zone 21 when desired. The products freed of catalyst are passed by way of line 26 to product separation wherein the products are separated into individual components by known separation procedures.

HYDRODESULFURIZATION

Chevron Research

A process developed by R.H. Kozlowski and B.F. Mulaskey; U.S. Patent 3,453,217; July 1, 1969; assigned to Chevron Research Company is a method for preparing a hydrotreating process catalyst comprising alumina and at least one Group VI metal sulfide hydrogenating component, and at least one Group VIII metal sulfide hydrogenating component. More specifically, the process comprises protecting the metal sulfides against reaction with oxygen by introducing in a liquid form into the pores of the catalyst a hydrocarbon protective material before exposure of the catalyst to an oxygen-containing atmosphere which has caused more than 5 weight percent of the metal sulfides to be converted to other compounds.

It is known that such sulfided catalysts are pyrophoric, and present storage problems when it is desired to store them for an extended period of time between their manufacture and their use in hydrotreating processes. Because of their pyrophoric nature they can react exothermically with oxygen in the atmosphere when exposed to the atmosphere, to the extent that a fire hazard can result. The exothermic reaction occurs between the metal sulfides and the oxygen, and even if does not occur to the extent of creating a fire hazard, it occurs at least to the extent of causing the generation of noxious fumes including sulfur dioxide. It is known to protect such sulfided catalysts against attack by oxygen in the atmosphere by such expedients as keeping them under blankets of inert gas such as CO_2 or nitrogen, by keeping them in closed drums, and by keeping the drums full. However, such expedients not only are unwieldy, but do not afford protection during the time when it is needed most, that is, when the catalyst is being transferred from storage drums into a hydrotreating reactor vessel.

The process will now be described in further detail with reference to the flow diagram in Figure 84. A mass 1 of particles of unsulfided catalyst, for example, comprising 7.5 weight percent nickel oxide, 24.0 weight percent molybdenum trioxide, 16.5 weight percent silica and 52.0 weight percent alumina, is passed from hopper 2 into sulfiding reactor vessel 3, where it is sulfided in an upper zone A at a temperature of 500° to 700°F. as it descends through reactor vessel 3 in countercurrent contact with a rising gaseous sulfiding agent introduced at a temperature of 100° to 300°F. into reactor vessel 3 through line 4. As the downwardly moving sulfided catalyst particles pass from zone A through zone B they are cooled by heat exchange with the rising gaseous sulfiding agent to a temperature of 300°F. or less, after which they pass from reactor vessel 3 through catalyst outlet 5.

The temperature of sulfiding zone A is maintained at the desired level by heating coils 6. Gaseous materials, including hydrogen, hydrogen sulfide, water vapor and carbon dioxide are removed from the upper portion of sulfiding zone A through line 7. The gaseous sulfiding agent may be, for example, a mixture of hydrogen sulfide with hydrogen or carbon dioxide. It is preferable to maintain the partial pressure of hydrogen sulfide at a low level so that the partial pressure of water vapor in sulfiding reactor vessel 3 will be maintained at a low level, to minimize any possible adverse effect of water vapor on the activity of the catalyst.

From catalyst outlet 5 the sulfided catalyst particles pass into a liquid body 10 of protective material, contained in container 11, and on to moving belt screen 12, moving in the direction shown to carry the protected catalyst particles out of liquid body 10 into catalyst storage container 13. As the catalyst particles emerge on moving belt screen 12 from liquid body 10 excess liquid protective material drains down and through belt 12 back into liquid body 10.

FIGURE 84:   APPARATUS FOR SULFIDING MOLYBDENUM-BASE HYDROTREATING PROCESS CATALYSTS

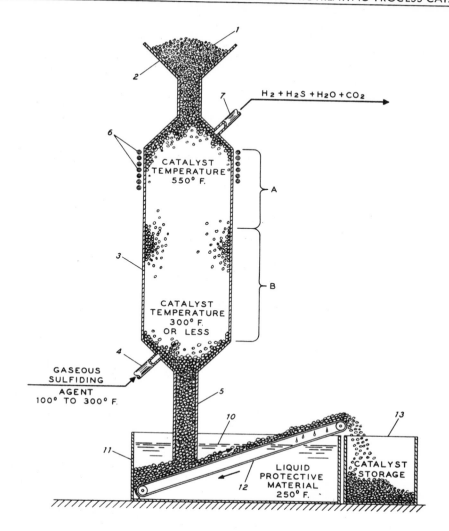

Source: R.H. Kozlowski and B.F. Mulaskey; U.S. Patent 3,453,217; July 1, 1969

The protective material used in the process consists essentially of hydrocarbons, including substituted hydrocarbons such as stearic acid. The protective material boils in the range of 410° to 1200°F., preferably 700° to 1100°F. Preferably it is a normally liquid hydrocarbon, including dodecane, catalytic cycle oils, and 700° to 1100°F. straight run petroleum distillates. However, it may also be a normally solid paraffin wax or microcrystalline wax.

Esso Research and Engineering

A process developed by C.E. Adams and W.T. House; U.S. Patent 3,509,044; April 28, 1970; assigned to Esso Research and Engineering Company for the desulfurization of petroleum residuum is carried out at nondestructive hydrosulfurization conditions by contacting the residuum with a high activity catalyst consisting essentially of a cobalt salt or nickel salt with a molybdenum salt or tungsten salt deposited on a support material consisting essentially of 1 to 6 weight percent alumina, the support being characterized by a maximum pore volume and surface area in pores 30 to 70 A. in diameter.

A process developed by C.E. Adams and W.T. House; U.S. Patent 3,531,398; September 29, 1970; assigned to Esso Research and Engineering Company involves the production of low-sulfur fuel oil by means of a continuous liquid phase hydrodesulfurization process for petroleum gas oil feed having an end point in the range of 750° to 1300°F. The process is applied to atmospheric gas oils, vacuum gas oils, and the like, containing about 0.1 up to 5.0 weight percent sulfur.

The high activity maintenance catalyst used in the process comprises a mixture of a salt of a Group VIII-B metal, e.g., cobalt or nickel, and a salt of a Group VI-B metal, e.g., molybdenum or tungsten, deposited on a support material consisting essentially of 1 to 6 wt. percent silica and 94 to 99 wt. percent alumina. The finished catalyst has a total pore volume of at least 0.25 cc/g. and a total surface area of at least 150 m.^{2}/g. Preferably the total pore volume is at least 0.4 cc/g. and the total surface area is at least 250 m.^{2}/g. In addition, a maximum of the total surface area, i.e., at least about 100 m.^{2}/g. of the surface area, is in pores having a diameter of 30 to 70 A.

Furthermore, since pores of the larger diameter seem to accelerate deactivation of the catalyst in the hydrodesulfurization of petroleum oils, it is preferred that the catalysts have a minimum of pores having a diameter of more than 100 A., i.e., less than 0.25 cc/g.

The process using such a catalyst is shown in the flow diagram in Figure 85. Referring to the drawing, a reduced crude oil feed is fed by line 1 through furnace 2 into vacuum distillation tower 3. Steam is fed by line 4 through the furnace into the vacuum unit. The feed is preheated in the furnace to a temperature in the range of 725° to 875°F. The vacuum unit is operated to maximize the recovery of a fraction amenable to continuous hydrodesulfurization. A stream comprising light ends and steam is recovered by line 5. Steam can be recovered and recycled by means not shown. A light gas oil fraction having a relatively low sulfur content can be recovered by line 6. A residuum fraction having an initial boiling point in the range of 1000° to 1150°F. is recovered by line 7. All or part of this material can be used as a fuel oil blending stock.

The prime desulfurization feed, a heavy vacuum gas oil, is passed via line 8, pump 9, lines 10 and 11, furnace 12 and line 13 to desulfurization reactor 14. The desulfurization feed may contain small quantities of residual material. Typical vacuum distillation conditions include a temperature in the range of 700° to 850°F. and a pressure in the range of 20 to 100 mm. Hg. Steam is added with the feed and to the bottom of the tower to enhance separation of distillable oil from the bottoms. This steam may amount to 1 to 20 pounds per barrel of oil feed. Velocity of the flow of vapors through the trays or other entrainment barriers above the flash zone is a serious limitation in operation of the vacuum tower. This velocity is normally in the range of 3 to 10 ft. per second. The velocity depends upon oil feed rate, temperature and pressure.

FIGURE 85:   PROCESS FLOW DIAGRAM FOR HYDRODESULFURIZATION OF HEAVY DISTILLATES

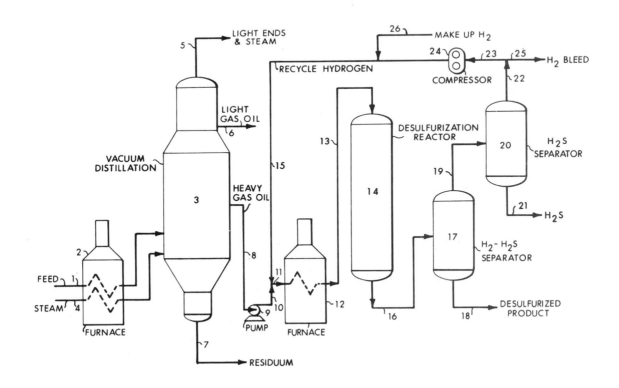

Source: C.E. Adams and W.T. House; U.S. Patent 3,531,398; September 29, 1970

At maximum temperature and minimum pressure with high feed rates, it can be said that the vacuum unit is being pushed and it is under these conditions that an occasional slug of residual petroleum oil is included in the deep cut desulfurization feed. The residual material contains a relatively high proportion of organo-metallic compounds, multiring aromatic hydrocarbons and other coke formers which rapidly deactivate a conventional hydrodesulfurization catalyst.

In furnace 12 the feed is preheated to a temperature in the range of 600° to 800°F. A hydrogen-containing gas containing 70 to 100 volume percent hydrogen is fed by line 15 into line 11 for mixing with the feed. A conventional liquid or mixed phase reactor is employed. The catalyst, previously described, is arranged in a series of stacked fixed beds in this embodiment. If desired, expanded beds, an ebbulating bed or slurry type operation can be employed. The oil and hydrogen are passed downwardly through the reactor but upflow techniques and countercurrent techniques can also be used. With typical gas oil feeds, the preferred reaction conditions include a temperature in the range of 650° to 800°F., a pressure in the range of 400 to 850 psig, and a hydrogen rate of 1,000 to 5,000 cu. ft./bbl.

A feature of the operating conditions is the pressure range. In the past it was felt that pressures above 1,000 psig, i.e., 1,200 to 1,500 psig, were essential for continuous 50 to 90% hydrodesulfurization of gas oils containing 2 to 3 wt. percent sulfur. Now it has been found that pressures below 1,000 psig are entirely suitable when the high activity maintenance catalyst disclosed herein is employed. Pressures as low as 400 psig are feasible and a temperature of about 700° to 750°F. is suitable throughout the run. These mild operating conditions coupled with the long run length provide numerous cost advantages in initial investment and in operating costs.

Desulfurization effluent is passed by line 16 to gas separator 17. Liquid product is removed by line 18. The treated product will usually contain less than 1% sulfur, i.e., 0.1 to 1.0 wt. percent sulfur, depending principally on the sulfur content of the feed and the severity of the reaction conditions. The gas stream is passed overhead by line 19 to H2S separator 20. H2S and light ends are separated by conventional means such as cooling and amine treating. H2S is removed by line 21. Hydrogen is recycled via lines 22 and 23, compressor 24 and line 15. A hydrogen bleed stream is removed by line 25. Makeup hydrogen is added as required by line 26.

Standard Oil (Indiana)

A process developed by R.J. Bertolacini and E.R. Strong, Jr.; U.S. Patent 3,393,148; July 16, 1968; assigned to Standard Oil Company (Indiana) yields a catalyst which is surprisingly useful for hydroprocessing heavy gas oils and hydrocarbon residua. This catalyst is a solid catalytic composition which comprises at least one hydrogenation component on a solid inorganic-oxide support comprising a large pore diameter alumina having a surface area within the range of about 150 to 500 square meters per gram and an average pore diameter within the range of about 100 to 200 A. Suitably, this catalyst comprises a hydrogenation metal of Group VI-A of the Periodic Table and a hydrogenation metal of Group VIII of the Periodic Table on a large pore diameter alumina having a surface area within the range of about 150 to 500 square meters per gram and an average pore diameter within the range of about 100 to 200 A. The support is further characterized in that a zeolitic molecular sieve is suspended in the matrix of the alumina.

Union Oil of California

A process developed by A.E. Kelley and F.C. Wood; U.S. Patent 3,644,197; February 22, 1972; assigned to Union Oil Company of California is a hydrofining process to reduce the nitrogen content of mineral oil feedstocks at relatively high-space velocities which comprises contacting the feedstock and added hydrogen with a conventional, substantially nonzeolitic, amorphous-based hydrofining catalyst, and contacting the effluent therefrom with a second hydrofining catalyst comprising a hydrogenating metal or metal sulfide supported on an active zeolitic cracking base.

Such hydrofining of mineral oil fractions is essential prior to their being subjected to catalytic hydrocracking. It is well known in the art that organic nitrogen has a deleterious effect on hydrocracking catalysts and that removal of the organic nitrogen in the feedstock is a necessary step in most hydrocracking processes. It is also well known that lower hydrocracking temperatures can be employed and longer run lengths can be realized in hydrocracking by reducing the nitrogen content of the feedstock to low levels, e.g., below 25 parts per million, and preferably below 10 parts per million. Conventional hydrofining catalysts can reduce the nitrogen content of the feedstock to these low levels; however, uneconomically low space velocities, high pressures and/or high temperatures are required.

According to the process, it has been found that the organic nitrogen content of a feedstock can be reduced to low levels at relatively low temperatures and pressures and at relatively high space velocities by a dual-catalyst system wherein the feedstock in the presence of hydrogen is first contacted with a conventional, substantially nonzeolitic hydrofining catalyst and is then contacted with a zeolite-based hydrofining catalyst comprised of a hydrogenating metal and/or sulfide thereof supported on a zeolitic cracking base having a cracking activity greater than that corresponding to a Cat-A Activity Index of about 40. Heretofore, it was considered undesirable to hydrofine with catalysts having a cracking activity higher than that corresponding to an Activity Index of about 25.

FIGURE 86:   INTEGRATED PROCESS FOR HYDROFINING AND HYDROCRACKING

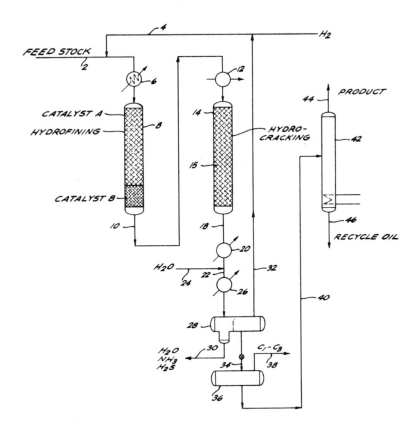

Source: A.E. Kelley and F.C. Wood; U.S. Patent 3,644,197; February 22, 1972

An integral hydrofining-hydrocracking process using such dual hydrofining catalysts is shown in Figure 86 above.   The initial feedstock, to be described later, is brought in via line 2, mixed with recycle and makeup hydrogen from line 4, preheated to incipient hydrofining temperatures in preheater 6, and is then passed directly into hydrofiner 8 and onto the conventional nonzeolitic hydrofining catalyst A, to be described later, where hydrofining proceeds under conditions substantially as follows:

	Hydrofining Conditions	
	Broad Range	Preferred Range
Temperature, °F.	600 – 850	650 – 800
Pressure, psig	400 – 4,000	800 – 2,500
LHSV, V/V-HR	0.5 – 10	0.9 – 7
H2/oil ratio, MSCF/B	0.5 – 20	2 – 12

The above conditions are suitably adjusted so as to effect partial denitrogenation of the feedstock.   The partially hydro-fined feedstock is passed from the conventional, nonzeolitic hydrofining catalyst A in hydrofiner 8 through the zeolite-based hydrofining catalyst B, where hydrofining proceeds even further under substantially the same conditions as those tabulated above, the conditions being correlated to carry the denitrogenation to the desired degree of completion.   The space velocity over catalyst B is preferably higher than the space velocity over catalyst A.   Due to the presence of sub-stantial quantities of organic nitrogen compounds in the catalyst B zone, little or no cracking of hydrocarbon components occurs and conversion to low-boiling materials is insubstantial.   Conversion to products, e.g., gasoline, boiling below the initial boiling point of the feed is normally less than about 25 volume percent, and preferably less than about 15 volume percent, and the "dry gas make", i.e., C1 – C3 yield, is usually less than about 60 SCF/barrel of feed, pref-erably less than about 30 SCF/barrel.   At this point, if further hydrocracking is not desired, the effluent from hydrofiner 8 may be withdrawn, cooled and fractionated to separate desired denitrogenated products.   However, in the modification illustrated, further hydrocracking is desired, and hence the total hydrofined product from hydrofiner 8, including ammonia and hydrogen sulfide formed therein, is withdrawn via line 10 and transferred via heat exchanger 12 to hydrocracker 14. Heat exchanger 12 is for the purpose of suitably adjusting the temperature of feed to the hydrocracking catalyst zone 15; this may require either cooling or heating, depending on the respective hydrofining and hydrocracking temperatures employed. The process conditions in hydrocracking zone 15 are suitably adjusted so as to provide about 20 to 70% conversion per pass to gasoline and/or other products, while at the same time permitting relatively long runs between regenerations, i.e., from about 2 to 12 months or more. For these purposes, it will be understood that pressures in the high range will normally be used in connection with temperatures in the high range, while the lower pressures will normally be used in conjunction with lower temperatures. The range of operative hydrocracking conditions is as follows:

Hydrocracking Conditions

	Broad Range	Preferred Range
Temperature, °F.	550 – 850	650 – 800
Pressure, psig	400 – 3,000	800 – 2,500
LHSV, V/V-HR	0.5 – 10	1 – 5
H_2/oil ratio, MSCF/B	0.5 – 20	2 – 12

It is important to observe that, if the hydrocracking catalyst is the same as the hydrofining catalyst B, a substantially lower average bed temperature should be maintained in the hydrocracking zone than in hydrofining catalyst bed B. This is to avoid overcracking which would otherwise take place in the hydrocracking zone in the substantial absence of organic nitrogen compounds. Normally in such cases, the average bed temperature in the hydrocracking zone should be about 10° to 80°F., preferably 30° to 70°F., lower than in hydrofining catalyst bed B. This will normally require cooling of the bed B effluent to below the inlet temperature of bed B.

Effluent from hydrocracking zone 15 is withdrawn via line 18 and partially cooled and condensed in exchanger 20 to a temperature of, e.g., 200° to 400°F., and mixed in line 22 with wash water injected via line 24. The resulting mixture is then further cooled in exchanger 26 to a temperature of, e.g., 50° to 200°F., and transferred into high-pressure separator 28 from which spent wash water containing dissolved ammonia and ammonium sulfide is withdrawn via line 30. Hydrogen-rich recycle gas is withdrawn via line 32 and recycled to line 2 via line 4. The condensed liquid product passes via line 34 into low-pressure separator 36, from which light hydrocarbon gases are exhausted via line 38. Low-pressure liquid condensate is then transferred via line 40 to fractionating column 42, from which desired product, e.g., gasoline, is taken overhead via line 44, and unconverted oil boiling above about 350° to 400°F. is withdrawn as bottoms via line 46. This unconverted oil may be recycled to hydrocracking zone 15 for further conversion to gasoline, or sent to a second hydrocracking stage (not shown) operated substantially in the absence of ammonia, whereby hydrocracking may be carried out at substantially lower temperatures than in zone 15.

It should be understood that the above-described process flow is for purposes of illustration only and that other flow schemes may be employed. For example, the effluent of hydrofiner 8 withdrawn via line 10 may be partially cooled and condensed and water-washed to remove ammonia and ammonium sulfide, and the washed condensate may then be fed to hydrocracker 14. The hydrogen contained in the effluent from hydrofiner 8 may be separated from the condensate and recycled back to hydrofiner 8. Also, any low-boiling products, e.g., gasoline, jet fuel, and/or diesel fuel, in the effluent of hydrofiner 8 may be separated from the high-boiling fraction, e.e., gas oil, therein so that such low-boiling products are not subjected to hydrocracking conditions in hydrocracker 14.

Additionally, hydrofining catalyst B may be located in the upper section of hydrocracker 14 in lieu of being located in the bottom section of hydrofiner 8, with the temperature of the feedstock and hydrogen entering the hydrocracking section of hydrocracker 14 being controlled by introducing quench hydrogen into the hydrocracking section. It is preferable, however, to locate hydrofining catalyst B in the bottom of hydrofiner 8 because it is usually more reliable and more efficient to control the temperature of the feed and hydrogen entering the hydrocracking section by heat exchanger 12 rather than by quench hydrogen.

As indicated above, the feedstock to hydrofining catalyst B is only partially hydrofined by the conventional, nonzeolitic hydrofining catalyst A, and the zeolitic hydrofining catalyst B completes the desired denitrogenation. The optimum amount of denitrogenation effected in each of the respective catalyst beds varies with the particular type of feedstock, tolerance of the hydrocracking catalyst to nitrogen poisoning, catalyst costs, vessel costs, and other similar factors. Generally, it is preferable to remove a substantial portion, e.g., 50 to 99%, preferably 75 to 98%, of the organic nitrogen in the upper catalyst bed B, and to utilize catalyst B to reduce the nitrogen content of the gas oil to the low level desired.

For example, with a feedstock containing about 2,000 to 4,000 ppm organic nitrogen and where it is desired to reduce the organic nitrogen content to below about 5 ppm, it is usually preferable to convert about 90 to 99% of the organic nitrogen in the upper catalyst bed comprising the conventional, nonzeolitic catalyst and utilize the zeolite-based catalyst B in the lower bed to complete the desired denitrogenation. If the feedstock contains only about 100 to 200 ppm of organic nitrogen and it is desired to reduce the nitrogen content to about 2 ppm, it is generally preferable to convert only about 50 to 80% of the organic nitrogen in the upper catalyst bed. It is generally advantageous to utilize the conventional, nonzeolitic catalyst to convert the less resistant organic nitrogen compounds, mainly the monocyclic and bicyclic compounds, and to utilize the zeolite-based hydrofining catalyst to convert the high-boiling, more resistant, organic nitrogen compounds having three or more rings, e.g., the carbazoles. Accordingly, it is preferable to effect relatively more denitrogenation in the lower bed containing the zeolite-based catalyst B where a large proportion, e.g., 30% of the organic nitrogen compounds are the highly resistant compounds than where only a small proportion, e.g., 5%, of the nitrogen compounds are the highly resistant compounds.

In broad summary, it may be said that as a general rule the degree of denitrogenation effected in the nonzeolitic catalyst A should be sufficient to reduce the organic nitrogen content of the feed to between about 5 and 250, preferably between 20 and 150, ppm, for at these organic nitrogen levels the zeolite-based catalyst B appears to exhibit its optimum selective denitrogenation activity.

The process conditions in each of the catalyst beds A and B are suitably adjusted to effect the above specified nitrogen removal in each of the beds. It is usually desirable to maintain both of the catalyst beds at substantially the same temperature, pressure and H2/oil ratio, and to vary the respective space velocities over each bed to effect the desired nitrogen removal by each of the respective catalysts. Accordingly, the space velocity with respect to catalyst B is generally 2 to 20 times, and preferably about 4 to 8 times, the space velocity with respect to catalyst A. The overall space velocity with respect to both catalysts is preferably between about 0.7 and 5.

As stated above, the catalyst in the upper bed A of hydrofiner 8 may comprise any of the conventional, substantially nonzeolitic, hydrofining catalysts, the base preferably having an ion exchange capacity of less than about 0.5 meq./g. Suitable catalysts include but are not limited to the transitional metals and sulfides thereof, and especially a Group VIII metal and/or sulfide (particularly cobalt or nickel) and/or a Group VI-B metal and/or sulfide (preferably molybdenum or tungsten). Such catalysts preferably are supported on an amorphous, nonzeolitic base in proportions ranging between about 2 and 25% by weight. Suitable bases include in general the difficultly reducible inorganic oxides which are amorphous and essentially nonzeolitic, e.g., alumina, silica, zirconia, titania, and clays such as montmorillonite, bentonite, etc.

Preferably the base should display little or no cracking activity, and hence bases having a cracking activity greater than that corresponding to a Cat-A Activity Index of about 25 are to be avoided. The preferred base is activated alumina, and especially activated alumina containing about 3 to 15% by weight of coprecipitated silica gel. The preferred hydrofining catalysts consist of nickel sulfide plus molybdenum sulfide supported on silica-stabilized alumina. Compositions containing between about 1 and 5% of Ni, 3 and 20% of Mo, 3 and 15% of SiO2, and the balance Al2O3, and where the atomic ratio of Ni/Mo is between about 0.2 and 4 are specifically contemplated.

Catalyst B in the lower portion of hydrofiner 8 comprises a minor proportion of at least one transitional metal hydrogenating component, preferably a Group VI-B metal and/or sulfide (preferably 5 to 30% molybdenum or tungsten) and/or a Group VIII metal and/or sulfide (preferably 1 to 10% of nickel or cobalt or 0.05 to 3% of palladium or platinum). The hydrogenating component is supported on a cracking base comprised of a zeolitic aluminosilicate, preferably a zeolite having relatively uniform crystal pore diameters of about 6 to 14 A., where the zeolitic cations comprise mainly hydrogen ions and/or polyvalent metal ions.

These zeolites may be used as the sole base, or they may be mixed with a minor proportion of one or more of the nonzeolitic bases such as silica-alumina cogel. Suitable zeolites include, for example, those of the X, Y, or L crystal types, mordenite, chabazite, and the like. Either crystalline and/or noncrystalline, amorphous zeolitic gases may be employed. In the case of both the crystalline and the noncrystalline bases, it is preferred that the ion exchange capacities thereof be greater than about 1.0 meq./g.

As stated above, it is desirable that the zeolitic base have a cracking activity greater than that corresponding to a Cat-A Activity Index of 40, and preferably greater than about 50. The Cat-A Activity Index of a catalyst is numerically equal to the volume-percent of gasoline produced in the standard Cat-A activity test as described in National Petroleum News, Aug. 2, 1944, vol. 36, p. R-537.

A particularly active and useful class of zeolite bases are those having a relatively high SiO2/Al2O3 mol ratio, e.g., between about 3 and 10. The preferred zeolites are those having crystal pore diameters between about 8 to 12 A., and where the SiO2/Al2O3 mol ratio is between about 3 and 6. A prime example of a zeolite falling in this preferred group is the synthetic Y molecular sieve.

The naturally occurring zeolites are normally found in a sodium form, an alkaline earth metal form, or mixed forms. The synthetic zeolites normally are prepared in the sodium form. In any case, it is preferred that most or all of the original zeolitic monovalent metals be ion exchanged out with a polyvalent metal, or with an ammonium salt followed by heating to decompose the zeolitic ammonium ions, leaving in their place hydrogen ions and/or exchange sites which have actually been decationized by further removal of water:

$$(NH_4^+)_x Z \xrightarrow{\Delta} (H^+)_x Z + xNH_3$$

$$(H^+)_x Z \xrightarrow{\Delta} Z + (x/2)H_2O$$

Both the hydrogen zeolites and the decationized zeolites possess desirable catalytic activity. Both of these forms, and the mixed forms, are designated here as being metal-cation-deficient. Hydrogen or decationized Y-sieve zeolites are more particularly described in U.S. Patent 3,130,006.

Mixed polyvalent metal-hydrogen zeolites may be prepared by ion exchanging first with an ammonium salt, then partially back exchanging with a polyvalent metal salt, and then calcining. Suitable polyvalent metal cations include the divalent metals such as magnesium, calcium, zinc, cobalt, nickel, manganese, and the like; the rare earth metals, e.g., cerium, or in general any of the polyvalent metals of Group I-B through Group VIII.

The hydrogenated metal, e.g., molybdenum, nickel, platinum and/or palladium, is then added to the base. The metal may be deposited on the zeolitic type cracking base by ion exchange. This can be accomplished by digesting the zeolite with an aqueous solution of a suitable compound of the desired metal as described, for example, in U.S. Patent 3,236,762.

Following the above procedures, the zeolite may be mixed with one or more of the nonzeolitic amorphous gels such as alumina and/or silica-alumina, then pelleted and calcined at temperatures between about 600° and 1200°F. The resulting catalyst is then presulfided, if desired.

ISOMERIZATION

Esso Research and Engineering

A process developed by E.M. Amir; U.S. Patent 3,484,385; December 16, 1969; assigned to Esso Research and Engineering Company yields an isomerization catalyst for polymethylbenzene formed by impregnating shapes of silica-alumina with soluble molybdenum compound and drying the impregnated shapes at a temperature below 650°F.

In the process shapes of amorphous silica-alumina containing from about 20 to 35% by weight of alumina and about 65 to 80% by weight of silica are formed into shapes such as pellets or pills, and the like, followed by heating the shapes to a temperature below about 950°F. after which the shapes are impregnated at ambient temperature with an aqueous solution of a molybdenum compound to provide a finished catalyst containing from about 3 to 10% molybdenum as MoO_3 on a dry basis. The impregnated shapes are then dried at a temperature below about 650°F. and preferably lower, and the so-formed catalyst may be then used in the isomerization of polymethylbenzene.

The temperature of the catalyst is adjusted to the desired isomerization temperature within the range from about 500° to 800°F. and the catalyst at this temperature is contacted with a polymethylbenzene containing feed under isomerization conditions in the presence of hydrogen to form an isomerized product. The isomerization conditions include a pressure within the range from about 100 to 500 psig, a feed rate of about 0.1 to 7 liquid v./v./hour and a hydrogen to feed mol ratio within the range from about 2:1 to 20:1.

The shapes may be heated to a temperature below about 950°F. for a period of time within the range from about 1 to 24 hours. Preferably the heating should be conducted for about 3 to 6 hours. The impregnated shapes should be dried over a period of time within the range from about 1 to 5 hours under the temperature conditions given. Superatmospheric pressure at which the shaped silica-alumina molybdenum-containing catalyst is heated may range from about 100 to 500 psig. A suitable pressure of about 250 psig may be used. These pressures may be maintained while the temperature of the catalyst is raised over the period of time given.

The step of heating the shapes to a temperature below about 950°F. may be omitted under some conditions where the shapes are formed without or in the absence of compaction and the shapes then impregnated without heating, following which the impregnated shapes are dried at a temperature below about 650°F. When the shapes are heated such as where extrusion, pilling, or compaction is used, the temperature must be maintained below about 950°F. within the preferred range from 800° to 900°F. Also, the impregnated shapes must be dried at a temperature below 650°F. (which may be within the range of about 250° to 450°F.) but preferably the impregnated shapes are dried at a temperature below 325°F. within a lower, more preferred range of 225° to 275°F. The shapes may be pellets or pills and may be formed by extrusion or by pelleting or by pressing.

The silica-alumina may be formed by coprecipitation or by mixing the separately precipitated silicic acid and aluminum hydroxide. Methods of forming synthetic amorphous silica-alumina are well known in the art.

It has been found that the efficiency of molybdenum-containing silica-alumina catalyst in the isomerization of poly-methylbenzenes is dependent on the method of preparation and especially the thermal history of the catalyst. Thus, it has been found that the isomerization of aromatic hydrocarbons such as the xylenes, the trimethylbenzenes and the tetramethylbenzenes, may be unexpectedly improved by the method of preparing the catalyst and the method of using the catalyst. It has been found that silica-alumina prepared either by coprecipitation or mixing of the separately pre-cipitated silicic acid and aluminum hydroxide should be formed into shapes such as pills or extrusions, and the like, following which the shapes such as pills may be heated to a temperature which must not exceed about 950°F.

Thereafter, the heated shaped silica-alumina is impregnated with a solution of a soluble molybdenum compound such as ammonium molybdate or molybdic acid. Thereafter, the impregnated shapes are dried by heating to a temperature below about 650°F. As stated later, the heating of the shapes prior to impregnation may be omitted, but when the heating operation is employed, the temperature may be within the preferred range from about 800° to 900°F. The drying must be below about 650°F. and may be within the range of about 225° to 650°F., more preferably in the range of 225° to 275°F. Where using catalysts containing silica in an amount of about 65 to 80% by weight and about 20 to 35% by weight of alumina, very desirable results are obtained in this procedure.

OXIDATION

Du Pont

A process developed by W.R. McClellan and A.B. Stiles; U.S. Patent 3,497,461; February 24, 1970; assigned to E.I. du Pont de Nemours and Company is one in which an improved bismuth molybdate on silica catalyst is made by mixing compounds to produce bismuth oxide, molybdenum oxide and, optionally, phosphorus oxide, in an aqueous slurry in the desired proportions, adding the slurry to an aqueous silica sol and then adding to the slurry ammonium carbonate or ammonium bicarbonate until the pH is in the range of 5 to 7.5, subsequently drying the calcining at 400° to 550°C. to produce the desired catalyst.

Bismuth phosphomolybdate or molybdate on silica catalysts are known in the art. They have been used for the oxidation of hydrocarbons, oxidative dehydrogenation of olefins, and also for the oxidation of olefin-ammonia mixtures to un-saturated nitriles. In general, such catalytic oxidations have required added water to obtain good conversions and yields.

Now it has been found that the catalytic activity of such catalysts can be improved by adding ammonium carbonate or ammonium bicarbonate to the catalysts during their preparation. Thus, during the conventional process of making such catalysts, prior to the drying step, ammonium carbonate or ammonium bicarbonate is added to the mixture of oxides in an amount to obtain a pH of 5 to 7.5. The composition is then dried and calcined in the usual manner.

Surprisingly, this addition creates such changes in the chemical and physical properties of the catalysts that the catalyst produced is homogeneous, finely divided, very stable, and has improved directivity or selectivity. Further, good con-versions and yields are obtained with these catalysts in many catalytic oxidations without the need for added water.

Knapsack-Griesheim

A process developed by K. Sennewald, K. Gehrmann, W. Vogt and S. Schafer; U.S. Patent 3,171,859; March 2, 1965; assigned to Knapsack-Griesheim AG, Germany is a process for the manufacture of unsaturated aldehydes or ketones from corresponding olefins by air oxidation with the use of a catalyst containing bismuth oxide and molybdenum oxide; more especially, for the manufacture of acrolein from propylene. In the course of time, acrolein has become a valuable inter-mediate for making, for example, acrylic acid derivatives, allyl alcohol, hexane diol, hexane triol, and other products. For this reason, attempts are being made to develop economical processes for making acrolein.

In known processes, acrolein is prepared by condensing formaldehyde and acetaldehyde, the acrolein being obtained in a yield of 70 to 80%. These processes use relatively costly and valuable starting materials. It has, therefore, been proposed that acrolein be prepared from cheap propylene and various such processes have already been described. Thus, for example, it has been proposed to oxidize propylene with the use of a catalyst consisting of bismuth oxide, phosphorus oxide, and molybdenum oxide; in another process, selenium is added to a copper oxide catalyst to improve the yield of acrolein.

In British Specification 655,210, for example, is described the preparation of acrolein with the use of a copper catalyst. In this process considerable proportions of expensive selenium must be added to a mixture of propylene-oxygen-inert gas in order to obtain a satisfactory yield of acrolein, so that the process is rendered uneconomical. Furthermore, only very dilute gases containing about 2% propylene are used. After a single passage, the yield of acrolein amounts to 77% and the total yield, after complicated recovery and return of unreacted starting material, amounts to 84%. In Belgian Patent 568,481, there is described a process for making unsaturated aldehydes and ketones, especially acrolein, where a catalyst is used to which no selenium, for example, must be added, but the yields obtained after single passage and the total yields of 57 and 72%, respectively, are relatively low. Still further processes also use copper oxide catalysts.

This process provides more especially for making unsaturated aldehydes or ketones from corresponding olefins by air oxidation with the use of a catalyst containing bismuth oxide and molybdenum oxide, where the olefinic hydrocarbon, advantageously after the addition of steam, is treated with air and/or oxygen at a raised temperature and in the presence of a catalyst of the empirical formula: $Fe_aBi_bP_cMo_dO_e$, in which a may be 0.1 to 12, b may be 0.1 to 12, c may be 0 to 10, d is about 12, and e may be 35 to 81. The elements bismuth, iron and molybdenum and optionally phosphorus are used in the catalyst in the form of their oxides. The catalyst may advantageously be applied onto a carrier, for example silicon dioxide or silica gel.

The catalyst is prepared by adding molybdenum oxide and, if desired, phosphoric acid to an aqueous solution of an iron salt and a bismuth salt, evaporating the resulting suspension together with colloidal silicon dioxide, for example, as carrier, and sintering the residue at an elevated temperature.

The process enables up to 99.5% of propylene to be converted and acrolein to be obtained in yields of up to 88%, calculated on the propylene, using a catalyst of the general formula $Fe_aBi_bP_cMo_dO_e$ without adding any activating or deactivating substance, by freeing the gas mixture leaving the reaction vessel from acrolein and recycling it into the reaction vessel after the amounts of propylene and oxygen consumed have been replenished. In view of the good yields obtained and the smooth reaction, none or only small amounts of carbon oxides are evolved during the reaction so that the proportion of off-gas is small. The off-gas only contains propylene in a proportion within the range of 0.5 to 1%, calculated on the propylene initially used. In addition thereto, the gas mixture used as starting material may contain propylene and oxygen in as high a concentration as about 15% of each substance. The reaction gas evolved therefore contains acrolein in a concentration of 6 to 8% by volume. Moreover, the catalysts are very active and furnish 150 to 400 grams per liter an hour of desired product, so that relatively small-dimensioned reaction vessels can be used.

Figure 87 shows the apparatus which may be used in the conduct of this process. The reaction gas which contains acrolein and escapes from reaction vessel 6 is conducted via hot line 7 to wash tower 8 where it is washed with water to remove acrolein, acetaldehyde and acrylic acid, the two latter substances being obtained in small proportions as by-products. Part of the heat content of the gas mixture may be utilized in a heat exchanger. The off-gas emanating from the water wash and flowing through cycle line 9 is supplied with measured quantities of fresh propylene and oxygen via supply

FIGURE 87: PLANT FOR THE MANUFACTURE OF ACROLEIN FROM PROPYLENE USING IRON BISMUTH PHOSPHOMOLYBDATE CATALYST

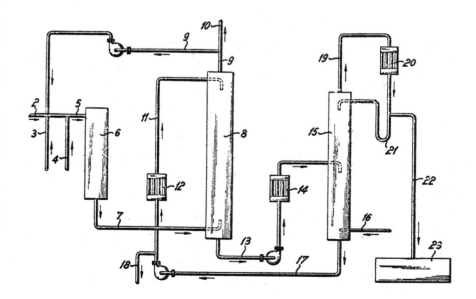

Source: K. Sennewald, K. Gehrmann, W. Vogt and S. Schafer; U.S. Patent 3,171,859; March 2, 1965

lines 2 and 3 to replenish the propylene and oxygen consumed. Steam is then added to the mixture via pipe 4 and the whole travels via collecting pipe 5 to reaction vessel 6 which is heated to a temperature of about 400°C. The additional carbon oxides obtained by propylene combustion are removed from the cycle via off-gas pipe 10 together with a small proportion of unreacted propylene.

In the optimum case, the loss of propylene amounts to only about 0.5%, calculated on the propylene initially used. The aqueous acrolein solution is removed at the bottom of wash tower 8 and conducted via line 13 and steam-heated preheater 14 to stripping column 15, the sump of which is maintained at 100°C. by means of steam line 16 and in which the acrolein together with acetaldehyde is obtained in the form of an azeotropic mixture with water as head product having a temperature of about 52°C. This product is withdrawn via head line 19 equipped with cooler 20 and is partially returned via return line 21 to the head of column 15. A part of the crude product is withdrawn via overflow line 22 and conveyed to collecting vessel 23. The crude product so obtained is already very pure contrary to the crude product obtained in conventional processes. The composition of the product is approximately as follows:

	Percent by Weight
Acrolein	94.57
Acetaldehyde	2.33
Propionaldehyde	0
Acetone	0
Water	3.1

The material removed at the bottom of column 15 travels through lines 17 and 11 equipped with cooler 12 back to wash tower 8 which is operated at a temperature of about 20°C. Part of the material is removed through wastewater line 18. In order to avoid polymerization, stripping column 15 is stabilized with a small proportion of an aqueous hydroquinone solution.

The crude product obtained in collecting vessel 23 is further treated in known manner by distillative purification and dehydration of the acrolein which finally has a degree of purity of at least 99.7%. Extractive distillation which would be normally required is omitted in view of the fact that the crude product obtained contains neither acetone nor propionaldehyde.

Shell Oil

A process developed by L.C. Fetterly, K.F. Koetitz and G.W. Conklin; U.S. Patent 3,190,913; June 22, 1965; assigned to Shell Oil Company involves the use of an arsenic-modified phosphomolybdic acid catalyst for the vapor phase oxidation of propylene to acrylic acid and isobutylene to methacrylic acid.

The arsenic component of the catalyst combination is preferably present in combination with oxygen, for example, as an oxide and/or acid form and/or salt of the acid form. The arsenic-modified phosphomolybdic acid catalysts employed in the process may be prepared by physical admixture of the phosphomolybdic acid with one or more arsenic compounds. They may be mixed in the dry state and the resulting physical admixture used as such; or use may be made of suitable carrying media in preparing the combination. Thus, the phosphomolybdic acid or the arsenic component, or both, may be dissolved or suspended in a suitable liquid medium and then combined; the carrying medium being thereafter removed by suitable means comprising one or more such steps as, for example, decantation, evaporation, filtering, centrifuging, and the like.

One or both components may be combined with a suitable carrying medium such as, for example, water, or any other suitable inert liquid, to form a paste before being admixed with each other. The resulting mixture is then dried and calcined. Comprised within the scope of the process is the pretreatment of the phosphomolybdic acid with arsenic or a compound thereof under conditions resulting in the decomposition of the arsenic component and/or the oxidation of arsenic etc., to result in a final mixture comprising phosphomolybdic acid in combination with an oxide of arsenic.

The arsenic content is maintained within the range of from about 0.1 to 10%, and preferably from about 0.1 to 5% by weight (calculated as elementary arsenic) of the catalyst. Particularly preferred are catalysts containing arsenic in the amount of from about 0.5 to 2% by weight of the catalyst, which range it has been found is definitely critical with respect to the obtaining of optimum and unexpected results.

Snam Progetti

A process developed by B. Notari, M. Cesari, G. Manara and G. Perego; U.S. Patent 3,492,248; January 27, 1970; assigned to Snam Progetti SpA, Italy yields a hydrocarbon oxidation catalyst which comprises molybdenum-bismuth-vanadium ternary compounds where the crystal lattice structure typifying monoclinic bismuth vanadates is distorted through the introduction of molybdenum atoms in place of vanadium atoms and the molar ratios of the components thereof are in the following range: Bi_2O_3/MoO_3 from 0.076 to 9.3; Bi_2O_3/V_2O_5 from 0.076 to 2.2; V_2O_5/MoO_3 from 0.076 to 9.3

Standard Oil (Ohio)

A process developed by J.L. Callahan, B. Gertisser and J.J. Szabo; U.S. Patent 3,248,340; April 26, 1966; assigned to The Standard Oil Company (Ohio) relates to an improved oxidation catalyst consisting essentially of oxides of the elements bismuth and molybdenum, and optionally, phosphorus, promoted by oxides of boron and bismuth, and to the catalytic oxidation of olefins to oxygenated hydrocarbons such as propylene to acrolein, and the catalytic oxidative dehydrogenation of olefins to diolefins such as butene-1 to butadiene, and tertiary amylenes to isoprene, and to the oxidation of olefin-ammonia mixtures to unsaturated nitriles such as propylene-ammonia to acrylonitrile using such catalysts.

POLYMERIZATION

Standard Oil (Indiana)

A process developed by J.W. Shepard, E.F. Peters and O.O. Juveland; U.S. Patent 3,352,795; November 14, 1967; assigned to Standard Oil Company (Indiana) relates to the polymerization of terminal vinyl olefins to produce normally solid polymers, and more particularly is concerned with providing an improved catalyst system for such polymerization.

In accordance with the process, terminal vinyl olefin monomers are polymerized to normally solid polymers by effecting polymerization in liquid phase with a catalyst system comprising: (a) a minor amount of an oxide of molybdenum in a submaximum valence state, supported on high surface area alumina, which has been contacted before use with a hydrogen halide in the vapor phase at multiple temperature stages within the range of about 150° to 500°C.; and (b) either sodium metal, lithium metal, or calcium metal.

It is essential that the metal oxide component of the catalyst system be molybdenum. Other metals of Group VI of the Periodic Table, viz. chromia and tungsten, do not give similar satisfactory results. Chromia hydrohalogenates too rapidly, and an excessive quantity of chromia, presumably in the form of a volatile and soluble oxyhalide, is removed from the catalyst. Indeed, hydrochlorination of unpromoted hexavalent chromia on silica kills its catalytic activity. Tungsten produces polymers of excessively high molecular weight, unusable for most molding applications. Further, it is required that the average valence of the molybdenum be below 6.0 (by dichromate titration in 50% H_2SO_4); this is sometimes referred to as a "submaximum valence state" or "subhexavalent" molybdena.

It is also important that the molybdenum be supported on a high surface area alumina base. The base or support may either consist of or substantially comprise alumina, but other supports such as silica, heretofore reported as more or less equivalent, are not satisfactory when used in conjunction with sodium promoter. Sodium, lithium or calcium promoters or components of the catalyst system bear a relationship to the molybdena on alumina catalyst not shared by other Group I and II metals. Potassium, for example, produces excessive amounts of low molecular weight greases and oils during polymerization; cesium and rubidium cause side reactions which reduce polymer yield and increase the difficulty of obtaining satisfactory polymer quality.

Undoubtedly the most distinctive aspect of the process is the effect of staged hydrohalogenation on subhexavalent molybdena on alumina. It has long been recognized in the patent literature that halogens have an empirical effect on polymerization catalysts; with some catalysts there is no disclosed tendency toward either enhancement or deactivation (U.S. Patent 2,944,049) while with others there is improvement, ranging from marginal (U.S. Patent 2,825,721) to good (U.S. Patent 2,725,374). On the other hand, with still other polymerization catalysts, halogens are said to be poisons (U.S. Patent 3,008,938).

A process developed by J.W. Shepard and O.O. Juveland; U.S. Patent 3,530,077; September 22, 1970; assigned to Standard Oil Company (Indiana) is one in which an olefin polymerization catalyst is prepared by distributing a minor amount of a transition metal oxide of Groups Va and VIa of the Periodic Table (such as molybdenum oxide) upon a major amount of alumina having a surface area within the range of 150 to 500 m.^{2}/g. and a pore diameter within the range of 100 to 200 A., then reducing the transition metal to an average valence about 1 less than maximum and chloriding the resulting combination with anhydrous hydrogen chloride to a chlorine content of about 2 to 5% by weight.

TUNGSTEN-CONTAINING CATALYSTS

HYDROCRACKING

Sinclair Research

A process developed by S.C. Haney and W.H. Decker; U.S. Patent 3,505,207; April 7, 1970; assigned to Sinclair Research, Inc. is one in which hydroconversion of raw, retorted shale oils by serial flow through a plurality of settled catalyst beds is improved by intermittently washing the initial catalyst bed with an aromatic hydrocarbon solvent. The wash is conducted under conditions which expand the catalyst bed and preferably in the presence of upflow hydrogen gas. The aromatic solvent can remove essentially completely the gummy, black, condensation products which form in the initial part of the first catalyst bed, causing excessive pressure drop and plugging the catalyst bed. The aromatic wash prolongs overall high catalyst activity by precluding or reducing the formation of coke by such deposits. Catalyst positioned subsequent to the initial bed can be washed as the need arises.

Under certain process conditions and catalyst system, shale oil may be hydroconverted to a product susceptible to a further refining using conventional petroleum processing to provide commercial products similar to those derived from petroleum crudes. Shale oil itself is not readily susceptible to such processing due to the complex cyclic structures which make up a substantial part of the raw oil. These complex materials tend to reduce the effectiveness of petroleum treating processes to an unattractive level. Further, raw shale oil contains considerable amounts of materials, such as sulfur and nitrogen compounds, metallic constituents, and the like, which rapidly poison some types of catalysts commonly used in petroleum processing. The expense of maintaining these catalysts active under such conditions often precludes the treatment of shale oils by processing techniques common to the petroleum industry. Hydroconversion of shale oils has been found to modify them to an acceptable degree to permit conventional petroleum processing of the hydrogenated shale oil products.

Retorted shale oil, when heated above a temperature of 250°F., in the presence of hydrogen, will form a gummy, black, adhesive precipitate which will deposit on the catalyst contacted in hydroconversion processes. These deposits accumulate over a period of processing and foul catalyst disposed as a fixed bed, causing excessive pressure drop through the bed and subsequent shut-down of the reactor for regeneration of the catalyst. It is thought that these deposits are caused by the rupture of the complex cyclic structures in the oil, partial dealkylation of some molecules, and condensation to form high molecular weight, resin-like materials. These reactions are further complicated by the presence in the shale oil of considerable amounts of metal oxides and sulfides, and nitrogen, sulfur, and oxygenated organic compounds. In the course of processing, the deposits will tend to further condense to form coke. These deposited materials impair the catalytic activity in the reactor by excluding the reactants from the necessary contact with the catalyst. As more material deposits, the catalyst bed eventually becomes plugged so that the feed is less and less able to pass through the reactor, creating an excessive pressure drop.

As the catalyst bed becomes fouled and the effectiveness of the catalyst declines, it is common to take the reactor off-stream, remove entrained oil from the reactor and then regenerate the catalyst by introducing air or other oxygen-containing gas at high temperature to burn deposits from the catalyst. Oxidative regeneration of the catalyst is expensive and time consuming. It is therefore highly desirable to provide a fixed catalyst bed shale oil hydrogenation process which is capable of extended operation without excessive fouling of the catalyst.

One method which will reduce the operational difficulties caused by such catalyst fouling involves periodic washing of the catalyst with an aromatic hydrocarbon solvent under conditions causing expansion of the catalyst bed. This treatment is also time consuming thereby representing an economic loss, especially when the catalyst bed is large or a plurality of beds are involved whether alternatively on stream or arranged for serial flow. However, raw shale oils exhibit the property of depositing the undesirable deposits to a much greater extent in the initial portion of the first hydroconversion catalyst with which the oils come in contact at an elevated temperature, than in subsequent portions of the catalyst.

This process takes advantage of this selective deposition to reduce the overall cost and inconvenience of fixed or settled catalyst bed systems used for hydroconverting raw shale oils with periodic washing of the catalyst by an aromatic hydrocarbon solvent. In this procedure an initial catalyst bed is provided subsequently contacted with the shale oil. Thus the raw shale oil first contacts an initial settled catalyst bed which is washed separately from remaining catalyst and the oil taken from the initial catalyst bed is subjected to further catalytic hydrotreatment.

The catalysts employed in the hydroconversion process can be the various sulfur resistant catalysts often employed in the treatment of heavy petroleum oils and also referred to as hydrogenation-dehydrogenation catalysts. Examples of suitable catalytic metal components include the members of Group VI-B in the Periodic Table, i.e. chromium, molybdenum and tungsten; vanadium; and the iron group metals of Group VIII, especially cobalt and nickel. Generally these metals are supported, e.g., on conventional refractory oxide catalyst supports such as alumina, silica, magnesia, zirconia, etc. and mixtures thereof and partially crystalline materials such as crystalline alumino-silicates of various types known in the art, and the metals are present in catalytically effective amounts, for instance 2 to 30 weight percent, in the form of the oxides, sulfides or other compounds thereof. Mixtures of these materials or compounds of two or more of the oxides or sulfides can be employed, for example, mixtures or compounds of the Group VIII metal oxides or sulfides with the oxides or sulfides of Group VI-B constitute very satisfactory catalysts.

Examples of such mixtures or compounds are nickel molybdate, tungstate or chromate (or thiomolybdate, thiotungstate or thiochromate) or mixtures of nickel and/or cobalt oxides with molybdenum, tungsten or chromium oxides. As the art is aware, these catalytic ingredients are generally employed while disposed upon a suitable carrier of the solid oxide refractory type, e.g., a predominantly calcined or activated alumina or silica-alumina, which can be mainly silica especially in hydroconverting catalysts. Commonly employed catalysts have 1 to 10% of a Group VIII metal and 5 to 25% of a Group VI-B metal (calculated as the oxide).

It has been found that macrosize catalyst particles of 1/64 to 1/2 or more inch in diameter preferably 1/16 to 1/4 inch in diameter, and lengths of similar dimensions or up to 1 inch or somewhat more, can be used in the catalyst beds. While particles of various shapes can be used and can be washed free of deposits, the spherical form has fewer objectional features, such as bridging, and is easier to set in a circulatory top-to-bottom bed motion during the bed wash cycle. Additionally, the spherical particles tend to form a relatively uniform packing upon settling, minimizing the void space between particles.

FIGURE 88: FIXED-BED HYDROCRACKING REACTOR WITH PROVISION FOR BENZENE WASH OF BED

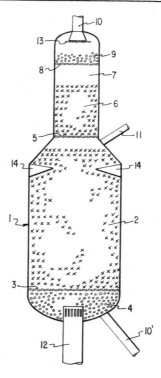

Source: S.C. Haney and W.H. Decker; U.S. Patent 3,505,207; April 7, 1970

Figure 88 shows a reactor 1, which is best described as a bottle reactor with features for washing the fixed catalyst bed. The main body of the catalyst is contained in the lower section of the "bottle" 2, supported on screen 3 which in turn rests on several feet of inert alundum spheres 4. The spheres are preferably 1/2 to 3/4 inch in diameter. Another screen 5, suitably supported is placed at the top of the main catalyst bed 2 and supports another bed of catalyst 6 in the "neck" portion of the "bottle". The upper catalyst section occupies from 1/2 to 4/5 of the volume of the upper section of the reactor, while 1/10 to 1/4 of the upper section is a void volume. At the top of the void space, screen 8 supports an additional layer of inert alundum spheres 9 which may range in size from 1/4 to 1 inch or even larger. The foregoing arrangement is particularly adapted to processing shale oils with intermittent washing of the top section of the reactor bed. The dimensional ratios shown below give satisfactory results, both in oil processing and washing of the catalyst bed.

	Range	Preferred
Vessel diameter, bottom : top	1.5 to 6 : 1	2 to 4 : 1
Length : diameter ratio:	2 to 20 : 1	
Top section	2 to 10 : 1	5 to 10 : 1
Bottom section	1 to 20 : 1	3 to 6 : 1
Catalyst volume ratio, bottom : top		3 to 10 : 1

During the oil process cycle, raw retorted shale oil enters the reactor through line 10, hits distribution plate 13, flows down through the top section of the catalyst 6, where the major part of the deposits form, hits redistribution plate 14, which extends around the periphery of the reactor, or any other suitable redistribution means to prevent oil flow directly down the wall of the reactor, and then passes downwardly through the main catalyst bed 2, and out through outlet line 12.

The wash cycle can be conducted in the following manner, still with reference to the figure: shale oil feedstock is stopped and the reactor is drained through line 12, aided by a purge flow of hydrogen or an inert gas if desired. Benzene, the preferred solvent, or other aromatic solvent as described above, is introduced through inlet line 10' and the reactor 1 is filled until it overflows into line 10, at which time the wash solvent flow may be halted. A hydrogen-rich gas, such as naptha reformer off-gas, is introduced through line 11, from which it flows upward through the upper catalyst bed 6. The catalyst bed in the upper section can be expanded to such an extent that particles from the top of the bed will strike screen 8, or, if desired or necessary and if adequate gas is available, the entire upper section may be agitated in such a manner that a top to bottom circulatory motion of the catalyst particles may be obtained.

The void upper section of the reactor provides necessary space for the motion of the particles and the consequent expansion of the upper catalyst bed 6, where the aforementioned gummy condensation products are deposited. After a suitable period of washing, the dirty solvent and gummy deposit are flushed through line 10 with fresh solvent. After 50 to 100% of the solvent is displaced, a second wash cycle is begun. The wash cycles may be repeated until the displaced solvent is clean. When essentially all the gummy material is removed, the solvent will appear as clear or a light straw color, indicating a clean catalyst. A continuous upflow of solvent at an appropriate mass velocity to agitate the catalyst bed may replace the hydrogen gas flow in part or completely. Other gases, e.g., nitrogen, carbon dioxide, flue gas and the like may also be utilized in place of hydrogen. Upon completion of the wash cycle, the solvent is drained from the reactor through line 10' or line 12. The used solvent may be recovered by decantation, fractionation, or any other of the methods known in the art.

In the washing cycle the upward flow of the wash solvent and the hydrogen or other gas at a suitable rate to expand the catalyst bed serves to promote contact between the solvent and the catalyst particles. In addition, the motion produces a scouring motion on the catalyst particles. The random movement of the catalyst causes the particles to rub against one another and against the wall of the reactor. This scouring action serves to loosen and remove the deposits from the catalyst and assist in dissolving or flushing away of these materials by the solvent. In this fashion, even the heaviest and least soluble materials deposited on the catalyst are flushed away.

HYDROTREATING

Chevron Research

A process developed by J. Jaffe; U.S. Patent 3,546,094; December 8, 1970; assigned to Chevron Research Company yields a hydrotreating catalyst comprising a layered synthetic crystalline aluminosilicate cracking component, preferably substantially free of any catalytic metal or metals, a silicaalumina gel component a Group VI hydrogenating component such as tungsten and a Group VIII hydrogenating component such as nickel.

It is known that a catalyst may comprise a crystalline zeolitic molecular sieve component associated with other catalyst components. It is also known that a least some of the other catalyst components may be in the form of a matrix in which the molecular sieve component is dispersed. It is also known that such catalysts may be used for such reactions as catalytic cracking, hydrocracking, and hydrodesulfurization. Representative prior arts patents disclosing one or more of the foregoing matters include: U.S. Patent 3,140,251; U.S. Patent 3,140,253; British Patent 1,056,301; French Patent 1,503,063, and French Patent 1,506,793. There has been a continuing search for further improvements in such catalysts, and in similar multicomponent catalysts, particularly for hydrocracking and hydrofining uses.

It is also known that a crystalline zeolitic molecular sieve cracking component, while relatively insensitive to organic nitrogen compounds and ammonia, has a well-ordered and uniform pore structure as a result of the crystal structure having bonds that are substantially equally strong in three dimensions. This provides definite limitations on the access of reactant molecules to the interiors of the pores.

It is also known, particularly from U.S. Patent 3,252,757, that a relatively new layered crystalline aluminosilicate clay-type mineral that has been synthesized has the empirical formula

$$nSiO_2 : Al_2O_3 : mAB : xH_2O$$

where the layer latices are comprised of silica, alumina, and the B compounds, and in which n is from 2.4 to 3.0 and m is from 0.2 to 0.6. A is one equivalent of an exchangeable cation having a valence not greater than 2, and is external to the latice. B is chosen from the group of negative ions which consists of F^-, OH^-, $1/2\ O^=$ and mixtures of these, and is internal in the lattice. At 50% humidity x is from 2.0 to 3.5. The mineral is characterized by a d_{001} spacing at 50% humidity within the range which extends from a lower limit of 10.4 A. to an upper limit of 12.0 A. when A is monovalent, to 14.7 A. when A is divalent; and to a value intermediate between 12.0 A. and 14.7 A. when A includes both mono-valent and divalent cations. The equivalent of an exchangeable cation, A, in the mineral may be chosen from the group consisting of H^+, NH_4^+, Li^+, K^+, $1/2Ca^{++}$, $1/2Mg^{++}$, $1/2Sr^{++}$, and $1/2Ba^{++}$, and mixtures of these.

The layered synthetic crystalline aluminosilicate mineral (hereinafter referred to for brevity as "layered aluminosilicate"), in the dehydrated form, is known from U.S. Patent 3,252,889 to have application as a component of a catalytic cracking catalyst; however, applications of the layered aluminosilicate, in either hydrated or dehydrated form, as a component of a hydrofining or hydrocracking catalyst have not been disclosed heretofore.

In view of the foregoing, objects of the process include providing an improved catalyst comprising a cracking component associated with other catalyst components that has, compared with similar prior art catalysts:

(1) improved hydrocracking activity,
(2) improved hydrodenitrification activity,
(3) improved hydrocracking stability,
(4) improved hydrodenitrification stability,
(5) a cracking component that is crystalline in structure, having pores
      elongated in two directions, contrary to the pores of crystalline
      zeolitic molecular sieves, and therefore having less reactant access
      limitations than the pores of such molecular sieves.

As shown in Figure 89, the catalyst of the process may be used on a once-through basis to concurrently hydrocrack and hydrodenitrify a hydrocarbon feedstock to produce more valuable products, some of which may be further upgraded by catalystic reforming or catalytic hydrocracking, if desired.

FIGURE 89:   FLOW DIAGRAM OF INTEGRATED HYDROCRACKING — HYDROTREATING PROCESS

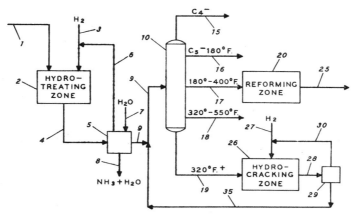

Source: J. Jaffe; U.S. Patent 3,546,094; December 8, 1970

A hydrocarbon feedstock, which in this case may boil above 400°F. and which may contain a substantial amount of organic nitrogen compounds, is passed through line 1 into hydrofining–hydrocracking zone 2, which contains the catalyst described above. The feedstock is hydrocracked in hydrocracking zone 2, in the presence of hydrogen supplied through line 3. Under these conditions, concurrent hydrodenitrification takes place to the extent that the feedstock is substantially denitrified. The effluent from zone 2 is passed through line 4 to separation zone 5, from which hydrogen separated from the treated feedstock is recycled through line 6 to zone 2. In zone 5, water entering through line 7 is used to scrub ammonia and other contaminants from the incoming hydrocarbon stream, and the ammonia, water and other contaminants are withdrawn from zone 5 through line 8.

From zone 5, the scrubbed, hydrocracked materials are passed through line 9 to distillation column 10, where they are separated into fractions, including a C4 fraction which is withdrawn through line 15, a C5 180°F. fraction which is withdrawn through line 16, a 180° to 400°F. fraction which is withdrawn through line 17, a 320° to 550°F. fraction which is withdrawn through line 18, and a 320°+F. fraction which is withdrawn through line 19. The C5 180°F. fraction withdrawn through line 16 is a superior–quality light gasoline. The 180° to 400°F. fraction withdrawn through line 17 is a superior catalytic reforming feedstock, which may be catalytically reformed in reforming zone 20, from which a superior catalytic reformate may be withdrawn through line 25. The 320° to 550°F. fraction withdrawn through line 18 is a superior–quality jet fuel. The 320°+F. fraction withdrawn through line 19 is a superior hydrocracking feedstock, which may be catalytically hydrocracked in hydrocracking zone 26 in the presence of a conventional hydrocracking catalyst and in the presence of hydrogen supplied to zone 26 through line 27.

From hydrocracking zone 26, an effluent may be withdrawn through line 28, hydrogen may be separated therefrom in separator 29, and hydrogen may be recycled to hydrocracking zone 26 through line 30. Alternatively, the 320°+F. fraction may be catalytically cracked in a catalytic cracking zone under conventional catalytic cracking conditions. From separator 29, hydrocracked materials may be passed through lines 35 and 9 to distillation column 10, where they may be separated into fractions, as previously described.

OLEFIN DISPROPORTIONATION

Phillips Petroleum

A process developed by R.L. Banks; U.S. Patent 3,631,118; December 28, 1971; assigned to Phillips Petroleum Company for preparing isoprene from ethylene is one wherein ethylene and butadiene are codimerized to produce a branched acyclic diene and the branched acyclic diene is contacted with an olefin disproportionation catalyst in the presence of ethylene to produce the isoprene. The butadiene can conveniently be provided by the olefin disproportionation of propylene to provide ethylene and butenes, and subsequent dehydrogenation of the butenes. Figure 90 illustrates the preparation of isoprene utilizing the olefin disproportionation and ethylene/butadiene codimerization steps.

FIGURE 90: FLOW DIAGRAM OF ISOPRENE PRODUCTION PROCESS USING OLEFIN DISPROPORTIONATION PLUS ETHYLENE–BUTADIENE CODIMERIZATION

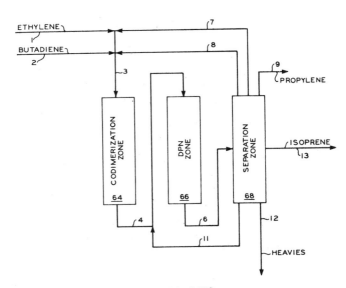

Source: R.L. Banks; U.S. Patent 3,631,118; December 28, 1971

As shown there, ethylene in line 1 and butadiene in line 2 are introduced into codimerization zone 64 by way of line 3. Therein, a suitable codimerization catalyst promotes the conversion of the feed materials to methylpentadiene codimer, such as 3-methyl-1,4-pentadiene. Within zone 64, other by-products including still different 1:1 addition products are formed such as 1,4-hexadiene, 1,3-hexadiene, cyclohexene, and linear butadiene dimers.

The effluent of the codimerization reaction is withdrawn via line 4 and passed to olefin disproportionation (DPN) zone 66. Unconverted ethylene from the codimerization reaction and unconverted butadiene are also subjected to the exhaustive ethylene cleavage reaction within zone 66. The olefin disproportionation of the effluent, containing a large proportion of ethylene, produces a variety of olefinic products including substantial amounts of isoprene. For best results, the molar ratio of ethylene to other olefins in the cleavage zone is at least 2:1 and preferably at least 4:1. The isoprene is the result of exhaustive ethylene cleavage of the branched polyenes such as the methylpentadiene codimer. The other principal product of this cleavage reaction is propylene. Within zone 66, substantial amounts of the other unbranched codimeriza-by-products such as cyclohexene and the linear hexadienes are reconverted to butadiene.

The effluent from disproportionation zone 66 is passed via line 6 to separation zone 68. Therein, ethylene and butadiene are separated and recycled via lines 7 and 8, respectively, to line 3 and to codimerization zone 64. By-product propylene is removed from the system via line 9. Unconverted codimerization products are returned to olefin disproportionation zone 66 by way of line 11. Heavier materials are removed from the process via line 12. Isoprene is passed from the separation zone 68 via line 13 and recovered as the product of the process.

In a modification of the process as depicted, some or all of the propylene which is recovered as a by-product of the process can be recycled to the olefin disproportionation zone 66 and utilized as a portion of the cleaving olefin. In this embodiment, butenes become a significant by-product of the reaction within the disproportionation zone 66 and are recovered from separation zone 68 in any suitable manner.

FIGURE 91: FLOW DIAGRAM OF ISOPRENE PRODUCTION PROCESS USING OLEFIN DISPROPORTIONATION AND ETHYLENE AS THE SOLE FEEDSTOCK

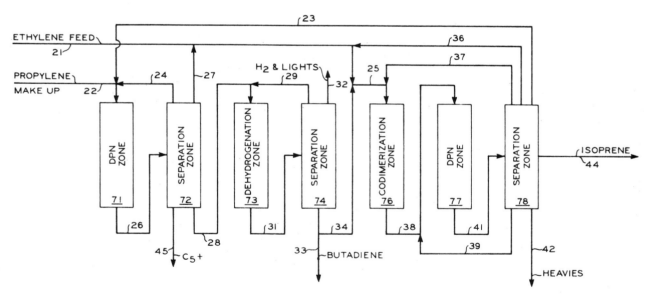

Source: R.L. Banks; U.S. Patent 3,631,118; December 28, 1971

Figure 91 shows a variation in the process wherein isoprene is prepared using ethylene alone as the feedstock. As shown there, ethylene in line 21 is fed via line 25 into a codimerization zone 76. In line 25, the ethylene is admixed with butadiene from line 34 which is prepared within butene dehydrogenation zone 73. The propylene disproportionation zone 71 receives make-up propylene from line 22 and recycle propylene from separation zones 72 and 78 via lines 23 and 24. Within disproportionation zone 71, the propylene is converted to ethylene and butenes according to:

$$2CH_2{=}CHCH_3 \longrightarrow CH_2{=}CH_2 + CH_3CH{=}CHCH_3$$

The effluent from the propylene disproportionation zone 71 is passed via line 26 into separation zone 72. From within zone 72, unconverted propylene is recycled to disproportionation zone 71 via line 24. Ethylene is carried overhead in line 27 into line 21 to provide a portion of the feed to the codimerization zone 76. Butenes are recovered in line 28 and passed to dehydrogenation zone 73. Any heavier hydrocarbons present in the effluent from the disproportionation zone 71 are passed from the system via line 45.

Within the dehydrogenation zone 73, the butenes are converted to butadiene. The effluent from dehydrogenation zone 73 is carried by line 31 to separation zone 74. Therein, hydrogen and light materials are recovered via line 32 and passed from the system. Unconverted butenes are separated and recycled via line 29 to zone 73. Butadiene can be recovered as valuable product of the process from separation zone 74 by way of line 33. Butadiene is also passed via line 34 into codimerization zone 76 wherein ethylene and butadiene are converted to codimers such as 3-methyl-1,4-pentadiene, 1,4-hexadiene, 1,3-hexadiene, cyclohexene, and butadiene oligomers as discussed above.

The entire effluent, containing the olefinic products and a substantial molar excess of ethylene from codimerization zone 76, is passed via line 38 into disproportionation zone 77 wherein the effluent is contacted with an olefin disproportionation-isomerization catalyst system. This reaction produces isoprene as a result of the ethylene cleavage of the methylpentadiene, while the exhaustive ethylene cleavage of the linear codimers, butadiene dimers, and other butadiene oligomers results in the regeneration of substantial amounts of butadiene. Another substantial product of the reaction within disproportionation zone 77 is propylene resulting from the ethylene cleavage reactions.

The effluent from the ethylene cleavage zone 77 is passed via line 41 into separation zone 78. Within separation zone 78, isoprene is recovered via line 44 and compounds heavier than butadiene oligomers are removed from the system via line 42 as heavies. The unconverted codimerization products are returned to olefin disproportionation zone 77 via line 39. Butadiene is recovered and recycled to the codimerization zone 76 via line 37. Ethylene is recovered and recycled to codimerization zone 76 via line 36. Propylene is recovered from separation zone 78 and passed via line 23 to line 22 and to propylene disproportionation zone 71 for conversion to ethylene and butenes.

The system as depicted provides an efficient way of preparing isoprene from ethylene alone. The quantity of propylene produced in the system is sufficient to satisfy the feed requirements for the olefin disproportionation reactor 71. Therefore, only makeup propylene need be introduced via line 22 to the system. In addition, the quantities of propylene, ethylene and butenes produced are sufficient that enough butadiene is prepared in dehydrogenation zone 73 to produce some butadiene as an additional product of this embodiment of the process.

The step requiring the dehydrogenation of butenes to butadiene can be carried out by any suitable catalytic process. For example, dehydrogenation processes employing catalysts such as the well known iron-potassia-chromia catalysts, or a catalyst such as lithium-treated tin oxide-tin phosphate catalysts can be used.

The olefin disproportionation steps of the process are carried out using suitable catalysts. Any catalyst having activity for olefin disproportionation reactions can be employed. These include solid (heterogeneous) and solution (homogeneous) catalysts, or combinations thereof. Suitable catalysts include those disclosed in U.S. Patent 3,261,879, (1963), U.S. Patent 3,365,513, (1968), and U.S. Patent 3,448,163, (1969), and U.S. Patent 3,558,518, (1971). Either solid or solution olefin disproportionation catalysts can be used for the propylene disproportionation steps. Solid catalysts are preferred for the ethylene cleavage disproportionation step. Some examples of preferred olefin disproportionation catalysts are WO_3/SiO_2, MoO_3/Al_2O_3, $Mo(CO)_3/Al_2O_3$, Re_2O_7/Al_2O_3, $WO_2/AlPO_4$, WO_3/Al_2O_3, and the like and mixtures thereof.

In that olefin disproportionation step of the process wherein the effluent from the codimerization reaction is exhaustively cleaved with ethylene, the olefin disproportionation catalyst is utilized in conjunction with or association with a suitable double bond isomerization catalyst such as, for example, magnesium oxide. The presence of the double bond isomerization catalyst greatly increases the conversion to the desired isoprene and butadiene. A particularly suitable combined catalyst for this step of the process is silica-supported tungsten oxide combination with magnesium oxide in the form of a mixed bed in which the MgO is present in amounts of 1 to 20 parts MgO per part WO_3/SiO_2 by weight.

OLEFIN HYDRATION

Japan Gasoline

A process developed by F. Sato, A. Okagami and T. Ueno; U.S. Patent 3,452,106; June 24, 1969; assigned to Japan Gasoline Co., Ltd., Japan is a process for the hydration of an olefin to the corresponding alcohol by passing the olefin together with liquid water over a catalyst at a temperature of 160° to 360°C., a pressure of 200 to 500 atmospheres gauge, an LHSV of olefin of 0.1 to 3.0 (m.^{3}/m.^{3} hr.) and a liquid velocity of over 1.0 (m.^{3}/m.^{2} hr.) The catalyst consists substantially of a blue oxide of tungsten and an oxide of chromium as a binding agent corresponding to a chromic oxide content of 5 to 50% by weight of the total weight of the catalyst. Figure 92 shows the application of such a catalyst to an olefin hydration process. The olefin which has been pressurized up to a desired reaction pressure is introduced into the system through a pipe 1, while similarly pressurized water is introduced into the system through a pipe 2, and they are passed through a heat exchanger 3 and a preheater 4 for being preheated, and thereafter, they are fed to a reactor 5. The reaction product withdrawn from the reactor 5 is separated into gas and liquid in a separator 6 which may be a high pressure separator, and the separated gas and liquid are depressurized at the valves 7 and 8, respectively. An unreacted olefin after separation may be recycled to the system. The liquid after being separated from the gas is introduced to a distillation column 9 where alcohol is separated from water. The separated water is withdrawn from the bottom of the column and is recycled to the system.

FIGURE 92:   DIRECT LIQUID PHASE OLEFIN HYDRATION PROCESS USING TUNGSTEN-CHROMIUM OXIDE CATALYST

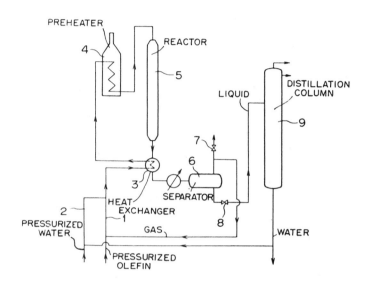

Source: F. Sato, A. Okagami and T. Ueno; U.S. Patent 3,452,106; June 24, 1969

MANGANESE-CONTAINING CATALYSTS

AUTOMOTIVE EXHAUST TREATMENT

Du Pont

A process developed by B.W. Howk and A.B. Stiles; U.S. Patent 3,216,954; November 9, 1965; assigned to E.I. du Pont de Nemours and Company is directed to mangano-chromia-manganite catalysts and their use for the catalytic conversion of components of automobile exhaust.

It is impractical to use catalysts that require a high exhaust gas temperature to initiate reaction because automobiles operate much of the time in city driving and at comparatively low exhaust temperatures. The catalysts light-off at relatively low temperatures, even after extended use, so that they are practical and effective under ordinary conditions of motor vehicle operation. The mangano-chromia-manganite catalysts can effectively be used to fuse the operation of catalysts which require a higher light-off temperature because the manganite catalysts begin operating at a low temperature and heat the catalyst bed and exhaust stream. Catalysts effective for treatment of such waste gases are ordinarily extremely sensitive to water. Liquid water, as well as water vapor, is found in the exhaust gases of motor vehicles and in other such combustion systems. The liquid water causes spalling of the catalysts and weakens them mechanically so much that after very short periods of operation they are no longer useful. Catalysts of the composition are quite resistant to damage from liquid water and water vapors.

Moderately effective systems have been devised for the handling of exhaust gases of motor vehicles providing the motors are operated on carefully selected fuels. A commercially satsifactory catalyst should be resistant to lead, sulfur, halogens, phosphates, boron, and their reaction products, and to hydrocarbon fuels and oils and their partial combustion products. Such a catalyst should also be resistant to manganese compounds which are sometimes used as antiknock agents. The catalysts of the process have remarkable resistance to such constituents often found in auto exhausts. They are also quite resistant to deactivation by high temperatures because of their composition and structure.

It is imperative that a satisfactory catalyst be effective for the oxidation or reduction of as many as possible of the numerous components in the waste gases being treated. The catalysts are effective in converting the nitrogen oxides, carbon monoxide, the hydrocarbons, and other exhaust gas components and are thus broadly effective. It will be noted that some of these are somewhat less effective in converting nitrogen oxides than others but they are nevertheless valuable for the conversion of carbon monoxide and hydrocarbons. It will be seen that exhaust gases can be converted in a plurality of stages. Thus the reduction of nitrogen oxides can occur in a first stage and then oxygen can be added in the second stage for the oxidation of carbon monoxide and hydrocarbons. The same or different catalysts can be used in the two or in the plurality of stages.

An air compressor driven by the motor can be used for supplying an excess of 30% or more over that stoichiometrically required to react with the hydrocarbons and carbon monoxide. Alternatively, a Venturi can be used. As mentioned earlier, the addition of air at a later stage may assist in reaching a light-off temperature since the air if introduced sooner would otherwise have a cooling effect. It is imperative that a satisfactory catalyst have a reasonably small volume so that it can conveniently be carried by a motor vehicle without requiring an unreasonable amount of room. Another practical consideration is that the catalyst should have a reasonably long service life, though this can be balanced somewhat against low cost. The catalysts are relatively inexpensive, have a surprisingly long life, and are effective in relatively small amounts.

Catalysts of the process are characterized by the use of mangano-chromia-manganite. This is composed of varying proportions of: MnO, MnO_2, Mn_2O_3, Mn_3O_4, Cr_2O_3, $Cr_2O_3 \cdot MnO_2$, $CrO \cdot Mn_2O_3$, $CrO_3 \cdot MnO$, and $Cr_2O_3 \cdot MnO$. In addition to the above oxides, a mangano-chromia-manganite catalyst quite possibly contains still other oxides of manganese, of chromium, and of chromium plus manganese which are not known compounds and which are not subject to positive identification by any means available. Thus Mn_2O_5 may be present when the state of oxidation of the catalyst is high and other oxides of similar unusual character may also exist in a transitory and unidentifiable form. The complex product of the process is chemically represented by the following

$$XCr_2O_n \cdot 2YMnO_m$$

in which the weight ratio of Mn:Cr is 3:0.5 to 3:30. Thus Y = 3, X = 0.5 to 30. Instead of the weight ratio, the atomic ratio can be used for it is almost the same. In the formula, n can be 2, 3, and 6 and m can be 1, 1.33, 1.5, 2, and 2.5. While any of the mangano-chromia-manganites as described can be used in preparing catalysts according to the process, it is much preferred to use those in which the ratio of Y:X is either above 3:3 or below 3:2. The weight ratio of Y:X, that is Mn:Cr, is preferably 3:0.5 to 3:1.5 and 3:3.5 to 3:30. Mangano-chromia-manganites thus constituted are far more valuable and thermally stable as catalysts than those having a ratio of 3:2 to 3:3 and can be used even without the interspersants which will be described hereafter. If the ratio is around 3:2 to 3:3 interspersants must be used.

While chemically the mangano-chromia-manganites can be designated as above, it is to be understood that they include ionic crystals in which the crystals have the spinel structure. Each chromium ion may be thought of as surrounded by 6 oxygen ions and each manganese ion by 4 oxygen ions at the apexes of regular polyhedra having the metal ions at their centers. The spinel crystal structure is isometric and the crystals may assume the various forms characteristic of the isometric system such as cube, octahedron, dodecahedron, tetrahexahedron, trisoctahedron, trapezohedron, hexoctahedron and combinations of these. The spinel structure generally can be represented as follows:

$$AB_2O_4$$

There are thus two kinds of cation sites: (A) those occupied by cations having an ionic radius of 0.7 to 0.9 angstroms, and (B) those occupied by cations having an ionic radius of 0.4 to 0.7 angstroms. In the partly ordered spinels of the mangano-chromia-manganites (A) is Mn^{++}, and (B) is Cr^{+++} and Mn^{+++}. The cations present in the (B) sites can be present in various proportions and in different valence states depending upon the temperature and upon whether the spinels are in an oxidizing, neutral, or reducing atmosphere as long as the total charges on the cations are equal to 8, thus preserving electrical neutrality. From theoretical considerations, though specific evidence is lacking, inverse spinels may also form in which (B) cations shift to (A) sites. Also, as the spinels of the mangano-chromia-manganites have defect structures, Mn^{++++} can go into the (B) or (A) sites.

In addition to the spinel represented by AB_2O_4, the mangano-chromia-manganites contain other manganese and chromium oxides. It is this whole complex of shifting chemical character that is herein termed a mangano-chromia-manganite. In the occupancy of the (A) and (B) sites in the spinel structure the ions which occupy the sites under one set of conditions may by ionic diffusion be replaced by others under changing conditions of chemical concentrations, temperature, oxidation and reduction. The labile and shifting character of the mangano-chromia-manganite is believed to lead to its high catalytic activity. The spinels are preferably in the labile isometric forms such as the cube, tetrahexahedron, trisoctahedron, and hexoctahedron. If the catalyst is heated to high temperatures the more stable forms develop such as the octahedron, dodecahedron, and trapezohedron. The cube is abundant in the preferred mangano-chromia-manganite catalysts from the process.

The mangano-chromia-manganites are preferably prepared in aqueous media by a reaction of a water-soluble manganese salt and a water-soluble chromium compound, preferably chromic acid anhydride. The manganese salts can be such compounds as manganese chloride, manganese nitrate, manganese acetate, manganese sulfate, and in general any salt of manganese. The chromium compounds can be such compounds as chromium nitrate, chromium sulfate, chromic acid, ammonium chromate, and ammonium dichromate. The manganese salt and the chromium salt are used in proportions to give the desired ratios as above described.

If ammonia or another precipitant is to be used which forms water-soluble chromates it will be evident that the water-soluble compounds of chromium, if in the hexavalent state, should not be used in excess of stoichiometric amount which will react with the manganese salt for where ammonia, or the like, is added later it will form a water-soluble ammonium chromate. If it is desired to obtain ratios higher than stoichiometric, these can be obtained by subsequent addition of the ammonium chromate salts. When the soluble compounds of manganese and chromium are brought together in aqueous solution they, of course, form a precipitate after sufficient ammonia has been added. Other precipitants can be used such as ammonium chromate and calcium hydroxide, magnesium hydroxide, barium hydroxide, and other bases which will not interfere with the action as a catalyst. However, ammonium hydroxide is inexpensive and is quite effective and is ordinarily satisfactory. The ammonia can readily be eliminated from the product by heating.

Preferred catalysts according to the process contain a second catalyst in addition to the mangano-chromia-manganite. This can be any catalyst or mixture which has value in treating the gases converted such as lead chromate, magnesium chromate, barium chromate, and strontium chromate. It is preferred to use base metal catalysts which are base metal chromites of copper, nickel, iron, zinc, cadmium, cobalt, tin, and bismuth. The mangano-chromia-manganite particles in catalysts have a tendency to grow in size. This growth occurs either by crystalline growth or by agglomeration, and the tendency is particularly great in use when the catalysts are heated to elevated temperatures. The mangano-chromia-manganite particles also have a tendency to agglomerate with base metal catalysts which may be included in the composition though this is not perhaps as great as the tendency to crystal growth.

To hinder such growth, preferred catalysts contain interspersants. These are comparatively inert particles of size comparable to that of the crystallites of the mangano-chromia-manganite. The interspersant, aluminum hydroxide, for example, is intimately associated with the manganite and chromate coprecipitate and being of a different crystal habit and size hinders agglomeration and crystallite growth in later phases of the catalyst preparation and use.

The interspersant has the function further of keeping the crystallites of the catalyst and the cocatalyst apart. The interspersant could somewhat less aptly be called a dissident and they are randomly distributed among the crystallites of the catalysts. Generally, it may be said that the interspersant can be any refractory material, the crystallites of which are similar in size to those of the mangano-chromia-manganite. The interspersed refractories should have a melting point above 1000°C. and more preferably above 1600°C.

Figure 93 represents a magnified view of one of the preferred compositions resulting from the process. There is shown an alumina support of high surface area which, as is well-known, carries upon its surface a myriad of alumina crystallites illustrated in the figure at 1. The mangano-chromia-manganite is illustrated showing at 2 a cube containing Mn and Cr lodged among the randomly disposed crystallites, 1. Manganese oxide as shown at 3, and typified by cubes of Mn2O3, is for the most part comparatively remote from the chromic oxide hexagonal crystallites illustrated at 4 because of the geography of the surface. Alumina is shown as an interspersant at 6 and 5 indicates chromia.

FIGURE 93: MAGNIFIED VIEW OF MANGANO-CHROMIA-MANGANITE CATALYTIC COMPOSITE

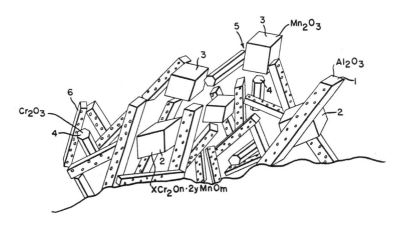

Source: B.W. Howk and A.B. Stiles; U.S. Patent 3,216,954; November 9, 1965

The following is a specific example of the preparation of such catalyst materials.

(1) Dissolve 110 pounds, 2 pound atoms, of metallic manganese as the nitrate in 75 gallons of water. Dissolve also in the solution 63 pounds, 1 pound atom, of copper as the nitrate.
(2) Slurry 200 pounds of alumina hydrate in the solution.
(3) Dissolve 300 pounds, 3 pound mols, of chromic acid anhydride in the solution.
(4) Add water to bring the volume to 800 gallons and heat to raise the temperature to 35°C.
(5) Agitate vigorously while adding vaporized anhydrous ammonia through a distributor at two pounds per minute until precipitation is complete as evidenced by further addition of ammonia producing no further precipitation.
(6) Stir the slurry for one hour and then filter. Wash the filter cake with water.
(7) Dry the filter cake at 135°C. for one hour.
(8) Raise the temperature of the dried product to 400°C. and calcine at that temperature for one hour to convert the manganate salts to mangano-chromic manganites having a crystallite size of about 50 angstroms. The alumina and copper chromite formed are of similar size.

(9) Charge a 100 pound lot produced as above, together with 65 pounds of water in which is dissolved 13 pounds of ammonium chromate into a sigma-arm mixer and knead until the mass is homogeneous and clay-like. Alternatively a similar product can be made using 33 pounds of ammonium chromate in this step.
(10) Dry and pulverize the product and mix with 1% of finely divided graphite. Pellet to form 3/6 by 3/16 inch cylindrical tablets.
(11) Heat in air at 400°C. for 3 hours.

HYDROCRACKING

Chevron Research

A process developed by S.M. Csicsery and J.R. Kittrell; U.S. Patent 3,558,473; January 26, 1971; assigned to Chevron Research Company yields a hydrocracking catalyst composed of a layered clay-type crystalline aluminosilicate cracking catalyst base and two hydrogenating components, one selected from the platinum group metals and compounds thereof and the other consisting of manganese and compounds thereof.

It is known in the art that a catalyst may comprise manganese and palladium, both exchanged into a crystalline zeolite molecular sieve. Mattox and Hammer U.S. Patent 3,329,604 disclose such zeolitic catalysts wherein at least about 25% of the original alkali metal is replaced by manganese. Even with the highly effective ion exchange techniques used by them, this catalyst requires about 2.5% manganese to be effective and, judging from the examples, manganese levels in excess of 5% appear to be preferred. With the catalyst support used in this catalyst, by contrast, a highly efficient and effective catalyst is obtained at much lower manganese levels, for example at levels of 0.5%. Indeed, high manganese concentrations surprisingly, actually inhibit the action of the catalyst support material.

Manganese also has been disclosed as a component for nonzeolitic catalysts for many different hydrocarbon conversion processes. For example, Meyers et al U.S. Patent 2,848,510 suggests that 0.1 to 5% platinum or palladium in addition to 1 to 20% manganese oxide may be used on acid-activated clay, silica-alumina, and many other supports. However, for a hydrocracking reaction, the acid-activated commercial clays are extremely ineffective with any level of the recited catalytic metals. Further, the layered, clay-type crystalline aluminosilicate of this catalyst results in higher catalytic activity and selectivity at low metal levels than any catalyst support comprehended by Meyers et al.

It is also known from the art to use manganese together with a platinum group metal for stabilization of the platinum group metal to crystallite formation. For example, Meyers et al U.S. Patent 2,911,357 discloses a catalyst consisting of at least one metal from the group Pt, Pd and Rh with at least one metal from the group Co, Ru, Mn, Cu, Ag and Au. The use of the metals from the latter group inhibits the crystallization of metals from the former group, as indicated by x-ray measurements. For the hydrocracking reaction a catalyst consisting of Pt, Pd or Rh together with a metal from the group Co, Ru, Cu, Ag and Au does not produce the superior results obtainable when Pt, Pd or Ir is used with Mn or Tc. Further, crystallization of Pd alone on the layered aluminosilicate of the catalyst does not appear to occur during hydrocracking. The catalyst of this process produces results not obtainable with the catalysts disclosed in the above references. The desirable attributes of an ideal hydrocracking catalyst may be listed as follows:

(1) Producing higher yields of gasoline, jet fuel or other liquid products and less butane and dry gas than are obtainable when operating with prior art catalysts at the same conditions and product cut point.
(2) Having a cost no greater than present commercial hydrocracking catalysts.
(3) Having a cracking component less sensitive to nitrogen poisoning than silica-alumina gel.
(4) Having a cracking component that is crystalline in structure, having pores elongated in two directions, contrary to the pores of crystalline zeolitic molecular sieves, and therefor having less reactant access limitations than the pores of such molecular sieves.
(5) Having a first hydrogenating component providing increased activity and stability to the catalyst, compared with a similar catalyst not containing the component.
(6) Having a second hydrogenating component providing additional activity and stability to the catalyst, compared with the same catalyst which contains the first hydrogenating component but not the second hydrogenating component.
(7) Having a high hydrocracking activity with economically low levels of the hydrogenating components.

It has been found that a catalyst comprising a layered clay-type crystalline aluminosilicate cracking component, for example the layered synthetic clay-type crystalline aluminosilicate mineral of Granquist U.S. Patent 3,252,757, a hydrogenating component selected from platinum and compounds thereof, palladium and compounds thereof, and iridium and compounds thereof, in an amount of 0.01 to 2.0 weight percent, calculated as metal and based on the cracking component, and a hydrogenating component selected from manganese and compounds thereof and technetium and compounds thereof, in an amount of 0.01 to 10.0 weight percent, calculated as metal and based on the cracking component, has all of the desirable catalyst attributes listed on the preceding page.

RHENIUM-CONTAINING CATALYSTS

HYDROCRACKING

Chevron Research

A process developed by J. Jaffe; U.S. Patent 3,535,233; October 20, 1970; assigned to Chevron Research Company yields a catalyst comprising a layered synthetic crystalline aluminosilicate cracking component and 0.01 to 3.0 weight percent, based on the cracking component and calculated as the metal, of a hydrogenating component selected from the group consisting of rhenium and compounds of rhenium. The rhenium hydrogenating component of the catalyst may be present in the final catalyst in the form of the metal, metal oxide, metal sulfide, or a combination thereof. The rhenium or compound thereof may be combined with the layered synthetic crystalline aluminosilicate mineral cracking component, or may be combined with other catalyst components in which the cracking component is dispersed, or both. The rhenium compound used in the impregnation or adsorption step generally will contain rhenium in anionic form. The compound should be one that is soluble in water, and that contains no ions that are known as contaminants in hydrocracking catalysts. Suitable rhenium compounds are perrhenic acid, $HReO_4$, and ammonium perrhenate, NH_4ReO_4. Impregnation also may be accomplished with an ammoniacal solution of rhenium heptoxide.

A process developed by J.R. Kittrell; U.S. Patent 3,558,472; January 26, 1971; assigned to Chevron Research Company yields a hydrocracking catalyst comprising a crystalline zeolitic molecular sieve cracking component, 0.01 to 2.0 weight percent, based on the cracking component and calculated as the metal, of a hydrogenating component selected from platinum, palladium and iridium and compounds thereof, and 0.01 to 2.0 weight percent, based on the cracking component and calculated as the metal, of a hydrogenating component selected from the group consisting of rhenium and compounds of rhenium.

HYDROGEN PRODUCTION

Shell Oil

A process developed by W.C.J. Quik and P.A. van Weeren; U.S. Patent 3,449,078; June 10, 1969; assigned to Shell Oil Company is one in which hydrogen is produced by the conversion of hydrocarbons in the presence of steam with a catalyst comprising rhenium supported on a carrier, with and without heavy metal promoters, stabilized by the addition of a small amount of alkali metal. It has been proposed to use rhenium catalysts rather than nickel for the preparation of hydrogen. The rhenium catalysts are highly active, which permits the use of relatively low temperatures, and are resistant to poisoning by sulfur. In fact, the sulfide form of rhenium usually has the highest activity. The rhenium catalysts have a serious disadvantage in that they are not stable, i.e., activity rapidly decreases with use.

It has now been found that rhenium catalysts having incorporated therein a minor amount of alkali metal are stable for the conversion of hydrocarbons in the presence of steam to produce hydrogen. The alkali metal content of the catalyst is less than 2% w. (calculated on the quantity of carrier material). As a rule, an optimum effect is obtained if the alkali metal content is below 1% w., preferably below 0.5% w. To obtain appreciable benefit, the amount of alkali metal is about 0.1% w. Good results are obtained with 0.2% w. For reasons of economy, preference is given to lithium, sodium and potassium or mixtures of these metals, i.e., in the form of alkali metal hydroxides or alkali metal carbonates. The rhenium can be present on the carrier in various forms, i.e., as metal and/or as an oxide. The sulfide form is preferred however, as it has been found that rhenium sulfide usually possesses the highest activity. It has further been found that activity and life of the rhenium catalysts can be increased by the use of promoters, such as metals from the right-hand column of Group I and/or metals of Group VIII of the Periodic System of Elements. Particularly suitable promoters are silver, nickel, cobalt, copper and gold, either as a metal or as a metal compound. Cobalt and/or silver are preferred promoters. The promoter is used in an amount of from 0.5 to 25% by weight. The rhenium catalysts are supported on a carrier. A preferred carrier is alumina (natural or synthetic). The presence of a certain amount of other oxides in the alumina is generally not detrimental. The alumina can contain up to 5% but preferably not more than 0.5% weight silica. Carriers with strongly acidic properties, such as silica-alumina cracking catalysts are generally not suitable. Generally carrier-supported catalysts will contain 1 to 25% w., and preferably 2 to 15% w., of rhenium. The rhenium catalysts can be conveniently prepared by impregnating the carrier material with a solution of the desired salt or salts. The metals can be impregnated separately or simultaneously from a common solution. The impregnated carrier is dried and calcined. After calcination, the rhenium which is in the oxide form can be reduced to the metal with hydrogen at 200° to 400°C. The rhenium can be sulfided by known means, such as with a mixture of hydrogen and hydrogen sulfide, or by a sulfur-containing hydrocarbon. The sulfiding operation can be applied directly to the oxide form if desired.

REFORMING

Chevron Research

A process developed by H.E. Kluksdahl; U.S. Patent 3,415,737; December 10, 1968; assigned to Chevron Research Co. involves reforming a sulfur-free naphtha in the presence of hydrogen with a catalyst composition of a porous solid catalyst support and 0.1 to 3 weight percent platinum and 0.01 to 5 weight percent rhenium.

Rhenium has been proposed for use in catalytic reforming as a substitute for more common catalytic components such as platinum. However, rhenium has been found to be extremely poor for reforming. Thus, rhenium alone supported on charcoal or alumina was found to possess only limited reforming activity and to require excessively large concentrations of the metal, above 5 weight percent, to obtain good activity. Rhenium has also been suggested for use with palladium for reforming; thus a catalyst comprising palladium and rhenium impregnated on alumina, which catalyst was presulfided, was shown to have better initial reforming activity than a presulfided palladium-alumina catalyst when used to reform a sulfur-containing naphtha. However, the activity of the palladium-rhenium presulfided catalyst decreased significantly after only limited use.

In development work leading to the process, there was prepared a catalyst comprising platinum and rhenium composited with alumina, but when tested for the reforming of a naphtha fraction, the catalyst caused excessive hydrocracking. The production of large amounts of light gases, such as methane and ethane, was significantly higher than the production of light gases with a catalyst comprising platinum alone on alumina. In other words, the platinum-rhenium catalytic composition was found to be less selective for the production of high octane gasoline products than a platinum-alumina catalyst. Also, upon introduction of the naphtha fraction to the reaction zone, a severe exotherm was observed in the catalyst bed. There was also prepared a catalyst comprising platinum and rhenium on alumina; the catalyst was sulfided and tested for the reforming of a sulfur-containing feed. The catalyst exhibited poor selectivity and activity. The yield of high octane products was low, thus making the reforming process unattractive economically. Thus, although rhenium has been suggested for use with noble metals such as palladium, it was found that such combination was not satisfactory under the above conditions.

It was found that when certain conditions of operations are followed and/or the catalyst subjected to certain pretreatments, a catalyst comprising platinum and rhenium supported on alumina possesses high activity, and, particularly, good selectivity and stability, for the reforming of sulfur-free feeds. It was especially unexpected to find that a supported platinum-rhenium catalyst would initially show undesirable hydrocracking, and then after reforming is continued until the hydrocracking becomes negligible. In fact, after the initial period, the catalyst comprising platinum and rhenium on alumina is so far superior to a catalyst comprising platinum alone on alumina that the initial poor reforming, which results in the production of high yields of light gaseous hydrocarbons, can be tolerated for the time needed to reduce the excessive hydrocracking activity of the catalyst. It is possible to get longer run lengths with higher yields of high octane products with the platinum-rhenium supported catalyst than with a platinum catalyst without rhenium.

A process developed by R.L. Jacobson, H.E. Kluksdahl and B. Spurlock; U.S. Patent 3,434,960; March 25, 1969; assigned to Chevron Research Company involves reforming a naphtha in the presence of hydrogen at low pressures to produce at least 98 F-1 clear octane gasoline with a catalyst composition of a porous inorganic oxide carrier containing 0.01 to 1.0 weight percent platinum and 0.01 to 2.0 weight percent rhenium. The pressure is maintained below 250 psig, and the feed rate is at least 2 LHSV. The hydrogen to hydrocarbon mol ratio is related to the distillation properties of the naphtha processed and to the octane number of the gasoline produced. The reforming process operates on stream for at least 2,000 hours with no greater than a 2 volume percent decline in the gasoline yield in 2,000 hours.

A process developed by R.L. Jacobson and B. Spurlock; U.S. Patent 3,449,237; June 10, 1969; assigned to Chevron Research Company involves starting up a reforming process conducted in the presence of a catalyst comprising platinum and rhenium associated with a porous solid catalyst carrier. It has been found that the high temperature excursion normally observed when a sulfur-free naphtha and hydrogen are initially contacted with a catalyst comprising catalytically active amounts of platinum and rhenium associated with a porous solid carrier can be decreased or substantially eliminated by contacting the catalyst in a reduced condition with a sulfur-free naphtha at reforming conditions and in the presence of an inert gas. Thus, the process for startup of a reforming process conducted in the presence of a reduced platinum-rhenium catalyst comprises introducing the catalyst into a treating zone, pressuring the treating zone with an inert gas, preferably nitrogen, to about 200 psig, heating the catalyst to 650°F., and contacting the catalyst with a substantially sulfur-free naphtha at a space rate of one volume of naphtha per volume of catalyst per hour (1 v./v./hr.). A naphtha-inert gas mixture withdrawn from the treating zone is passed into a gas separation zone to separate inert gas from the naphtha, and the separated inert gas is recycled to the treating zone. Over a period of 2 to 3 hours, the temperature of the catalyst in the treating zone is raised at a generally uniform rate from 650° to 900°F. and the naphtha feed rate to the treating zone is gradually increased at a generally uniform rate from a space velocity of 1 v./v./hr. to a space velocity of at least 2 v./v./hr. The inert gas flow into the treating zone is gradually decreased over the period of 2 to 3 hours while building up autogenous pressure of produced hydrogen, e.g., by dehydrogenation of naphthenes. The operation of the process will now be described in further detail with reference to Figure 94.

FIGURE 94:   FLOW DIAGRAM FOR A PLATINUM-RHENIUM CATALYST REFORMING PROCESS

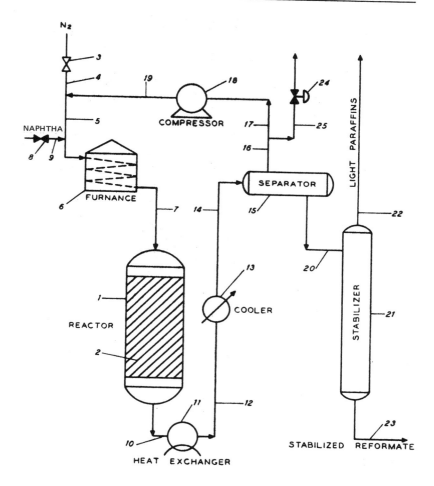

Source: R.L. Jacobson and B. Spurlock; U.S. Patent 3,449,237; June 10, 1969

A suitable reforming reactor 1 is filled with a platinum-rhenium catalyst 2. An inert gas, in this case nitrogen, is passed through open valve 3, through lines 4 and 5 into the furnace 6. The nitrogen is heated in furnace 6 to an elevated temperature and passed through line 7 to reactor 1 into contact with the bed of catalyst 2. Valve 8, through which naphtha feed will enter into line 9, then to the furnace 6 and reactor 1, is closed. Nitrogen is withdrawn from reactor 1 through line 10 for recirculation to the reactor. The nitrogen withdrawn through line 10 passes through heat exchanger 11 through line 12 to a cooler 13. Thereafter, nitrogen passes through line 14 to gas liquid separator 15, through line 16, line 17, compressor 18 and line 19 to line 5 where the recycled nitrogen mixes with nitrogen introduced through line 4. The reforming system is pressured to about 200 psig with nitrogen, and the catalyst temperature raised to 650°F. by circulating nitrogen heated in furnace 6 through the catalyst bed 2. Thereafter, valve 3 is closed, and valve 8 opened to permit naphtha into line 9 and then into line 5 and into furnace 6. The naphtha passes through line 7 into contact with the bed of platinum-rhenium catalyst 2 at a space velocity of about 1 v./v./hr. At the low temperature of 650°F. and in the presence of nitrogen very little reforming of the naphtha occurs. A small amount of dehydrogenation of the naphthenes does occur with the accompanying production of hydrogen.

The mixture of nitrogen and naphtha and any produced hydrogen is withdrawn from reactor 1 through line 10 into heat exchanger 11. In the heat exchanger 11, the nitrogen-naphtha mixture generally passes in indirect contact with a fresh naphtha (not shown); the heated fresh naphtha stream is then sent through line 9 and line 5 to furnace 6. The nitrogen-naphtha mixture next passes from heat exchanger 11 through line 12, cooler 13, line 14 into separator 15. The nitrogen and any produced hydrogen are separated from the naphtha in separator 15; the naphtha is removed through line 20 and passed to stabilizer 21. It may be desirable during the initial stages of contact of the naphtha with the catalyst, that is, during the period of time that little or no reforming of the naphtha occurs, to recycle naphtha from separator 15 to furnace 6 and to reactor 1 for further reforming. Means for recycling naphtha are not shown in the figure. After the reforming process is on stream any reformate removed from separator 15 is sent to stabilizer 21, the light paraffins being removed through line 22 and the stabilized reformate removed through line 23. The nitrogen and hydrogen removed from separator 15 pass through line 16, line 17, into compressor 18. The compressed gases are then recycled to furnace 6 through lines 19 and 5.

The catalyst temperature in reactor 1 is gradually increased from 650° to 900°F. over a period of 2 to 3 hours. The naphtha feed rate is gradually increased from 1 v./v./hr. to at least 2 v./v./hr. over the same period. As the temperature is gradually increased, hydrogen is produced in greater amounts; hydrogen withdrawn from reactor 1 is recycled to the reactor. The nitrogen concentration in the recycle gases is gradually decreased over a period of 2 to 3 hours by bleeding off or discharging part of the recycle stream from the reforming system through line 25 by opening valve 24. The rate at which the gases are withdrawn is not permitted to exceed the rate at which hydrogen is produced. The hydrogen pressure is allowed to build up over a period of 2 to 3 hours while maintaining a total pressure in the reaction zone of at least 200 psig. Normally, at the end of the 2 to 3 hour period of bringing the catalyst temperature and reactor pressure to the desired level for reforming, substantially all the nitrogen should have been discharged from the system and reforming of the naphtha in the presence of hydrogen should be occurring. By means of the startup procedure described, a temperature excursion in the catalyst bed will have been substantially eliminated.

A process developed by H.E. Kluksdahl; U.S. Patent 3,496,096; February 17, 1970; assigned to Chevron Research Co. is one in which a catalyst comprising a platinum group component and a rhenium component in association with a porous inorganic oxide carrier has become deactivated from exposure to a hydrocarbon feed under reforming conditions and has carbonaceous matter accumulated thereon. Such a catalyst may be regenerated by the steps of (1) contacting the catalyst with a regeneration gas containing oxygen at a partial pressure of from 0.1 to 2.5 psia, at a temperature below 800°F. to remove substantially all of the carbonaceous matter, (2) contacting the catalyst with a regeneration gas containing oxygen at a partial pressure of from 1.0 to 2.5 psia, at a temperature of from 800° to 900°F., and then (3) contacting the catalyst with a regeneration gas containing oxygen at a partial pressure greater than 2.5 psia, at a temperature above 900°F., and (4) finally contacting the catalyst with a hydrogen-containing gas at a temperature above 600°F. Halide is introduced into the regeneration gas at a temperature greater than 800°F. to provide the finished catalyst with a halide content of at least 0.6 weight percent.

IRON-CONTAINING CATALYSTS

AMMONOXIDATION

Nitto Chemical Industry

A process developed by T. Yoshino, S. Saito, Y. Sasaki and K. Moriya; U.S. Patent 3,591,620; July 6, 1971; assigned to Nitto Chemical Industry Co., Ltd., Japan involves the production of acrylonitrile by contacting a mixture of propylene, molecular oxygen and ammonia in the vapor phase at an elevated temperature with a catalyst having the empirical formula $Fe_{10}Sb_{5-80}V_{0.01-2}X_{0-2}O_{22-186}$ where X represents P or B.

The catalyst mentioned above is distinguished over the promoted or unpromoted iron oxide-antimony oxide catalyst disclosed in the prior art, in the following respects. (1) It comprises a base catalyst system having a specific atomic ratio of Fe/Sb and a particular promoter (vanadium) being present in an extremely small amount. (2) It exhibits an improved conversion of propylene to acrylonitrile, particularly in the case where high conversion is achieved and the amount of residual oxygen is very small.

DEHYDROGENATION

Shell Oil

A process developed by G.J. Hills and R.C. Siem; U.S. Patent 3,360,579; Dec. 26, 1967; assigned to Shell Oil Co. is one where a catalyst particularly effective in the dehydrogenation of ethyl benzene to styrene is obtained by calcining a mixture of monohydrated yellow iron oxide, chromium oxide, potassium carbonate and water. The dehydrogenation reaction most usually employed is to pass the alkyl aromatic reactant diluted with 2 to 30 mols of steam per mol of reactant over a dehydrogenation catalyst at a temperature of 580° to 700°C. at a low pressure, e.g., subatmospheric to 25 psig.

The class of catalyst found to be most effective for this process is an alkalized iron oxide containing a small amount of another heavy metal oxide more difficultly reducible than iron oxide. The catalyst is usually a potassium carbonate promoted iron oxide (calculated as Fe_2O_3) containing a small amount of chromium oxide (Cr_2O_3) to improve the stability of the iron oxide. Catalysts containing various ratios of iron oxide, chromium oxide and potassium carbonate have been used. One of the major advantages of these catalysts is that they are autoregenerative under reaction conditions in the presence of steam.

This obviates the necessity of interrupting the process and regenerating the catalysts, such regeneration including the burning of the carbon deposits off the surface of the catalyst particles, which is necessary with other dehydrogenation catalysts. There is, however, over a long period of time, e.g., 250 to 300 days, a noticeable decline in the activity of the catalyst requiring higher temperatures and pressures than initially used to maintain constant conversion. This decline in activity is noted by a drop in the selectivity (mols of desired product per mol of reactant reacted) while maintaining a constant conversion (percent reactant reacted).

The selectivity to desired product of this reaction varies inversely with the conversion. Thus, there is a point at which the sum of the conversion and selectively, commonly referred to as the CSV (conversion-selectivity value) reaches a maximum. Any improvement that increases either the selectivity or conversion without lowering the other results in an increased CSV. The iron oxide used in catalyst preparation is extremely critical in terms of catalyst performance. The iron oxide employed in catalyst preparation is usually a synthetically produced powdered red, red-brown or black pigment. The red or red-brown pigments are highly pure ferric oxide while the black pigment is the magnetite form ferrosoferric oxide (Fe_3O_4), which is usually the form found in the catalyst under reaction conditions. These oxides are prepared by various methods, e.g., oxidation of iron compounds, roasting, precipitation, calcination, etc. Whatever method is used the final step is usually the application of substantial heat to remove all or substantially all of the water from the oxide, thereby producing an oxide red or red-brown in color. It is known that the most active and selective catalysts are those having an available surface area below 10 square meters per gram and in many cases below 5 square meters per gram.

Since iron oxides have surface areas in excess of this requirement many methods have been employed to reduce their available surface area. One of the most common is the precalcination of the iron oxide at a temperature in excess of 700°C. for a period of time ranging from one-half hour to several hours. Other methods are involved in catalyst preparation. One method adds Portland cement to the catalytic components followed by calcination at 600°C. thereby lowering the surface area by plugging pores in the catalyst by cement as well as by calcination. Another method reduces the available surface area by calcination of the catalyst at 800° to 950°C.

While most of the above methods result in catalysts having the desired surface area they also result in catalyst having a relatively high density. It has been found that catalysts having a highly porous structure and a low surface area are highly active for the catalytic dehydrogenation of alkyl aromatics such as ethyl benzene. Various methods have been employed in an attempt to form highly porous catalysts. For example, combustible materials such as sawdust, carbon, etc., have been added during catalyst formation and then burned out after the pellet had been formed.

Wood flour has also been added to the catalyst mixture and the catalyst has been prepared by adding an excess of water to the mixture before extruding into pellets and then drying. The catalysts formed by these methods often suffer disadvantages in that the pellet crushing strength is low or the catalysts shrink thereby losing any increase that might have been gained in porosity. If too much water is added before catalyst extrusion the paste is too thin to extrude efficiently.

It has been found that substantially improved catalysts particularly effective in catalyzing the dehydrogenation of the ethyl benzene to styrene are obtained by using a yellow iron oxide as a starting material in catalyst preparation instead of the red or red-brown iron oxide usually used in preparing iron oxide-chromium oxide-potassium carbonate dehydrogenation catalysts. While it is not known just what effect the yellow iron oxide has upon the catalyst that makes it different from prior art catalysts of the same elemental composition, it is known that the catalyst prepared from yellow iron oxide produces superior results in the dehydrogenation of ethyl benzene to styrene.

Yellow iron oxide appears in a variety of colors ranging from a light yellow to a deep yellow-orange. The yellow iron oxides are the hydrated form of ferric oxide and are substantially monohydrates. The water content is usually 13% by weight. They may be prepared by the controlled oxidation of iron in ferrous sulfate solutions such as disclosed in U.S. Patent 2,111,726, to G. Plews dated March 22, 1938. It is believed that the water of hydration may have a significant effect. It is thought that in drying the catalyst after extrusion, the water added to mix the catalyst components leaves much more readily than the water of hydration of the iron oxide. Upon additional heating or calcining, the water of hydration is driven off resulting in a catalyst of increased porosity thereby allowing greater diffusion of the gaseous reactant throughout the catalyst particle.

The manner in which the catalysts are prepared is not extremely critical and depends to a large extent upon the composition of the catalyst being prepared. For catalysts containing a relatively low potassium carbonate content, about 13.5% by weight, the potassium carbonate is dissolved in a small amount of water and the solution is thoroughly mixed with powdered yellow iron oxide and chromium oxide to form a paste. The paste is extruded and pelleted. The pellets are pan dried at 175° to 235°C. for one hour and subsequently calcined for 1/2 to 2 hours at 800° to 1000°C.

The amount of water added to form a paste is much more critical when using yellow iron oxide than with the red iron oxides due to the water of hydration. If too much water is added, the paste will be too thin and extrusion into pellets or formation of tablets difficult. Calcination of the catalyst is necessary to strengthen it and to reduce the available surface area to the desired limits which is usually below 5 square meters per gram and is preferably below 3 square meters per gram. The percentages of catalytic components may vary. As stated before, catalysts containing from 80 to 90% by weight iron oxide, calculated as Fe_2O_3, a potassium compound equivalent to 9 to 18% by weight potassium carbonate and 1.5 to 5.0% by weight chromium oxide calculated as Cr_2O_3 are preferred. Catalysts made from 84% by weight iron oxide, 13.5% by weight potassium carbonate and 2.5% by weight chromium oxide have been found especially useful.

The weight percentages of yellow iron oxide and potassium carbonate are especially critical in controlling the physical features of the catalyst produced. If the potassium content rises substantially above the preferred amount resulting in a corresponding reduction of iron oxide, the catalyst has a much higher density, larger surface area and small pore volume. The size of the pellets produced may vary, catalysts having a diameter of from 1/8 to 1/4 inch and from 1/8 to 5/8 inch in length being the most common. The small diameter catalysts are more active but are not as strong. Generally speaking, catalysts having a diameter of 3/16 of an inch are usually employed. It is preferable that the catalyst contain only iron oxide, potassium carbonate and chromium oxide as mentioned above. However, if desired, one may include other additives such as phosphates, silica, cement, etc.

Universal Oil Products

A process developed by <u>K.D. Uitti; U.S. Patent 3,515,763; June 2, 1970; assigned to Universal Oil Products Company</u> provides an improved method for the conversion of ethylbenzene to styrene via catalytic endothermic dehydrogenation reaction which comprises:

(a) passing a feed mixture containing ethylbenzene and steam in outward radial flow direction through a plurality of annular dehydrogenation reaction zones, each zone containing particulate catalytic material, under conversion conditions including a relatively high temperature;

(b) withdrawing effluent containing stream from the first zone at a relatively low temperature;

(c) admixing the low temperature effluent with added superheated steam in a mixing zone prior to introduction of the admixture in outward radial flow manner through the next succeeding reaction zone under conditions sufficient to increase the temperature of the effluent to a predetermined relatively high level;

(d) introducing the admixture into the next succeeding reaction zone in the manner indicated; and

(e) recovering styrene from the effluent of the last reaction zone.

The catalyst employed for the dehydrogenation reaction is preferably an alkali-promoted iron catalyst. Typically, such a catalyst may consist of 85% by weight ferrous oxide, 2% by weight of chromia, 12% by weight of potassium hydroxide, and 1% by weight of sodium hydroxide. Other catalyst compositions include 90% by weight iron oxide, 4% by weight chromia, and 6% by weight potassium carbonate. The process may be described in further detail with reference to the flow diagram in Figure 95.

FIGURE 95: FLOW DIAGRAM OF PROCESS FOR STYRENE MANUFACTURE USING ALKALI-PROMOTED IRON CATALYST

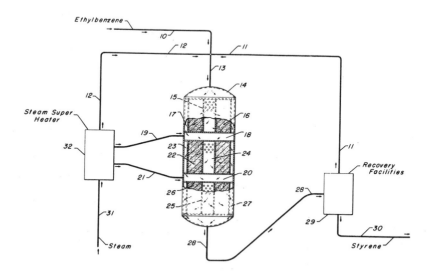

Source: K.D. Uitti; U.S. Patent 3,515,763; June 2, 1970

With reference now to the accompanying drawing, an ethylbenzene-containing feedstock enters the process through line 10 being admixed with recycle ethylbenzene in line 11, the source of which is hereinafter described. In addition, the ethylbenzene in line 10 is admixed with from 5 to 15% by weight of the total amount of steam utilized in the overall process and entering through line 12 at a temperature of 1400°F. The steam-ethylbenzene mixture at a temperature of 1200°F. passes via line 13 into reactor 14 which contains three annular fixed beds of catalyst, 16, 22, and 26, respectively. (As defined in the prior art, the dehydrogenation of ethylbenzene is generally effected at a reaction temperature from within the range of 932° to 1292°F.).

As the steam and ethylbenzene mixture passes into reactor 14, the admixture proceeds into concentrically placed conduit 15 which has perforations therein. These perforations, of course, are of such a size and shape that the catalytic mass 16 cannot pass through the perforations. The admixture next proceeds in outward radial flow manner through catalyst bed 16 and is subsequently withdrawn at the outer perimeter of catalyst bed 16 through passageway 17. This first effluent is at a temperature of 1100°F. and is then channeled from passageway 17 into mixing zone 18.

In mixing zone 18, superheated steam 19 is introduced at a temperature of 1500°F. in an amount previously described sufficient to raise the temperature of this first effluent to substantially reaction temperature. It is to be noted that the outlet of mixing zone 18 is concentrically located to the inlet of the next succeeding reaction zone. Therefore, the reheated first effluent plus added steam pass through conduit 24 in outward radial flow manner through a second annular catalytic mass 22 where additional conversion of ethylbenzene to styrene takes place. A second effluent is withdrawn from the outer perimeter of catalyst bed 22 and is thereafter channeled into mixing zone 20 in a manner identical to the manner previously described for mixing zone 18.

Additional superheated steam at a temperature of 1600°F. is introduced into mixing zone 20 via line 21 in an amount previously described sufficient to raise the temperature of the second reaction zone effluent to substantially reaction temperature. In similar manner, the reheated second reaction zone effluent passed through concentrically located conduit 25 in outward radial flow manner through third catalyst bed 26. A third reaction zone effluent is passed through passageway 27 and out of reactor 14 via line 28 and sent to product recovery facilities 29.

The total amount of steam for the process enters the process via line 31 and passed into steam superheater 32 of a conventional type and design well known to those skilled in the art. By proper operation of steam superheater 32, the temperature of the superheated steam in lines 12, 19, and 21 may be substantially the same, but preferably are at a succeeding higher temperature, to wit: the temperature of the superheated steam in line 21 is greater than that of the steam in line 19 which in turn is greater than the temperature of the steam in line 12.

Product recovery facilities 29 are conventional in nature and usually comprise distillation facilities for separating the unreacted ethylbenzene from the product styrene and/or distillation facilities for recovering products made in the reaction, such as benzene and toluene, from the desired products. Preferably, ethylbenzene in high concentration is recovered from facilities 29 and returned to the reaction zone via line 11 in the manner mentioned previously. Styrene in high concentration and high purity is withdrawn from the process via line 30.

OXYDEHYDROGENATION

Badger

A process developed by J.R. Ghublikian; U.S. Patent 3,502,737; March 24, 1970; assigned to The Badger Company, Inc. involves contacting ethylbenzene, steam and an oxygen-containing gas in the presence of an iron oxide catalyst, with the amount of oxygen-containing gas being carefully controlled to maintain a selected substantially isothermal condition in the reaction zone by preferential reaction with carbon, hydrocarbon-containing material and/or hydrogen to provide heat by combustion. The amount of steam used is materially less than would be required in the absence of oxygen.

The selective dehydrogenation combustion catalyst is an oxide of iron, preferably promoted by being admixed with alkali such as sodium carbonate, and activated by being heated with steam for several hours at the reaction temperature. A preferred catalyst is 2% potassium oxide, 5% chromium oxide and 93% ferric oxide which is distributed on an inert carrier such as magnesia and activated by heating with steam at 1100°F. for two hours. The preferred and other commercial iron oxide catalysts are available from Shell, Girdler, and others. The essential features of the process are shown in Figure 96.

An elongated reactor 10, enclosing a fixed bed of the preferred catalyst described above, is fed by ethylbenzene vapors entering at the upper end through a line 12 controlled by a valve 16. Simultaneously, elemental oxygen-containing gas (which may be pure oxygen but more usually is ordinary air) enters through a line 20 controlled by a valve 18, and is combined with superheated steam, at the selected temperature as stated above, which enters through a line 14 controlled by a valve 19. With this arrangement, the oxygen-containing gas is preheated to the temperature of the steam as they jointly enter the reactor. The oxygen-containing gas and the steam then pass together into the upper end of the reactor 10 in controlled quantity, commingle with the ethylbenzene upon the catalyst, and pass downward through the catalyst body. Alternately, it is obvious that the piping may be arranged so that the ethylbenzene is mixed with the steam before entering the reactor to combine with the oxygen-containing gas. In this case, the oxygen-containing gas can be passed directly and independently into the upper part of the reactor by way of a line 26 controlled by a valve 24. The preliminary quantity input of oxygen-containing gas is controlled by either valve 18 or 24. The reaction product is withdrawn continuously via a line 42 and passed to any conventional recovery system (not shown).

FIGURE 96:   OXYDEHYDROGENATION REACTOR FOR PROCESS USING IRON OXIDE CATALYST

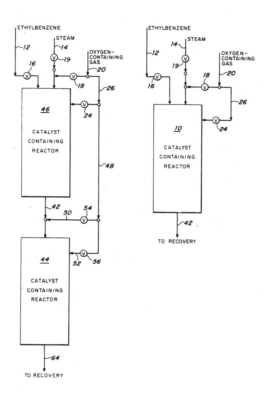

Source: J.R. Ghublikian; U.S. Patent 3,502,737; March 24, 1970

Petro-Tex Chemical

A process developed by M.Z. Woskow; U.S. Patent 3,513,216; May 19, 1970; assigned to Petro-Tex Chemical Corp. involves dehydrogenation of organic compounds by reacting hydrogen removed from the organic compound with oxygen to form water. Oxygen is supplied by a solid oxidant comprising a ferrite of magnesium, manganese, cobalt, nickel, zinc, cadmium and mixtures with boron which releases oxygen and is thereby converted to a composition diminished in oxygen.

COBALT-CONTAINING CATALYSTS

HYDROFORMYLATION

Toa Nenryo Kogyo

A process developed by S. Usami, T. Kondo, K. Nishimura, and Y. Koga; U.S. Patent 3,378,590; April 16, 1968; assigned to Toa Nenryo Kogyo Kabushiki Kaisha, Japan is a process for producing oxygen-containing compounds, such as aldehydes and alcohols, in which an olefinic compound, carbon monoxide and hydrogen are reacted in the presence of a cobalt catalyst and a promoter selected from the group consisting of palladium supported on carbon and platinum supported on carbon, in a hydroformylating reaction zone. The resulting reaction product can be introduced without change into a hydrogenating reaction zone and hydrogenated in the presence of hydrogen, and the cobalt catalyst is separated by decomposing the cobalt carbonyl in the hydrogenated product and by depositing it on the promotor.

The process is shown in the flow diagram in Figure 97. A mixed liquid of a hydroformylating catalyst solution and a liquid prepared by suspending the palladium on carbon carrier or the platinum on carbon carrier in a proper medium is fed through pipe 3 from a catalyst storing tank 202. A material having an olefinic double bond is fed through pipe 2 from an olefinic raw material storing tank 201. Further, a mixed high pressure gas of carbon monoxide and hydrogen is fed through a pipe 1. They are passed through a heat exchanger 301, are thus heated to a fixed temperature and are then introduced into a hydroformylating reaction zone 203.

The reaction product produced after they are kept in the reaction zone for a fixed period enters a heat exchanger 302 through a pipe 4, is thus cooled and is then led to high pressure receiver 204, medium pressure receiver 205 and low pressure receiver 206 so as to be separated into gas and liquid. The gas separated in such case is returned to the pipe 1 through a pipe 7 and is recirculated and used again. The separated liquid is fed to a hydrogenating reaction zone 207 through a pipe 5 and a heat exchanger 303.

FIGURE 97:   COMBINED HYDROFORMYLATION AND HYDROGENATION PROCESS USING COBALT CATALYST

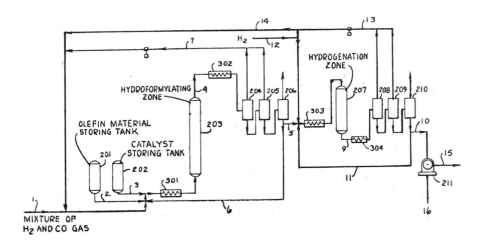

Source: S. Usami, T. Kondo, K. Nishimura and Y. Koga; U.S. Patent 3,378,590; April 16, 1968

However, as required, a part of the liquid from the low pressure receiver 206 can be recirculated to the hydroformylating reaction zone through a pipe 6. Hydrogen to be used in the hydrogenating reaction zone is fed through a pipe 12. However, the gas produced in the case of the gas-liquid separation after the hydrogenating reaction may be recirculated and fed through a pipe 13. But, in such case, when an amount larger than a certain fixed amount of carbon monoxide produced by the decomposition of a metallic carbonyl has come to be accumulated in the gas fed through the pipe 13, the circulation to the hydrogenating reaction zone 207 should be stopped and the circulation to the hydroformylating reaction zone should be made through a pipe 14 so that the gas may be used to adjust the concentration of the incoming gas.

Further, in the hydrogenating reaction zone, as described above, the metallic carbonyl dissolved in the liquid will be decomposed to produce the metal which will be deposited on the palladium on carbon carrier or platinum on carbon carrier suspended in the liquid and the suspension will enter a heat exchanger 304 through a pipe 9. After the suspension is thus cooled, it is led to high pressure receiver 208, medium pressure receiver 209 and low pressure receiver 210 and the gas separated therefrom. The liquid is then fed to a catalyst separator 211 which may be of a centrifugal separation type or a filtration type. The liquid part separated by the catalyst separator is fed through the pipe 15 to the next step, which may be a distillation procedure. The solid part is taken out of the system at 16, is then fed to a catalyst preparing procedure and is again circulated and used.

HYDROGENATION

Catalysts and Chemicals

A process developed by H.W. Fleming; U.S. Patent 3,155,739; November 3, 1964; assigned to Catalysts and Chemicals Incorporated involves the conversion by hydrogenation of acetylenic and diolefinic compounds in gas mixtures whereby a high degree of selectivity is achieved by reason of the catalyst utilized.

An ideally selective catalyst is one in which the highly unsaturated hydrocarbons are hydrogenated to monoolefins rather than to paraffins. In other words, in a gas stream containing 1 to 2% methyl acetylene, vinyl acetylene or propadiene, a truly selective catalyst will convert these impurities to propylene with essentially no hydrogenation of the propylene or ethylene in the gas composition. It has previously been thought that nickel sulfide was unique in selectively hydrogenating alkyl acetylenes and alpha,beta diolefins to olefins. However, a catalyst comprising cobalt sulfide is significantly more selective over a wide range of conditions than is nickel sulfide. Therefore, a gas mixture containing predominantly C_2 olefins and trace amounts of sulfur may be treated with such a catalyst to produce a purified gas containing acetylene in a concentration of less than 10 ppm.

Cities Service Research and Development

A process developed by L.M. Rapp; U.S. Patent 3,541,002; November 17, 1970; assigned to Cities Service Research and Development Company involves semicontinuously withdrawing, regenerating, and replacing particulate catalyst in a continuously operating high pressure hydrogenating treatment process, particularly a process utilizing an expanded or ebullated catalyst bed. The process comprises intermittently withdrawing spent particulate catalyst from the ebullated bed, washing and stripping the withdrawn catalyst, accumulating the stripped catalyst particles in a first regeneration zone, and regenerating the accumulated catalyst in the first regeneration zone while simultaneously accumulating stripped catalyst particles in a second regeneration zone. Regenerated particulate catalyst is semicontinuously fed to the hydrogenation zone from the first regeneration zone as required to maintain the level of the operating catalyst bed, while accumulated catalyst in the second regeneration zone is regenerated and the process repeated.

Figure 98 shows schematically the operations involved in such a process. Referring to the figure wherein a pair of reactor vessels 12 and 14 respectively for catalytically treating petroleum and hydrocarbon oils with hydrogen, are shown. The reactor vessels 12 and 14 are capable of operating at the high pressure (e.g., 1,000 to 5,000 psig) and high temperatures, i.e., above 400°F. which are normally associated with the treatment of petroleum oils by contacting the oils with hydrogen in the presence of catalyst. While any petroleum or hydrocarbon oil suitable for catalytic treating with hydrogen may be used, those boiling in the range between 350° and 1100°F. are preferred, and may include e.g., virgin or thermal gas oils, coker distillate, vacuum gas oils, deasphalted gas oils, and other fractions derived from crude oil or from naturally occuring sources or from synthetic crudes produced from natural tar, shale oil, tar and/or coal.

Relatively heavy hydrocarbon oils boiling above about 650°F. are especially suitable for treating with hydrogen in the presence of a particulate catalyst. A suitable hydrocarbon oil feed and hydrogen gas mixture is fed to reactor vessel 12 from a source not shown, and the first hydrotreating stage effluent passed out of the reactor vessel 12 through effluent conduit 16 to the reactor vessel 14 for second stage hydrotreating with the resultant product stream being passed out of vessel 14 for further processing as required. The reactor vessels 12 and 14 each have a catalyst bed 18 and 20 respectively supported on a conventional bubble cap tray 22 and 24 respectively. The catalyst beds 18 and 20 are each made up of a mass of catalyst particles larger than 300 angstrom units and are generally described as macrospheres in contrast to the

FIGURE 98: FLOW DIAGRAM OF HIGH PRESSURE COBALT-CATALYZED HYDROGENATION PROCESS FEATURING SEMICONTINUOUS CATALYST REGENERATION

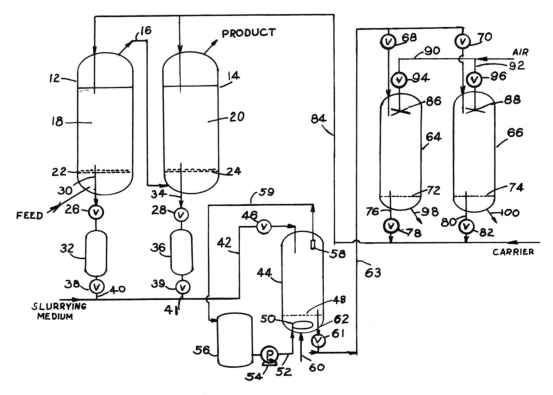

Source: L.M. Rapp; U.S. Patent 3,541,002; November 17, 1970

smaller microspheres which are commonly employed in fluid bed processes. Catalyst particles having a diameter on the order of from about 1/64 inch to about 1/2 inch and being in the form of elongated extrudes are particularly preferred. The particulate catalyst may be composed of any of the conventional and well-known constituents. Those catalyst components which are suitable for use in hydrogenation process such as that described herein are, for example, metals or combinations of the metals from groups VI and VIII of the periodic table in various proportions, which are supported on an alumina, silica or silica-alumina base. Thus suitable catalysts for use on such bases include cobalt, iron, molybdenum, paladium, nickel, tunsten etc. as well as combinations of the same. They may be used in the oxide or sulfide forms either alone or together with other suitable catalysts. Finally the newer synthetic catalyst particles such as zeolites or molecular sieves may be utilized if desired.

The type of catalyst bed particularly applicable to the process is that described as an expanded or ebullated bed and is disclosed in detail in U.S. Patent Reissue 25,770. The ebullated bed is maintained by passing liquids or a gaseous liquid mixture up through the mass of catalyst particles at a sufficient velocity to expand the volume of the catalyst bed and cause random movement of catalyst particles within the bed. In general the expansion of the catalyst bed in such instances is normally between 10 and 300% based upon the settled volume of the particulate catalyst mass.

Resulting catalyst concentrations obtained in the treatment zone are at least 15 pounds of catalyst per cubic foot of reactor volume while concentrations of more than 25 pounds per cubic foot are preferred. The random movement of the catalyst particles in the bed allows periodic removal of a portion of the bed including spent catalyst, while operation of the hydrogenation treatment is continued. There is therefore no necessity of periodically shutting down the reactor in order to replace or regenerate spent catalyst. Consequently in the hydrogenation process the withdrawn spent catalyst particles must either be discarded or regenerated and the catalyst beds 18 and 20 maintained at operating level by replenishment with fresh or regenerated catalyst.

Particulate catalyst is removed from the reactor vessel 12 and 14 by opening up withdrawal valves 26 and 28 respectively controlling catalyst withdrawal from each reactor vessel. A catalyst withdrawal conduit 30 extends up into the bed of the first stage reactor vessel 12 and is connected through withdrawal valve 26 to a catalyst withdrawal pot 32. Likewise a second catalyst withdrawal conduit 34 extends up into the bed 20 of the second stage reactor vessel 14 and is connected through valve 28 to a second catalyst withdrawal pot 36. Catalysts dumping valves 38 and 39 are each respectively connected to the aforesaid withdrawal pots 32 and 36 and control the dumping of withdrawn particulate catalyst from the pots through conduits 40 and 41 to a withdrawn catalyst transport conduit 42.

Spent particulate catalyst is withdrawn from either of the reactor vessels 12 and 14 by opening the respective withdrawal valve 26 or 28 while the respective dumping valves 38 and 39 remain closed or are closed. After an amount of spent catalyst has been collected in the withdrawal pot, withdrawal is terminated by closing the respective withdrawal valve. Subsequently withdrawn catalyst is dumped from the withdrawal pot into the transport conduit 42 by opening the respective dumping valve 38 or 39. Such a procedure allows withdrawal of catalyst from the high pressure reactor vessel without interference in the operation of the reactor vessel. In the transport conduit 42, the particulate catalyst is slurried in a light oil stream from a source not shown and carried to a washing and stripping vessel 44. Flow in the transport conduit 42 is controlled by a flow valve 46.

The washing and stripping vessel 44 is a chamber similar in construction to the reactor vessels 12 and 14 and had a transverse grid 48 with a multiplicity of ball check caps, not shown, to allow upward flow of fluid through the strip vessel 44 while particulate catalyst is retained on the grid 48. Feeding wash oil into the bottom of the stripping vessel 44 is a wash oil distributor ring 50 which is supplied with wash oil through conduit 52 and pump 54 from a storage tank 56. Wash oil is removed from the stripping vessel 44 by solids-liquid separator 58 and returned through return conduit 59 to the storage tank 56. Also located in the bottom of the stripping vessel 44 is a stripping medium feed line 60 through which steam or preferably hydrogen or an inert gas such as nitrogen from a source not shown, is supplied to the vessel 44 to act as a stripping medium.

The passage of wash oil from the distributing ring upwardly through the stripping vessel 44 expands and also tends to randomly move the particulate catalyst mass. As the expansion and particle movement within the stripping vessel is not identical to that found in the ebullated-bed treatment process such a descriptive name will not be applied to the washing and stripping process. After the wash oil has been passed upwardly through the stripping vessel 44 for a time sufficient to remove adsorbed heavy and residual hydrocarbon oils, the oil wash is terminated, i.e., by shutting down the pump 52, and the stripping treatment is commenced. A stripping gas such as superheated (dry) steam or more preferably hydrogen gas is passed upwardly in the stripping vessel 44. Stripping of the washed particulate catalyst with hydrogen is continued for at least one hour and results in a substantially dry particulate catalyst (i.e., substantially free of adsorbed hydrocarbon fluid).

After completion of the stripping operation, the dry particulate catalyst is removed from the stripping vessel 44 by opening a valve 61 thereby causing the stripped dry catalyst to pass through a conduit 62 into regenerator feed line 63. The dry particulate catalyst is carried by a substantially inert gas such as hydrogen or nitrogen acting as a conveying or transporting medium in the regenerator feed line 63 to one of two similar regeneration vessels indicated by reference numbers 64 and 66, respectively, as a first and a second regeneration vessel or zone. Valve 68 controls flow from the feed line 63 into the top of the first regeneration vessel 64 while a similar valve 70 controls dry particulate catalyst feed to the second regeneration vessel 66.

Mounted transversely at the bottom of each of the regeneration vessels 64 and 66 is a grid denominated by reference numbers 72 and 74 respectively. The grids 72 and 74 allow gases and liquids to pass therethrough while retaining the mass of particulate catalyst. A regenerated catalyst removal conduit 76 extends into the lower portion above the grid 72 in regeneration vessel 64 and is provided with a valve 78. Similarly another regenerated catalyst removal conduit 80 extends into regeneration vessel 66 and is provided with a valve 82 for controlling removal of regenerated catalyst. The two removal conduits 76 and 80 are communicatingly connected to a regenerated catalyst feed line 84 through which light oil or an inert gas from a source not shown is passed to act as a carrier medium for the regenerated catalyst. The regenerated catalyst feed line 84 connects through valves, not shown, to each of the reactor vessels 12 and 14.

A pair of gas distributors, 86 and 88 respectively, are located in the upper part of each of the regeneration vessels 64 and 66, and are connected by suitable piping 90 and 92, and valves 94 and 96, respectively, to a source, not shown, of regeneration gases preferably air and inert gases such as superheated steam. Finally combustion gas removal conduits 98 and 100, respectively, are communicatingly connected to the bottom of each of regeneration vessels 64 and 66 and provide for removal of gases evolved during the regeneration of the dry particulate catalyst. In operation, a spent particulate catalyst and liquid reactant mixture is withdrawn about once a day from each of the operating high pressure reactor vessels 12 and 14. About 0.01 to 1.0 lbs. catalyst per bbl. feed is removed each day and passed into each of the withdrawal pots 32 and 36.

After termination of the withdrawal from the reactor vessels, the withdrawn particulate catalyst is passed as a slurry in a light oil medium to the washing and stripping vessel 44. Wash oil is first recirculated through the washing and stripping vessel for a period sufficient to remove substantially all the heavy and residual oil adsorbed on the particulate catalyst. After washing, hydrogen gas is passed upwardly through the washing and stripping vessel 44 for at least one hour to remove any light or wash oil remaining on the catalyst. The upward flow rates of both the wash oil and the hydrogen drying gas are at velocities which expand the mass of particulate catalyst and impose a random motion to the catalyst particles as hereinbefore described. A liquid velocity of 2 to 100 gal./min./ft.^{2} is sufficient. The dry stripped particulate catalyst is then passed to the first regeneration vessel 64 via an inert gas carrying medium and accumulated in the regeneration vessel 64 until a full charge of spent particulate catalyst is accumulated. This may amount to about seven days accumulation of withdrawn spent catalyst from the reactor vessels. However also contemplated as part of the process, is the partial charging of the regeneration vessel 64 with a portion of fresh newly produced particulate catalyst, so that the accumulated dry particulate catalyst may be composed of both fresh and spent catalyst particles. The inclusion of fresh catalyst particles with the spent catalyst provides an economic and advantageous break-in treatment for the fresh catalyst.

The regenerative treatment for a particular hydrogenation catalyst, specifically a cobalt-molybdenum catalyst Aero HDS-3A is as follows. After the regeneration vessel has been fully charged with hot dry particulate catalyst, steam at a pressure of about 100 psig and a temperature of 600° F. mixed with air, is introduced into the regeneration vessel 64 through gas distributor 86. The steam rate is about 1 lb. of steam per hour per pound of catalyst, while air is present in a concentration equivalent to about 0.5 mol percent oxygen. After coke burnoff is initiated, temperature of the flame front in the regeneration zone is maintained at as close to 750° F. as possible. (An increase of approximately 0.1 mol percent oxygen results in a 25° F. temperature rise above inlet temperature.)

Coke burnoff is completed by bringing the flame front temperature to about 800° F. if possible. After coke burnoff is completed as indicated by a lowering temperature, the steam flow is terminated. Air flow is continued to cool the burnoff particulate catalyst mass to about 350° F. and the reactor vessel 64 is then purged with nitrogen. If desired the regenerated particulate catalyst may be sulfided by adjusting the reactor vessel pressure to about 250 psig with recycled hydrogen gas containing hydrogen sulfide at a temperature of about 400° F. until a copious breakthrough of hydrogen sulfide is observed. The regenerated particulate catalyst is stored in the regeneration vessel 64 and fed to the reactor vessels 12 and 14 periodically, to maintain the catalyst beds 18 and 20 respectively at the desired levels.

After a full charge of dry particulate catalyst has been accumulated in the first regeneration vessel 64, accumulation of dry particulate catalyst is commenced in the second regeneration vessel 66, and the regeneration process described above in reference to the first regeneration vessel 64, is repeated utilizing the second regeneration vessel 66. Periodic withdrawal of spent particulate catalyst and continual periodic replenishment of the reactor vessels with regenerated particulate catalyst allows alternate use of the first and second regeneration vessels, 64 and 66, to provide a semicontinuous catalyst regeneration process.

OXIDATION

Teijin

A process developed by Y. Ichikawa; U.S. Patent 3,525,762; August 25, 1970; assigned to Teijin Limited, Japan involves recovery of the cobalt-containing catalysts used in liquid phase oxidation processes. More particularly, the process relates to a method of purification and recovery of the trivalent catalysts used in the production of oxidation products by oxidizing hydrocarbons and/or their oxidized derivatives with molecular oxygen, in the presence of cobalt salts of lower aliphatic monocarboxylic acids of 2 to 4 carbons as catalysts.

Recent developments in petroleum chemical industries afford rich supplies of aliphatic, alicyclic and aromatic hydrocarbons, and as the result, liquid phase oxidation of these hydrocarbons or their oxidized derivatives using molecular oxygen, such as air, to form ketones, alchols and carboxylic acids, etc. has come to be the object of industrial concern and has been widely practiced. A typical process includes, for example, the production of acetic acid from acetaldehyde, of benzoic acid from toluene, of cyclohexanol and cyclohexanone from cyclohexane, of adipic acid from these products, and of terephthalic acid from p-xylene.

Such liquid phase oxidation processes using molecular oxygen are normally practiced in the presence of catalysts. As such catalysts, salts of valence-variable metals such as cobalt and manganese are preferred, particularly cobalt salts of lower aliphatic monocarboxylic acids of 2 to 4 carbons. The cobalt salt catalysts of lower aliphatic monocarboxylic acids of 2 to 4 carbons are used with particularly satisfactory results in oxidation of hydrocarbons and/or their oxidized derivatives in a solvent comprising a lower aliphatic monocarboxylic acid of 2 to 4 carbons or an aqueous solution thereof of no more than 30 mol percent of water content.

When the product, or that and the side-produced water are removed from the reaction mixture and the remaining mother liquor is recycled as for further oxidation of hydrocarbons or their oxidized derivatives, the catalytic activity of the catalyst (a cobalt salt of an aliphatic monocarboxylic acid of 2 to 4 carbons) is lowered as the number of times of the recirculation increases, and the purity of the product is also lowered as impurities tend to be mixed therein. Therefore, for liquid phase air-oxidation using a liquid media, effective treatment of the mother liquor containing the catalyst, remaining after the removal of the object oxidized product from the reaction mixture is of industrial importance.

As one of such treatments, it is known in the production of terephthalic acid by oxidation of p-xylene with molecular oxygen in the presence of bromine and a cobalt salt, that the mother liquor remaining after the removal of crude terephthalic acid from the reaction mixture may be oxidized with nitric acid, so that the intermediate oxidation products and unreacted p-xylene in the liquor may further be converted into terephthalic acid, and the cobalt salt in the liquor may be converted to cobalt nitrate to be recovered. However, while such a process allows the recovery of the unreacted material and the intermediate oxidation products as terephthalic acid, it suffers the disadvantage that the catalyst cannot be directly recycled and reused because it is recovered in the form of cobalt nitrate. Again, since in such a process nitric acid oxidation is practiced in the presence of a bromine-containing compound, the equipment is heavily corroded, which is industrially very objectionable.

On the other hand, it was recently proposed to obtain aromatic carboxylic acids, particularly terephthalic acid, with good yields under relatively mild conditions at lower reaction temperatures, using cobalt-containing catalysts, together with other additives such as methylenic ketones, ozone and aldehydes. However, no advantageous means of recovery of the cobalt-containing catalyst has yet been found with respect to that process.

An exemplary process for the recovery of oxidation catalysts is such as illustrated, for example, in U.S. Patent 2,964,559 to Burney et al. The process is one involving the recovery of a heavy metal oxidation catalyst used in an oxidation process carried out with the use of bromine as a promotor. In accordance with such an oxidation method, the oxidation is ordinarily carried out at a temperature above 150°C. and the used cobalt catalyst is present in the reaction mixture in a divalent form. It is indicated that the distillation bottoms from the reaction mixture can be extracted with water or glacial acetic acid to provide for the recovery of the heavy metal catalyst. As indicated previously, however, such a process as described in U.S. Patent 2,964,559 is disadvantageous in view of the use of the bromine promotor which complicates the system and provides distinct disadvantages with respect to the type of apparatus which can be utilized due to the corrosive nature of the promotor.

The environment of the development of the process differs from that illustrated, for example, in U.S. Patent 2,964,559 in that the oxidation process employing a cobalt salt catalyst is not conducted in the presence of bromine or a bromine-type promotor, and, accordingly, the reaction mixture is not the same as that obtained in accordance with the previously described process. Thus, in accordance with the environment in which the process was developed, the cobalt salt catalysts to be recovered are primarily or substantially in the trivalent form.

Until the development of the process no satisfactory process had yet been developed for recovering such trivalent form of the cobalt salt catalysts from an oxidation process conducted in the absence of bromine or a bromine-type promotor. In accordance with the process, however, such a satisfactory process has been developed whereby the cobalt salt catalysts resulting from such oxidation process is heated to a temperature within the range of 50° to 300°C. in the presence of an aqueous solution of a lower aliphatic monocarboxylic acid of 2 to 4 carbon atoms wherein the aliphatic monocarboxylic acid and water are present in at least certain amounts per gram-atom of the cobalt metal present and the aliphatic monocarboxylic acid is present in the aqueous solution in a mol percent of at least 0.2 to a mol percent of not less than 80.

It has been discovered that when hydrocarbons and/or their oxidized derivatives are oxidized in liquid phase with molecular oxygen in the presence of a cobalt salt of an aliphatic monocarboxylic acid of 2 to 4 carbons as the catalyst, besides the object oxidation product and the intermediate oxidation products, complicated side products are formed. It is presumed that these side products contain the substances which prevent the oxidation reaction. For example, oxidation products of phenolic substances side-produced during the oxidation of p-xylene are such objectionable side products.

Therefore, if the catalyst system containing these side products is recycled and continuously used for the oxidation reaction without any refining treatment, as the number of recycles increase, its catalytic activity is lowered, which also causes lowering of the purity of the product. These objectionable side products cannot be easily separated and removed from the mother liquor of the reaction mixture from which the object oxidation product has been removed, since they are soluble in the mother liquor. This is probably because the objectionable substances form certain types of chelate bonds with the metal cobalt.

However, when a suitable amount of water is added to the mother liquor to form an aqueous solution and the aqueous solution is heated to 50° to 300°C. followed by cooling, a solid precipitate is formed. When the solid is removed, the remaining liquor is apparently free from the oxidation reaction-preventing substances. As the liquor comprises the cobalt salt of the aliphatic monocarboxylic acid of 2 to 4 carbons dissolved in the aqueous solution of the monocarboxylic acid, it may be recycled, as it is or after being subjected to distillation to have some of the water content removed. If necessary, the mother liquor treated as in the above from which the solid precipitate has been removed may be subjected to further treatments such as distillation or drying so that the cobalt salt of the aliphatic monocarboxylic acid of 2 to 4 carbons may be recovered in the form of a concentrated solution isolated solid, which can be reused as the catalyst.

STEAM REFORMING

Pullman

A process developed by J.P. Van Hook and T.H. Milliken; U.S. Patent 3,385,670; May 28, 1968; assigned to Pullman Incorporated utilizes a catalyst composition consisting of elemental cobalt, cobalt oxide or mixtures thereof an a zirconia carrier and in which the content of cobalt, expressed as cobalt oxide, is from 0.5 to 15 weight percent of the total weight of the composition.

Steam reforming of hydrocarbons is a process by which a hydrocarbon and steam are contacted with a catalyst to produce gaseous product comprising hydrogen and carbon oxides and is well-known to the art of hydrogen production. Catalysts which are used commerically for this reaction usually comprise nickel as the catalytic agent supported on or diluted with an oxidic refractory support comprising alumina such as aluminous cement containing magnesia and calcium oxide as binding agents.

While such two-component catalysts may successfully be employed for steam reforming of normally gaseous paraffins at relatively low steam-to-carbon ratios, they are usually unsuitable under the same operating conditions for the conversion of feedstocks containing significant amounts of unsaturated compounds such as olefins and/or heavier, higher molecular weight paraffinic compounds because of considerable carbon deposition on the catalyst. The problem of carbon deposition during steam reforming of olefinic feedstocks such as refinery gases and coke oven gases and of normally liquid feeds is one well-known to those skilled in this art. Although considerable effort has been expended to improve nickel catalysts so as to overcome the carbon deposition problem and yet provide a process which can be operated at feasible levels of steam very little attention has been devoted to the study of catalytic agents other than nickel since it is generally believed that nickel is the most suitable agent from the standpoint of catalytic activity.

However, in the development work leading to the process, a unique composition was developed consisting of cobalt on zirconia. Figure 99 shows a correlation of hydrocarbon conversion and steam-to-carbon ratios for various supported cobalt catalysts, as follows:

> A - cobalt on zirconia
> B - cobalt on titania
> C - cobalt on titania-alumina
> D - cobalt on alumina
> E - cobalt on aluminous cement

Such cobalt on zirconia catalysts operate without serious carbon deposition problems as well.

FIGURE 99: GRAPH SHOWING RELATION OF HYDROCARBON CONVERSION TO STEAM-CARBON RATIO FOR VARIOUS SUPPORTED COBALT CATALYSTS

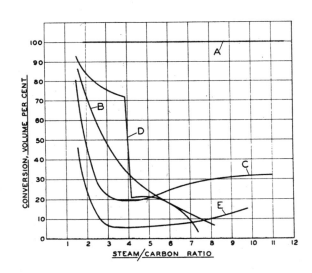

Source: J.P. Van Hook and T.H. Milliken; U.S. Patent 3,385,670; May 28, 1968

RHODIUM-CONTAINING CATALYSTS

HYDROFORMYLATION

British Petroleum

A process developed by M.T. Westaway and G. Walker; U.S. Patent 3,617,553; November 2, 1971; assigned to The British Petroleum Company Limited is one in which transition metal complexes are separated from mixtures containing organic components of lower molecular weight by contacting the mixture with one side of a cellulosic membrane. The applied pressure is greater than the pressure on the opposite side of the membrane and the pressure differential is greater than the osmotic pressure of the system. The permeate has a reduced concentration of complex. The membrane is preferably a cellulose acetate membrane. Applied pressures are from 500 to 2,500 psig and complexes separated may be those of rhodium, nickel, vanadium and other metals of groups VIIa and VIII of the Periodic Table. Organic components may be hydrocarbons or mixtures from the dimerization of olefins or from the hydroformylation of propylene.

Monsanto

A process developed by F.E. Paulik, K.K. Robinson and J.F. Roth; U.S. Patent 3,487,112; December 30, 1969; assigned to Monsanto Company is a hydroformylation process which uses a supported rhodium catalyst, with olefins being charged in the vapor phase to react to aldehydes. A preferred catalyst is chlorocarbonylbis(triphenylphosphine)rhodium (I) and the carrier is selected from the group consisting of pumice, alumina, silica, silica-alumina, magnesia, diatomaceous earth, bauxite, titania, zirconia, clays, attapulgite, lime, magnesium silicate, silicon carbide, carbons and zeolites.

Hydroformylation processes are well known in the art and have been directed to the production of reaction mixtures comprising substantial amounts of aldehydes and alcohols by the reaction of olefins with carbon monoxide and hydrogen at elevated temperatures and pressures in the presence of certain catalysts dissolved in liquid reaction media. The prior art teaches the use of dicobaltoctacarbonyl or its various modified forms as well as carbonyls of other group VIII metals such as rhodium, ruthenium, iridium, etc. which may also be modified by ligands comprised of organic compounds of group V elements such as triaryl- and trialkyl-phosphine, arsine, etc. Certain disadvantages present in hydroformylation processes described in the prior art are catalyst instability, lack of product selectivity, low levels of catalyst reactivity and because of the liquid phase present, the need for larger and costly processing equipment for product isolation, as well as for catalyst recovery, catalyst regeneration and catalyst recycle to the reactor.

One particular disadvantage of hydroformylation processes of the prior art is their dependence upon the use of liquid phase catalysts comprised of metal carbonyls or certain modified metal carbonyls including dicobaltoctacarbonyl, tetracarbonyl-cobalt hydride and organophosphine substituted cobalt carbonyls, which necessitate the use of high pressures to remain stable under the high reaction temperatures employed.

Dicobaltoctacarbonyl requires a very high partial pressure of carbon monoxide to maintain catalyst stability under hydroformylation conditions. These partial pressures of carbon monoxide are often in excess of several hundred psig at moderate temperatures in the range of 50° to 100°C., and range as high as 1,000 to 3,000 psig carbon monoxide partial pressure under normal hydroformylation conditions. Organophosphine substituted complexes of dicobaltoctacarbonyl are often more stable in a liquid phase catalyst, in that they require considerably lower partial pressures of carbon monoxide and consequently can be used in a lower pressure hydroformylation process. However, even these catalysts are generally not stable enough to withstand the severe operating conditions employed for product isolation and catalyst recovery in the hydroformylation process.

One particular disadvantage of hydroformylation processes of the prior art is their dependence upon the use of liquid phase catalyst systems comprised of metal carbonyls or certain modified metal carbonyls dissolved in a liquid reaction medium. In liquid phase systems, additional processing steps are necessary for separation of products from the catalyst solutions, and there are always handling losses of the catalyst. This handling of the catalyst solution in liquid phase processes requires the use of large and costly processing equipment for separation of the product, and for catalyst recovery and catalyst recycle to the reaction zone. Also, the losses of the metal complex due to handling of the catalyst solutions are costly because the metal complexes per se are very expensive.

For example, the prior art hydroformylation reaction utilizing a cobalt carbonyl catalyst in liquid phase suffers a serious disadvantage due to substantial losses of the cobalt carbonyl catalyst during the necessary handling for separation of product from the catalyst solutions. It is well known that such solutions readily decompose during the processing steps for product separation so that the cobalt plates out as an inactive coating on the walls of the reactors and piping. This is a serious defect since significant losses of catalyst occur. The catalyst solution thus loses its effectiveness and the precipitated cobalt metal is difficult to remove and recover.

Furthermore, separation of products from catalyst solutions of the conventional liquid phase processes, such as the cobalt catalyzed process, results in the cobalt carbonyl catalyst being carried into the ultimate product, from which the cobalt must be separated by processing in a "decobalting" operation. Here the cobalt is precipitated as the metal or as a salt which must then be dissolved and again transformed to the active cobalt carbonyl catalytic compound in the liquid phase. Significant losses of the catalyst metal complex occur during all these handling and processing steps.

The process using a solid catalyst in the absence of a liquid phase overcomes the above difficulties inherent in the handling and processing of a liquid reaction medium by providing a system for continuous separation of products from the solid catalytic phase. These advantages are even more important when utilizing a catalyst comprised of the more expensive noble metal complexes where utilization in industrial applications requires an absolute minimum of catalyst losses.

Still another object of the process is the provision of a vapor phase hydroformylation process enabling the efficient single stage production of aldehydes by reaction of olefinic hydrocarbons with carbon monoxide and hydrogen in the presence of an improved and more stable catalyst enabling the use of lower temperature, lower pressure and shorter contact time than generally possible heretofore, and eliminating product isolation, catalyst recovery and recycle steps, and also solvent recovery.

Figure 100 illustrates the application of the improved catalyst to a vapor phase process wherein the reactant olefin (e.g., propylene) and the hydroformylated product (e.g., butyraldehyde) are removed in the vapor effluent from the reactor. In this operation propylene feed is introduced into reactor 3 through line 1, while the hydrogen and carbon monoxide feed (e.g., synthesis gas) is introduced through line 2.

FIGURE 100: FLOW DIAGRAM FOR RHODIUM-CATALYZED HYDROFORMYLATION PROCESS

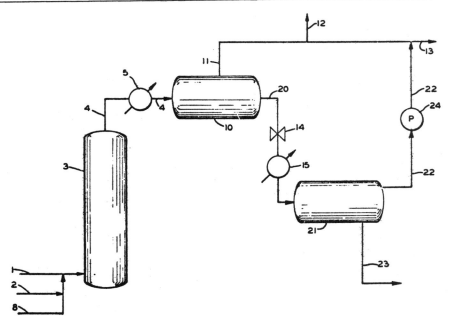

Source: F.E. Paulik, K.K. Robinson and J.F. Roth; U.S. Patent 3,487,112; December 30, 1969

A recycle gas stream consisting principally of hydrogen and carbon monoxide with a small quantity of propylene is introduced through line 8. Catalytic reactor 3 contains the supported catalyst which is present, for example, as a fixed bed of catalyst. The typical catalyst loading is about 0.35 weight percent (based on metal content), and comprises the rhodium complex chlorocarbonylbis(triphenylphosphine) rhodium dispersed on an inert support such as 10 to 30 mesh, low surface area, porous alumina. The reactor operates at about 125°C. and 500 psig total pressure (carbon monoxide partial pressure about 240 psig); the normal gas rate is 150 lb./min. propylene, 200 lb./min. carbon monoxide, and 16 lb./min. hydrogen.

The gaseous reactor effluent, containing about 100 lb./min. propylene, 86 lb./min. butyraldehyde, 166 lb./min. CO, and 14 lb./min. H_2 exits through line 4 and is partially condensed in condenser 5 at about 10°C. The condensed phase consisting of greater than 90% butyraldehyde is separated from the noncondensables in a high pressure separator 10. The gases exit the separator through line 11, and after removal of a purge stream through line 12 to prevent buildup of inert gases, the remaining gases are vented at line 13, or preferably are recycled. The condensed phase is removed from high pressure separator 10 through line 20, let down in pressure through valve 14, cooled in exchanger 15, and then separated in vessel 21. The gases which contain dissolved propylene, liberated on reduction in pressure, exit through line 22, are repressured in compressor 24, and combine with line 11 gases to form the gaseous stream available at 13 for recycling. The liquid from separator 21 represents the crude butyraldehyde which passes out through line 23 to the aldehyde purification area.

NICKEL-CONTAINING CATALYSTS

DEHYDROGENATION

Progil

A process developed by R. Gac and L. Zeppieri; U.S. Patent 3,336,399; August 15, 1967; assigned to Progil, France is a process for the manufacture of heavy phenols from corresponding oxygenated cycloaliphatic compounds and, more particularly, for the production of naphthols by the dehydrogenation of alcohols and ketones derived from more or less hydrogenated naphthalene and alpha-naphthol, from 1-hydroxy-1,2,3,4-tetrahydro-naphthalene or 1-oxo-1,2,3,4-tetra-hydro-naphthalene.

In the past, various methods have been proposed for the preparation of heavy phenols by the dehydrogenation of the corresponding cycloaliphatic compounds. None of these methods has given complete satisfaction because, for instance, these known methods result in a low rate of transformation and in poor yields. Furthermore, the known processes generally give rise to the formation of considerable quantities of secondary products. For instance, when a catalytic metal is employed in a very active form, such as fresh Raney nickel or platinum sponge, for the dehydrogenation of 1-hydroxy- or 1-oxo-1,2,3,4-tetrahydro-naphthol-1, which is very difficult to separate later on from the alpha-naphthol. If the conventional dehydrogenation catalysts (made up of metal oxides) are used, the speed of dehydrogenation is too low and this permits a side reaction to take place to an excessive degree. The same disadvantage is found when using catalysts formed by the deposit of finely divided metal on carriers generally employed for catalytic members, particularly active carbon, clays, and oxides of elements of Groups III to VIII of the Periodic Table, particularly Al_2O_3, TiO_2, SiO_2, and Cr_2O_3, all compositions containing such oxides.

This process, then, is based on the discovery of special catalysts to be used in the dehydrogenation of oxygenated cyclo-aliphatic compounds for the purpose of obtaining the corresponding heavy phenols. It has been found that, in order for the dehydrogenation of the aliphatic ring to take place selectively, the catalytic mass should not contain any "acid foci". In other words, it should have a completely neutral or basic character. The "acid foci" (that is to say, compounds or groups which can have an acid character, such as those found in active carbon, active silica containing silicic acid, as well as known catalyst carriers, such as titanium dioxide, aluminum oxide or chromium oxide) are undesirable in the de-hydrogenation process. The process consists, in the first place, of subjecting oxygenated cycloaliphatic compounds in the liquid state to heating in the presence of a dehydrogenation catalyst made up of one or more finely divided catalytic metals deposited on an oxide support of a basic or entirely neutral character and not containing any composition of an acid character in the free state. Catalytic metals which can be used may be one or more of the metals selected from the group consisting of nickel, copper, iron, cobalt, chromium, platinum, or other metals of the platinum group, such as iridium, palladium, and rhodium.

Carriers which can be used with the catalyst include the basic oxices, particularly those of metals of Groups I and II of the Periodic Table, especially calcium oxide, magnesium oxide, copper oxide, strontium oxide, barium oxide, and zirconium oxide. Some oxides, such as silica, may be used, but only in a perfectly neutral form, such as kieselguhr, which does not contain any acid foci. Catalysts may be prepared in accordance with this process by using 20 to 100 parts by weight of the finely divided catalytic metal per 100 parts of carrier. In the preferred embodiment, the catalytic metal makes up 50 to 100 parts per each 100 parts by weight of carrier. A preparation of these catalysts can be carried out in the usual manner. For instance, a mixture of a desired metal oxide with a selected support powder may take place by co-precipitation, by tabletting, or by granulation and reduction in hydrogen, preferably at a temperature equal to or in the vicinity of that at which the catalyst is to be used.

Another important feature of the process is that the dehydrogenation take place at temperatures which are in the vicinity of the boiling point of the reaction liquid. It is particularly advantageous to work under boiling conditions at atmospheric pressure or at any pressure from 1 to 5 atmospheres. When the oxygenated cycloaliphatic compound which is to be treated is 1-oxo-1,2,3,4-tetrahydro-naphthalene, the best temperature to use for obtaining alpha-naphthol is in the range from 200° to 350°C. The velocity of the reaction and the rate of conversion increases with the temperature and the results are particularly favorable at the boiling point of the reaction mixture, which lies between 256° and 290°C. at atmospheric pressure. If the process is carried out at a slightly higher pressure in the range from absolute pressures from 1 to 3 kg./cm.2 the favorable temperature range would be in the range from 305° to 340°C.

The process can be carried out in two different ways: with a fixed catalytic bed, or with catalyst particles in suspension in the reaction medium. When liquids containing principally 1-oxo-1,2,3,4-tetrahydro-naphthalene or 1-hydroxy-1,2,3,4-tetrahydro-naphthalene are dehydrogenated to naphthol by passage through a fixed catalytic bed, it is recommended that the catalyst be in the form of granules or particles of a size from 0.5 to 10 mm. and, preferably, in the range from 2 to 5 mm. Larger dimensions lead to a poor utilization of the catalyst, while smaller dimensions increase the loss of charge and risk the formation of gummy residues involving agglomeration of the granules. The linear rates of flow for the reaction liquid through the catalytic bed are preferably from 4 to 100 meters per hour, and, preferably, from 10 to 40 meters per hour, calculated for the theoretically empty apparatus.

When the process employs the catalyst in suspension in the reaction liquid, the preferred dimensions of the catalytic granules are less than that used in a fixed bed. They can be in the range from 0.1 to 5 mm., but preferably are in the range from 0.1 to 1 mm. In fact, below 0.1 mm. in size separation of the catalyst from the treated liquid after the reaction can be difficult, while granules which are greater than 1 mm. size are not readily maintained in suspension and lead to a slow rate of dehydrogenation. According to the nature of the liquid and of the catalyst, the suspension can contain 10 to 40 parts by weight of the latter per 100 parts of liquid. However, the preferred proportions are from 20 to 30 parts per 100. Figure 101 shows a suitable form of apparatus for the conduct of the process.

FIGURE 101: PROCESS FLOW SHEET FOR DEHYDROGENATION OF CYCLOALIPHATIC COMPOUNDS TO PHENOLIC COMPOUNDS

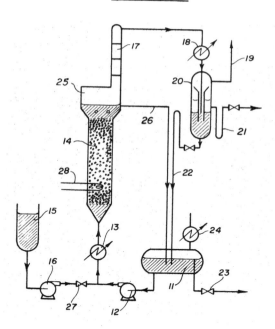

Source: R. Gac and L. Zeppieri; U.S. Patent 3,336,399; August 15, 1967

Referring to the accompanying drawing, it can be seen that the apparatus is provided with a flask 11 which serves for the intermediate storage of liquid obtained by a dehydrogenation which takes place in a reactor 14 of the starting material obtained from a reservoir 15. A supply pump 16 causes the liquid to pass from the reservoir 15 into the bed of catalyst in the reactor 14. At the same time, the bed receives the treated liquid from the flask 11 by way of a circulation pump 12. The liquid passes through a preheater 13 and arrives at a catalytic bed at the desired temperature, for instance, 250°C. The reactor 14 is provided with a heater 28 which permits the catalyst bed in the reactor 14 to be preheated. At its upper part, the reactor is provided with an enlargement 25 provided with an overflow pipe 26 which leads back to the flask 11. The top of the enlargement 25 is provided with a column 17 through which the vapors evolved by the reaction pass and, eventually, reach a condenser 18. At the bottom of the condenser 18, a separator 20 is provided in which water recovered separates out and is removed through the pipe 21. The hydrogen escapes through a vent 19 at the top of the separator. The decanted liquid from the separator 20 returns to the flask 11 through a pipe 22. A vent 24 at the top of the flask 11 is provided with a cooler for avoiding the escape of naphtholic vapors into the atmosphere. Withdrawal of the finished product is made by way of a takeoff valve 23 and a valve 27 regulates the flow of liquid from the pump 16 to the reactor 14. The reaction can also be carried out with a single passage of the reaction mixture through a tubular apparatus of sufficient length but which is divided into several sections for eliminating the hydrogen produced in each section.

However, in order to achieve high rates of flow of liquid, it is generally more advantageous to recirculate the reaction mass in a closed circuit through the catalyst in order to obtain the desired rate of conversion. The apparatus shown in the drawing is particularly suitable for this type of operation. A predetermined volume of the primary material is withdrawn from the reservoir 15 and the pump 16 forces it into the reactor 14. The valve 27 is then closed and operation takes place using the circulation pump 12. The liquid circulates around the various circuits, while the hydrogen formed in the reactor 14 is released through the pipe 19. When the flask 11 contains the desired naphthol (or other heavy phenol) content, the material is withdrawn through the valve 23 and a new charge of starting material from the reservoir 15 is introduced into the apparatus.

HYDROCRACKING

Chevron Research

A process developed by B.F. Mulaskey; U.S. Patent 3,487,007; December 30, 1969; assigned to Chevron Research Co. involves hydrocracking using a catalyst containing nickel and tin, or compounds thereof, in an amount from 2 to 50 combined weight percent metals, the nickel to tin weight ratio being from 0.25 to 20, associated with a porous inorganic oxide at hydrocracking conditions. The process is controlled at a temperature from 400° to 750°F., by varying the sulfur concentration in the hydrocracking zone. The aromaticity of the product is increased by increasing the sulfur concentration in the reaction zone and is decreased by decreasing the sulfur content.

Another process developed by B.F. Mulaskey; U.S. Patent 3,542,696; November 24, 1970; assigned to Chevron Research Company involves the production of a catalytic composition which comprises nickel, or compounds thereof, in association with a coprecipitated composite of tin, or compounds thereof, and a siliceous oxide. The nickel and tin, or their compounds, are present in an amount of 2 to 50 combined weight percent metals, with a nickel to tin weight ratio of 0.25 to 20. As a specific embodiment, the catalytic composition of matter comprises a coprecipitated composite of nickel and tin, or their compounds, and a siliceous oxide. The coprecipitated composite is preferably prepared by the coprecipitation or cogelation of a mixture of compounds of the hydrogenating metals, that is, compounds of nickel and tin, and a compound of silicon.

As a further specific embodiment, a catalytic composition of matter has been found comprising a crystalline zeolite aluminosilicate thoroughly admixed with a composite comprising nickel, or compounds thereof, and a coprecipitate of tin, or compounds thereof, and a siliceous oxide. The combined weight percent of nickel and tin is from 2 to 50, based on the finished catalyst, with the nickel to tin weight ratio of from 0.25 to 20. The crystalline zeolitic aluminosilicate preferably has uniform pore dimensions of greater than 6 A.

A process developed by J.R. Kittrell; U.S. Patent 3,546,096; December 8, 1970; assigned to Chevron Research Company is a hydrocracking process utilizing a catalyst consisting essentially of an ultrastable crystalline zeolitic molecular sieve component, a component selected from nickel and compounds thereof, and a component selected from tin and compounds thereof. The ultrastable crystalline zeolitic molecular sieve-containing catalyst of the process is an effective hydrocracking catalyst of itself. It is not contained in a matrix of other catalyst components. It is surprising to find that by using a critical combination of an ultrastable crystalline zeolitic molecular sieve component and a nickel hydrogenation component augmented with a tin component, excellent results can be achieved without dispersing the molecular sieve component in a matrix of other catalyst components.

The ultrastable crystalline zeolitic molecular sieve component of the catalyst may be an ultrastable form of any type of crystalline zeolite molecular sieve that is known in the art as a useful component of a conventional hydrocracking catalyst. An ultrastable form of faujasite, that is, one having a sodium content below 3 weight percent, calculated as Na_2O a unit cell size below about 24.65 A., and a silica/alumina ratio above about 2.15, in the ammonia or hydrogen form, is especially preferred. It has been found that the presence of tin, in the metal, oxide or sulfide form, in the catalyst of the process, results in higher hydrocracking activity and higher hydrogenation activity than would be exhibited by a catalyst that is identical except that contains no tin. Further, the presence of tin permits the hydrogenation activity to be controlled in an essentially reversible manner by varying the amount of sulfur present in the hydrocarbon feed.

HYDROGEN PRODUCTION

Universal Oil Products

A process developed by <u>D.E. McCartney and H.A. Hauser; U.S. Patent 3,314,761; April 18, 1967; assigned to Universal Oil Products Company</u> involves an improved method for operating a continuous catalytic hydrogen producing system in a manner to maintain an active surface on the catalyst particles. More particularly, the process is directed to means for introducing a metal carbonyl or a metal carbonyl-halogen compound into contact with the catalyst being used in the system and effecting the decomposition of such carbonyl compound to provide a fresh deposition of the activating metal or metal compound on the catalyst whereby activity thereof is maintained at a high level.

In carrying out a hydrogen producing system by the cracking or decomposing of methane or other hydrocarbon charge stream, it is preferable and substantially necessary from the commercial and economic aspects of the operation to maintain a continuously operating system capable of providing a high yield of hydrogen with a high degree of purity. Maintaining a high yield and purity seems to be directly related to maintaining a high activity level for the catalyst in the system. For example, in the operation of a fluidized catalytic hydrogen producing unit, wherein heated particles of a nickel-alumina catalyst are contacted by methane in the reaction zone to produce hydrogen and carbonized particles, and wherein the latter are subsequently contacted with air in a regeneration zone to effect the gasification and removal of the carbon, as well as to effect the reheating of the catalyst particles, it has been found that there is a gradual deactivation of the catalyst with an accompanying loss of equilibrium conversion. However, the same unit utilizing a fresh catalyst charge will provide substantially equilibrium conversion of methane to hydrogen and carbon with such optimum conversion continuing until there is a gradual deactivation of the catalyst.

Figure 102 shows the essential features of a fluidized catalyst hydrogen producing system with means to introduce a metal carbonyl compound into contact with catalyst particles circulating in the system and effect maintenance of the catalyst activity. There is shown a reactor 1 adapted to receive heated catalyst particles and a charge stream from a riser line 2. The catalyst and charge stream effect a desired intermixing and contact within the interior of the reactor zone such that there is a decomposition of the charge stream, indicated herein as a methane stream, to produce a hydrogen rich product stream and carbonized catalyst particles. The hydrogen rich stream with entrained particles is passed from the upper end portion of the reactor zone 1 by way of line 3 and particle separator 4. The latter serves to recover entrained catalyst particles and return them to the unit by way of dip leg or standpipe 5 while the hydrogen rich product stream passes overhead by way of line 6. Carbon containing catalyst particles are continuously withdrawn from the lower portion of reactor 1 by way of line 7 and control valve 8, to be subsequently entrained in an air or free oxygen containing stream within line 9 for transfer into regeneration chamber 10.

FIGURE 102: FLOW DIAGRAM OF REGENERATIVE NICKEL-CATALYZED PROCESS FOR PRODUCING HYDROGEN FROM METHANE

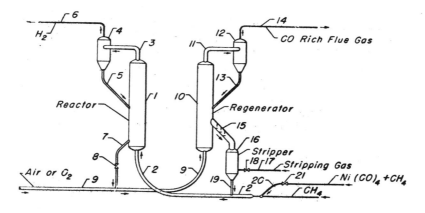

Source: D.E. McCartney and H.A. Hauser; U.S. Patent 3,314,761; April 18, 1967

The regenerator is generally maintained, in a preferred operation, under controlled oxygen content conditions to provide the gasification and removal of carbon from the catalyst particles to in turn effect the desired production of heat and reheating of catalyst particles as well as carbon removal to a desired lower carbon level. Again, the oxygen and carbonized catalyst contacting is indicated as being carried out in a fluidized bed within the regeneration zone such that a resulting carbon monoxide-carbon dioxide flue gas stream and entrained particles will pass overhead by way of line 11 to a separator

12. Collected particles from separator 12 are returned by way of line 13 to the fluidized bed in regenerator 10 while a carbon monoxide rich flue gas stream passes overhead by way of line 14. A continuous stream of heated catalyst particles having a lower carbon content are withdrawn from the regenerator by way of line 15 and stripping zone 16. In the latter there may be introduction of a suitable stripping gas such as hydrogen, methane or nitrogen, such that entrained and occluded carbon oxides are stripped from the particles prior to their reintroduction into the reaction zone. The stripping gas stream is indicated as being introduced by way of line 17 and valve 18. Preferably the stripping stream comprises a portion of the hydrogen product stream or other reducing gas such that the catalyst particles are not returned to the reaction zone in an oxidized state which may lead to the production of greater quantities of carbon oxides therein. The resulting stripped particles pass from stripper 16 by way of line 19 into the transfer line 2 which carries methane and heated particles into the conversion zone 1.

In accordance with the process, there is an introduction of a metal carbonyl compound that is compatible with the activating metal component of the catalyst system into contact with a portion of the catalyst particles, whereby there is a deposition of fresh activating metal or metal oxide on such particles. Thus, there is indicated line 20, having control valve 21, providing for the introduction of the carbonyl compound into admixture with catalyst in line 2. In an illustrative example, nickel carbonyl in admixture with methane passes in a vaporized stream into line 2 such that there is a coating of at least a part of the particles passing from the regenerator to the reaction zone. Since the catalyst particles are at a high temperature which may be of the order of 1600°F. or higher, there will be a coating and a subsequent decomposition of the carbonyl compound onto the particles to provide a fresh nickel deposition for use in effecting further conversion of the charge stream in the downstream end portion of line 2 and within the reactor 1.

HYDROGENATION

Chevron Research

A process developed by B.F. Mulaskey; U.S. Patent 3,480,531; November 25, 1969; assigned to Chevron Research Co. is one in which unsaturated hydrocarbons, particularly aromatics, are converted to more saturated hydrocarbons, without substantial cracking, by contacting the hydrocarbons in the presence of hydrogen at hydrogenation conditions with a catalyst comprising nickel and tin, or their compounds, associated with a porous solid carrier.

Phillips Petroleum

A process developed by J.L. Groebe; U.S. Patent 3,178,373; April 13, 1965; assigned to Phillips Petroleum Company relates to an improved method for activating or reducing metal hydrogenation catalysts, especially nickel-kieselguhr hydrogenation catalysts. In the hydrogenation of unsaturated compounds, and particularly in the hydrogenation of aromatic hydrocarbons, such as benzene, metal catalysts are used. These metal catalysts are usually received from the manufacturer as metal oxide or other compound on a support such as kieselguhr. The metal oxide or other compound must be reduced, at least in part, to the metal before it is active for hydrogenation. Normally, this reduction step is accomplished with the catalyst in place in the hydrogenation reactor, but before the reactor is placed onstream for hydrogenation. This reduction step has been ordinarily carried out in the prior art by employing pure hydrogen as a reducing medium.

However, pure hydrogen is extremely expensive to use. Also, in some plant practices steam is added to a hydrogen-containing reducing stream as a diluent to control the temperature rise during reduction of the catalyst. However, this has also been found to be undesirable since the catalyst life is somewhat less when employing steam as a diluent compared with a catalyst reduced with commercial pure cylinder hydrogen. According to the process, it has been found that metal hydrogenation catalysts, particularly nickel hydrogenation catalyst, can be effectively and efficiently activated or reduced prior to use in the absence of added steam without the prior art disadvantages by a two-step process comprising contacting the catalyst first with an inert gas and then with a hydrogen-containing stream.

In somewhat more detail, the process comprises the steps of first heating the catalyst with an inert gas until the catalyst has reached a temperature of at least about 500°F. and is substantially dehydrated, and then contacting the catalyst with an impure hydrogen-containing stream at a temperature not exceeding about 750°F. until the desired degree of reduction of the catalyst is achieved. Figure 103 shows the essentials of the operation of this process. Referring in detail to the drawing, the system shown essentially comprises a furnace 10, catalyst chamber 15a and 15b containing a bed of a metal catalyst to be reduced, preferably a nickel-kieselguhr catalyst, a cooler 22, a separator 25 and compressors 30a and 30b.

An inert gas such as nitrogen is first utilized to flush the system of undesirable materials and then to bring the catalyst beds in the catalyst chambers up to the desired reduction temperature. In actual operation, the system is first evacuated, nitrogen is then introduced by way of line 11 into the system and then the system evacuated one or more times. After the system is freed of undesirable materials, the whole system is pressured with nitrogen introduced by way of line 11. The nitrogen is cycled through cooler 22, separator 25, compressors 30a and 30b and passed through furnace 10 wherein it is heated to a a temperature of at least 500°F. The heated nitrogen effluent from the furnace is cycled through the catalyst chambers

15a and 15b, cooler 22, separator 25, compressors 30a and 30b, furnace 10, and then back through the catalyst chambers until the outlet temperature of the catalyst chambers reaches a level above the water dew point of the system and the temperature of the beds of catalyst have reached a temperature of at least 500°F. During the time the catalyst is heated to a reduction temperature of at least 500°F., water of hydration of the catalyst is condensed in cooler 22 and removed from the system in separator 25 by way of conduit 17. The amount of this water can be measured to determine when the catalyst has been substantially completely dehydrated.

FIGURE 103: FLOW SHEET SHOWING ACTIVATION PROCEDURE FOR NICKEL-KIESELGUHR HYDROGENATION CATALYSTS

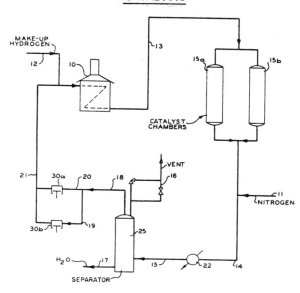

Source: J.L. Groebe; U.S. Patent 3,178,373; April 13, 1965

When the catalyst bed has reached a temperature of at least 500°F., and the bed is substantially dehydrated, the inert gas, preferably nitrogen, is gradually bled from the system through vent 16 and a make-up hydrogen-containing stream is introduced into the system by way of line 12 to replace the inert gas being removed. The make-up hydrogen stream together with circulating nitrogen is passed through furnace 10 wherein it is heated, through conduit 13 and passed through catalyst chambers 15a and 15b. The make-up hydrogen stream preferably contains demethylizable materials that are demethylizable at a temperature in the range of 500° to 750°F. The demethylizable hydrocarbons on contacting the heated catalyst demethylate and thus increase the rate of heating of the catalyst bed due to the demethylation reaction (which is exothermic) until a reduction temperature in the range of 650° to about 750°F. is reached.

The reduction effluent removed from the catalyst chambers is comprised principally of hydrogen, methane, and decreasing amounts of nitrogen as the cycling continues. The effluent is passed by way of conduit 14 through cooler 22, line 15 and then to separator 25. The temperature of the reduction gas is reduced to a temperature of about 100°F. in cooler 22. Condensed water is removed from the lower portion of separator 25 by way of conduit 17 and bleed gas, principally nitrogen (but also some hydrogen and methane), is removed overhead from separator 25 by way of vent 16. A gas stream for recycle comprising hydrogen and methane is removed from separator 25 by line 18, introduced into compressors 30a and 30b by way of lines 20 and 19, respectively, and then passed by way of line 21 to furnace 10 and then to the catalyst chambers as previously described.

Siemens

A process developed by E. Weidlich and G. Kohlmuller; U.S. Patent 3,489,694; January 13, 1970; assigned to Siemens AG, Germany involves a method of stabilizing and increasing the activity of Raney catalysts, particularly Raney nickel. It is known that Raney catalysts are produced by dissolution of inactive components from a Raney alloy, for example, aluminum from a nickel-aluminum alloy, by alkali solutions or acids. The nobler metal is thereby precipitated as a fine grained microporous powder with large surface. During this process, a great number of holes simultaneously form in the metal lattice of the nickel which renders the latter exceptionally active and well qualified for binding, either chemically or physically, the hydrogen resulting from the dissolution of the nobler metal. It is also known that Raney catalysts, which are stored, hydrated or used for long periods decrease in activity and after a time become useless.

237

The stabilization and activation treatment involves treating the catalysts with aqueous hydrochloric acid or with hydrochloric acid containing ferric chloride. $FeCl_3$, and thereafter eliminating the inactive component by dissolving out, using an alkali liquor. During the alkali aftertreatment of the Raney metal, activated with hydrochloric acid or with hydrochloric acid containing ferric chloride, the surface of the Raney metal is etched with simultaneous dissolving out of any hydroxides or oxides present. Concomitant with the above, an adsorption of iron occurs at the surface.

Varta

A process developed by M. Jung and H. von Doehren; U.S. Patent 3,573,038; March 30, 1971; assigned to Varta AG, Germany is one in which catalyst powders, such as platinum, palladium or Raney metals, are rendered insensitive to air or oxygen by treatment with an oxygenated chlorine, bromine or iodine compound, such as potassium or sodium iodate, chlorate or bromate. The powder obtained is essentially hydrogen-free, nonpyrophoric and reactivatable. The powder may be molded by heat and pressure to a formed porous body.

OLEFIN ISOMERIZATION

Dow Chemical

A process developed by C.R. Noddings and R.G. Gates; U.S. Patent 3,297,778; January 10, 1967; assigned to The Dow Chemical Company is based on the findings that a nickel phosphate prepared under conditions to maintain the pH between about 6 and 9 will yield a catalytic material which will isomerize alpha-alkenes to their corresponding beta-alkenes in near quantitative yields. The catalyst is prepared by mixing together a water-soluble metal salt of nickel with a water-soluble form of the orthophosphate in an aqueous medium under conditions such that the pH of the complexing environment is within the range of from 6 to 9. Further, while not critical, but desirable, the phosphate moiety is employed in a slight excess over that which is theoretically necessary to combine with the metal ions to form a metal orthophosphate. It is to be understood that the pH may, but does not have to, be maintained within the operative range during mixing but can be adjusted, after mixing, by addition of a base or acid as necessary to the reaction mixture to bring the solution within the desired range thereby causing precipitation of a catalytic material.

The contacting and mixing of the reactants in accordance with the above can be carried out in several manners, such as simultaneously, stepwise or intermittently, each either in a batchwise or continuous manner. Examples of salts which may be used as starting materials in preparing the catalyst are the chlorides, bromides, nitrates, and acetates, etc., of nickel. Examples of soluble phosphates that may be employed as starting materials are disodium phosphate, trisodium phosphate, dipotassium phosphate diammonium phosphate, etc. The catalyst can also be prepared in either a batchwise manner or a continuous manner by feeding separate streams of an alkali, preferably aqueous ammonia, although other bases can be employed as well as two different bases, and as either a single or as a separate stream an aqueous solution of a nickel salt, and either a separate or as a part of the aforesaid streams a dissolved orthophosphate, into a reaction chamber.

The relative rates of flow being adjusted such that the resultant mixture will achieve continuously or upon completion of the mixing a pH between 6 and 9. It is desirable, in a continuous or stepwise operation, to retain within the reaction zone a portion of the nickel phosphate which forms and precipitates. This is conveniently achieved by adjusting the outflow of the nickel phosphate precipitated to retain a portion of the flocculent material which settles rapidly to form, as a lower layer of the resultant mixture, an aqueous nickel phosphate slurry that contains 2% by weight or more, usually from 7.5 to 10%, of the nickel phosphate. The reaction mixture, or preferably the settled lower layer thereof, may be filtered to obtain a filter cake which contains 10% or more, usually about 22% of the nickel phosphate.

After the reaction is complete the precipitate is separated from the liquor by filtration or decantation and is washed with water, decanting or filtering after each washing. The washing should be carried out so as to remove as thoroughly as possible readily soluble compounds from the product, since such impurities have a disturbing and erratic action on the thermal decomposition of hydrocarbons. Of particular attention is the unreacted chloride or by-product chloride which, if retained in the catalyst, tend to deactivate the latter. The catalyst is, at this stage in its preparation, a solid or gel-like substance which is apparently amorphous.

After being washed with water, the product is dried, usually at temperatures between 60° and 150°C. The dried product is a hard gel usually of yellowish color. The gel may be crushed or otherwise reduced to granules, or small lumps, and can be used directly as an isomerization catalyst. However, it is preferably pulverized, e.g., to a particle size capable of passing a 28 mesh screen, and the powdered product is treated with a lubricant and is pressed into the form of pills, tablets, or granules of size suitable for use as a catalyst, e.g., into the form of tablets of from 1/16 to 1/2 inch diameter. The lubricant serves to lubricate the particles during the operation of pressing them into pills and its use permits the formation of pills of greater strength and durability than are otherwise obtained. As the lubricant, a substance capable of being removed by vaporization or oxidation from the product, e.g., a substance such as graphite, a vegetable oil, or a hydrocarbon oil, etc. is used.

The following is a specific example of the conduct of the process which will be outlined with reference to the block flow diagram in Figure 104. In the manner shown in the drawing, 15 g. mols of nickel chloride as in a 22 weight percent aqueous solution thereof was mixed in a vessel with 10.3 g. mols of phosphoric acid as a 75.5 weight percent aqueous solution and the resulting mixture is diluted with water to a total volume of 65 gallons. Upon completion of the addition of the above enumerated chemicals to the vessel reactor, an aqueous 13.6 weight percent ammonium hydroxide solution was, or had been, added. In some instances, the aqueous ammonium hydroxide was added together with the reactants, in others after addition of all of the reactants, and in still others the phosphoric acid and ammonia were first mixed and then admixed with the other reactants. The reaction mass was continuously stirred and base or acid added to produce and maintain a pH of the system between 5.8 and 8.5. In the specific instance 43.3 g. mols of ammonium hydroxide were required to maintain the pH at 7.8 at the end of 2.3 hours of reaction.

The reaction was considered complete when the final pH remained constant. Thereafter the reaction mass was allowed to settle overnight after which the supernatant liquid above the precipitate was drawn off (approximately 60 gallons decanted) and the resulting thick slurry filtered and washed with water. The filtrate was discarded. In the specific instance the slurry was washed with water three times until chloride free, then removed and dried at 100°C. in a rotary drier. The dry powder was recovered to the extent of 99% of the theoretical yield, based on the starting materials used, and was crushed mixed with 2% by weight of a lubricant grade graphite and expressed into pellets about 1/4 inch in diameter and 1/4 inch long. The graphite was burned off by treating the pellets with air and steam at about 650°C. for about 6 hours. The resulting catalyst pellets were tested as an isomerization catalyst at 200°C., 100 v./v. hr. (volumes of gas per unit volume of catalyst per hour) (STP) of butene-1 utilizing a mixture of 54.3% butene-1 and 45.7% butene-2 at 1 atmospheres pressure. Of the butene-1 fed to the reactor, there was obtained a 33% conversion to butene-2 representing a 100% yield of butene-1 converted.

FIGURE 104: BLOCK FLOW DIAGRAM OF PROCESS FOR PREPARING NICKEL PHOSPHATE OLEFIN ISOMERIZATION CATALYST

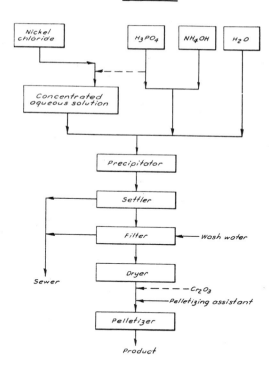

Source: C.R. Noddings and R.G. Gates; U.S. Patent 3,297,778; January 10, 1967

Pure Oil

A process developed by N.L. Carr; U.S. Patent 3,108,974; October 29, 1963; assigned to The Pure Oil Company relates to a method of hydroisomerizing low-molecular-weight hydrocarbons and more particularly to a method for hydroisomerizing normal paraffinic hydrocarbons having 4 to 7 carbon atoms in the molecule. Low-molecular-weight, normal paraffinic hydrocarbons having 4 to 7 carbon atoms, and particularly normal pentane and hexane, can be isomerized to branch-chain paraffins by contact with solid catalysts at temperatures of the order of 650° to 800°F. and elevated pressures of the order of 200 to 1,000 pounds per square inch. Catalysts which are effective in selectively converting normal paraffins to iso-paraffins are composed of an acidic silica-alumina cracking catalyst as support, impregnated with a small amount of a Group VIII metal, such as nickel, platinum, palladium, iridium and rhodium, or combinations thereof. Although nickel supported on a silica-alumina cracking catalyst is an effective isomerization catalyst, its selectivity is improved when incorporated in the form of nickel molybdate. In the decomposition (activation) of metal-oxygen compounds, such as nickel molybdate, the metal oxide is usually produced, and then is subsequently reduced in the presence of hydrogen at elevated temperatures. It is possible that reduction of the metal oxide is not the only criterion for activity; activating reactions such as the breaking of metal-support complexes and hydrogen chemisorption also occur and are affected by time and temperature.

Isomerization catalysts are prepared in active state by a process of oxidation with a substantially dry, oxygen-containing gas, preferably gas having a water vapor partial pressure below 15 mm. of mercury, followed by reduction with hydrogen-containing gas. The activity of the catalyst is dependent on a careful control of the oxidation and reduction conditions. For example, in the case of nickel and nickel molybdate supported catalysts, it is important to maintain the partial pressure of water vapor in the reducing gas not higher than about 25 mm. of mercury in order to obtain maximum activity.

In somewaht more detail, the process comprises the steps of oxidizing the catalyst with substantially dry oxygen-containing gas at a temperature between approximately 750° and 1000°F. for a period of time sufficient to convert the hydrogenation component to the oxide form and thereafter reducing the catalyst in a substantially oxygen-free and water-free hydrogen-containing atmosphere at a temperature of 750° to 1000°F. for a minimum period of time determined by the equation:

$$Y = 0.07XZ$$

in which Y is the minimum reduction time in hours, X is the length of the oxidation period in hours, and Z is the partial pressure of oxygen in the oxygen-containing gas in millimeters of mercury.

OLEFIN OLIGOMERIZATION

Sentralinstitutt for Industriell Forskning

A process developed by N. Bergem, U. Blindheim, O.-T. Onsager and H. Wang; U.S. Patent 3,558,736; January 26, 1971; assigned to Sentralinstitutt for Industriell forskning, Norway is one in which ethylene is converted to 3-methyl-pentene using a catalyst system comprising a compound of a metal from Group VIII of the Periodic Table, a Lewis acid compound of a metal from Group II and/or III of the Periodic Table, and a Lewis base compound of the elements of Group V and/or VI of the Periodic Table. The reaction product consists essentially of normal butene and 3-methylpentene, the 3-methylpentene is recovered, and all or a part of the normal butene is recycled to the reaction zone.

A number of processes for oligomerization of ethylene have been developed. These processes, however, do either give dimerized products, normal butenes, or they give a reaction product which mainly contains a mixture of higher olefins with a high content of normal (straight-chained) olefins. From U.S. Patent 3,087,978 it is known to prepare 3-methyl-pentenes by reaction of ethylene and butene-1 or butene-2 by using alkali metal catalysts. Due to the relatively low catalytic activity it is necessary to use pressures from at least 13 atmospheres and up to 130 atmospheres and a reaction temperature of from about 100° to 325°C.

The preparation of 3-methylpentenes is also known from Belgian Patent 651,596 employing catalysts based on π-allyl-nickel compounds. In the catalyst systems described, use is made of components which are unstable and difficult to obtain; the synthesis and use of the catalyst systems are therefore connected to great operational difficulties. The catalytic activity of these systems are stated to be caused by presence of the unstable and easily decomposable π-allyl-nickel bonds. Due to the development of increasingly larger cracking units for higher hydrocarbons, ethylene has become a cheap and easily available raw material for chemical industries.

The method for the manufacture of 3-methylpentene capitalizes on this availability and is characterized by the following scheme:

$$\text{ethylene} \xrightarrow{\text{catalyst}} \text{3-methylpentenes + butenes}$$

Particularly the process pertains to the simultaneous conversion of ethylene to ethylene-trimer and the reaction of ethylene with ethylene-dimer, a normal butene, with the formation of a reaction product predominantly containing hexenes and normal butenes, the hexenes being mainly 3-methylpentenes. The 3-methylpentenes are recovered as reaction product with at least a part of the normal butenes being recycled to the reaction step.

The preferred metals from the Group VIII of the Periodic Table for incorporation in the catalysts of the process are nickel or cobalt. The best results are obtained by using nickel compounds. The Group VIII metal compounds effective for use in the selective production of 3-methylpentenes are characterized by an electron configuration in which the outer electron shell is not filled, i.e., the metal atoms have an outer electron shell deficiency of 2 or 3 electrons as compared with the nearest noble gas.

Preferably, the Group VIII metal compounds are soluble in the system at least to the extent of 0.001 mmol. per liter of the reaction mixture. An important property of the Group VIII metal compounds which act as effective catalyst components in the production of 3-methylpentenes, is that they must be sufficiently stable against reduction to metallic form under the given reaction conditions, so that they are not reduced to the corresponding metal to such an extent that they no longer act as catalyst. Typical nickel compounds meeting the above requirements and being effective catalyst components in the selective production of 3-methylpentenes are illustrated by the following five types of compounds, it being understood that the corresponding compounds of other Group VIII metals may also be employed:

In the types I to V above D| represents a donor equivalent having one free electron pair as for example in the groups:

$$|O<, \quad |S<, \quad |N^- \quad \text{and} \quad |P^-$$

|D—D| is a bifunctional donor ligand, X is an equivalent of an organic or inorganic acid or a hydrocarbon group of the type alkyl or aryl containing from 1 to 10 carbon atoms, X—D| is a chelate ligand containing both a donor and an acid equivalent, and cyclodiene and cyclodienyl is a cyclic hydrocarbon with at least two double bonds, and a radical thereof respectively, containing 3 to 12 carbon atoms. Illustrative of the Type I compounds are: tetramethylcyclobutadiene-nickel-dichloride and tetraphenylcyclobutadiene-nickel-dibromide. Illustrative of the Type II compounds are: cyclopentadienyl-triphenyl-phosphine-nickel-monochloride and cyclopentadienyl-tri-n-butylphosphine-nickel-monoethyl. Illustrative of the Type III compounds are: (triphenylphosphine)2-nickel-nitrate, (tri-n-butylphosphine)2-nickelsulfate, (tricyclohexylphosphine)2-nickel-rhodanide, (tri-di-n-butylamino-phosphine)2-nickel-chloride, (tri-isopropyl-phosphine)2-nickel-iodide, (tricyclohexylphosphine)2-nickel-chloracetate, (2,4,6-trimethylpyridine)2-nickel-chloride, (piperidine)2-nickel-chloride, (tricyclohexylphosphine)2-nickelethylate, and (triiospropylphosphine)2-nickel-naphtholate.

Illustrative of the Type IV compounds are: nickel-dimethylglyoxime, nickel-acetylacetonate, nickel-benzoyl-acetonate, nickel-(8-hydroxychinolite), and nickel-phthalocyanine. Illustrative of the Type V compounds are: ethylenediamine-nickel-chloride, 1,2-bis-diethylphosphino-ethane)-nickel-bromide, and 2,2'-dipyridyl-nickel-diethyl. The catalytically active nickel compounds of the Types III and V can readily be formed by reaction of a nickel salt of an organic or inorganic acid with a Lewis base. Thus nickel compounds such as $NiCl_2$, $NiBr_2$, $Ni(NO_3)_2$, $NiSO_4$, $Ni(OOCCH_3)_2$, $Ni(OC_2H_5)_2$ and the like may be prereacted with a Lewis base to form a compound of the above mentioned type, or reacted in situ with a corresponding Lewis base in the reaction mixture to form the active catalyst components. Illustrative Lewis bases for these reactions are compounds containing trivalent nitrogen, phosphorus, arsenic or antimony, or divalent oxygen or sulfur.

Figure 105 is a flow diagram of the process. As shown there, the catalytic components and the solvent are premixed in zone 4. The nickel component is added from zone 1 and the aluminum component plus the Lewis base are added from zone 2. Solvent is introduced from zone 9. The catalyst solution is introduced continuously into reaction zone 5 from the mixing zone 4 via line 15, in an amount sufficient to compensate for the amount of catalyst consumed. A recycle stream of solvent, catalyst and some higher olefins is passed from separation zone 6 via line 16 to reaction zone 5. Fresh ethylene feed is passed from zone 3 via line 17 to the reaction zone 5 while unreacted ethylene and ethylene-dimer is passed from separation zone 7 via lines 18 and 17 to reaction zone 5.

In zone 5, the ethylene and ethylene-dimer undergo catalytic reaction to form a reaction mixture consisting essentially of from 70 to 99% of normal butenes and hexenes, the hexenes being essentially, from 60 to 98% 3-methylpentenes, primarily in the form of 3-methylpentene-2. The average residence time in zone 5 is adjusted to give the above composition of the product, and the reaction mixture is continuously withdrawn via line 19 and passed to separation zone 6. In zone 6 the hexenes and the low temperature boiling components are separated from the reaction mixture. The distillation conditions in the upper part of the distillation column are suitably chosen in the temperature range of 20° to 100°C. at pressures from 0.15 atmospheres to 2.0 atmospheres.

The overhead fraction is passed via line 20 to separation zone 7 wherein the hexene fraction is separated from the lower boiling components and passed via line 21 to zone 8 for storage. The overhead fraction from separation zone 7 consisting of C$_2$ and C$_4$ olefins is passed via lines 18 and 17 back to reaction zone 5. The distillation conditions in the upper part of column 7 are suitably chosen in the temperature range from -20° to 90°C. at pressures from 0.15 to 5.0 atmospheres. The bottom fraction from distillation zone 6 is returned via line 16 to zone 5. A small part of this fraction is withdrawn and passed via line 22 to distillation zone 9, wherein the solvent is distilled off and passed via line 23 to zone 4 for reuse.

In the example illustrated on the flow sheet is used a solvent having a boiling point lower than the lowest boiling C$_8$ in the reaction mixture. An intermediate distillate fraction consisting essentially of C$_8$ olefins may then be recovered from zone 9 via line 24 to storage zone 11. By selection of a solvent having a boiling point higher than the C$_8$ fraction the solvent may suitably be recovered as an intermediate fraction and passed to zone 4 as mentioned above. Used catalyst together with higher olefins are passed via line 25 to zone 10. This catalyst may be recovered for regeneration and reuse, or it may be discarded. However, the consumption of catalyst in the process is very low; thus by use of pure raw materials a conversion of ethylene in the range of 10^4 kg./kg. nickel compound may be obtained.

FIGURE 105: PROCESS FOR THE PRODUCTION OF 3-METHYLPENTENES FROM ETHYLENE

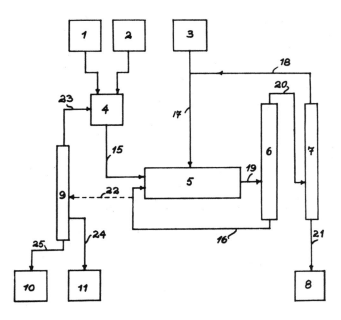

Source: N. Bergem, U. Blindheim, O.-T. Onsager and H. Wang; U.S. Patent 3,558,736; January 26, 1971

STEAM REFORMING

Esso Research and Engineering

A process developed by W.F. Taylor; U.S. Patent 3,410,661; November 12, 1968; assigned to Esso Research and Engineering Company is one in which a hydrogen gas product is produced by reaction of liquid hydrocarbons with water in the presence of a nickel reforming catalyst at low reaction temperatures, e.g., 110° to 500°F., and at atmospheric pressure and above. The liquid phase reforming of emulsions of hydrocarbons with water in the presence of a solid catalyst for producing a gas product rich in hydrogen is a distinct achievement. In contrast to vapor phase reforming, the liquid phase reforming conserves large quantities of heat needed for vaporization in vapor phase reforming and thus improves the thermal efficiency of the process.

Highly active catalysts useful for the low temperature partial conversion of the hydrocarbons to obtain principally hydrogen are typified by mixed nickel-alumina and nickel-silica catalysts which have nickel contents from 10 to 75 weight percent, preferably 40 to 45 weight percent and these mixed catalysts may be promoted by certain metals, e.g., barium, strontium, cesium, cerium, lanthanum, yttrium, iron, potassium, and copper, present as oxides, carbonates or both oxides and carbonates. The proportion of promoter may be between 0.5 to 12 weight percent of the catalyst. In general, the highly active nickel catalysts have high nickel surface areas, i.e., 20 to 30 m.^{2}/g. They are obtained by coprecipitations of nickel with aluminum as hydroxides, carbonates, or basic carbonates from aqueous solutions of nitrate salts by use of NH$_4$HCO$_3$ at low temperature (200° to 400°F.), drying of the precipitates, and low temperature (400° to 900°F.) calcining of the dried precipitates in air and low temperature (600° to 900°F.) activation of the calcined precipitates by hydrogen. The promoters are admixed as decomposable compounds, e.g., hydroxides, carbonates or nitrates with the precipitates. Similarly mixed catalysts of nickel with silica may be prepared using a metasilicate and kieselguhr, in place of aluminum compounds to have the nickel interspersed with SiO_2 instead of Al_2O_3. The catalyst granules may be 1 to 5 mm. particles or be compressed into pellets or be extruded. Figure 106 shows the general overall element involved in the conduct of the process.

FIGURE 106: FLOW DIAGRAM OF LIQUID-PHASE NICKEL-CATALYZED STEAM REFORMING PROCESS

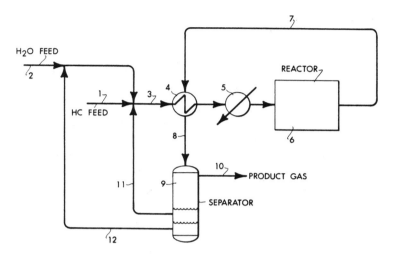

Source: W.F. Taylor; U.S. Patent 3,410,661; November 12, 1968

There is shown a conduit 1 for the admission of the liquid hydrocarbon oil feed to the system, conduit 2 for admission of liquid water feed, conduit 3, wherein the hydrocarbon and water feeds are mixed to form an emulsion which is passed into heat exchanger 4 where the emulsion is warmed. The warmed emulsion may be passed through a heater 5, which supplies heat for the desired reaction temperature, then into the reactor 6. Reactor 6 contains one or more catalyst beds, i.e., beds of active nickel-containing catalyst particles, or surfaces coated with catalyst for the reforming reaction. The product gases from the reforming reaction are removed from reactor 6 by means of conduit 7 and are passed through heat exchanger 4 where the product gases and unconverted hydrocarbon and water are cooled by indirect heat exchange with the liquid emulsion feed. The product gases and condensate are removed from heat exchanger 4 at a lowered temperature and are passed to gas-liquid separator 9 by conduit 8.

The gas-liquid separator 9 permits removal of the product gases which comprise primarily H_2, carbon dioxide, and some methane or hydrocarbon gas by way of conduit 10. The unreacted liquid hydrocarbon condensate forms an upper liquid phase which is withdrawn through conduit 11 from separator 9 for recycling to conduit 1. The lower phase of liquid water collected in separator 9 is withdrawn through conduit 12 and recycled to water-feed conduit 2. The product gases removed from the separator 9 by conduit 10 may then be separated into component parts by liquefaction and fractionation or they may be used as they are removed from the separator. The gases rich in hydrogen may be used effectively as a fuel for a fuel cell. Gaseous hydrocarbon components separated from the H_2 product may be used as a heating fuel, e.g., to undergo combustion for heating the feedstock going into the reactor 6. Various known solvent absorption or adsorption processes may be used to remove carbon dioxide and gaseous hydrocarbons from the hydrogen.

In the basic process carried out as shown in the flow diagram, the principal features of the process are mixing of liquid hydrocarbon oil feed with water feed to form an aqueous emulsion under pressure, heating of the emulsion under pressure to the desired reaction temperature while maintaining the hydrocarbon and water principally in the liquid phase upon contact of the emulsion with the solid catalyst in the reactor in a manner which permits the gas to be removed as it is formed with a partial conversion of the hydrocarbon, and passing the unreacted liquid hydrocarbon and water together with the gas product through a cooler 4 to a separator 9 where the gas product is separated from unreacted liquids. Using the catalyst in a plurality of beds in reactor 6 each of the beds may be in a separate tube surrounded by a heating medium so that heat is imparted by indirect heat exchange from the heating medium to the emulsion flowing through the catalyst beds and the gas formed in each of the beds is made to flow with unreacted emulsion out of the tubes in reactor 6 through conduit 7 to the heat exchanger 4 for cooling, then to the separator 9.

By feeding 16.5 mols of liquid heptane and 159 mols of liquid water into conduits 1 and 2 shown, it is possible to obtain a hydrogen-rich gas product with 0.1 to 15% conversion of the liquid heptane per pass using a temperature of 350°F. and a pressure of at least 1,350 psia so that the gas product contains less than 10% water vapor.

A process developed by W.F. Taylor, J.H. Sinfelt and H. Berk; U.S. Patent 3,449,099; June 10, 1969; assigned to Esso Research and Engineering Company produces gases containing methane, hydrogen, and carbon dioxide by reaction of paraffin hydrocarbons with steam in the presence of a highly active catalyst containing nickel interspersed with alumina or silica to give the catalyst a nickel surface area of about 20 to 60 m.2/g. of catalyst when freshly prepared and activated by treatment with hydrogen. The hydrocarbon feed-steam mixture is first passed through a bed of the catalyst which has been deactivated by use in the reaction but retains residual nickel surface area for removing sulfiding and coking contaminants from the reactant mixture before the reactant mixture is passed through a bed of the freshly prepared and activated catalyst under reaction conditions for producing the desired gas products.

Institute of Gas Technology

A process developed by A.R. Kahn; U.S. Patent 3,416,904; December 17, 1968; assigned to Institute of Gas Technology is a one-stage process wherein hydrocarbons having an end boiling point up to 500°F. are converted by steam reforming to hydrogen-rich gas which may be used for any applicable chemical processes, or can be selectively methanated for use in an acid fuel cell. The hydrocarbon feedstocks may contain a relatively high proportion of normal olefins and cyclo-olefins and aromatics such as benzene. The vaporized hydrocarbons are steam reformed in a reactor in the presence of a nickel-alumina-aluminum catalyst at pressures preferably from 1 to 5 atmospheres and at temperatures ranging from about 700° to 1100°F. The typical catalyst contains 44 to 56 weight percent nickel, 22 to 38 weight percent alumina and the remainder aluminum. Figure 107 shows the essentials of the process.

FIGURE 107:   FLOW DIAGRAM OF VAPOR PHASE STEAM REFORMING PROCESS USING NICKEL-ALUMINA-ALUMINUM CATALYST

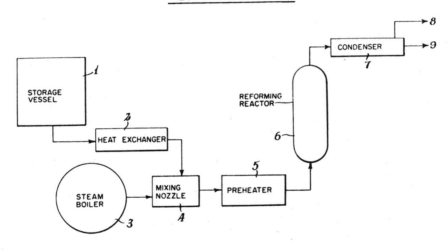

Source: A.R. Kahn; U.S. Patent 3,416,904; December 17, 1968

The numeral 1 represents a storage vessel wherein the hydrocarbon feedstock is stored. Preferably the feedstock is paraffinic hydrocarbons but may contain aromatics and olefins. The hydrocarbon feed is pumped through a heat exchanger 2 wherein it is vaporized and then blended with steam from boiler 3 in a mixing nozzle 4. The mixture is maintained at a pressure between 1 to 5 atmospheres depending on the operating conditions, nature of feed and desired product gas. The stream of intimately mixed hydrocarbon vapors and steam is then passed through a preheater 5 wherein it is preheated to initial reaction temperature ranging from 700° to 1100°F., depending on the operating conditions, nature of feed and desired product gas. The mixutre of hydrocarbons, vapors and steam then pass into the reactor 6 through a bed of catalyst, as hereinabove described. The gasification reactions occur here and the hydrocarbons are essentially completely gasified. The resulting effluent which is primarily a mixture of hydrogen, methane, carbon monoxide, carbon dioxide and unreacted steam exits from the reforming reactor, is cooled in heat exchanger 7 wherein steam is removed as liquid, as shown diagrammatically at 9, leaving a gas rich in hydrogen exiting the condenser at 8.

A process developed by H.L. Feldkirchner, H.R. Linden, A.R. Khan and F. Todesca; U.S. Patent 3,433,610; March 18, 1969; assigned to Institute of Gas Technology is a steam reforming process for gaseous hydrocarbons or liquid hydrocarbons or mixtures thereof which can be fed into a reactor in vapor form at operating conditions to produce a high methane content gas useful as a peak shaving pipeline gas partially or completely interchangeable with natural gas. The steam reforming is run in the presence of a nickel-alumina-aluminum catalyst at superatmospheric pressures in the range between above about 5 to 30 atmospheres and at temperatures ranging from about 650° to 1050°F. Typical hydrocarbon feed stocks are liquefied petroleum gases, petroleum naphthas, natural gasoline, kerosene and the like. For such feedstocks, the product gas after removal of carbon dioxide and water contains an excess of 70 mol percent methane. The steam-to-hydrocarbon weight ratio of feed material is maximally about 4.5:1, and minimally 2.6:1.

Phillips Petroleum

A process developed by R.R. McMullan; U.S. Patent 3,027,237; March 27, 1962; assigned to Phillips Petroleum Co. is a process for reactivating catalyst used to produce hydrogen from a mixture of steam and hydrocarbon. In one of its more specific aspects, it relates to an on-stream catalyst reactivation procedure for use in the primary reformer of a synthesis gas unit where steam and natural gas are reacted to produce hydrogen for ammonia manufacture. The preferred catalyst in such a process is nickel oxide which is reduced to nickel prior to use by reaction with hydrocarbon. As the reaction proceeds the activity of the catalyst tends to decline, probably because of carbon deposition or sulfur poisoning. Carbon deposition causes the catalyst to disintegrate which in turn greatly increases the pressure drop through the catalyst bed.

Although troublesome carbon deposits can be avoided, even a small decline in catalyst activity results in an increase in methane content in the synthesis gas which in turn reduces the efficiency of the ammonia synthesis reaction. An increase of methane in the synthesis gas reduces the capacity of the compressors used to compress the hydrogen and nitrogen prior to synthesis and also necessitates a purge of the synthesis plant cycle gas with a resultant loss of hydrogen and nitrogen. It is estimated that an increase of 1% methane in the synthesis gas decreases the potential output of ammonia by about 6%. It is customary to reactivate the catalyst when an activity decline becomes troublesome either by raising the catalyst temperature or by oxidizing the catalyst with steam or a mixture of steam and air. The latter method is preferred because of lower fuel consumption; however, even this method has necessitated diverting the natural gas feed and removing the unit from production for the duration of the oxidation and subsequent reduction procedure.

According to this process, a procedure is provided whereby a gas reforming catalyst can be reactivated without removing the catalyst bed from production. It has been found that a gas reforming catalyst used to convert a mixture of hydrocarbon and steam to hydrogen gas and carbon oxides can be reactivated by gradually increasing the steam to hydrocarbon feed ratio to a predetermined elevated value, holding this feed ratio at the elevated value for a period of time sufficient to reactivate the catalyst and thereafter returning the feed ratio to about its original value. In general, the reactivation is complete when a decline in the methane content of the reactor effluent or of the synthesis gas is no longer evident.

Pullman

A process developed by J.F. McMahon; U.S. Patent 3,417,029; December 17, 1968; assigned to Pullman Incorporated involves contacting a hydrocarbon and steam in the presence of an alkalized catalyst comprising nickel and a refractory material under conditions such that hydrogen-rich gas is produced. Numerous advantages are realized by the use of such catalysts to effect conversion of hydrocarbons to hydrogen-rich gas in the presence of steam. One advantage is that steam reforming of hydrocarbons is accomplished at ratios of steam-to-hydrocarbon which are significantly lower than those required when standard steam reforming catalysts are employed. Even at these lower steam requirements, carbon deposition is substantially avoided. These advantages are particularly important when olefin-containing feeds, and normally liquid and heavy hydrocarbon feedstocks are employed, since steam reforming of such feedstocks is thereby rendered a commercially feasible process. Thus, by this process and relatively inexpensive feedstocks which heretofore could not be used because of the necessity of using prohibitively high steam requirements are rendered useful for steam reforming. In addition, heavy hydrocarbon feedstocks may be converted to hydrogen-rich gas at steam requirements which are significantly lower, thereby improving the general economics of the process. The reduction of steam requirements is particularly marked and outstanding when the catalyst contains an added alkali metal compound.

The catalysts contain nickel including elemental nickel or a compound of nickel such as nickel oxide, and mixtures thereof. The nickel content of the catalysts may range between about 4 and about 40 weight percent based on the total weight of the catalyst, the high nickel catalysts usually being preferred, such as, for example, those containing between about 10 and 30 weight percent. A second ingredient of the catalyst is an added alkaline compound of which alkali metal compounds including those of sodium, lithium, and potassium, are preferred. Typical examples of such compounds are the alkali metal salts of an oxygen-containing acid such as the carbonates, bicarbonates, nitrates, sulfates, silicates, oxalates and acetates of sodium, potassium and lithium; the alkali metal oxides or an alkali metal salt capable of yielding the oxides or an alkali metal salt capable of yielding the oxides at elevated temperatures including the aforesaid salts, as well as the alkali metal hydroxides.

Particularly efficacious catalysts are those to which an alkali metal carbonate or hydroxide has been added, especially sodium carbonate and sodium hydroxide. The catalysts are prepared so as to provide a concentration of added alkali of at least 0.5 weight percent calculated as the metal, preferably at least 2.0 weight percent. The concentration of added alkali may be as high as 30 weight percent, and usually an amount below 20 weight percent calculated as the metal, is employed.

Societe Chimique de la Grande Paroisse

A process developed by J. Housset, J. Quibel, P. Honore and R. Pidoux; U.S. Patent 3,457,192; July 22, 1969; assigned to La Societe Chimique de la Grande Paroisse, Azote et Produits Chimiques, France involves catalyst compositions suitable for use in the steam reforming of hydrocarbons heavier than methane and optionally unsaturated, comprising nickel as the active metal between 1 to 40% by weight calculated as nickel oxide and a mixture of silicon dioxide, magnesium oxide and zirconium dioxide as the support, the magnesium oxide to zirconium oxide ratio by weight being between 1.7 and 2, and the magnesium oxide to silicon dioxide ratio by weight being between 4.5 and 5. These compositions are disposed inside the reforming zone, in the direction of flow of the gas, in several layers having an increasing content of nickel oxide adapted to the evolution of reaction.

PALLADIUM-CONTAINING CATALYSTS

HYDROCRACKING

Chevron Research

A process developed by S.M. Csicsery and J.R. Kittrell; U.S. Patent 3,632,500; January 4, 1972; assigned to Chevron Research Company yields a hydrocracking catalyst comprising a layered clay-type crystalline aluminosilicate cracking component, 0.01 to 2.0 weight percent, based on the cracking component and calculated as the metal, of a hydrogenating component selected from platinum and compounds thereof, palladium and compounds thereof, rhodium and related compounds, ruthenium and compounds thereof, iridium and related compounds, and nickel and compounds thereof, and 0.01 to 5.0 weight percent, based on cracking component and calculated as the metal, of a hydrogenating component selected from the group consisting of iron and compounds thereof. The following is one specific example of the conduct of this process.

Example: A catalyst consisting of iron, palladium, and a layered clay-type crystalline aluminosilicate was prepared in the following manner. These starting materials were used: (1) 500 g. of a layered synthetic crystalline aluminosilicate mineral as described in Granquist U.S. Patent 3,252,757; and (2) 1,250 ml. of an aqueous solution containing 7.21 g. of tetraammino palladium nitrate [Pd(NH3)4](NO3)2 and 18.05 g. of hydrated ferric nitrate, Fe(NO3)3·9H2O. The mineral, a lumpy powder form, was mixed with the aqueous solution to form a pasty mass. The pasty mass was dried in a vacuum at 200° to 250°F. The resulting material was broken into small pieces, and then calcined. The calcined catalyst contained 0.5 weight percent palladium, calculated as metal and based on the cracking component, and contained 0.5 weight percent iron, calculated as metal and based on the cracking component.

HYDROGENATION

Engelhard Minerals & Chemicals

A process developed by C.D. Keith, P.L. Romeo, Sr. and O.J. Adlhart; U.S. Patent 3,489,809; January 13, 1970; assigned to Engelhard Minerals & Chemicals Corporation involves the use of unitary ceramic block catalyst supports with active sites, particularly sites containing Group VIII metal catalysts. With such supported catalysts purifications are carried out at higher linear velocities than common with conventional packed small particle catalyst beds. Improved selectivity is obtained in reactions in which the impurity is more strongly adsorbed than the product. One example of the purifications involved concerns selective hydrogenation in olefinic gas streams of acetylenes and diolefins.

Palladium is generally recognized as a superior catalytic metal for selective purification of olefin and diolefin streams because of its high selectivity and activity. Supported palladium has been found particularly suitable and the palladium has been used on particulate carriers such as pellets. The palladium concentrations in these particulate catalysts are very low, frequently 0.01 to 0.001% Pd based on the weight of the catalyst metal plus support. This is necessary to obtain the desired selectivity.

Whereas these particulate catalysts function well at low impurity levels, with higher concentrations, especially with acetylene concentrations of 1% and higher, they are not satisfactory. Because of the very low catalytic metal concentration, they are readily poisoned and have short life. They are deactivated particularly by polymers produced in side reactions and require frequent regeneration thereby causing unwanted difficulty and increased expense during ethylene purification. Additionally, the design of equipment used in conjunction with conventional particulate catalysts is limited to relatively large diameter reactors in order to avoid high pressure drop in the apparatus. This problem increases with increasing space velocity. Inasmuch as commercial high temperature naphtha cracking processes produce ethylene streams having typically 1.75 to 2.50% acetylene and because most of the newer cracking operations provided ethylene with high acetylene concentrations, it has become increasingly important to obtain catalysts which can adequately cope with the new demands for ethylene purification. In development work leading to the process, it has been found that unitary ceramic honeycomb type supports for Group VIII metal catalysts provide a structure uniquely qualified to greatly increase the efficiency of processes for removing acetylenic and diolefinic contaminants from normally gaseous olefins, and for removing acetylenes from diolefins. Palladium when deposited on such honeycomb type supports is substantially superior to palladium catalysts known previously.

Good selectivity is obtained with these honeycomb type catalysts even at high precious metals concentrations. The improved catalyst structures provide greater overall removal of contaminants at less cost with smaller equipment, greater throughput and improved selectivity. The catalyst support is made up of a honeycomb type skeletal structure having gas-flow paths throughout which permit operation of the purification processes at high linear velocity, at high space velocity, and with high selectivity. Moreover the process is effected with long catalyst life, and problems of back pressure as well as catalyst regeneration are minimized.

The process is particularly effective for reducing the amount of acetylene in ethylene streams from 1 to 3% by volume and higher to 0.2% and less. The unitary ceramic honeycomb type catalyst may be used in single stage or multistage processes which may incorporate both honeycomb type and particulate catalysts. In a preferred embodiment for achieving an overall selective removal of acetylene from ethylene streams containing 1 to 3% by volume acetylene to less than 10 ppm, a multistage process is used in which the unitary ceramic honeycomb type catalyst containing 0.01 to 10% Pd is used at a volume hourly space velocity of 50,000 to 4,000,000 in one or more stages to remove the acetylene to a level of less than 0.5%.

In a final stage a particulate catalyst containing 0.5 to 0.001% Pd is used at a volume hourly space velocity of 3,000 to 50,000 to remove the acetylene concentration to less than 10 ppm. In the preferred multistage process, a high degree of purity can be achieved in ethylene streams containing high levels of acetylene contamination with overall advantages over prior art processes with respect to process conditions, reactor design, catalyst life, and efficiency.

FMC

A process developed by T.M. Jenney, D.H. Porter and E.M. Zdrojewski; U.S. Patent 3,112,278; November 26, 1963; assigned to FMC Corporation relates to the production of hydrogen peroxide by the catalytic reduction and subsequent oxidation of an anthraquinone working material, and more particularly to the regeneration of the catalyst used in the reduction step. Heretofore, the anthraquinone process has been carried out by dissolving an alkylated anthraquinone, e.g., 2-ethyl anthraquinone or mixtures of various alkylated anthraquinones, in selected solvents and alternately reducing and oxidizing the mixture or working solution. Hydrogen peroxide is produced during the oxidation step and is separated from the remainder of the working solution. This process is fully described in U.S. Patent 2,657,980, issued to Sprauer on November 3, 1953.

A catalytic reduction is required in this process in order to add hydrogen to the anthraquinone working compound at a rate sufficient for commercial operation. The catalyst is selected to selectively hydrogenate the quinone form of the working compound to the hydroquinone form, without unduly attacking the compound to produce degradation products, which are incapable of producing hydrogen peroxide. All known catalysts produce some degradation, and maximum selectivity is a desired property.

One drawback to the catalytic hydrogenation is that the catalyst loses its activity in a relatively short time and either must be replaced with fresh catalyst or alternately must be treated to restore its activity. Prior workers have employed a variety of methods for regenerating the catalyst. In U.S. Patent 2,692,240, issued to Sprauer on October 19, 1954, the catalyst is regenerated with an oxidizing agent, while in British Patent 788,340, issued to La Porte, the catalyst is regenerated using alkali.

Some of the drawbacks of these prior regeneration methods are that they are relatively expensive, in most cases require special equipment to carry out the regeneration treatment, and can cause spalling of the metallic catalyst from its support. The object is to regenerate a hydrogenation catalyst in situ in a way which improves activity and selectivity of the catalyst.

It has been found that the noble metal catalysts supported on inert carriers can be regenerated to both restore their activity and increase their selectivity by treating the catalyst with wet steam at temperatures between 80 and 200°C. for periods as low as 4 hours. The steam should contain at least 10% water, and 2.0 kg. of this wet steam are required to regenerate a liter of catalyst. The use of wet steam is necessary if all the improvements in selectivity and activity are to be realized. Regeneration with dry steam does not produce the results described.

The process can be effectively employed in fixed bed types of catalyst chambers. In this type of operation, the catalyst is suspended in a fixed bed within the catalytic chamber and rests on a perforate support. The working solution to be hydrogenated is passed through the bed along with hydrogen, and reduction takes place as the working solution and hydrogen contact the catalyst bed. While other catalyst systems, i.e., fluid beds containing suspended catalyst, can also be regenerated by the process, it has been found that it is particularly suitable for fixed bed operations because regeneration of the catalyst can be carried out in place without removing the catalyst from the bed.

The catalysts which may be treated by the process are generally the noble catalysts such as palladium, rhodium, and platinum. These noble metals may be carried on inert carriers such as carbonates, alumina, carbon, alkaline earths, silica gel, etc. The preferred noble metals which have been found suitable for regeneration by the process, and which are extensively used in the anthraquinone process, are palladium and platinum.

Great Lakes Carbon

A process developed by L.A. Bryan; U.S. Patent 3,458,576; July 29, 1969; assigned to Great Lakes Carbon Corporation is one in which phenylnitropropenes are hydrogenated under mild conditions, e.g., at a temperature under 100°C. and a pressure of less than 75 psig, in the presence of a slurry of a catalyst in the hydrogenation solvent which has been treated at 45°C. for 4 minutes prior to the hydrogenation. Palladium, platinum and mixtures of these metals with nonpyrophoric nickel are the catalysts used.

There are numerous methods outstanding in the art for reducing nitro groups to amino groups and ethylenic linkages to saturated bonds. When these methods are brought to bear on certain materials such as arylnitroalkenes however, there arise difficulties that have yet to be overcome. For example, cathodic or sodium amalgam treatment have been used to reduce 1-phenyl-2-nitropropene to 1-phenyl-2-aminopropane, a compound commonly known as amphetamine. These particular reductions as taught by Alles, U.S. Patent 1,879,003 have failed to gain wide use.

An explanation for this is offered by Susie and Hass, U.S. Patent 2,233,823, who acknowledge that it has been found extremely difficult to effect satisfactory reduction of, e.g., 1-phenyl-2-nitropropene; only low yields are secured by electrolytic reduction and only high molecular weight polymeric products are formed on catalytic hydrogenation. Susie and Hass meet these problems by employing a two-stage reduction which consists in first submitting the arylnitropropene to the reducing action of iron in the presence of water in order to obtain the oxime of the corresponding arylpropanone, then treating this product with high pressure hydrogen gas, 1,700 psig, in the presence of the metal catalysts, e.g., platinum, palladium, or nickel.

A 50% overall conversion of the arylnitropropene to the arylaminopropane is obtained in this manner. Pyrophoric nickel has been used as a catalyst with yields of up to 60% of the amine calculated on the nitro olefin. Tindall proposes this method in U.S. Patent 2,636,901 in lieu of noble metal catalysts which he considers poor prior art for various reasons. It is an object of this process to provide a one-step high yield method for reducing arylnitroalkenes to arylaminoalkanes. Another object is to provide a method that requires the use of very moderate pressures. A further object is to avoid the use of easily flammable materials such as pyrophoric catalyst.

These and other objects have been accomplished by subjecting the arylnitropropene to the action of hydrogen gas in the presence of a noble metal catalyst or of a mixture of noble metal and nonpyrophoric nickel catalysts, the catalysts having previously been subjected to a mild heat treatment.

HYDROISOMERIZATION

Shell Oil

A process developed by H.A. Benesi; U.S. Patent 3,527,835; September 8, 1970; assigned to Shell Oil Company is a process for preparing a paraffin hydroisomerization catalyst containing highly dispersed platinum or palladium homogeneously distributed into a synthetic crystalline alumino-silicate mordenite support. Dispersion of metal throughout the support is achieved by competitive cationic exchange in an ammoniacal solution having an ammonium ion to metal complex ion ratio of 100 to 1 or more. Dispersion of metal is maintained by controlling the metal reduction temperature. The catalyst thus prepared is used to isomerize normal paraffins in the C_4 to C_7 range in the presence of hydrogen at elevated temperature and pressure.

OXIDATION

Farbwerke Hoechst

A process developed by W. Riemenschneider, K. Dialer, O. Probst and O.-E. Bander; U.S. Patent 3,301,905; Jan. 31, 1967; assigned to Farbwerke Hoechst AG, Germany is a process for conversion of an olefinic hydrocarbon to a carbonyl compound selected from the group consisting of aldehydes and ketones by oxidizing an olefinic carbon atom of the olefinic hydrocarbon to a carbonyl group by contacting the olefinic hydrocarbon and oxygen at a temperature between 50° and 160°C. and at a pH between 0.5 and 6 with water and a catalyst of (a) a salt of a noble metal selected from the group consisting of palladium, iridium, ruthenium, rhodium and platinum and (b), as a redox system, an inorganic salt of a metal showing several valence states under the reaction conditions applied. The process may be carried out in the presence of a diluent gas selected from the group consisting of carbon monoxide and hydrogen. The essential flow scheme of such a method is shown in Figure 108.

FIGURE 108: FLOW DIAGRAM OF PROCESS FOR PRODUCTION OF ACETALDEHYDE FROM ETHYLENE

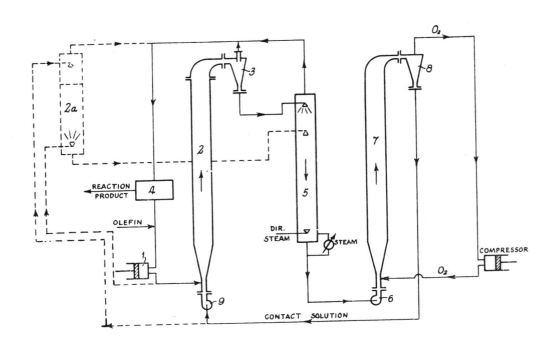

Source: W. Riemenschneider, K. Dialer, O. Probst and O.E. Bander; U.S. Patent 3,301,905; January 31, 1967

A current of an olefinic gas mixture is introduced through a compressor 1 into reactor 2 or 2a in concurrent or counter-current to the catalyst liquid, and the gas is finely distributed in the catalyst solution by means of a suitable device, for example a frit, a mixing nozzle, an oscillatory sieve, a vibrator, a rapid agitator or the like. The olefin is converted in this solution to a carbonyl-containing reaction product leaving the reactor together with unreacted olefin and diluting gas, if desired via a cyclone 3.

The reaction product is separated in a separating device 4 from the remainder of the olefinic gas mixture, which latter substance may then be recirculated together with a fresh amount of olefin into the reactor via compressor 1 if it still contains substantial amounts of olefin, or may be used for other purposes. The reaction liquid is then conducted to a stripper 5, where it is treated with steam to be directly or indirectly freed from olefin and residues of reaction product. The gases obtained by stripping are conducted, if desired, to separating device 4. The stripped reaction solution is then introduced into regenerator 7 by means of a pump 6 or a static incline.

The stripping stage may, however, be omitted, and the reaction solution is then directly introduced into regenerator 7. The solution is intimately contacted in generator 7 with oxygen or gases containing oxygen. The regenerator may be designed so that the contact liquid flows in a countercurrent to the oxygen-containing gases, which leave the regenerator through cyclone 8, and may be returned into the cycle by means of a compressor. The contact liquid is then communicated in the regenerated state to the reactor 2 or 2a by means of pump 9. All partial operations may be carried out individually or together at a raised pressure, at a reduced pressure, or at atmospheric pressure.

The following is one specific example of the operation of the process.

Example: 50 cc of a catalyst solution containing, per liter of water, 1 g. of PdCl$_2$, 100 g. of CuCl$_2$·2H$_2$O, and 5 cc of concentrated hydrochloric acid, are heated to 80°C., and 1.5 liters of gas are passed through per hour. The gas used consists of 50% of ethylene and 50% of carbon monoxide and is mixed with half its volume of oxygen prior to the reaction. The formed acetaldehyde is separated from the reaction mixture by washing. The conversion, calculated upon the ethylene used, is above 40%.

W.R. Grace

A process developed by E.W. Stern and M.L. Spector; U.S. Patent 3,479,392; November 18, 1969; assigned to W.R. Grace & Co. involves reacting ethylene and acetic acid in the presence of palladium chloride and disodium hydrogen phosphate under substantially anhydrous conditions to produce vinyl acetate. The process will be described with reference to the flow diagram in Figure 109.

FIGURE 109: FLOW DIAGRAM OF PROCESS FOR DIRECT PRODUCTION OF VINYL ACETATE FROM ETHYLENE AND ACETIC ACID

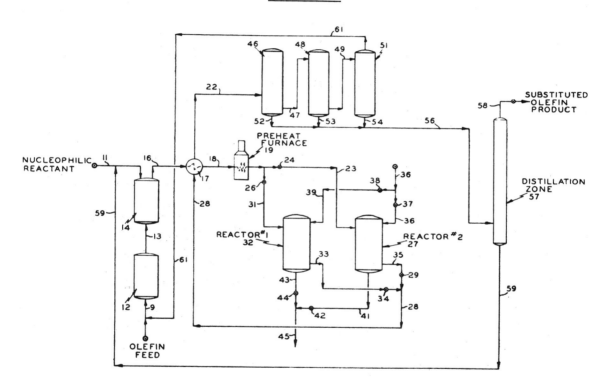

Source: E.W. Stern and M.L. Spector; U.S. Patent 3,479,392; November 18, 1969

As shown in the drawing, ethylene is passed along line 9 through drier 12 where any water is substantially removed by conventional methods and is then passed into vaporizer 14 along line 13 where it is admixed with acetic acid introduced by means of line 11. The vaporized mixture is then passed along line 16 through heat exchanger 17 where heat is picked up from a product stream passing through line 28. The mixture is then conducted along line 18 into preheat furnace 19 where it is heated to a temperature between 150° and 250°F. (e.g., 200°F.) at a pressure between 5 and 100 psig (e.g., 20 psig).

The reactants are passed into reactor 27 by means of line 23, opening valve 24 and closing valve 26. Reactor 27 contains palladium chloride and is maintained at a temperature between 150° and 250°F. (e.g., 200°F.) and a pressure between 5 and 100 psig (e.g., 20 psig). The reactants are passed through reactor 27, removing vinyl acetate product and any unreacted ethylene or acetic acid through line 35 and then into line 28. When substantially all of the palladium chloride has been converted to palladium or at any desired point before such time, valve 24 on line 23 is closed and valve 26 on line 31 is opened to allow the ethylene acetic acid mixture to pass through line 31 into reactor 32 which also contains palladium chloride and is maintained under substantially the same conditions as reactor 27.

While the reaction is thus proceeding in reactor 32, palladium chloride is regenerated in reactor 27 by introducing oxygen and hydrogen chloride thereto along line 36 having valve 37 thereon. When reactor 27 is thus being used as the regeneration zone, it is maintained at a temperature between 300° and 400°F. (e.g., 375°F.) and a pressure between 5 and 100 psig (e.g., 20 psig). The water vapor produced as a result of this reaction, as well as unconsumed oxygen and hydrogen chloride gases, is allowed to pass from zone 27 by means of line 41 having valve 42 thereon and exits from the system by means of line 45.

The gases may then be passed through conventional drying apparatus and the dried gases recycled for further regeneration. When it becomes necessary to regenerate the palladium chloride in reactor 32, the same regenerative procedure described above may be effected by cutting off the flow of reactants thereto by closing valve 26 and passing the oxygen-hydrogen chloride mixture along line 39 by opening valve 38, removing water vapor and gases from reactor 32 by means of line 43 having valve 44 thereon, exiting from the system by means of line 45.

When the reaction between ethylene and acetic acid is being conducted in either or both of reactors 27 and 32, vinyl acetate product is removed therefrom by means of lines 35 and 33, respectively, and carried along line 28 through heat exchanger 17 and line 22 to a conventional condenser system represented by condensers 46, 48 and 51, interconnected by lines 47 and 49, distillate from each being removed by means of lines 52, 53 and 54, respectively, into line 56 by means of which the crude vinyl acetate product is carried to fractional distillation zone 57. Overhead, comprising unreacted ethylene, is removed from the condenser system by means of line 61 and is recycled for further reaction to drier 12. In zone 57, vinyl acetate product is removed as overhead by means of line 58 and unreacted acetic acid is removed as bottoms by means of line 59 and is recycled to vaporizer 14.

Union Oil of California

A process developed by W.D. Schaeffer; U.S. Patent 3,260,739; July 12, 1966; assigned to Union Oil Company of California for the manufacture of vinyl acetate comprises contacting ethylene with an anhydrous catalyst solution at a temperature between 25° and 250°C. the solution comprising acetic acid solvent; between 0.001 and 1.0 weight percent palladium ion; between 1 and 25 weight percent cupric ion; between 0.1 and 5 weight percent of chloride ion; and between 0.5 and 30 weight percent of acetate ion.

When the adsorption of ethylene substantially ceases, the contacting is discontinued and a crude product recovered containing vinyl acetate and spent catalyst solution. The crude product is treated to recover the vinyl acetate and the spent catalyst solution is regenerated by treatment with oxygen and a soluble nitrite or nitrogen oxide. The regenerated solution is then dehydrated and returned to contact with ethylene.

The essential elements of the process are illustrated in the flow diagram in Figure 110. An olefin, typically ethylene, enters the process at 1 and is contacted with the catalyst solution in a suitable reaction zone 3 which can be a packed tower through which the olefin is passed upwardly countercurrent to the downward flow of the catalyst solution. More simply, the reaction zone can contain a liquid reservoir of catalyst into which the olefin is passed. Preferably, an excess of olefin is used, the excess being removed and recycled through line 2.

In general, reaction temperatures between 25° and 250°C. can be used, preferably between 80° and 150°C. The reaction pressure is sufficient to maintain the catalyst solution in liquid phase, generally between 1 and 15 atmospheres being sufficient. The reaction period, of course, depends on the rate of ethylene and inventory of catalyst metal ion, secondary metal ion, etc., in the solution. Generally, contacting times between a few minutes to 2 hours can be used, the actual reaction time necessary being preferably determined by the rate of ethylene absorption by the solution, no adsorption or a very slight adsorption indicating that the solution is spent and must be regenerated. The ethylene absorption rate can of course be determined by observation of the reactor pressure or of the recycle flow of ethylene in a manner apparent to those skilled in the art.

The reactor effluent is passed to a suitable separator 4 and the pressure reduced to remove unreacted olefin which can be recycled by pump 5, line 6, to the olefin feed stream at 1. The catalyst solution containing the ester product passes through line 6 into distillation tower 7. When necessary, the liquid stream can be heated at 8 to effect the distillation. If desired, the reactor effluent can be flashed directly into tower 7. Distillation in tower 7 is performed to separate the ester product from the catalyst solution which contains the reduced form of the catalyst metal and cuprous and/or ferrous salts. Preferably, the solvent used in the process has a higher boiling point than the desired ester product and the ester can thus be recovered in the overhead from the distillation zone while the catalyst solution is recovered as the residue of the distillation. The overhead from the distillation tower passes through line 9 and cooler 10 and the resultant

FIGURE 110:   FLOW DIAGRAM OF ALTERNATIVE PROCESS FOR PRODUCTION OF VINYL ACETATE FROM ETHYLENE
AND ACETIC ACID

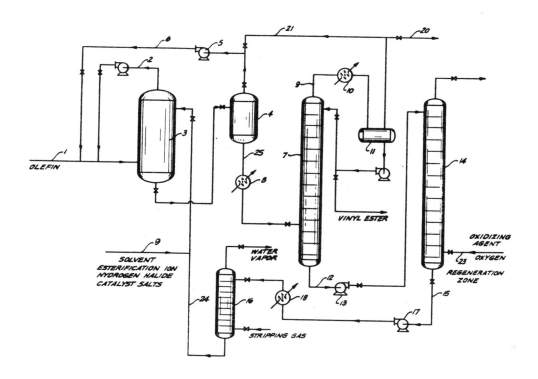

Source: W. D. Schaeffer; U.S. Patent 3,260,739; July 12, 1966

condensate passed into distillate drum 11. The distillate is withdrawn from this drum as ester product for further purification or treated as desired. The condensate is also used as reflux for the distillation tower in a conventional manner. The
gas separated in drum 11 can be vented at 20 or the unreacted olefin contained therein recovered by recycle of the gas
through line 21 to pump 5.

Preferably, the product distillation is effected at atmospheric pressure, however, subatmospheric or superatmospheric
pressures can be used if desired. Although separate vessels are illustrated for the reaction and product distillation zones,
it is apparent that the vessel 7 can be superimposed onto reactor 3 or, more simply, the reaction could be performed in
the lower portion of distillation tower 7. This technique would permit continuous removal of the ester product as it is
formed. Excess ethylene could be recovered from the distillate drum 11 and recycled to the reaction.

The residue of the distillation passes by line 12, pump 13, to a regeneration zone 14 where the spent catalyst solution
is caused to flow downwardly countercurrent to the upward flow of an oxygen-containing gas. If desired, batch processing could also be used in this step simply by accumulating a pool of spent catalyst in 14 and thereafter introducing the
oxidant beneath the liquid level of the pool. Although, as previously mentioned, the presence of copper or iron salts
in the catalyst solution permit the oxygen or air oxidation of the reduced catalyst metal, it is preferred to employ various
oxidation promoters to accelerate the rate of oxidation. The regeneration can be conducted at temperatures between 50°
and 250°C., preferably between 100° and 200°C.

Of the oxidation promoters, ozone, nitrogen oxides and hydrogen peroxide are preferred as they do not introduce extraneous materials into the solution. If desired, other strong oxidizing agents can be used such as manganic or cobaltic
ions, bromide ions, chromate or dichromate ions. These oxidizing agents can be introduced into the solution as various
organic or inorganic salts, e.g., cupric or ferric bromides; manganic or cobaltic halides, particularly chlorides or
bromides, nitrates, manganic or cobaltic organic salts such as the acetates, butyrates, etc.; lithium chromates or di-
chromates, etc. The catalyst solution can then be contacted with air or oxygen to effect the generation. The nitrogen
oxides, including nitric oxide, dinitrogen trioxide, nitrogen dioxide, dinitrogen tetraoxide, dinitrogen pentaoxide and

nitrogen trioxide are preferred and are simply admixed with the oxygen-containing gas supplied at 23. In general, between 1×10^{-6} and 1×10^{-1} parts of an oxidation promoter are employed per part by weight of solution, the bulk of the oxidation being performed by oxygen. With the preferred nitrogen oxides, between 1×10^{-6} and 1×10^{-2} parts of nitrogen oxide per part of solution are employed.

After the regenerative oxidation, the catalyst solution must be treated to remove the water formed during regeneration. Preferably, the water can be distilled off in the tower 14 by conducting the regeneration at temperatures above the vaporization temperature of water, or the regenerated solution can be passed to stripping vessel 16 where an inert stripping gas, e.g., air, carbon dioxide, nitrogen, etc., is employed to strip the water vapor from the solution. In some instances, it may be desirable to further heat the solution with heater 18 to effect the complete removal of water vapor from the solution. If desired, a suitable desiccant, e.g., carboxylic acid anhydrides such as acetic, propionic, butyric, valeric, benzoic, etc., can be added to effect the water removal. The regenerated catalyst solution is then passed through line 17 for introduction into the reactor. Makeup chemicals, e.g., solvents, organic acid salts, halides, etc., are added by line 19 to provide the desired catalyst formulation.

The catalyst metal ion is included in the catalyst solution as a soluble salt, preferably as the halide, although the catalyst metal salt of any of various esterifying ions can also be employed. Generally, however, the halides are preferred, and of the halides, the chloride is most preferred. Salts of any of the following metals, which are catalysts for the reaction, or admixtures thereof can be used: platinum, rhodium, gold, ruthenium, vanadium, palladium and iridium. Of these metals, palladium is preferred because of its greater reactivity. Other metals, of course, can be preferred for other reasons; for instance, rhodium can be preferred because in its reduced state it exists as a soluble rhodious ion rather than as the free metal. As a result, it remains in solution, permitting facile regeneration. In general, between 0.001 and 5.0, preferably between 0.001 and 1.0 weight percent of the solution, comprises the catalyst metal ion.

Also included in the solution is a reservoir of secondary metal ions which serve to extend the effective life of the catalyst metal. These secondary metal ions are the cupric and ferric ions which are added as the halides, or other soluble salts, preferably as salts of the aforementioned esterifying ions, e.g., cupric acetate, ferric propanoate, cupric valerate, etc. Cupric is the preferred secondary metal ion because of its greater activity and solubility. In general, between 1 and 25, preferably between 3 and 15 weight percent of the solution comprises a cupric or ferric ion as a dissolved or suspended salt. The cupric or ferric salt can be used in excess of that soluble in the solution since even suspended salts are effective in reoxidizing the reduced form of the metal catalyst.

A process developed by K.L. Olivier and W.D. Schaeffer; U.S. Patent 3,461,157; August 12, 1969; assigned to Union Oil Company of California involves the oxidation of olefins in a liquid reaction medium comprising a platinum group metal catalyst and a redox agent. The process features an improved method of regenerating the catalyst by removing a portion of the reaction medium, separating a tarry fraction from the medium, contacting the tarry fraction with oxygen in a combustion chamber and returning the inorganic ash product to the reaction zone. Figure 111 shows the application of this technique to the production of vinyl acetate from ethylene.

As shown there ethylene is introduced into the process through line 10 by compressor 12 and mixes with the recycle gas in line 14. The mixed gases are introduced into the reactor 16 through a suitable control valve 18 which is set to provide the desired flow rate of the oxidizable material. An oxygen plant is provided at 20 to provide a stream of oxygen through line 22 which is introduced into reactor 16 through flow control valve 24 which is controlled by gas analyzer 26. This gas analyzer determines the elemental oxygen content of the gases in the vapor space of the reactor or in the vaporous effluent removed therefrom.

Preferably, this controller is set to control the rate of introduction of oxygen so as to prevent the oxygen content of the gaseous effluent from exceeding 2 or 3 volume percent and most preferably to maintain the elemental oxygen content less than 1 volume percent. The reaction zone is maintained under substantially anhydrous conditions and sufficient pressure is employed to maintain a liquid phase within the reactor.

The reaction conditions within the reactor are relatively mild; temperatures from 30° to 300°C. and preferably from 50° to 125°C. are employed. The reactor pressure is maintained from 10 to 1,000 atmospheres and the use of higher pressures is preferred because the increased pressure favors solubility of the olefin in the reaction solvent. Preferably, reaction pressures from 50 to 750 atmospheres are employed. The reaction zone contents are maintained under agitation by a suitable mixer such as the propeller mixer 28 which is rotated by power means such as motor 30 that is mounted outside the vessel.

The olefin is supplied in a large excess and this excess of olefin serves to sweep the oxidized products out of the reactor through overhead line 32. The products are cooled in condenser 34 and passed to a high pressure separating zone 35 from which the gases, i.e., excess olefin, and the fixed gases such as carbon oxides, are removed through line 38 and recycled to the reactor through lines 40 and 14. A portion of the recycle gases is diverted to the recycle gas purification zone where it is contacted with an aqueous solution of monoethanolamine or other conventional adsorbent for the removal

FIGURE 111: PROCESS FOR VINYL ACETATE FROM ETHYLENE FEATURING PALLADIUM CATALYST REGENERATION

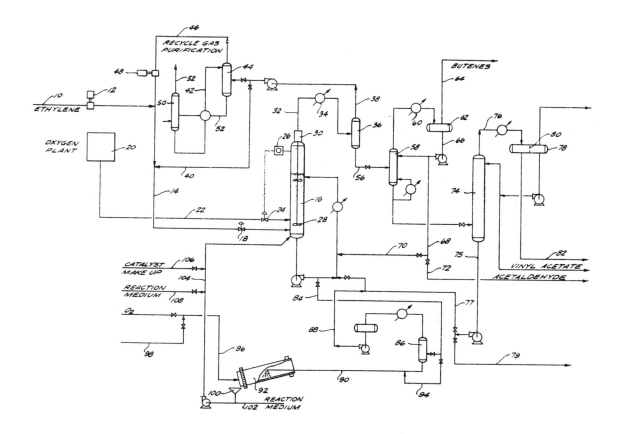

Source: K.L. Olivier and W.D. Schaeffer; U.S. Patent 3,461,157; August 12, 1969

of carbon dioxide. The contacting is performed countercurrently in adsorption zone 44 and the purified recycle gases are removed overhead through line 46 and returned for recycling by recycle gas compressor 48. The solution of monoethanolamine is regenerated in zone 50 and the carbon oxides are removed through line 52. The regenerated solution is returned through line 42 in heat exchange relationship with the rich adsorbent solution removed from zone 44 through line 52.

The condensed liquids in zone 36 are removed through line 56 and passed to fractionation zone 58 where the low boiling by-products are removed. These low boiling by-products are condensed in cooler 60 and passed to product receiver 62. In this product receiver the low boiling by-products such as the butenes, methyl acetate, etc., are removed overhead as gases for further processing through line 64 and the condensed liquid phase which is chiefly acetaldehyde is removed through line 66 and used to reflux the fractionation zone 58.

A portion of the product acetaldehyde is withdrawn through line 68 and all or a portion can be returned to the oxidation zone through line 70 for oxidation in the reaction zone to acetic acid to replenish the acetic acid consumed in the formation of the vinyl acetate product. If desired, all or a portion of this acetaldehyde could also be removed through line 72 as a product of the oxidation zone or as an intermediate to be passed to a separate reaction zone where it is oxidized to acetic acid. The acetic acid is then returned to the oxidation process to replenish the acetic acid consumed in the formation of the vinyl acetate.

The bottoms from the fractionation zone 58 are then further fractionated in product fractionator 74 where the desired vinyl acetate product is recovered in a high degree of purity through overhead line 76, condensed and collected in product receiver 78. The vinyl acetate is employed to reflux fractionation zone 74 and a portion thereof is recovered as vinyl acetate product. Product receiver 78 is preferably combined with a phase separator by the positioning of a weir 80 within the vessel to permit separation of an aqueous and organic phase. The aqueous phase is removed through line 82 as a vinyl acetate rich aqueous stream which is passed to further processing for the recovery of all vinyl acetate values therefrom.

The particular concern of this process comprises the maintaining of the catalyst activity within reaction zone 16. During the processing of the olefin there occurs an accumulation of high boiling residue which is chiefly polyvinyl acetate. To prevent the accumulation of this high boiling residue to an objectionable level, a portion of the contents of the reaction zone are withdrawn through line 84 and passed to the catalyst recovery where the organic portion is burned and an ash containing the catalyst metals is recovered and returned directly to the reaction zone.

This is illustrated by passing all or a portion of the withdrawn reaction medium to a separation zone 86 which can comprise an atmospheric or vacuum distillation zone where a tarry residue is separated from a clear, lower boiling solvent. The lower boiling components are returned through line 88 to the reactor and the tarry residue is passed through line 90 to a combustion zone 92. As previously mentioned, separation in zone 86 is optional and if desired the withdrawn reaction medium can be passed directly to the oxidation zone through line 94.

The particular oxidation zone illustrated comprises a rotating kiln where the tarry residue is sprayed onto a bed of ash and where oxygen is introduced through line 96 to support combustion of the tarry residue. The combustion is maintained at the surface of the ash. If necessary, a suitable fuel gas can be introduced through 98 to maintain the necessary combustion temperature. The ash is discharged from the rotating kiln into hopper 100 and a suitable carrier liquid, preferably fresh reaction medium, is introduced through line 102 to provide a pumpable slurry of the ash which is then returned through line 104 to the reaction zone. The necessary quantities of makeup catalyst components such as the halides, alkali metal salts, etc., are supplied through line 106 and the necessary quantities of fresh reaction medium and acetic acid reactant are introduced through line 108.

The high boiling fraction from the product fractionation zone 74 is removed through line 75 and a portion thereof can be recycled to the reaction zone through line 77. The majority of this high boiling fraction comprises acetic acid which was removed with the product from reaction zone 16. This acetic acid also contains quantities of high boiling by-products such as propionic acid and vinyl propionate acid and vinyl propionate. All or a portion of this liquid stream can be passed to further processing through line 79 for the recovery of these by-products to obtain a purified acetic acid which is then returned to the reaction zone.

Figure 112 shows an alternative version of the process which is directed to the production of various carbonylation products of ethylene. In this oxidative carbonylation an olefin such as ethylene is introduced through line 110 and is admixed with the recycle gas in line 112 and passed to the reactor 114. The rate of introduction of the olefin into reactor 114 is controlled by flow control valve 116. An oxygen plant 118 provides a suitable source of oxygen through line 120 and the rate of introduction of oxygen into the reaction zone is controlled by flow control valve 122 which responds to the free oxygen content of the gases in the reactor or the vapor effluent removed therefrom as determined by gas analyzer and controller 124.

Preferably, the rate of oxygen introduction is controlled to limit the free oxygen content of this vapor stream to less than 3 and most preferably less than 1 volume percent. A gas rich in carbon monoxide can be obtained from various natural gas reforming furnaces and this gas is introduced through line 126. The gas is contacted countercurrently in adsorption zone 128 with the lean cuprous acetate solution introduced through line 130 and a pure hydrogen stream is removed through line 132. The rich adsorbent is removed through line 134, heated in heat exchanger 136 and passed to stripping zone 138 where the carbon monoxide is volatilized and removed from the cuprous acetate solution. The lean cuprous acetate is then removed and returned to the adsorption zone 128. The carbon monoxide stream is then passed by compressor 140 through line 142 and flow control valve 144 into reactor 114.

Reaction 114 is maintained under substantially anhydrous conditions and at relatively mild temperatures between 30° and 300°C.; preferably between 50° and 150°C. This temperature is maintained with the exothermic reaction by circulating all or a portion of the reaction medium through cooler 146 or, if desired, a cooling coil can be positioned internally of the reactor itself. The reaction zone is maintained under liquid phase conditions and the liquid contents and the liquid phase is stirred by propellers 148 which are rotated by motor 150 that can be mounted externally of the vessel.

The products are swept from reactor 114 by circulating a large excess of ethylene and carbon monoxide therethrough. The products are removed as a vaporous effluent through line 152, cooled in condenser 154 and passed to a high pressure accumulator from which the gases are removed through line 156 and returned to the reaction zone through recycle line 158. A portion of the recycle gas stream can be diverted through line 160 to a recycle gas purification zone which is any conventional system for the recovery of carbon dioxide from gas streams.

This is illustrated as an aqueous solution of monoethanolamine which is introduced through line 162 into countercurrent

FIGURE 112: PROCESS FOR VINYL ACETATE FROM ETHYLENE FEATURING ALTERNATIVE PALLADIUM CATALYST REGENERATION PROCEDURE

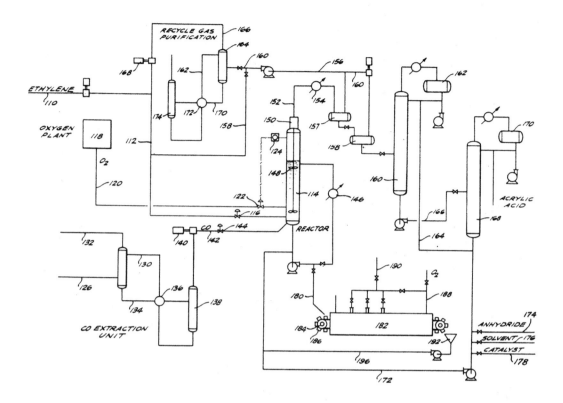

Source: K.L. Olivier and W.D. Schaeffer; U.S. Patent 3,461,157; August 12, 1969

contact with the recycle gas in adsorber 164. After that the purified gas is removed through line 166 and returned through recycle gas compressor 168 to the reaction zone for further processing. The enriched monoethanolamine solution is withdrawn through line 170, heated through heat exchanger 172 and passed to stripper 174 where the carbon dioxide is removed and the lean monoethanolamine solution is removed for return to the adsorption zone. The liquid contents of high pressure accumulator 157 are passed to a second accumulator 158 at a lower pressure and the dissolved ethylene contained in this stream is removed and returned for further processing to the reaction zone through compressor 160.

The liquid contents of low pressure accumulator 158 are passed into fractionation zone 160 where the acetic acid is vaporized and recovered in product receiver 162. This acetic acid is returned to the reaction zone through line 164. The liquid residue from this fractionation zone 160 is passed through line 166 to fractionation zone 168 where the desired product, i.e., acrylic acid, is recovered as product in product receiver 170 and the high boiling fractions such as beta-acetoxy propionic acid, high boiling acids and anhydrides are recycled to the reaction zone through line 172. The necessary makeup components of the reaction medium are introduced into this recycle line such as acetic anhydride through line 174, fresh solvent acetic acid or pivalic acid through line 176 and fresh catalyst components such as additional halide or any of the necessary catalyst salts is introduced through line 178.

The particular improvement provided by the process comprises the maintaining of the activity of the reaction medium in reactor 114 by removing a portion of the reaction medium from the remainder of this medium through line 180 and passing this material to a suitable combustion zone 182 where the organic portion of the removed reaction medium is burned to produce an ash containing the catalyst components that can be recovered and directly returned to the oxidation zone. If desired, the withdrawn portion can be distilled at atmospheric or reduced pressures to recover the lower boiling components which are returned to the reaction zone and the distillation residue is passed to zone 182.

Combustion zone 182 is illustrated as a furnace having a conveyor belt 184 with open pans 186 positioned therein. The reaction medium can be introduced into the open pans and passed into the furnace where it is heated to combustion temperatures and contacted with an oxygen-containing gas that is introduced through line 188. This oxygen can be pure or diluted oxygen, e.g., air or mixtures of air with inert gases. If necessary to maintain the desired temperature, a fuel gas can be introduced through line 190. The ash produced in furnace 182 is discharged therefrom into hopper 192 and a suitable carrier liquid such as fresh reaction medium is admixed with the ash and pumped through line 196 for return to the reaction zone.

PLATINUM-CONTAINING CATALYSTS

AMMONIA OXIDATION

Engelhard Industries

A process developed by C.D. Keith; U.S. Patent 3,428,424; February 18, 1969; assigned to Engelhard Industries, Inc. is one in which the oxidation of ammonia to produce nitric oxide is effected in the presence of a catalyst block comprising a porous inert unitary refractory skeletal structure having gas flow channels therethrough and a platinum group metal deposited thereon, especially on the surfaces of the gas flow channels. This process is applicable to the production of nitric acid.

In the commercial production of nitric acid, ammonia gas is mixed with an oxygen-containing gas, e.g., air, and the gas mixture passed through or over platinum metal gauze catalyst maintained at an elevated temperature in a converter to obtain nitric oxide. The effluent gas from the ammonia converter is then cooled and introduced together with additional oxygen-containing gas into absorption towers wherein the nitric oxide is oxidized to nitrogen dioxide and the nitrogen dioxide is absorbed in water to form nitric acid. The platinum metal gauze is usually of a platinum-rhodium alloy and in the form of a fine gauze packed in numerous layers, typically 10 to 30 gauze sheets packed together.

The platinum metal gauze is heated at high temperatures of 650° to 1000°C. and higher for the ammonia oxidation and the pressure for the oxidation varies from atmospheric pressure to 110 psig and higher. Further, the catalyst in the converter for the ammonia oxidation of high capacity plants may be subjected to high gas flows as high as 1 million cubic feet and higher of the gas mixture per cubic foot of catalyst gauze per hour. Under such severe conditions of the ammonia oxidation, there is considerable loss of the expensive platinum group metal due to physical and chemical attack by the gases and the loss of catalyst is greater for converters operating at higher pressure and/or temperatures. Indeed the losses of expensive platinum metal catalyst are not infrequently as high as 2.2 troy ounces per 100,000 pounds avoirdupois of ammonia oxidized in the converter.

Further, with the catalytic metal gauze packed in numerous layers in the reaction zone, material gas back pressures tend to develop which are undesirable. Moreover, it is difficult to maintain constant temperature in the gauze. The shell of such catalyst unit is heat conductive and a temperature gradient is set up in the bed. The difficulty of temperature control presents problems with respect to the expensive Pt-Rh catalyst. If the temperature is too high, precious metal losses increase sharply even under atmospheric pressure. Hot spots in the gauze cause holes through which the gases channel unconverted. In operation, the gauze darkens appreciably and quite large excrescences of low mechanical strength form on the wires. These excrescences cause increased back pressure in the gauze and are partly carried away by the gas stream with loss of the expensive precious metal.

In addition to the abovementioned problems, usually the gauze catalysts start losing their activity in a relatively short time. It is estimated that the gauze units, particularly in high pressure systems, must be regenerated every 3 to 8 weeks. Further, the gauze has a tendency to pry up, tear and sag. Replacement, repair and regeneration of the conventional catalysts causes expensive shut-downs in plant operation.

In accordance with the process, is has now been found that ammonia can be oxidized to produce nitric oxide by a catalytic oxidation which is considerably more economical and efficient and with appreciably less loss of catalytic metal than with the prior art process utilizing the platinum gauze catalyst, and with elimination of the material gas back pressures occurring especially after prolonged use of the metallic gauze catalyst with the gauze in numerous layers. This process involves passing gaseous ammonia together with an oxygen-containing gas present in an amount sufficient to supply molecular oxygen in at least the stoichiometric amount required to react with the ammonia to produce nitric oxide, at a reaction temperature through a plurality of gas flow channels extending in the direction of gas flow through a supported catalyst comprising a porous inert unitary solid refractory skeletal structure as support, and a platinum group metal as catalyst on surfaces of the flow channels and of superficial macropores communicating with the channels. During its passage through the flow channels, the gaseous mixture of ammonia and oxygen–containing gas contacts the platinum group metal on the flow channel surfaces and also on the surfaces of the accessible superficial macropores, with the result that the ammonia is oxidized to nitric oxide. The accessible superficial macropores are predominantly of size in excess of 200 angstrom units. The platinum group metal required by this process is reduced to a low level and to as small an amount as 1% of that required by the prior art process utilizing the platinum gauze.

Further, the instant process eliminates the undesirable growth of excrescences occurring on the Pt–Rh gauze of the prior art process and therefore the loss of the expensive catalytic metal, and the channeling of the gases through the resulting openings in the gauze with attendant by-passing of the catalyst by unconverted gases is also eliminated by the process. The catalytic platinum group metal alloy herein is preferably deposited as a thin continuous or substantially continuous layer on the surfaces of the flow channels and of the superficial macropores communicating therewith. Figure 113 is a flow diagram for the production of nitric acid according to the process.

FIGURE 113:   FLOW DIAGRAM OF NITRIC ACID MANUFACTURING PROCESS WITH DETAIL OF AMMONIA OXIDATION CATALYST STRUCTURE

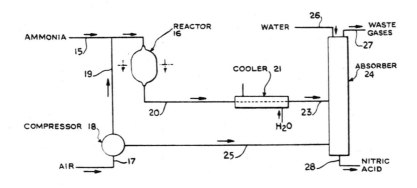

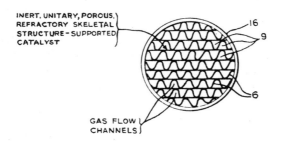

Source: C.D. Keith; U.S. Patent 3,428,424; February 18, 1969

As shown in Figure 113, the ammonia gas is passed through conduit 15 into reactor 16, preheated air being supplied into admixture with the ammonia gas via conduit 17, compressor 18 and conduit 19. The air is passed into admixture with the ammonia in conduit 15 in an amount sufficient to provide an amount of molecular oxygen in the admixture which is at least that required to stoichiometrically react with the NH_3 to form nitric oxide and water. The gaseous mixture of ammonia and air passes in contact within reactor 16 with the platinum group metal catalyst deposited on the surfaces of gas flow channels 9, as shown in the detail at the base of the figure, and of macropores communicating therewith of the unitary refractory skeletal structure or "honeycomb" support 6 disposed within reactor 16 wherein the ammonia is oxidized to $NO + H_2O$. The temperature of the catalyst within reactor 16 is between 650° and 1000°C. or higher due to the exothermic reaction and the pressure from atmospheric to 110 psig and higher.

The space velocity of the gaseous admixture through the flow channels of the supported catalyst in reactor 16 may range up to 1 million cubic feet of gas per cubic foot of catalyst per hour. However, there is but minimal pressure drop due to the unobstructed gas-flow channels of the catalyst.

The gas exiting at elevated temperature from reactor 16 through conduit 20 comprises a mixture of primarily N_2, NO, water vapor and NO_2, and this gas is passed into cooler 21 wherein its temperature is lowered appreciably by indirect heat exchange with cooling water or other suitable coolant. Some NO_2 is converted to HNO_3 in cooler 21 by reaction with condensed water therein in accordance with the reaction $3NO_2 + H_2O \longrightarrow 2HNO_3 + NO$. Further, some NO may be converted to NO_2 in cooler 21 by the reaction $2NO + O_2 \longrightarrow 2NO_2$.

From cooler 21, the mixture is passed through conduit 23 and introduced into a lower portion of absorber column or tower 24. Air is supplied from compressor 18 through conduit 25 and is also introduced into a lower portion of absorber 24. Absorber column 24 is equipped with trays and bubble caps or alternatively packed with acid-resistant packing, e.g., stoneware. Absorber 24 and other elements and vessels of the apparatus herein handling the corrosive acidic gases and liquid HNO_3 are constructed of an acid-resistant material, for instance stainless steel. Water is introduced into the upper portion of absorber 24 through conduit 26.

Nitric oxide is oxidized in absorber 24 to form NO_2 by reaction with excess oxygen from the air. The nitrogen dioxide passes upward within absorber column 24 in intimate countercurrent contact in the region of the bubble cap trays or packing (not shown) with the liquid water flowing downward therewithin, whereby the nitrogen dioxide is absorbed by the water forming nitric acid and releasing additional nitric oxide. The waste gases from absorber 24 are withdrawn through conduit 27 and, after preferably being first sent through a mist separator, may then be heated by indirect heat exchange with the hot effluent gases from reactor 16 in a suitable gas to gas heat exchanger (not shown) where the hot gases can then be utilized for power recovery, for instance as motive fluid for operation of a gas turbine which in turn is coupled with the air compressor 18 for compressing the incoming air. Nitric acid of typically 60% acid concentration is withdrawn from column 24 through conduit 28 and, after preferably being first bleached by contact with a countercurrent stream of air, is passed to storage.

Phillips Petroleum

A process developed by O.W. Hill and J.C. Thomas; U.S. Patent 3,093,597; June 11, 1963; assigned to Phillips Petroleum Company relates to regeneration of an ammonia oxidation catalyst. In one of its aspects, the process relates to the regeneration of platinum gauze pads used as catalyst in the oxidation of ammonia to produce nitric acid by subjecting the catalyst to ultrasonic vibrations. In another of its aspects, the process relates to a method as described, wherein catalyst fines are removed from beds of catalyst, using a combination of steps including one or more of the following: mechanically cleaning the catalyst by brushing and shaking, oxidizing contaminants with a hydrogen torch, treating with hydrochloric acid to remove alloys, iron, etc., treating with hydrogen fluoride to remove sand, silica, etc., treating with sulfuric acid to remove organic matter, etc., but always placing the gauze under circumstances to subject the same to ultrasonic vibration, as in a detergent bath, to dislodge tiny particles, usually platinum, from the catalyst. In a further aspect of the process, particles which are dislodged can be recovered from the detergent by filtration and the detergent reused.

A typical gauze pad contains 120 ounces of "platinum catalyst" comprising 90% platinum, 5% rhodium and 5% palladium. The pad is composed of about 16 gauze units made of three to four individual gauzes per unit. The individual gauzes in a unit are woven from wire 75 microns in diameter with 80 meshes per inch and are welded together at the edges. The pads are retained in the converter on a support grid. A catalyst gauze of the type described is made of smooth wire. When such a gauze is placed in ammonia oxidation service, metallic protrusions develop on the surfaces of the wire and greatly increase the total surface area of platinum in contact with the reacting gases. As a result, when a new gauze pad is placed in ammonia oxidation service, the initial conversion efficiency of ammonia to nitrogen oxides is very low. Within an hour, however, the conversion efficiency reaches a level of about 90% and after 3 days, reaches the maximum level of 93 to 96%.

After a period of time, the conversion efficiency drops to 90%, at which time the catalyst gauze is removed from service and cleaned. When the conventional cleaning steps, hydrogen and acid treatments, are used, the maximum period of service before efficiency drops to 90% is about 25 days. Each time a gauze pad is cleaned, it is usually necessary to replace several layers of gauze with new gauze. Losses of catalyst also occur during service in the oxidation process and during cleaning operations. The total loss of catalyst normally averages 0.0374 ounce per ton of nitric acid produced, when conventional cleaning steps are used.

In contrast, for catalyst gauzes treated by the improved process, catalyst life increased from the usual 25 days to an average of 51 days. Catalyst loss decreased from the normal 0.0374 to an average of 0.0316 ounce per ton of nitric acid produced. Based on a plant production rate of 400 tons per day, these catalyst savings represent a savings of better than $200.00 per day when the catalyst gauze sells for $87.00 per ounce.

Universal Oil Products

A process developed by L.C. Hardison; U.S. Patent 3,467,491; September 16, 1969; assigned to Universal Oil Products Company is one in which ammonia is removed from ammonia-containing gases by conversion thereof to nitrogen and water without production of noxious nitrogen oxides by contacting the gases and air with a platinum-alumina catalyst at a temperature from 400° to 450°F. In particular, there is provided means for substantially eliminating the discharge of harmful ammonia and nitrogen oxides into a confined work space from printing equipment using a diazo paper.

Various "dry" or "diazo-type" printers in turn utilize slightly different systems for obtaining color or printing. However, most of them have either aqueous ammonia or anhydrous ammonia to develop the lines or color of a print from a diazo compound, a coupler and a stabilizing acid which are impregnated into the light sensitive copy paper. The coated or impregnated papers are generally referred to as "diazo paper" even though different suppliers will use different chemical impregnations. Although ammonia vapors may be released within the printing equipment to contact but one face of the copy paper, such vapors are generally carried throughout the interior of the entire printer and will be released into the printing room unless expensive and elaborate exhaust systems are incorporated into or are attached to the printer installation. Certain data released by the National Safety Council indicates that for some people the detectable threshold for ammonia is approximately 20 parts per million and that some people notice slight eye irritation at about 40 parts per million. Concentrations of the order of 1500 parts per million are highly irritating and may be fatal. Preferably, the exhaust gases that vent from a printing machine should be treated to reduce the ammonia concentration to less than 20 parts per million for use in a well ventilated space and to provide substantially lower levels for installations in confined or poorly ventilated work spaces.

Prior associated work carried out in effecting the catalytic combustion of ammonia has invariably produced substantial quantities of the various nitrogen oxides, even though the usual high temperature catalytic conversion of ammonia in the presence of oxygen will result in primarily water vapor and nitric oxide, the latter being colorless and odorless. On the other hand, if the combustion treatment is carried out at room temperature, then there is a high combination of nitric oxide and oxygen to give the brown nitrogen dioxide. The latter is of course quite harmful with only amounts of less than 5 parts per million being considered safe. The odor of nitrogen dioxide is detectable in amounts of the order of 5 parts per million, although usually no discomfort is felt at this level. It is generally stated that at about 62 parts per million of nitrogen dioxide there will be some discomfort; however, with respect to both nitric oxide (NO) and nitrogen dioxide (NO$_2$) the harmful effects are particularly insidious because there is no immediate reaction and a person who has been exposed to the atmosphere may not experience sensations of poisoning until several hours after the exposure.

It is thus a principal object of this effort to provide a controlled catalytic process for treating an ammonia-containing vent stream whereby both the ammonia and nitrogen oxides levels will be below that which is considered harmful for a person working in a confined space. The process will now be further described with reference to the equipment diagram in Figure 114 shown on the following page.

Referring to the drawing, there is indicated diagrammatically a diazo printer 1 with a vent line 2 connecting with a catalytic reactor chamber 3. It is to be understood that the drawing is not to scale and that the reactor means 3 might well be of a size and design which is readily adaptable for installation within the cabinet portion of printer equipment, as well as exteriorly thereof, such that the ammonia containing vent stream collected from within the cabinet portion of the printer may be drawn through the catalytic converter portion and a resulting treated, harmless vent gas stream be discharged from the cabinet portion of the printer 1.

Specifically, with respect to reactor 3 of the drawing, there is an inlet section 4 having an inlet port 5 and an internally positioned electric heating coil 6. The latter is shown as being supplied a suitable electric power source by way of lines 7 and 8. The heater coil 6 may be formed of a conventional type of resistance wire or may comprise one or more sections of "calrod" type of heater unit. In any event, the heater coil 6 shall be suitably placed within the inlet section 4 such that the vent gases passing therethrough will be preheated to a temperature of 400°F. prior to entering the shell portion 9 which in turn holds a catalyst bed 10. Although not shown, suitable thermocouple or other temperature sensitive means may be positioned downstream from the heater coil 6 and in turn made connective with automatic switching means from the electrical power source to the coil such that there may be automatic control of the temperature leaving the section 4.

The catalyst section 10 is shown as being of annular shape or as having a hollow cylinder-like configuration, with catalyst particles being held in a fixed bed between an outer screen or perforate member 11 and an inner screen or perforate plate member 12. Screen 11 is placed a spaced distance inward from the housing wall 9 such that there is an annular space 13 for the ammonia containing stream to surround the catalyst 10 and pass in a radial inward flow therethrough to a central passageway 14. The latter is blocked upstream by a suitable nonperforate plate 15 but is open at a downstream end to connect with a suitable outlet conduit 16. There is of course, an internal opening or passageway 17 between the inlet section 4 and the catalyst housing 9 whereby the heated vent gas stream may readily enter the space 13 and surround the catalyst bed 10. The ammonia vapors within the vent stream will be contacting catalyst particles 10, such as alumina with a small percentage of platinum activation, whereby there will be substantially complete conversion and elimination

FIGURE 114: APPARATUS FOR CATALYTIC TREATMENT OF VENT GASES CONTAINING AMMONIA

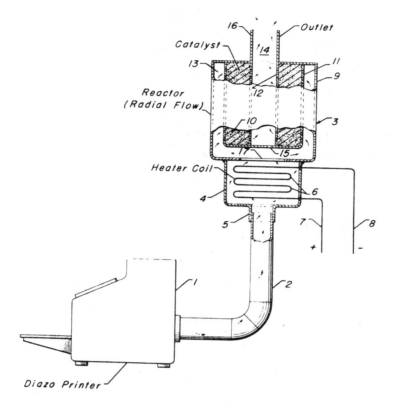

Source: L.C. Hardison; U.S. Patent 3,467,491; September 16, 1969

of the ammonia in the stream to permit the discharge of a stream from outlet 16 with less than 5 parts per million of the ammonia remaining in the stream. In addition, where the catalytic conversion is carried out at a controlled temperature within the range of 400° to 450°F., there will be substantially no nitrogen oxides formed during the conversion, which in turn could provide poisoning to such persons that may be constantly near the printer equipment in a confined work space.

In one specific instance, the catalyst within bed 10 may comprise 1/16" to 1/8" round spherical type alumina particles impregnated with platinum to have approximately 0.5 to 1.0% platinum by weight of the alumina base; however, lesser or greater amounts may be utilized to provide optimum conversion for a particular ammonia vapor vent stream treatment such that the resulting discharge will have ammonia vapors and nitrogen oxides in only small permissible quantities of a few parts per million, or merely fractional parts per million.

AUTOMOTIVE EXHAUST TREATMENT

Universal Oil Products

A process developed by E. Michalko; U.S. Patent 3,259,589; July 5, 1966; assigned to Universal Oil Products Company involves preparing a catalytic composite of alumina and a platinum component, which method comprises commingling the alumina with chloroplatinic acid and from 0.13 to 0.70% by weight, based upon the weight of the alumina, of an organic acid selected from the group consisting of dibasic acids and derivatives thereof having the following structural formula,

$$HO-\underset{\underset{O}{\parallel}}{C}-\left[\underset{\underset{R'}{|}}{\overset{R}{\underset{|}{C}}}\right]_a-\underset{\underset{O}{\parallel}}{C}-OH$$

where R is selected from hydrogen, hydroxyl and alkyl groups; R' is selected from hydrogen, alkyl and carboxyl groups; and n is within the range of 0 to 6, drying the resulting mixture at a temperature within the range of from 100° to 250°F., and thereafter subjecting the mixture to an atmosphere of hydrogen at a temperature from 900° to 1800°F.

A process developed by J. Hoekstra; U.S. Patent 3,409,390; November 5, 1968; assigned to Universal Oil Products Co. provides a method for converting a noxious exhaust gas stream to less objectionable components by contacting the stream at conversion conditions with a catalytic composite of low density porous alumina particles, platinum and an alkaline earth component prepared by commingling the alumina with chloroplatinic acid to provide from 0.01 to 1% platinum by weight of the composite, drying the resulting mixture at a temperature from 100° to 250°F., subsequently subjecting the mixture to an atmosphere of hydrogen at a temperature from 200° to 1800°F., thereafter impregnating the resulting material with a solution of a component selected from the group consisting of barium, strontium and calcium, and then drying the resulting catalyst composite.

A process developed by V. Haensel; U.S. Patent 3,503,715; March 31, 1970; assigned to Universal Oil Products Company utilizes a unitary apparatus for effecting the catalytic oxidation of engine exhaust gases, where they may be both high and low quantitites of hydrocarbon emissions with such gases, with one catalyst layer comprising platinized alumina particles and a next adjacent layer comprising platinized alumina particles containing a barium, calcium or strontium component. Such a device is shown in Figure 115.

FIGURE 115:   SECTIONAL VIEW OF EXHAUST GAS CONVERTER USING TWO DIFFERENT PLATINUM
CATALYST LAYERS

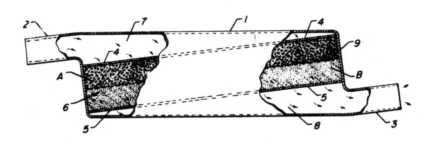

Source: V. Haensel; U.S. Patent 3,503,715; March 31, 1970

Referring to Figure 115, there is shown an enclosed converter-muffler chamber 1 having an inlet port 2 and outlet port 3. The internal portion of chamber 1 is provided with an upper screen or perforate member 4 spaced from a lower screen or perforate member 5. The spaced perforate members 4 and 5 provide a catalyst retaining section 6 therebetween, as well as define an inlet manifold section 7 which connects to inlet port 2, while partitioning member 5 defines in combination with outlet 3 an outlet manifold section 8.

In accordance with the process, a physical mixture or arrangement of two different portions of catalyst composites are maintained within the catalyst section 6 such that any given exhaust gas stream passing through the converter-muffler must contact each of the types of catalyst. There is diagrammatically indicated an upper portion of catalyst A which is adapted to accommodate a lower temperature, low emission engine exhaust gases and effect a high degree of conversion. Downstream from the catalyst portion A is a second stage catalyst portion B which contains barium oxide, or other suitable alkaline earth component adapted to give high temperature resistance and stability thereto. Such portion is thus particularly adaptable for use in contacting high temperature exhaust gas streams from high emission engines. The unitary device, as shown, is thus capable of handling exhaust gases from any kind of vehicle engine, or from an engine which may change from being a low emission engine to a high emission engine.

CRACKING

Mobil Oil

A process developed by V.O. Bowles; U.S. Patent 3,412,013; November 19, 1968; assigned to Mobil Oil Corporation involves the utilization of high activity crystalline alumino-silicate containing cracking catalysts of orderly lattice structure and uniform pore diameter of at least 6 A. and which have been ion exchanged and/or rare earth exchanged to produce desired active sites.

The thus modified catalytically active crystalline alumino-silicates may be used alone or these alumino-silicates may be combined with one or more hydrogenating and/or hydrocracking promoting components. Furthermore, the catalytically active crystalline alumino-silicate either with or without a hydrogenating component, physically combined, deposited on or in the crystalline alumino-silicate may be used alone, mechanically admixed with or composited with synthetic or naturally occurring components, such as clays, hydrous oxide gels, gels of mixed hydrous oxides and the like, which may be inert or may themselves possess catalyst activity. A specific example of the preparation of such a catalyst is as follows.

Example: A platinum-zeolite 5A complex was prepared by adding a solution of platinum amine chloride to a solution of sodium aluminate, mixing with a solution of sodium metasilicate, the aluminum and silicon components being in proportions to produce a zeolite of the A pattern, and the mixture refluxed at 100°C. for 7 hours, the solid filtered out, washed, ion exchanged to substitute a portion of calcium for the sodium, washed, dried and calcined at temperatures high enough to assure than platinum is interspersed in the crystalline lattice of the zeolite. The platinum content was 0.31% by weight.

Separately, a sodium Y zeolite was treated at 180°F. by continuous contact with a 25% ammonium chloride solution at conditions such that 60 lbs. of this solution were contacted in the base exchange operation over a period of 48 hours. The zeolite was then water washed to chloride free, oven dried for 20 hours at 230°F., and calcined in air for 10 hours at 100°F. Proportions were such as to provide 500 ml. of finished catalyst. This provided an acid or H form of Y catalyst with approximately 99% of the original sodium replaced by hydrogen, as determined by analysis for residual sodium. Equal amounts of the platinum 5A catalyst and the HY catalyst were ball-milled together and pelleted.

DEHYDROGENATION

Chevron Research

A process developed by E. Clippinger and B.F. Mulaskey; U.S. Patent 3,531,543; September 29, 1970; assigned to Chevron Research Company is one utilizing composites of a Group VIII noble metal, tin and an inorganic, solid, refractory oxide carrier which have excellent dehydrogenation activities and little or no isomerization and cracking activities.

It is known in the art that composites of noble metals of Group VIII of the Periodic Table are multifunctional in that they exhibit substantial and concurrent activites which include isomerizing, cracking and dehydrogenation activities in their catalytic action with vaporized hydrocarbons at elevated temperatures. It is also known in the art that isomerization and cracking activities can be inhibited by the use of several expedients. These, however, suffer from certain drawbacks which impair their usefulness including limited catalyst life, catalyst regeneration problems, migration of the modifier from the catalyst during use and the inconvenience and risks involved from the employment of materials such as tellurium, selenium, arsenic and the like which are particularly hazardous to man.

It has been found that composites of a Periodic Table Group VIII noble metal, tin and a solid, inorganic refractory metal oxide carrier are improved hydrocarbon dehydrogenation catalysts. For each 100 parts by weight of the carrier, the composite should contain from 0.05 to 5 parts of the noble metal, and for each atom of the noble metal from 0.001 to 3 atoms of tin. The tin inhibits isomerization and cracking activities normally concurrently experienced in the use of unmodified noble metal composites in the processing of hydrocarbon feeds in the vapor phase at elevated temperatures. The subject tin-containing composites are stable and may be used for long periods of time. Ostensible loss of dehydrogenation activity through slow accumulation of deposits on these catalysts is readily recovered by the use of conventional carbon burn-off cycles. The catalysts are particularly useful where it is desired to dehydrogenate a hydrocarbon feed with little or no concurrent isomerization or cracking of the feed or product. In this case the refractory metal oxide carrier should be nonacidic and preferably a lithiated alumina.

On the other hand, where it is desired to favor dehydrogenation and yet concurrently effect a moderate degree of isomerization and cracking of a hydrocarbon feed, a composite of a Group VIII metal, containing a relatively minor amount of tin, e.g., from 0.001 to 0.1 atom per atom of the noble metal, is desirably employed. In this case the refractory oxide support may be nonacidic or acidic depending upon the relative emphasis desired, in particular with regard to the cracking function of the catalyst in the hydrocarbon processing.

Halcon International

A process developed by R.S. Barker; U.S. Patent 3,361,818; January 2, 1968; assigned to Halcon International Inc. is a process for the preparation of aminated benzenes by dehydrogenating aminated cyclohexanes at temperatures from 180° to 500°C. in the presence of a Group VIII metal at a liquid hourly space velocity of from 0.1 to 20 hr.$^{-1}$. Yields are improved by conducting the dehydrogenation in the presence of hydrogen, and aminating agent.

Aminated benzenes are commercial chemicals of great industrial importance. For example, the varied uses of the simplest aminated benzene, aniline, include application as a rubber accelerator, antioxidant, dye intermediate, drug intermediate

intermediate, explosive and fuel. Conventionally, aniline is prepared by the reduction of nitrobenzene with iron filings or borings and 30% hydrochloric acid; by reaction of chlorobenzene with aqueous ammonia at 200°C. and 800 psi; or by catalytic vapor phase reduction of nitrobenzene with hydrogen. In the prior art, attempts to convert cycloaliphatic amines to aminated benzenes were unsuccessful. For example, Moss et al., Nature, 178, 1069 (1956) found that benzene rather than aniline was formed by the dehydrogenation of cyclohexylamine.

The dehydrogenation catalyst used in the process is preferably a metal of Group VIII of the Periodic Table, or mixtures of two or more such metals, i.e., metals of the platinum group or the palladium group as well as cobalt and nickel. Copper, iron, molybdenum, chromium or mixtures thereof with each other or any of the foregoing Group VIII metals may be used as promoters for the dehydrogenation catalyst metal. Best yields are obtained using a catalyst supported on neutral materials, such as carbon, silica, silicon carbide, zircon, spinels, or alpha-alumina of the porous type. As used herein, by a neutral support is meant one which either of itself or when neutralized, e.g., by treatment with an acid or a base, will not deaminate the starting material when tested in the absence of catalyst under the reaction conditions. The support may be neutralized at any state in the preparation of the catalyst. It may, for example, be neutralized prior to catalyst impregnation, during catalyst impregnation, or subsequently to catalyst impregnation. Self-supporting catalysts such as the "forminates" may also be employed. These catalysts are described in detail in Groggins, Unit Processes in Organic Synthesis, McGraw-Hill Book Co., Inc. (New York), 1958, edition 5, pp. 434-438.

HYDROCRACKING

Socony Mobil Oil

A process developed by H.L. Coonradt and W.E. Garwood; U.S. Patent 3,169,107; February 9, 1965; assigned to Socony Mobil Oil Company, Inc. is an improved catalytic hydrocracking process wherein a high boiling petroleum hydrocarbon or hydrocarbon mixture is subjected to cracking in the presence of hydrogen and a solid porous catalyst composed of a metal of the platinum series deposited on a metal oxide and stabilized by intimate admixture thereof with silica in the form of a hydrogel or gelatinous precipitate in an amount corresponding to at least 25 weight percent of the resulting dry composite.

In another embodiment, the process is concerned with hydrocracking selected petroleum stocks in the presence of a catalyst consisting essentially of a composite of 25 to 75% by weight silica, 0.05 to 10% by weight of a metal selected from the group consisting of platinum and palladium and remainder alumina prepared by impregnating the alumina component with the metal selected from the aforementioned group, mechanically mixing the resulting impregnated product with the silica component under conditions of intimate contact such as to effect chemical interaction between the same and drying and treating the resulting composite at an elevated temperature.

HYDROGEN CYANIDE MANUFACTURE

W.R. Grace

A process developed by P.M. Brown and H.C. Duecker; U.S. Patent 3,577,218; May 4, 1971; assigned to W.R. Grace & Co. is one in which a mixture of ammonia and methane is reacted over platinized mullite catalyst to give high yields of hydrogen cyanide. The catalyst is prepared by depositing platinum on either natural or synthetic mullite; it possesses improved stability and avoids the problem of carbonitridation frequently encountered in use of alumina as a catalyst support in the production of hydrogen cyanide.

Hydrogen cyanide is generally prepared by conversion of a nitrogen containing gas and a hydrocarbon with or without a source of oxygen over a suitable catalyst. The bulk of the hydrogen cyanide is prepared by one of two catalytic processes. In the first of these processes, hydrocarbon and a source of nitrogen such as ammonia are passed through ceramic tubes containing platinum as the catalyst. No air is added to the reaction mixture. In the second of these processes, air is added with the reaction mixture and the reactants are heated to a temperature of 1000°C. in the presence of platinum gauze as a catalyst. This method is disadvantageous in that the product produced contains water. In addition, large quantities of platinum are required with the attendant high cost.

The reaction in the absence of oxygen using platinum on ceramic tubes as a catalyst has certain advantages over the process where the catalyst is platinum sponge in that small quantities of platinum are required and the product is anhydrous. However, problems are encountered in this process in that the catalyst tends to become carbonized at high temperatures in the absence of oxygen. The alumina tubes slowly convert to aluminum nitride due to carbonitridation. The nitride formed acts as a catalyst for the ammonia decomposition prior to its contact with the hydrocarbon gas such as methane. As a result, high quantities of ammonia are required to prevent coking. Coking may be diminished by employing a high surface area highly active catalyst which operates at lower temperatures.

Platinum on high surface area alumina give satisfactory results. However, the alumina tends to sinter with occlusion and loss of platinum. In addition, the alumina again tends to be converted to the aluminum nitride.

It has been found that a very satisfactory catalyst for the preparation of hydrogen cyanide can be prepared by depositing platinum from chloroplatinic acid on granular mullite and converting the platinum to the metal. The platinum may be alloyed with other metals in an amount of up to 20% of the metals' content if desired. The mullite base is a very stable base. It does not have the problems inherent in use of alumina as the catalyst base. The alumina in the mullite is bound into the structure in such a manner that there is no tendency for nitride formation. In addition, it retains high surface area over prolonged periods of time without sintering and encapsulating the platinum. When a mixture of ammonia and methane is passed over a catalyst comprising a base coated with the platinum or platinum alloys, the yield of hydrogen cyanide from ammonia, as observed over many days of continuous operation, reaches a peak shortly after the start of the operation and thereafter continues to fall off over a period of several days until it is no longer economical to continue with the catalyst in the spent condition.

Accordingly, the reaction is normally stopped at some predetermined level of yield. The spent catalyst is reactivated by burning off the coke so that when operations are renewed, the yield of hydrogen cyanide will again obtain a high level. Because of the high stability of the mullite, the platinum catalyst on the mullite base does not become deactivated as rapidly as the catalysts of the prior art due to lower operating temperature and resistance of nitride formation. Figure 116 shows a reactor design employing such a catalyst composition.

FIGURE 116: CONVERTER DESIGN FOR HYDROGEN CYANIDE MANUFACTURE USING PLATINIZED MULLITE CATALYST

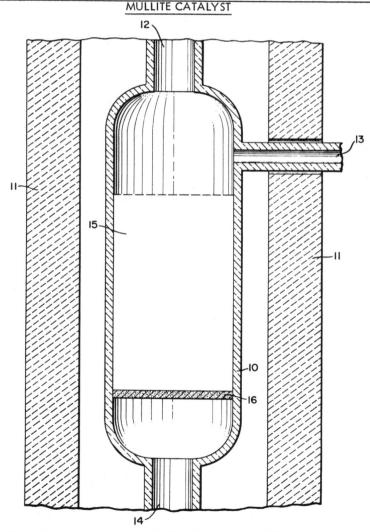

Source: P.M. Brown and H.C. Duecker; U.S. Patent 3,577,218; May 4, 1971

The reactor 10 is positioned in a furnace 11. The reactor is a cylindrical structure having gas inlets 12 and 13 and an outlet 14. The catalyst is suspended in the reactor at 15 over a suitable porous disk 16. In operation, the reactor is brought up to temperature in the furnace and preheated methane, or nitrogen and methane if a diluent is required, are fed into the reactor at 12. Gaseous ammonia is fed into the reactor at 13. The reaction takes place in the catalyst bed 15 and the gaseous product passes out the exit tube 14 to a suitable collection vessel.

When the ammonia gas is passed into the system immediately above the catalyst bed, the yield of hydrogen cyanide remains constant over a protracted period of time. In addition, best results are obtained when the amount of methane is carefully controlled to assure there is an adequate amount of ammonia present to react with the methane at all times. If these precautions are followed, the tendency for the methane to be reduced to carbon is alleviated with the resultant increase in catalyst life. The reactants are maintained in contact with the catalyst for periods of 0.01 to 10 seconds, preferably about 4 seconds.

Although mullite is a naturally occurring mineral, it is desirable for obvious reasons to prepare the base synthetically. In one process, solutions of sodium silicate, aluminum nitrate and sodium aluminate are prepared and intimately mixed. The mixture is spray dried and exchanged to remove sodium to a suitably low level. The spray dried product is then calcined to prepare the base. The mullite base is then impregnated with a solution of a platinum salt. The preferred salt for the impregnation is chloroplatinic acid although other salts such as chloride, bromide, fluoride and sulfate may be used. A sufficient quantity of the solution is added to prepare the catalyst having a platinum content of 0.01 to 2%, preferably about 0.1 to 0.6%. In the final step, the platinum is reduced to the metal by heating in a hydrogen atmosphere. Suitable reduction is achieved if the catalyst is heated to 250° to 500°C. for 1 to 6 hours, preferably 500°C. for 4 hours.

Monsanto

A process developed by B.Y.K. Pan; U.S. Patent 3,371,989; March 5, 1968; assigned to Monsanto Company is an improved process for the production of hydrogen cyanide by the vapor-phase reaction of ammonia, natural gas and air in the presence of a platinum gauze catalyst. It comprises activating the catalyst by passing the reactants in specified mol ratios and temperatures through it as a specific mass flow rate ranging from 1.28 to 1.45 lb./hr./layer of gauze/sq. in. of gauze surface until a maximum yield of hydrogen cyanide is obtained.

In the reaction of ammonia, natural gas and air over platinum gauze catalysts, there is a period of time after the reaction is initiated with a new gauze catalyst before maximum catalytic activity is achieved. This catalyst-maturing or activation period varies in length or duration and may last as long as 12 days. During the activation period, the conversion of reactants to hydrogen cyanide is lower than optimum and the yield slowly increases to the normal optimum. The lowered conversion and yield in a plant-scale operation during this period represent significant losses in utilization of raw materials as well as loss in production output. It has been found however, that the conversion of raw materials to hydrogen cyanide and the yield of this product can be significantly increased, the life of the catalyst can be lengthened, and the time required for activation of the catalyst can be shortened considerably by controlling the specific mass flow rate of the reactant mixture through the reactor during the activation period at a particular level.

More specifically, the activation step consists of passing a reactant mixture of ammonia, natural gas and air in such proportions that the mol ratio of ammonia to natural gas is in the range from 0.70 to 1.0 and the mol ratio of natural gas to air is in the range from 0.170 to 0.200 through the gauze catalyst at a specific mass flow rate in the range of 1.28 to 1.45 lb./hr./layer of gauze/sq. in. of gauze surface at a temperature in the range from 1100° to 1200°C. for the period of time required to give a maximum yield of hydrogen cyanide and thereafter passing a reactant mixture of ammonia, natural gas and air through the catalyst under conditions known to give hydrogen cyanide.

ISOMERIZATION

Phillips Petroleum

A process developed by J.W. Myers; U.S. Patent 3,449,264; June 10, 1969; assigned to Phillips Petroleum Company is a process for improving the activity of alumina and platinum-alumina catalysts to increase their capacities for isomerizing and hydrocracking hydrocarbons.

The most widely used isomerization catalysts are Friedel-Crafts catalysts, of which the preferred one is aluminum chloride. However, other catalyst have been developed and are used in certain instances. The isomerization of hydrocarbons over platinum-on-alumina, promoted with noble metals, has been practiced. The halogens, chlorine and fluorine, have been added in minor amounts to these catalysts to enhance the activity thereof for isomerization processes. The temperatures utilized in the isomerization of hydrocarbons with the aforesaid catalyst have been above 600°F., e.g., 600° to 900°F. The process is concerned with activating alumina and platinum-alumina catalysts, which may contain aluminum chloride and/or halogen to isomerize hydrocarbons with high efficiency and high yield, below isomerization temperature, 600°F.

In somewhat more detail, the process comprises treating active alumina-containing catalysts with at least one treating agent selected from the group consisting of Cl, HCl, Br, HBr, mono-, di-, tri- and tetrachloromethane, and mono-, di-, tri-, and tetrabromomethane, in the range of 400° to 1500°F. so as to activate the catalyst. Anhydrous HCl and HBr must be used. The activation process is carried out by heating the active alumina, platinum metal-alumina, or platinum metal-halogen-alumina catalyst at a temperature of 400° to 1500°F., preferably 1050° to 1400°F., for a period of at least 10 minutes and up to 100 hours or more, preferably 1/2 to 6 hours, at least 10 minutes, preferably 1/2 hour of heating being in the range of 900° to 1500°F. and at least 10 minutes, preferably, 1/2 hour being in contact with the treating agent. The treating agent may be used alone, or it may be carried in a stream of carrier gas such as nitrogen, methane, ethane, or other gas essentially inert in the treating process. The treating agent may comprise 1 to 100% of the treating gas stream. The activating gas pressure is not critical, but is usually about atmospheric or a convenient higher pressure. The treatment is preferably carried out in a muffle furnace or equivalent apparatus at a temperature in the prescribed range.

By drying the catalyst at a temperature in the range of 900° to 1500°F. for at least 1/2 hour in a dry ambient, the temperature of treatment with the treating gas may be as low as 400°F. Also, the drying and treating may be effected in one step at temperatures in the range of 900° to 1500°F., preferably 1200° to 1500°F. for at least 0.5 hour.

Texaco

A process developed by K.D. Ashley and J.H. Estes; U.S. Patent 3,551,516; December 29, 1970; assigned to Texaco Inc. is one in which a catalyst for isomerization of paraffinic hydrocarbons is prepared by contacting a composite of platinum and alumina with an activating agent selected from the group consisting of carbon tetrachloride, chloroform, methylene chloride, phosgene and trichloroacetyl chloride.

REFORMING

American Cyanamid

A process developed by R.W. Hagy, Z.V. Morgan and R.L. Northcraft; U.S. Patent 3,518,207; June 30, 1970; assigned to American Cyanamid Company is a process for preparing platinum-alumina reforming catalyst of improved activity having platinum uniformly distributed therein which comprises continuously contacting calcined formed particles of alumina with a carbon dioxide saturated platinum-containing solution. In practice, the continuous contacting of the formed particles is carried out by continuously feeding gaseous carbon dioxide into a platinum-containing solution while the platinum-containing solution is being continuously recirculated through a bed of formed alumina particles.

A substantial problem in the preparation of platinum-alumina catalyst, especially catalyst compositions prepared by the impregnation of solid particulate or formed alumina particles, is the achievement of uniform distribution of the platinum within the alumina structure. It has long been recognized that uniform distribution of platinum is important to the efficient utilization of this costly catalytic promoter. Thus, such distribution is important to expose as great an area of platinum surface as possible to make full use of the alumina support and to protect the platinum against loss by attrition.

Typical of the patent literature concerned with procedures for achieving uniform distribution of platinum in platinum-alumina reforming catalyst are U.S. Patents 2,840,514; 2,840,532 and 2,884,382. The first two of these patents are concerned with the addition of various salts as impregnating aids to a platinum-containing solution to assist in obtaining more uniform impregnation. Typical to such impregnating aids are various aluminum salts such as aluminum chloride, aluminum nitrate and the like. With regard to the latter patent mentioned, the process therein disclosed comprised pretreating a porous alumina support material, while in the dry state with gaseous carbon dioxide in such a manner that all of the air in the pores of alumina is replaced by the gaseous carbon dioxide. Thereafter, while maintaining the thus treated alumina out of contact with air, the alumina is impregnated with a platinum solution and the resulting material dried and calcined.

The times required to achieve improved or enhanced impregnation in accordance with U.S. Patent 2,884,382 are substantial, most of the examples describing pretreatments with gaseous carbon dioxide for 30 or more minutes and then while the alumina is in contact with the gaseous carbon dioxide and out of contact with the air, the alumina impregnation with a platinum solution such as chloroplatinic acid, is carried on for substantial periods of time, typically 24 hours or more at room temperature.

Of the methods employed to achieve more uniform distribution of platinum in the alumina particles, that described in U.S. Patent 2,884,382 is most closely related to that of this process. While the method of obtaining uniform impregnating of platinum described in U.S. Patent 2,884,382 has demonstrated merit, it is time consuming and therefore costly, does not consistently achieve a high degree of uniform impregnation, nor does it lend itself to large scale commercial production. The operation of the process may be further illustrated with reference to Figure 117 on the following page.

FIGURE 117: FLOW DIAGRAM FOR THE PRODUCTION OF PLATINUM-ALUMINA CATALYST BY IMPREGNATION

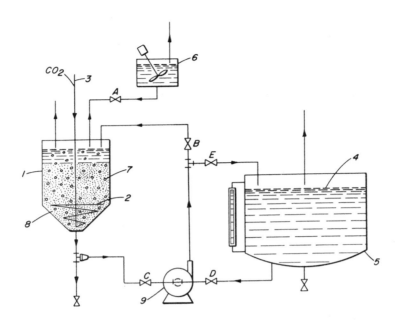

Source: R.W. Hagy, Z.V. Morgan and R.L. Northcraft; U.S. Patent 3,518,207; June 30, 1970

In Figure 117, 1 represents an impregnation tank having a capacity of 110 gallons, containing a sparger or cone shaped carbon dioxide source 2 positioned in the cone shaped bottom thereof and connected with a carbon dioxide source 3. In operating the exemplified system 60 gallons of demineralized water 4 from holding tank 5 is fed to impregnation tank 1 through valves D and B after which 5 gallons of a chloroplatinic acid solution containing 1.96% of platinum by weight and 374 g. $AlCl_3 \cdot 6H_2O$ is fed from tank 6 to the demineralized water. The demineralized water and platinum solutions are mixed and gaseous carbon dioxide is fed from carbon dioxide source (not shown) through line 3 to sparger 2 to saturate with carbon dioxide 7 the impregnating solution, i.e., the demineralized water–chloroplatinic acid mixture.

As indicated above, sparger 2 is cone shaped, constituted by a series of concentric circles of a tubing material, such as polyethylene or polyvinyl chloride typically of rounded cross-section, the smallest circle of which would be at the bottom of the cone and in contact with carbon dioxide source line 3. Each circle will have a number of perforations along its surface, whereby gaseous carbon dioxide may be dispersed uniformly upward through the platinum solution throughout the entire solution. To the carbon dioxide saturated impregnated mixture is added 200 pounds of 1/16" calcined alumina extrudates 8 having an average length of 4 to 7 mm., a surface area of 160 to 220 m.^{2}/g., and a pore volume of 0.5 to 0.7 ml./g.

Simultaneously with the introduction of the extrudates 8 circulation of the carbon dioxide saturated platinum solution is commenced through the bottom of impregnator 1 through line C by means of circulating pump 9, through line B and back into impregnator 1. Circulation of the platinum solution at steady state is preferably such that the volume of impregnator 1 is completely recirculated every 0.8 minute. During steady state operations gaseous carbon dioxide is continuously fed to the platinum solution in the manner indicated so as to continuously maintain the platinum solution saturated with carbon dioxide. It will be seen that while in operation the carbon dioxide is bubbled against or countercurrent to the recirculating flow of the platinum solution.

Normally, impregnation in the above manner is complete in about 1 hour and will result in a volume impregnation determination of about 100%. In the above exemplification, the final catalyst will contain 0.42% platinum and 0.6% chloride, the balance being essentially alumina. When the impregnation step is complete the impregnated catalysts are removed through the bottom of impregnator 1, dried and calcined to form the final product. The solution from which the platinum and chloride content is substantially exhausted is delivered to a storage tank facility for recovery of accumulated trace amounts of platinum.

Chevron Research

A process developed by R.L. Jacobson, H.E. Kluksdahl and B. Spurlock; U.S. Patent 3,487,009; December 30, 1969; assigned to Chevron Research Company involves reforming a naphtha in the presence of hydrogen at low pressures to produce at least 98 F-1 clear octane gasoline with a catalyst composition of a porous inorganic oxide carrier containing 0.01 to 0.3 weight percent platinum, 0.01 to 0.3 weight percent rhenium and 0.001 to 0.1 weight percent iridium. The pressure is maintained below 300 psig, and the feed rate is at least 1.5 LHSV. The hydrogen to hydrocarbon mol ratio is related to the distillation properties of the naphtha processed and to the octane number of the gasoline produced. The reforming process is characterized by the capability of operating on-stream for at least 2,000 hours with no greater than a 2 volume percent decline in the gasoline yield in 2,000 hours.

A process developed by B. Spurlock and R.L. Jacobson; U.S. Patent 3,507,781; April 21, 1970; assigned to Chevron Research Company is a reforming process using a catalyst containing platinum and iridium on a porous solid carrier. The process is started up by contacting the naphtha with the catalyst in the presence of an inert gas, for example, nitrogen. The pressure in the reforming zone should be about 200 psig and the catalyst temperature about 650°F. when the naphtha is first contacted with the catalyst at a space velocity of about 1 v./v./hr. Thereafter, the catalyst temperature is increased to 900°F. over a 2 to 3 hour period while building up autogenous pressure of produced hydrogen.

During the startup period of a platinum-iridium catalyst, that is, when the catalyst is initially contacted with hydrogen and naphtha at reforming conditions, the catalyst causes excessive hydrocracking. Thus, large yields of hydrocarbon gases, for example, methane and ethane, are produced during the early stages of reforming. During this initial reforming period, the selectivity of the catalyst for the production of high octane gasoline is accordingly low. After an initial period of reforming, the hydrocracking activity, or demethanation activity as it is sometimes referred to, of the catalyst decreases to an acceptable level, i.e., to a level at least equivalent to that of platinum catalyst without iridium.

As a consequence of the high demethanation activity of platinum-iridium catalysts, a temperature excursion, or heat front, travels through the catalyst as naphtha is initially contacted with the catalyst in the presence of hydrogen and at reforming conditions. Although the temperature excursion only exists during the initial perid of contact with the naphtha feed, such an excursion could be the cause of a temperature runaway in a commercial reforming plant. The temperature in the bed may increase as high as several hundred degrees above the temperature of the naphtha introduced to the reaction zone. Obviously such a severe temperature increase can damage the reactor and/or catalyst.

One method of controlling the hydrocracking activity or the demethanation activity of the platinum-iridium catalyst would be to add a small amount of sulfur to the feed during the startup period. However, the platinum-iridium catalyst, as indicated previously, is very sensitive to the presence of sulfur. Thus, a sulfur startup procedure would not be advantageous in this case.

It has been found however, that the high temperature excursion normally observed when naphtha and hydrogen are initially contacted with a catalyst comprising catalytically active amounts of platinum and iridium associated with a porous solid carrier can be decreased or substantially eliminated by contacting the catalyst in a reduced condition with a substantially sulfur-free naphtha at reforming conditions and in the presence of an inert gas. Thus, the process for startup of a reforming process conducted in the presence of a reduced platinum-iridium catalyst comprises introducing the catalyst into a treating zone, pressuring the treating zone with an inert gas, preferably nitrogen, to about 200 psig, heating the catalyst to 650°F., and contacting the catalyst with a substantially sulfur-free naphtha at a space rate of about 1 v./v./hr. A naphtha-inert gas mixture withdrawn from the treating zone is passed into a gas separation zone to separate inert gas from the naphtha, and the separated inert gas is recycled to the treating zone. Over a period of 2 to 3 hours, the temperature of the catalyst in the treating zone is raised at a generally uniform rate from 650° to 900°F. and the naphtha feed rate to the treating zone is gradually increased at a generally uniform rate from a space velocity of 1 v./v./hr. to a space velocity of at least 1.5 v./v./hr. The inert gas flow into the treating zone is gradually decreased over the period of 2 to 3 hours while building up autogenous pressure of produced hydrogen, e.g., by dehydrogenation of naphthenes.

Pullman

A process developed by E.F. Schwarzenbek; U.S. Patent 3,230,179; January 18, 1966; assigned to Pullman Incorporated relates to an improved method of preconditioning a platinum catalyst which has been previously regenerated with an oxygen-containing gas, and more particularly, it pertains to a method of prereducing a regenerated platinum catalyst whereby the life of the catalyst is prolonged significantly.

Platinum catalysts which have been regenerated by means of an oxygen-containing gas are prereduced to eliminate from the catalyst any adsorbed oxygen or to convert any oxide of platinum to the metallic form. In either case, the elimination of oxygen results in the production of water, and platinum catalysts, in general, are sensitive to the presence of water at elevated temperatures. Further, it is known that as a result of regenerating the platinum catalyst there is a permanent loss of activity and, therefore, it is important to conduct the regeneration of catalyst under conditions resulting in a minimum loss of activity. Accordingly, it is the purpose of this process to provide a method of regeneration whereby the life of the catalyst is prolonged significantly. By means of the process, it is proposed treating the platinum catalyst which was previously treated with an oxygen-containing gas with a hydrogen-containing gas in a zone maintained at a hydrogen partial pressure of 0.05 to 1.0 psia, based on the inlet conditions, and at an elevated temperature.

Shell Oil

A process developed by R.H. Coe and H.E. Randlett; U.S. Patent 3,278,419; October 11, 1966; assigned to Shell Oil Company relates to platinum catalysts, and more particularly to the treatment of deactivated platinum catalysts to restore the activity and selectivity thereof.

During the course of the catalytic reforming reaction, catalyst activity gradually declines owing to a build-up of carbonaceous deposits on the catalyst and/or a depletion of halogen from the catalyst. Eventually it becomes necessary to regenerate the catalyst by subjecting the catalyst to an oxidizing atmosphere to remove carbonaceous deposits by burning. Halogen can be added to the catalyst during the regeneration procedure or by the addition of a volatile decomposable halogen compound to the feed during operation. Generally, however, carbon burn and/or halogen replenishment fails to restore the catalyst to initial activity and selectivity, or if so, only temporarily, and activity and selectivity decrease at an increased rate during subsequent use of the catalyst. This decreased activity, even with regenerated and halogenated catalyst, is attributed to agglomeration of platinum crystallites. Consequently, the "spent" platinum catalyst is processed for the extraction, separation and recovery of the platinum which is then used for fresh catalyst. This is of course, an expensive operation because of the platinum recovery charges and the cost of manufacturing the catalyst.

It has been proposed to reduce the size of the agglomerated platinum by subjecting the catalyst, after being burned substantially free of carbon, with an oxygen-containing gas under certain conditions of time, temperature and oxygen partial pressure. This procedure, generally referred to as an "air soak", is often only partially effective to favorably alter the size of the platinum crystallite.

This process provides an improved method for restoring the activity and selectivity of "spent" platinum group metal catalyst in a fixed, moving or fluid bed catalytic reforming process. The method comprises the combined steps of (1) pretreating the catalyst during regular operation, (2) regeneration and (3) reactivation. More particularly, the reactivation method comprises the sequential steps of (a) pretreating the catalyst under normal operating conditions, and preferably during such normal operations, with chlorine, (b) contacting the chlorine-treated deactivated carbonized catalyst with an oxygen-containing gas to burn off carbonaceous deposits thereon, and (c) reactivation of the chlorine-treated and decarbonized catalyst by "air soaking". This treatment restores the chlorine content of the catalyst, effects a redispersion of the crystallites of catalytic metal (platinum or palladium) and thus restores the catalyst's activity.

Sinclair Research

A process developed by W.H. Decker; U.S. Patent 3,248,338; April 26, 1966; assigned to Sinclair Research, Inc. is particularly useful in regenerating platinum-metal-alumina or platinum-type metal catalysts containing about 0.01 to 2 weight percent of platinum or other platinum group metal, preferably 0.1 to 0.75 weight percent.

Petroleum refiners have attempted more or less unsuccessfully to develop catalysts having long life or aging stability in order to successfully operate high severity reforming processes without the necessity of frequent regeneration. Since known regeneration systems have been intrinsically expensive in terms of materials and manufacture, there has resulted the need for the development of a regeneration system which meets the requirements of low cost and simplicity of operation.

Also, a regeneration system that eliminates or reduces the danger of carbon monoxide poisoning of reforming catalysts, particularly the platinum-metal-containing catalysts commonly used in reforming, is highly desirable, for such poisoning results in rapid deactivation and subsequent loss in the performance of these catalysts. Conventional regeneration systems normally utilize either gas- or oil-fired generators to produce the inert combustion gas that is circulated through the reactor beds during the regeneration operation. While operating conditions of these units are adjusted to minimize the amount of carbon monoxide present in the inert gas, experience has indicated that even with the most exacting controls, significant amounts of carbon monoxide are present and damage the catalyst. These generators, furthermore, produce an inert gas which is unfortunately saturated with water at high temperatures. This gas must, therefore, be cooled and dried in order to reduce the water content to an acceptable level, inasmuch as water has a deleterious effect on catalyst supports, such as alumina. Accordingly, it is the purpose of this process to produce a regeneration method which alleviates these problems and provides a process which is economical and has simplicity of design.

In accordance with the process, a reactor having a deactivated catalyst bed is isolated from the system and positioned in a circulatory regeneration system whose path includes a heater, the isolated reactor, a flash drum and a gas compressor. The reactor in question is evacuated of hydrocarbons and hydrogen and an inert gas from a storage drum is introduced into the regeneration system and circulated to further denude the reactor of hydrocarbons and hydrogen. The inert gas mixture is then evacuated from the regeneration system and heated inert gas is again circulated through the system until the cooled catalyst in the reactor is at a temperature in excess of 550°F. Liquid hydrocarbons are condensed from the circulating gases by action of the flash drum. Oxygen-containing gas is then introduced and circulated to burn off carbon deposits on the catalyst in the reactor. During the initial stages of this combustion period inert gas from the flash drum is returned to the storage drum. The oxygen-containing gas is then evacuated from the system and the reactor containing regenerated catalyst is returned to the conversion system.

Socony Mobil Oil

A process developed by L.M. Capsuto; U.S. Patent 3,137,646; June 16, 1964; assigned to Socony Mobil Oil Company, Inc. involves the regeneration of particle-form solid platinum-group metal reforming catalyst and, more particularly, preventing deterioration of particle-form solid platinum-group metal reforming catalyst resulting from contact of the catalysts with sulfur dioxide during regeneration by combustion in free oxygen-containing gas of carbonaceous material deposited during an on-stream period.

It is known that all petroleum naphthas contain sulfur. The corrosive effect of hydrogen sulfide on ferrous alloys has also been recognized. Consequently, when designing a reforming unit the sulfur content of the feed to be treated is an important factor in determining the steel to be used in fabrication. Initially, it was solely a question whether it was economically advantageous to fabricate the piping, heaters, and reformers from stainless steel or to treat the reformer feed to remove the sulfur and fabricate the piping, heaters and reactors from low alloy steel. A compromise was effected and when the sulfur content of the naphtha to be treated was less than 200 ppm (parts per million) the naphtha was not pretreated and high alloy steel was used for the reforming reactors, heaters and piping. On the other hand, when the sulfur content of the naphtha to be treated exceeded 200 ppm, provision was made for hydrodesulfurizing the naphtha feed prior to reforming.

Subsequently, it was recognized that, from the aspect of the on-stream time of the reforming unit between regenerations, the nitrogen content of the feed was far more important. As a consequence, naphthas having a nitrogen content in excess of 1 ppm are either diluted with another naphtha to reduce the nitrogen content of the mixture to not more than 1 ppm or the naphtha is hydrodenitrogenated. However, substantially all reformer feeds are hydrodecontaminated by contact in the presence of hydrogen with a hydrogenating catalyst having hydrodesulfurizing and hydrodenitrogenizing capabilities. Nevertheless, the tubes of at least the heater up-stream of the reforming reactor are coated with a sulfide film which, when inert gas containing free oxygen flows therethrough during regeneration of the platinum-group metal reforming catalyst, is partly converted to iron oxide and sulfur dioxide and partly blown through the piping and deposited on the top and through the upper portion of the static catalyst bed in the reactor down-stream of the heater.

To overcome the difficulties attendant upon the deposition of sulfide scale on and in the upper portion of a static bed of platinum-group metal catalyst, the submersion of foraminous baskets in the upper portion of the static bed of platinum catalyst is indicated in U.S. Patent 2,884,372. U.S. Patent 2,792,337 indicates that the oxygen-containing gas be introduced into the catalyst bed without prior contact with parts of the reactor and feed inlet line and to pass part of the gas back through the forepart of the catalyst bed and the feed inlet line of the reactor and the remainder forward through the catalyst bed without recycling any part of the gas through the catalyst bed. U.S. Patent 2,873,176 discloses that difficulties can be avoided by not exposing the sulfide scale in the heater tubes to free oxygen. It suggests passing inert carrier gas through the heater and injecting sufficient oxygen to produce combustion of the carbonaceous material, usually designated as coke, into the carrier gas between the heater and the reactor. U.S. Patent 2,923,679 recommends that the heated oxygen-containing regeneration gases flow through the reforming unit in a direction which is the reverse of the flow of naphtha and hydrogen-containing gas.

The process provides another solution to this problem which for simplicity of operation appeals to operators of reforming units and for economic reasons is very attractive. Whether the naphtha is hydrodecontaminated prior to reforming or contains so little sulfur and nitrogen to justify no hydrodecontamination, the method of avoiding or of substantially preventing the deleterious effect of sulfur dioxide upon platinum-group metal reforming catalyst is useful. Experience has shown that even when low-sulfur, low-nitrogen naphthas are reformed without prior hydrodecontamination in a reforming unit having a plurality of reactors little hydrogen-sulfide corrosion of the heater tubes occurs in the heaters subsequent to the first reactor. Accordingly, the process provides for converting the sulfide scale in at least the heater up-stream of the first of a plurality of reforming reactors to sulfur dioxide and iron oxide and venting the gases to either the inert gas generator stack, to the refinery flare, or to a unit in which the sulfur dioxide can be absorbed or converted to a useful material. The choice of the means of disposing of the sulfur dioxide-containing gases is dependent upon local anti-pollution ordinances and economic considerations.

A process developed by J.W. Payne; U.S. Patent 3,140,263; July 7, 1964; assigned to Socony Mobil Oil Company, Inc. involves rejuvenating spent hydrocarbon conversion catalysts consisting essentially of platinum deposited on a porous solid support. This method comprises rejuvenating a spent platinum supported catalyst, initially containing 0.35 to 1 weight percent platinum deposited on a solid porous support and which has been employed in a hydrocarbon conversion operation, by burning carbonaceous deposit from the spent catalyst in a combustion supporting atmosphere at at elevated temperature, cooling the catalyst after burning to below the boiling point of the platinum compound impregnating solution employed, contacting the cooled catalyst with the impregnating solution of sufficient platinum compound concentration to deposit at least 0.2 weight percent platinum (dry basis) on the catalyst, drying the catalyst by passing a heated gas therethrough and reheating to the hydrocarbon conversion reaction temperature.

Universal Oil Products

A process developed by A.R. Greenwood and K.D. Vesely; U.S. Patent 3,647,680; March 7, 1972; assigned to Universal Oil Products Company is a continuous reforming-regeneration process comprising one or more reforming reactors and employing a platinum catalyst wherein catalyst activity is maintained at a predetermined level by continuous regeneration thereof without the removal of any reactor from the process stream.

Prior art reforming processes generally comprise one of two types of operation, i.e., a nonregenerative type and a regenerative type. In the nonregenerative type of operation, the catalyst is maintained in continuous use over an extended period of time, i.e., from 5 months to 1 year or more depending on the quality of the catalyst and the nature of the feed stock. Following this extended period of operation, the reforming reactor is taken off stream while the catalyst is regenerated or replaced with fresh catalyst. In the regenerative type of operation, the catalyst is regenerated with greater frequency utilizing a multiple fixed bed reactor system arranged for serial flow of the feed stock in such a manner that at least one reactor can be taken off stream while the catalyst is regenerated or replaced with fresh catalyst, one or more companion reactors remaining on stream, or going on stream, to replace the off stream reactor. Subsequently, the regenerated fixed bed reactor is placed on stream while another is taken off stream and the catalyst bed is regenerated or replaced with fresh catalyst in like manner.

It is apparent from the brief description of prior art nonregenerative and regenerative reforming processes that both means of operation embody certain undesirable features. For example, in the nonregenerative type of operation, the entire plant is usually taken off stream to effect regeneration or replacement of the catalyst with a resultant significant loss in production. Further, the nonregenerative type of operation is characterized by a continuing decline in catalyst activity during the processing period requiring an operation of increasing severity to maintain product quality, usually at the expense of product quantity. In the regenerative type of operation utilizing a multiple fixed bed reactor system, or "swing reactor" system, similar problems are encountered although to a lesser degree. However, the start up and shut down procedures relating to insertion and removal of a reactor in the process stream are unduly complicated and require a complex system of valves, lines and other equipment to accomplish reactor changer over with a minimum loss of process time.

It is therefore desirable to provide a reforming process which would substantially obviate the undesirable features of prior art nonregenerative and regenerative type reforming processes. More specifically, it would be desirable to have a reforming process whereby a predetermined high level of catalyst activity and stability is maintained without resorting to the removal of the reforming reactor from the process stream and a consequent loss of process time.

In the preferred system using stacked reactors, the reactor system has in effect a common catalyst bed moving as a substantially unbroken column of particles from the top reactor to the bottom reactor. Thus, used catalyst is withdrawn from the bottom reactor while regenerated catalyst is added to the top reactor, and the catalyst of the reactor system is regenerated with all reactors remaining on stream at reforming conditions. The flow diagram in Figure 118, on the following page, shows the essential elements of the process.

A straight-run gasoline fraction boiling in the 200° to 400°F. range is charged to the process through line 1 at a liquid hourly space velocity of 2.0, entering a heater 2 in admixture with a hydrogen-rich gas stream recycled through line 3 from a product separator (not shown). In a conventional reforming operation, a considerable excess of hydrogen is admixed with the hydrocarbon charge to minimize carbon formation on the catalyst. Typically, hydrogen is employed in a 10:1 mol ratio with the hydrocarbon charge. However, the method wherein the catalyst is subjected to frequent regeneration, affords a substantial reduction in hydrogen consumption. Preferably, the hydrogen-hydrocarbon mol ratio is from 1:1 to 5:1. Thus, in this illustration, the heated combined stream comprises hydrogen and hydrocarbon in a mol ratio of 3:1, the heated combined stream being withdrawn from the heater 2 by way of line 4 and charged into the upper portion of reforming reactor 5.

Reforming reactor 5 is shown in vertical alignment with reforming reactor 11, with an intermediate heater 10. The reforming catalyst charged to reactor 5, as hereinafter described, is comprised of spherical 1/16" diameter particles containing 0.375 weight percent platinum and 0.9 weight percent combined chlorine, the remainder being alumina. The reactor temperatures are maintained in the 850° to 950°F. range and the pressure at about 200 psig. Reforming reactor 5 is shown with catalyst confined in an annular moving bed 6 formed by spaced cylindrical screens 7. The reactant stream is passed in an out-to-in radial flow through the catalyst bed, the reactant stream continuing downward through the cylindrical space 8 formed by the annular bed 6 and exiting to heater 10 by way of line 9.

FIGURE 118: FLOW DIAGRAM OF CONTINUOUS PLATINUM REFORMING–REGENERATION PROCESS

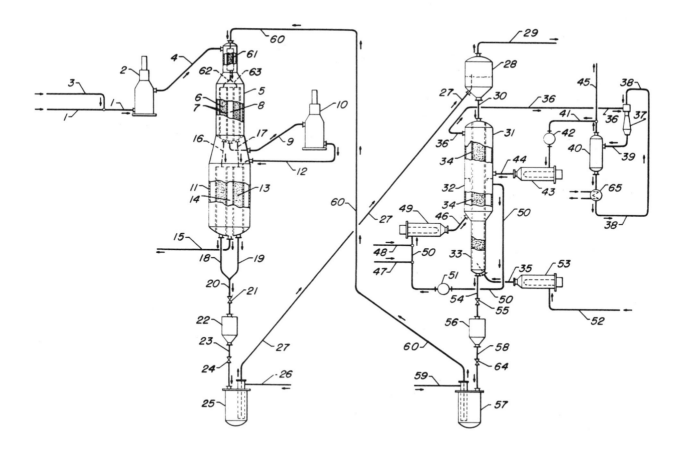

Source: A.R. Greenwood and K.D. Vesely; U.S. Patent 3,647,680; March 7, 1972

Since the reaction is endothermic, the effluent from the reactor 5 is reheated in heater 10 and thereafter charged to re-actor 11 through line 12. Again, the reactants stream is passed in an out-to-in radial flow through the annular catalyst bed 13 substantially as described with respect to reactor 5, the flow passing downward through the cylindrical space 14 and passing from reactor 11 by way of line 15. The reactor effluent withdrawn through line 15 is passed to conventional product separation facilities for recovery of high octane product, for example, a reformate having a clear octane number rating of about 95, and recovery and recycle of a hydrogen-rich gas stream to the reactor system.

The catalyst particles descending through the reactor 5 as an annular moving bed 6, are continued to the annular moving bed 13 of reactor 11 by way of catalyst transfer conduits 16 and 17. Conduits 16 and 17 represent a multitude of catalyst transfer conduits permitting passage of the catalyst between the annular beds 6 and 13 and effecting a suitable pressure drop whereby substantially all of the reactant stream from reactor 5 is directed through the heater 10 by way of line 9 with only a minimal amount by-passing the heater 10 and passing directly to reactor 11 together with the catalyst flow through conduits 16 and 17. Thus, in effect, the reactor system has a common catalyst bed moving as a substantially un-broken column of particles through the top reactor 5 and the bottom reactor 11.

In the example, the used catalyst is withdrawn through lines 18 and 19 at a rate such that the catalyst inventory of the reactor system is replaced in 30 day cycles. The catalyst is withdrawn intermittently through line 20 and by means of control valve 21 to effect a moving bed type of operation, the catalyst being discharged into a lock-hopper 22 for sepa-ration of residual hydrocarbon therefrom. The used catalyst is subsequently transferred through line 23 and control valve 24 to a lift engager 25 to be lifted in a nitrogen stream to a disengaging hopper 28 by way of line 27. The catalyst is lifted to hopper 28 by a flow of nitrogen charged to the lift engager 25 through line 26 from an external source (not shown). Nitrogen is charged to the lift engager 25 at the rate of 531 scfh and at a temperature of 100°F. The used catalyst de-posited in the disengaging hopper 28 comprises 0.7 weight percent combined chloride and 2 to 5 weight percent carbon.

An overhead line 29 is provided to vent the disengaging hopper 28 to the atmosphere or to recycle the nitrogen. Catalyst particles from the disengaging hopper 28 are fed through line 30 to a catalyst regenerator comprising a carbon burn–off zone 31, a chlorination zone 32, and a drying zone 33. The catalyst particles are processed downward, as a moving bed or column in a confined regeneration zone 34, passing from the carbon burn–off zone 31 to the chlorination zone 32. Chlorinated and substantially carbon–free particles are then continued downward through the drying zone 33 and contacted therein with a hot, dry air flow to effect the separation of excess adsorbed gaseous components from the catalyst.

In this instance, the catalyst particles are passed from the disengaging hopper 28 to the catalyst regenerator at an average rate of 200 pounds per hour. The catalyst particles are processed downward through the carbon burn–off zone 31 at a rate to establish an average residence time therein of 2 hours.

In the carbon burn–off zone 31, the catalyst particles are heated in contact with an oxygen–containing gas including hot recycle gases charged to the carbon burn–off zone 31 at a gaseous hourly space velocity of 4700 by way of line 44. The oxygen–containing gas is derived from air charged to the drying zone 33 by way of line 35, the air becoming admixed with steam, chlorine, and HCl on passing upward through the chlorination zone 32, and the gaseous mixture continuing upward through the carbon burn–off zone 31 in contact with the catalyst particles contained therein. The resulting gaseous products, including oxides of carbon and sulfur, are withdrawn from the carbon burn–off zone 31 as flue gases through line 36.

In some cases, it may be desirable to separate sulfur components from the flue gases prior to recycle. In that event, the flue gases are charged to a scrubber 37 wherein they are admixed with a caustic stream recycled from a caustic settling tank 40 through a cooler 65 and line 38. The mixture is then passed to the caustic settling tank via line 39. The resulting flue gases, substantially free of halogen and oxides of sulfur, are recovered overhead from the settling tank 40 through line 41 and charged by means of a blower 42 through a heater 43 and line 44 to the carbon burn–off zone 31 as aforesaid. The heater 43 is provided for start–up operation. The gases recycled to the carbon burn–off zone 31 comprise 0.7 weight percent oxygen to effect a controlled burning at a temperature of from 830° to 930°F. in the burn–off zone. An overhead vent line 45 is provided to discharge excess flue gas from the process.

The catalyst particles, substantially free of carbon are processed downward from the carbon burn–off zone 31 through the chlorination zone 32 at conditions to effect an average residence time therein of about 1 hour. In the chlorination zone 32, the catalyst particles are brought in contact with a 2 : 1 mol ratio of steam and chlorine charged thereto through line 46, the steam and chlorine being admixed with air passing upward from the drying zone 33 as previously mentioned. Steam is charged to the system at a temperature of 450°F. and at a rate of 2.4 pounds per hour through line 47. The steam is passed through line 50 together with recycle vapors contained therein and admixed with chlorine from line 48, the chlorine being charged at a rate of 1.45 pounds per hour. The steam–chlorine mixture is heated to 930°F. in heater 49 and charged to the chlorination zone 32 through line 46 at a gaseous hourly space velocity of 4700. A recycle stream comprising excess steam and chlorine is withdrawn from the chlorination zone 32 by way of line 50 and circulated by means of a blower 51 as a portion of the steam–chlorine charged to the chlorination zone 32.

From the chlorination zone 32, the catalyst particles are continued downward through the drying zone 33 whereby vaporous components are stripped from the catalyst by a flow of dry air. The air is charged to the system through line 52 from an external source (not shown) and heated to 800° to 1000°F. in heater 53 prior to being charged to the drying zone 33 through line 35. The air is charged at a gaseous hourly space velocity of 150.

The catalyst particles are withdrawn from the regenerator at regular intervals by way of line 54 through control valve 55 and collected in lock–hopper 56. The catalyst is subsequently transferred to a lift engager 57 through line 58 and control valve 64 to be conveyed to the reforming reactor 5. The catalyst particles are carried through line 60 by a dry, pure hydrogen stream introduced to the lift engager 57 by way of line 59 at 3300 standard cubic feet per hour and at a temperature of about ambient. The hydrogen is subsequently used as a reducing gas and as a portion of the hydrogen feed to the reforming reactor 5.

Prior to direct contact with the reactant stream in reforming reactor 5, the catalyst, in admixture with hydrogen, is processed in a dense phase through a reducing zone 61 to effect an indirect heat exchange with hot reaction gases charged to the reactor. The catalyst is processed downward through the reduction zone 61 at a rate to establish residence time of about 2 hours therein at a temperature of 950° to 1000°F. The resulting reduced catalyst is thereafter added to the catalyst bed through lines 62 and 63 replacing that withdrawn from the reactor system through lines 18 and 19 for regeneration. The method finds particular application with respect to low pressure reforming. While low hydrogen partial pressures favor the main octane–improving reactions, e.g., dehydrogenation of paraffins and naphthenes, a principal objection to low pressure reforming is in the excessive formation of carbon resulting from condensation and polymerization reactions also favored by low hydrogen partial pressures. However, the continuous reforming–regeneration process obviates this objection, and the relative catalyst instability resulting from carbon formation is no longer a limiting factor to a successful low pressure reforming operation. An added advantage derived from the process is in the increased and continued supply of hydrogen available for hydrogen–consuming refinery operations, such as hydrocracking.

FUTURE TRENDS

The trends in the catalyst field are as follows:

In manufacture, the trend is to the production of catalyic shapes particularly adapted to process conditions. This may mean honeycomb catalysts particularly adapted to automotive exhaust treatment; it may mean the production of mechanically-stable microspheres particularly adaptable to a moving-bed catalytic process.

In recovery, the trends are toward minimizing recovery procedures on the basis of reducing investment and operating costs. On the other hand, the trend is to maximizing the efficiency of recovery procedures for scarce and precious catalyst components and also to maximize removal of contaminants from plant waste streams such as heavy metals from catalysts.

In use, the trend to more and more specific catalytic process continues. Aside from chemical synthesis, the major growth area for catalyst application lies in the field of pollution control with particular emphasis on control of automotive air pollution.

COMPANY INDEX

The company names listed below are given exactly as they appear in the patents, despite name changes, mergers and acquisitions which have, at times, resulted in the revision of a company name.

INVENTOR INDEX

U.S. PATENT NUMBER INDEX

NOTICE

Nothing contained in this Review shall be construed to constitute a permission or recommendation to practice any invention covered by any patent without a license from the patent owners. Further, neither the author nor the publisher assumes any liability with respect to the use of, or for damages resulting from the use of, any information, apparatus, method or process described in this Review.

Complete copies of the patents described in this book may be obtained at a cost of $0.50 per patent prepaid. Address order to the Commissioner of Patents, U.S. Patent Office, Washington, D.C. 20231.

1

TOBACCO SMOKE FILTERS 1972

by Dr. N. E. Bednarcyk

Cigarette smoke is an aerosol maintained in a gaseous phase made up from air and the combustion products of tobacco and paper wrapping.

For a smoke filter to achieve commercial production, it must fulfill several stringent criteria. It must have a low unit cost, and it must be capable of being applied to the cigarette at a high and fast rate. In the performance area it must have an easy draw, be efficient, and not permit substances removed from the first few puffs to be eluted by the last puffs. Also, it must have a minimal effect on the flavor of the smoke.

While filter-tipped cigarettes have been on the market for many years, research toward producing more effective and selective filters has increased substantially since the U. S. Surgeon General's 1964 report. That the 184 patents covered in this book were issued over a little more than a five-year period, is evidence of this activity.

A partial and much abbreviated table of contents follows here. The wealth of information provided by this publication is easily seen by reference to the numbers shown in parentheses. These numbers indicate the number of processes covered for each filter type, method, or component.

PART I—NONSELECTIVE FILTERS AND ADDITIVES
1. MECHANICAL FILTERS
   Removal of Smoke Components by Impingement (8)
   Membrane-Containing Filters (7)
   Ventilation Filters (5)
   Miscellaneous Mechanical Devices (12)
2. POLYMERS AND CELLULOSE DERIVATIVES
   Polymers and Polymer Fibers (7)
   Polymer Foams (4)
   Cellulose Derivatives (7)
   Plus Other Components (4)
3. OTHER MATERIALS OF CONSTRUCTION
   Glass-Micropore and Fibers (4)
   Natural Materials (5)
   Various Fibers (3)
4. FRANGIBLE CAPSULES IN FILTERS
   Improved Filtration Effects (6)
   Encapsulated Fluid Release (5)
   Other Effects from Capsule Fluids (5)

5. CHARCOAL BONDING AGENTS (9)
   Polyolefin
   Ethylene Copolymer
   Polyoxyethylene Glycol
   Polyalkylene Oxide
   Polyalkylene Glycol
   Cyanamide-Cellulose Acetate Tow
   Water-Soluble Resins
   Vinylidene Polymer Plus Thermosetting Aminoplast
   Cellulose Crystallite Aggregates
6. VARIOUS FILTER ADDITIVES
   Sugar Esters (3)
   Others (4)
   Activated Carbon
   Expanded Polystyrene
   Honey and Other Viscous Liquids
   Beeswax and Similar Materials
7. TOBACCO ADDITIVES AND SUBSTITUTES
   Pure Additives (5)
   Substitutes (3)

PART II—SELECTIVE FILTERS AND ADDITIVES
8. INORGANIC SALTS
   Phosphates, Phosphites, Carbonates (8)
   Other Salts (3)
9. METALS AND METAL OXIDES
   Zinc Compounds (3)
   Oxides of Other Metals (4)
10. NITROGEN COMPOUNDS
    Amines and Amides (2)
    Other Nitrogen Compounds (7)
11. IONIC & ELECTROSTATIC FILTERS
    Ion Exchange Resins (4)
    Electrostatic Filtering Media (2)
    Use of Electrets (2)
12. VARIOUS SELECTIVE ADDITIVES
    Alcohols and Esters (3)
    Nitrogen, Phosphorus, Sulfur (3)
    Organic Compounds (5)
    Multicomponent Filters (3)

PART III—FILTER MANUFACTURE
13. FILTER BODIES (8)
    Bonding Agents (4)
    Processes for Materials
14. INCORPORATION OF ADDITIVES (8)

263 pages

$36

FINE DUST AND PARTICULATES REMOVAL 1972

by H. R. Jones

Pollution Control Review No. 11

This book concerns itself with the control of airborne particles which measure from 0.01 to 2.0 microns. One micron is one millionth (10^{-6}) part of a meter. Submicron particles in polluted air are controlled hardly at all by gravitational forces. Once they enter the atmosphere, their residence time is likely to range from a month to several years.

While the generation of natural fine dusts is just as large, particulates from man's activity contribute significantly to all the major aspects of air pollution. How to reduce such emissions and how to remove the particulates at the source, is described in this book which is based mainly on government reports and on recent U.S. patents.

A partial and abbreviated table of contents follows here.

Introduction
1. SOURCES OF EFFLUENTS
2. EFFLUENT CHARACTERISTICS
   Particulate Characteristics
   Carrier Gas Characteristics
   Particulate Sampling
   Size Measurements
   Adverse Effects on Health
   Effects on the Atmosphere
3. REMOVAL EQUIPMENT
   Cyclones
   Wet Scrubbers
   Electrostatic Precipitators
   Fabric Filters
   Mist Eliminators
   Afterburners
4. ECONOMICS OF CONTROL
   Removal and Disposal Systems
5. ELECTRIC POWER GENERATION
   Emission Sources and Rates
   Effluent Characteristics
   Fine Particle Emissions
   Control Practices and Equipment
6. FOREST PRODUCTS INDUSTRIES
   Sawmill Operations
   Plywood and Particleboard Plants
   Pulp Industry
   Bark Boilers
   Recovery Furnaces

7. INORGANIC CHEMICAL INDUSTRY
   Lime Manufacture
   Rotary Kilns
   Crushing and Screening
   Sulfuric Acid Manufacture
   Phosphoric Acid Manufacture
   Fertilizer Manufacture
   Phosphate Fertilizers
   Ammonium Nitrate Fertilizers
   Aluminum Silicate Pigment Manufacture
   Magnesium Chloride Concentration Methods
   Sodium Carbonate Manufacture
8. ROCK AND CLAY INDUSTRIES
   Crushed Stone
   Sand and Gravel
   Cement Manufacture
   Emission Rates
   Secondary Sources
   Clay Products
   Refractories
9. IRON AND STEEL INDUSTRIES
   Sinter Plants
   Blast Furnaces
   Open Hearth Furnaces
   Basic Oxygen Furnaces
   Electric Furnaces
   Iron Foundries
   Control Processes
10. NONFERROUS METALS INDUSTRIES
    Cloth Filters
    Centrifugal Separators
    Electrostatic Precipitators
    Primary Lead Smelting
    Primary Zinc Smelting
    The Aluminum Industry
11. PETROLEUM REFINING
12. CONCLUSIONS AND RESEARCH RECOMMENDATIONS
13. FUTURE TRENDS
    Major Pollutant Sources
    Improved Control Devices
    Currently Installed Equipment
    Projection Methods
    Results
    Projections of Fine Dust Emissions

307 pages

$36

POLISHING COMPOSITIONS AND MATERIALS 1972

by J. Partridge

A review of 127 U.S. patents on polishing materials and covering many formulations.

The word "polish" as used here and in much of the literature often signifies a composition which cleans and coats a surface, in addition to polishing it, which usually means to make it smooth and shiny by abrasion or chemical action.

Consumption of polishes is high and growing substantially.

A partial table of contents follows:

1. AUTOMOBILE POLISHES
   SILICONES (5 Processes)
   Polyalkylsiloxane, Thickener, Propellant, Montan Wax, Microcrystalline Waxes and Liquid Silicones
   AMINES (7)
   Cationic and Nonionic Surfactants
   Monoesters with 6 to 12 Carbon Atoms, Acetic Acid, Salts of Fatty Amines, Hydrotrope, Alcohol and Water
   PARAFFIN OIL (1)
   FLUOROCARBON MATERIALS (1)
2. FLOOR POLISHES
   PHENOLIC COMPOUNDS (2)
   POLYETHYLENE (2)
   Paraffin Wax, Oxidized Wax, and Polyethylene
   HALOGENATED ORGANIC COMPOUNDS (3)
   Perfluoroalkyls
   Polychloroalkylbenzene Waxes
   AMINES (4)
   Volatile Amines + Triethanolamine
   Polyvinyl Ether Wax, Anionic and Nonionic Emulsifiers plus Resin
   SILICONES (1)
   ACRYLATES (2)
   ORGANIC PHOSPHORUS COMPOUNDS (4)
   Polycarboxylic Resin, Alkaline Earth Salt, Phosphonate Chelating Agent and NH₄OH
   POLYVALENT METAL CHELATES (1)
   PRESSURIZED POLISHES (2)
   OTHERS (6)
   Trimellitic Anhydride + Alkylene Oxide Resin
   Lactic Acid, Methanol, Hydrogen Peroxide, and Ammonia

3. GLASS AND GEM POLISHES (13)
   Hydrous α-Ferric Oxide
   Bastnaesite (Mineral Fluocarbonate)
   Zirconium Compounds
   Fatty Acid Esters of Glycols
4. METAL POLISHES
   SILVER AND COPPER (17)
   Elemental Sulfur
   Organic Sulfides
   Thioethers
   Polishing Cloths
   CHROMIUM AND STEEL (4)
   Gersthofen Wax + Silicone Oil
   Waxes, Solvents, Triethanolamine, Oleic Acid, Ammonia, Abrasives
   OTHER METALS (3)
   Graphite, Pumice + Bodied Grease
   Aluminum Oxide
5. LEATHER AND PLASTICS (6)
   Paraffin Waxes
   Alkylamines
   Polysiloxanes
6. SEMICONDUCTOR POLISHES (5)
   Chelating Agents
   Silica and Anatase
7. FURNITURE—HOUSEHOLD POLISHES (5)
   Silicone Fluids
   Polyvinyl Ether Waxes
   Dextran
   Fine Particle Abrasives
8. MULTIPLE USE POLISHES
   AMINES, AMIDES, NITRILES (8)
   SILICONES (5)
   ORGANOPHOSPHORUS COMPOUNDS (2)
   COMPONENTS AND COMPOSITIONS (18)
   Bacteriostatic Hydroperoxides
   Microcrystalline Waxes
   Paraffin Waxes and Solvents
   Ethyl Cellulose Latex
   Polymers of Glycidyl Ethers
   Distillation Residues from Straight Chain Alcohols
   Esters of High Molecular Weight Carboxylic Acids
   Vinyl Stearate-Maleic Anhydride Copolymers
   Amine-Modified Polyolefin Wax Adducts
   Wax Acids from Guerbet Alcohols
   Strewable Waxes
   Paraffin Wax Emulsion Creams

204 pages

$36

PLASTIC CONTACT LENSES 1972

by S. Summerville

A major fraction of U.S. patents on plastic lenses, issued within the past few years, deals with contact lenses. Many of these discuss work in connection with flexible lenses or the so-called "soft" lenses, among which are the hydrophilic lenses placed on sale in 1971. According to one business publication, "soft" contact lenses is one of the hottest fields Wall Street has seen in years.

Compared to glass, plastic is lower in weight and will not shatter when broken, among other desirable properties. For contact lenses conventional hard plastics have two of glass' shortcomings, i.e. difficulty in obtaining a proper fit and lack of permeability.

Flexible contact lenses may, however, be the solution to the problem of fit and hydrophilic polymers may solve the problem of water permeability as well as fit. Gas permeability, particularly oxygen and carbon dioxide permeability, is another and equally important problem in the use of contact lenses.

It is estimated that present wearers of contact lenses represent less than 10% of the potential U.S. market.

108 U.S. patents are reviewed in this book. A partial list of contents follows. The numbers in ( ) indicate the number of processes for each topic.

1. HYDROPHILIC MATERIALS (3)
   Hydrophilic Silicone Rubber
   Hydratable Polymers
2. FABRICATION OF HYDROPHILIC LENSES (11)
   Shaping of Xerogels
   Polymerization in Rotating Mold
   Aspherical Convex Surface
   Shaping by Temporary Deformation
   Opaque Interlayer
   Hard Insert in Hydrophilic Body
3. BIFOCALS AND MULTIFOCALS (7)
   From Monolithic Mass
   Lens-Orienting Metal Insert
   Fused Bifocal Segment

4. DESIGN CONSIDERATIONS (6)
   Bifocal Lenses with Colored Zones
   Fusion of Materials with Different Refractive Indexes
5. MOLDING AND POLYMERIZATION OF SINGLE FOCUS LENSES (8)
   Plastic-Filled Pocket Between Films
   Poly (4-Methyl-1-Pentene) Lens with Hydrophilic Rim
   Nontransparent Annular Zone
6. FINISHING OF SINGLE FOCUS LENSES (8)
   Reduction of Stresses
   Continuous Grinding
   Deformable Rotating Surface
   Shadow Picture Control
7. MOLDING APPARATUS (3)
   Eye Molds
   Blanks from Sheet Plastic
8. CUTTING, GRINDING, POLISHING (13)
   Lap Formation from Eye Impression
   Ellipsoidal Grinding
   Altering Lens Power
   Edging with Optical Comparator
9. LENS MATERIALS (4)
   Fluorescent Lenses
   Dimethylvinylsilyl Copolymers with Filler
   Material Isotonic with Tears
   Polyelectrolyte Resins
10. FLEXIBLE LENSES (7)
    Fluid-Filled Lenses
    Pliable Scleral Portion
    Radial Fluid Passages
    Hydrophilic Rim on Hydrophobic Body
11. CURVATURE TECHNIQUES (13)
    Recessed Optical Zones
    Conicoids for Both Surfaces
    Symmetrical Inner and Asymmetric Outer Zones
12. INTENTIONAL COLORATION OR OCCLUSION (4)
13. LENSES WITH HOLES, DUCTS, INDENTATIONS, FLANGES AND PROTUBERANCES (7)
14. DEVICES (14)
    Telescopic System for Subnormal Vision

268 pages

$36

COPY PAPERS 1972
by S. G. Weiss

Some 10 years after World War II, when science and industry were hampered by insatiable demands for information, xerography and a well-developed thermography entered the scene. Within a short time it became surprisingly easy and fast to copy the printed or written page, and to obtain a clear and distinct image. The need for old-fashioned carbon paper virtually vanished. Advances in image copy techniques sometimes make it difficult to differentiate between original and copy—with the copy showing improved contrast over the original.

There has been no noticeable slackening in the pace of technological development of copying methods. It is estimated that within 5 years the office copying market will be double its present size.

A review of 275 U.S. patents. A partial table of contents follows. Numbers in ( ) indicate a plurality of processes per topic.

1. PHYSICAL THERMOGRAPHY
Procedures (7)
IR and UV Radiation Systems
Heat-Sensitive Sheets
Transparentization (19)
Fusible Copolymer Films
Change of Phase or Form (15)
From Crystalline to Tacky Amorphous
States
Thermal Opacification
Diffusion Processes (11)
Vapor Transfer from Microcapsules
Permeability Variations

2. CHEMICAL THERMOGRAPHY
Silver Soaps and Reducing Agents (13)
Basic System
Modified Toners
Oxidation-Reduction Systems (8)
Carbohydrates and o-Anisidine
Amine Molybdates
Coupling Reactions for Dyes (12)
Metal Color Forming Agents (14)
Phenolic Compounds (12)
Pyran and Pyrrole Derivatives (6)
Amines (5)
Ninhydrin + Amine Ligands
Indan Derivatives

Polyhydroxy Compounds (3)
Decomposition Reactions (6)
Reduction of Decomposition Temperatures
Thermal Decomposition of Clathrates

3. ELECTROGRAPHY
Electrostatic Image Development (10)
Dry Powder Processes
Liquid Processes
Voiding Electrostatic Images (3)
Non-copyable Photochromic Papers
Organic Photoconductive Coatings (11)
Quaternary Electroconductive
Compounds (9)
Water-Based Coatings (11)
Zinc Oxide-Based Coatings (7)
Unusual Electrographic Methods (3)
Latent Free Radicals
Electro-Gas-Dynamic Image
Chemical Frost Image

4. DIAZOTYPES
Thermal Development (16)
Coatings (7)
Heat-Coalescible Polymers
Boron Compounds for Precoating
Image Speed and Stabilization (6)
Encapsulation Techniques (4)
Transparentization (2)

5. DRY PHOTOSENSITIVE SYSTEMS
Dry Development Systems (8)
Leuco Dye Bases (5)
Polymers and "Blushing" (4)
Image Stabilization (3)

6. PRESSURE-SENSITIVE SYSTEMS
Contact Imaging (13)
Microencapsulation (4)
Stabilization (5)

7. VESICULAR SYSTEM
Vesicular Sheets & Materials (4)

8. ELECTROSENSITIVE SYSTEMS
Sheets & Recording Materials (3)

9. DIFFUSION & MICROWAVE SYSTEMS
Various Diffusion Processes (3)
Microwave Process (1)

266 pages

$36

PRINTING INKS 1972
by A. Williams

Printing ink production in the U.S.A. now exceeds one billion pounds per year, supplying an extremely complex market. The printing industry is one of the largest consumers of pigments due to an increasing demand for color.

The mounting emphasis on pollution control has required many formulation changes and particularly the use of water-based systems in flexographic inks. Solvent-free systems are being developed, and "curing" by high energy radiation and ultraviolet is being studied.

Specialty inks (magnetic inks, conductive and fluorescent inks, etc.) constitute ca. 20% of the U.S. market, while the bulk of the supply goes into newsprint (letterpress), lithographic, rotogravure, and flexographic inks.

159 Processes based on 184 U.S. patents. Numbers in ( ) indicate numbers of processes per heading in the partial table of contents that follows.

1. FLEXOGRAPHIC INKS
General Formulations (7)
Polyester Overprint Varnishes
Bank Check Ink
Safety Inks
Polyamides (7)
Polymeric Fatty Acids
Polyamide-Polyimide Resins

2. GRAVURE & MOISTURE SET INKS
Gravure Inks (7)
Rosin-Based Inks
Water-Color Inks
Simultaneous Colors
Moisture Set Inks (8)
Shellac Base
Limed Rosin
Nonionic Surfactants

3. LITHOGRAPHIC VARNISHES
Drying Oil Vehicles (11)
Rosin Salts
Cobalt Borate Drier
Tung Oil Copolymers
Heat-Set Varnishes (5)
Gelled Linseed Oil
Zein Vehicle
Metallic Ink
Other Vehicles (3)

4. NEWSPRINT INKS
Carbon Black (8)
Pelletized Furnace Black
Sulfonated Carbon Black
Pigmented Masterbatches
Additives (5)
Antimisting Agents
Dispersing Agents
Rub-Off Control
Emulsions (2)
Tall Oil Monoglycerides

5. INK ADDITIVES
Pigments (13)
Azo Compounds
Naphthindolizinediones
Stabilized Phthalocyanines
Triaryl Methane Dyes
Cr and Co Alcohol-Soluble Salts
Colorforming Compositions
Crystal Violet Lactone
Trihalostannates and Germanates
Treated Pigments (6)
Coated Kaolinite Particles
Amine-Treated Clays
Hydrophobic Silica
Titanium Hydrate Treatment
Silane-Treated Pigments
Dispersing Aids (4)
Dialkyl Sulfosuccinates
Thiophosphatosuccinates
Anti-Offset Agents (4)
Starch + Drying Oils
Polysaccharides + Fluorocarbons
Various Additives (7)
Antiskinning Agents
Control of Souring
Nonabrasive Inks

6. SPECIALTY INKS
Hectographic Transfer Inks (13)
Magnetic Inks (5)
For Printing on Polyolefins (8)
For Printing on Fluorocarbons (3)
Stencil Inks (7)
Conductive Inks (4)
Fluorescent Inks (4)
Textile Printing Inks (5)
Various Specialties (13)

293 pages

$36

FOOD CANNING TECHNIQUES 1972
by M. Gutterson
Food Processing Review No. 26

Canning, as opposed to frozen food preservation, makes use of metal and glass containers almost exclusively and relies on heat processing of food in hermetically sealed containers. Much research has been done toward flavor preservation and elimination of harmful microorganisms, making long term nonrefrigerated storage a present-day reality.

107 Patent-based processes. A partial table of contents is given here. Numbers in ( ) indicate numbers of processes per subject.

1. GENERAL TECHNIQUES
CANNING SYSTEMS (8)
High Temperature Radiation
Heating in Fluidized Bed
Hydrostatic Cooking
Radio Frequency Sterilization
Microwave Sterilization
ASEPTIC CANNING (3)
Chemical Bactericides
Superheated Steam
Infrared Radiation
CLOSURES (4)
Thermoplastic Caps
Easily Opened Containers
Vacuum Seals
Reclosable Seals
CAN COATINGS (4)
Butadiene + α -Methylstyrene
Palladium for O₂ Removal
DEOXYGENATION (4)
Oxygen-Permeable Barriers
Bisulfites
Water Synthesis Catalyst
Enzymatic Oxygen Removal
CONTAINERS (3)
CHEMICAL PRESERVATION (4)
Alkyl p-Hydroxybenzoates
Benzohydroxamic Acid
Hydrochloric Acid

2. MEAT AND POULTRY
WHOLE PIECES & CHUNKS (7)
Eliminating Air Pockets
Inserting Meat into Cans
COMMINUTED MEAT (6)
Use of Ascorbic Acid
Plus Sodium Nitrate
Alginate and Mg Salts

SMOKE & SMOKE FLAVORS (2)
MEAT RELEASE AGENTS (1)
SEMICOOKED MEAT (1)
BARBECUED RIBS (1)
POULTRY (2)
Breaded Fried Chicken

3. SEAFOOD
PREPARATION OF FISH (4)
Degassing
Storage under CO₂
PRECOOKING (3)
FISH CANNING SYSTEMS (4)
FISH PRODUCTS (3)

4. VEGETABLES, FRUITS, JUICES
CANNING SYSTEMS (2)
TOMATOES (4)
Aseptic Filling
Acid Treatment
CITRUS PRODUCTS (4)
Low Pressure Treatment
Flavor Improvement
APPLES (1)
CREAMED CORN (1)
PICKLES (4)
PEAS—Color Preservation (1)
OLIVES (2)

5. DAIRY PRODUCTS
MILK (8)
Prevention of Thickening
Elimination of Cooked Flavor
Viscosity Control Agents
Infants' Formulations
Condensed Cream (2)
CHEESES (2)

6. BEVERAGES
BEER AND ALE (2)
SOFT DRINKS (2)
Air-Free Packaging
Prevention of Can Corrosion
CANNING WATER (1)
Retaining Freshness

7. DOUGHS & BAKED GOODS (4)

8. VARIOUS PRODUCTS (5)
Aerosol Containers

238 pages

$36

LIQUID FUELS
FROM OIL SHALE AND TAR SANDS 1972
by J. McDermott
Chemical Process Review No. 65

Synthetic crude oil, obtained by pyrolysis of the kerogen fraction of oil shale, is expected to add 200,000 barrels per day or more to our petroleum supply by 1980.

Vast reserves of oil are also found in tar sands, such as the Athabasca deposits in Canada. Oil from tar sands must be mined with huge shovels and separated by water-based techniques.

All recovery and conversion processes require a high conservation of energy to obtain oil competitive with conventional crudes.

This book describes 101 processes related to the retorting and refining of oil shale and the separation of oil from tar sands. A partial table of contents follows. Numbers in ( ) indicate a plurality of processes per topic.

I. OIL SHALE RETORTING
Gas Combustion Processes (23)
Utilizing Flue Gases
Control of Fines
Multistage Retorting
Hydrogen as By-Product
Controlled Retort Atmosphere
Ceramic Hot Ball Systems (6)
Heat Transfer Balls
Inert Ball Heater
Heat from Pyrolysis Products
Attrition and Heat Transfer
Control of Entrained Solids
Granular Heat Carrier
In Situ Processes (4)
Laser Beam Heating
Nuclear Detonation
CO₂ as Heat Carrier
Perforated Pipe
Others (7)
Travelling Grates
Agglomeration of Fines
Electrothermal Pyrolysis
Hot Shale Ash for Heat
Use of Nuclear Reactor
Use of Sonic Energy
Recovery of Aluminum
Pipeline Transportation
Mining Deep Deposits

II. SHALE OIL REFINING
Hydrogenation (8)
Using H₂ and H₂O
With Synthesis Gas
Simultaneous Retorting and
Hydrogenation
Pyrolysis and Hydrogenation
Coarse and Fine Catalysts
Impregnation with Gasoline
Treatment with Solvents
Other Processes (7)
Retorting + Cracking
Separation + Cracking
Zeolite Catalysts
Asphaltite Treating
Pour Point Reduction
HCl Treatment of Crudes

III. TAR SAND SEPARATION METHODS
Hot Water Processes (8)
Phosphates + Surfactants
Oil from Middlings
Addition of Oil to Slurry
Fluidized Beds
Oil from Bituminous Emulsion
Micellar Dispersions
Centrifugal Separation
Cold Water Processes (11)
Non-Aqueous Processes (9)
Other Processes (3)

IV. TAR SANDS—RETORTING & REFINING
Retorting and Coking (10)
Separation While Cracking
Internal Combustion Retorting
Recycling Hot Solids
Compacted Tar Sands
Compaction + Retorting
Slurry Process
Hot Sand Recycle
Coking and Froth Product
Fluid Coking Processes
Filtration of Suspended Material
Hydrogenation (3)
Hydrofining Process
Hydrogen Donor Diluents
Jet Fuels from Coked Tar

276 pages

$36

AGGLOMERATION PROCESSES IN FOOD MANUFACTURE 1972
by N. D. Pintauro
Food Processing Review No. 25

In an agglomeration process several particles are caused to adhere to each other, resulting in a porous, open structure aggregate with new characteristics, such as increased flowability, wettability, and dispersibility. 108 Patent-based processes.

1. Nonfat Dry Milk
2. Chocolate Drink Powders
3. Whole Milk and Other Dairy Products
4. Sugars and Other Sweeteners
5. Soluble Coffee and Tea
6. Flour and Cake Mixes
7. Other Agglomeration Processes
Beverage Mix Products
Dried Egg Products
Monosodium Glutamate in Agglomerates
Colored Agglomerates
Dispersibility with Surfactants
All Purpose Processes

270 pages. $36

SAUSAGE PROCESSING 1972
by Dr. E. Karmas
Food Processing Review No. 24

A patent-based review of sausage production technology, with numbers of specific processes indicated in ( ).

Cured Color (18)
Additives (3)
Forestalling Rancidity (4)
Fermented Flavors (3)
Texturizing Agents (13)
Emulsion Stabilizers (5)
Defatting (6)
Demeating Bones (4)
Composition Control (2)
Ingredients Identification (2)
Stuffing (10)
Linking & Tying (13)
Forming & Shaping (3)
Smoking & Pumping (4)
Hot Air Cooking (6)
Cooking in Liquids (3)
With Heat Exchange
Electrical Cooking (14)
Casing Release & Removal (5)
Fresh Pork Sausage (3)
Dry Sausage (8)
Various & Novelty Products (4)

136 Processes. 218 pages. $36

ENZYMES IN FOOD PROCESSING AND PRODUCTS 1972
by H. Wieland
Food Processing Review No. 23

Commercial availability of enzymes has increased considerably. Enzymes applicable to food processing are now plentiful, and the alert food processor is urged not to miss this opportunity to improve his products in many ways. 101 Processes. The numbers in ( ) indicate the number of processes allocated to each topic.

FRUIT & VEGETABLE PROCESSING (13)
STARCH & SUGAR CONVERSION (9)
BAKED GOODS APPLICATIONS (12)
CHEESE MAKING (11)
MEAT TENDERIZATION (18)
SPECIAL APPLICATIONS (13)
FLAVORS THROUGH ENZYMES (13)
DEOXYGENATING AND DESUGARING (6)
ENZYME STABILIZATION (6)

269 pages. $36

SEAFOOD PROCESSING 1971
by M. Gillies
Food Processing Review No. 22

Describes 84 processes based on U.S. patents issued since 1960. Numbers in ( ) denotes numbers of processes in each chapter.

1. PRESERVATION (13)
On Fishing Vessels
Chemical Methods
Edible Coatings
Various Preservatives
2. CANNING PROCEDURES (11)
Tuna and Similar Fish
Sardines
Forestalling Struvite
3. PROTEIN CONCENTRATES (14)
Mechanical Means
Chemical Means
Biological Means
Stickwater Proteins
4. MOLLUSKS & SHELLFISH (18)
Squid
Bivalves
Crustaceans
5. CONSUMER PRODUCTS (17)
6. ANIMAL FOODS (11)

206 pages. $36

FRUIT PROCESSING 1971
by M. Gutterson
Food Processing Review No. 21

All 140 processes (mostly developed since 1960) were selected with the purpose of providing fruits and fruit products retaining the characteristics of freshly picked fruit, yet capable of being shipped the world over and being highly acceptable by organoleptic tests. Preventing the growth of microorganisms with a minimum of chemicals and processing equipment was another goal.

1. GENERAL TECHNIQUES (34 processes)
Heat Treatments, Increasing Cellular Permeability, Inhibiting Discoloration, Ripening, Dehydration
2. TREATMENT OF POMES (23)
3. CITRUS FRUITS (18)
4. BERRIES (20)
5. DRUPES (16)
6. DRIED FRUITS (10)
7. OTHER FRUITS (19)
Flavor Improvement, Delaying Senescence, Use of Enzymes and Freezing Techniques. 223 pages. $36

POULTRY PROCESSING 1971
by G. H. Weiss
Food Processing Review No. 20

Poultry is the most efficient and effective means for converting grain to protein. This book discusses in detail the different methods devised to assure excellent flavor, texture and tenderness concomitant with easy preservation, maximum storage time, easy handling, and consumer acceptance. 55 processes in 8 chapters:

1. Preservation (11 processes)
2. Chilling and Freezing (6)
3. Enhancing Palatability (10)
4. Stuffed Products (3)
5. Molded Rolls and Loaves (19)
6. Batter-Coated Products (2)
7. Cooking Procedures (2)
8. Poultry Concentrates (2)

168 pages. $24

SOLUBLE COFFEE MANUFACTURING PROCESSES 1969
by Dr. N. Pintauro
Food Processing Review No. 8

This book describes significant manufacturing processes for producing soluble coffee, and offers a wealth of detailed practical information based primarily on the U.S. patent literature. Describes 114 specific processes in this field with substantial background information.

Introduction: Roasting, Extraction, Filtration and Concentration, Recovery of Aromatic Volatiles, Spray Drying and Other Dehydration Processes, Freeze Drying Processes, Aromatization of Soluble Coffee Powder, Agglomeration Techniques for Soluble Coffee, Decaffeinated Soluble Coffee, Packaging of Soluble Coffee. Illustrations, Indexes, 254 pages. $35

ALCOHOLIC MALT BEVERAGES 1969
by M. Gutcho
Food Processing Review No. 7

The traditional brewing process is a batch operation, costly and time consuming. There would be economic advantages to improved continuous processes which would require less capital investment for plant and equipment, give savings in labor, better use of raw materials, shorter processing time, and a more uniform product.

Detailed descriptive process information is found in this review, based on 157 U.S. Patents in the brewing field, issued since 1960. The 157 processes are organized in 7 chapters which tend to follow the steps in the brewing process.

Contents: Malting, Wort, Hops, Fermentation, Freeze Concentration and Reconstitution of Beer, Chillproofing, Preservation against Microbiological Spoilage, Foam, Indexes. Illustrations. 333 pages. $35

CONFECTIONARY PRODUCTS MANUFACTURING PROCESSES 1969
by M. Gutterson
Food Processing Review No. 6

This book is of technological significance in that it details over 200 processes for producing confections, based on the U.S. patent literature since 1960.

Based solely on new technology, this book offers substantial manufacturing information relating to this field. The wide scope of detailed data can be seen by the chapter headings indicated below:

Candy
Chocolate Products
Whipped Products
Icings
Gels
Coatings and Glazes
Gums and Stabilizers
Egg Products
Marshmallows and Meringues
Puddings
Frozen Confections
Chewing Gum
Other Confections
Indexes

Illustrations. 321 pages. $35

EDIBLE OILS AND FATS 1969
by Dr. N. E. Bednarcyk
Food Processing Review No. 5

This book describes in detail 225 recent process developments.

Shortenings; Fluid, Plastic, Miscellaneous: Margarine and Spreads; Margarine Oils, Highly Nutritional Oil Blends, Antispattering Agents, Fluid and Whipped Margarines, Flavor, Color, and Texture Modifications, Low Calorie Spreads: Salad Oils, Mayonnaise and Emulsified Dressings; Crystallization Inhibitors, Emulsified Dressings, Flavored Salad Oils, Low Calorie Dressings: Frying and Cooking Oils; Equipment, Breakdown Inhibitors, Antispattering Additives, Other Additives: Hard Butters; Preparation by Fractional Crystallization, Preparation by Ester Exchange, Miscellaneous: Oil Processing; Antioxidants and Stabilizers; Emulsifiers and Emulsions; Mixed Ester Emulsifiers, Dried Emulsion, Miscellaneous: Peanut Butter and Spreads; Chocolate Products; Indexes. Illustrations. 404 pages. $35

SNACKS AND FRIED PRODUCTS 1969
by Dr. A. Lachmann
Food Processing Review No. 4

The sales of snack foods in the U.S. may reach the two billion dollar mark soon. Many companies are actively working on new snack foods or on improved processes. The patent literature on french fried potatoes, potato chips, corn chips and other crisps is continually growing and it is the purpose of this book to present this literature in easy readable form.

French fried potatoes and their methods of production are described in the second chapter. The next chapter deals with potato chips, still the most popular product of the snack food industry. The U.S. market for potato chips is estimated to be approximately 600 million dollars in 1969. In Chapter Four the processes for corn chips are covered; in Chapter Five, apple crisps. Chapter Six describes processes for expanded chips and some specialty items; and the last chapter deals with batter mixes. Many illustrations. 181 pages. $35

PROTEIN FOOD SUPPLEMENTS 1969
by R. Noyes
Food Processing Review No. 3

The 126 Processes in this book are organized in 8 chapters by raw material source including the important newer processes for producing protein by fermentation of hydrocarbons. Another chapter on textured foods describes in detail a number of processes for producing these products that simulate meat. Indexes by company, inventors and patent number help in providing easily obtainable information.

This book is based upon the patent literature and serves a double purpose in that it supplies detailed technical information and can be used as a guide to the U.S. Patent literature on processes to obtain protein materials.

Contents: Hydrocarbon Fermentation, Fish-Based Protein, Soybeans, Cottonseed, Other Oilseeds and Legumes, Wheat and Gluten, Milk-Based Protein, Textured Foods, Miscellaneous, Indexes. Many illustrations. 412 pages. $35

VEGETABLE PROCESSING 1971
by M. Gutterson
Food Processing Review No. 19

Shipping vegetables from one continent to another has become a normal means of supply. But such transporting is possible only by adequate processing of the perishable vegetable goods. Many of the 184 process descriptions in this book are concerned with just such treatments. Processes for improving the stability of vegetables in regard to time, temperature and moisture are numerous. So are those where the emphasis is in making vegetables more digestible and more acceptable to children and adults:

1. General (27 processes)
2. Potatoes (59)
3. Other Roots (12)
4. Bulbs (13)
5. Leaves & Stems (8)
6. Tomatoes & Others (30)
7. Corn (6)
8. Legumes (23)
9. Olives & Mushrooms (6)

335 pages. $36

FLAVOR TECHNOLOGY 1971
by Dr. N. D. Pintauro

Scientific and trade journals contain only limited information on practical and applied flavor research and technology. Industry and commercial operators wish to keep such information confidential. This book reviews such technology from U.S. patents since 1960. There are 99 processes in 9 chapters:

1. Spice Technology (11)
2. Peppermint & Citrus (11)
3. Fruit Essences (9)
4. Dairy Flavors (6)
5. Bread Flavors (8)
6. Vanilla (9)
7. Meat Flavors (17)
8. Meat Seasonings (10)
9. Fixation (18)

228 pages. $35

MILK, CREAM AND BUTTER TECHNOLOGY 1971
by G. Wilcox
Food Processing Review No. 18

In these days of heightened consumer awareness, the alert dairy processor cannot afford to bypass the latest developments in his field. Much emphasis now is on modified milk products which contain numerous proteins of high nutritional value. The even distribution of essential amino acids bestows on modified milk a great enticement for use with otherwise deficient diets. Methods for low sodium milk products, hypoallergenic dietary prepns. and infant milk products are given special attention. 181 Processes are described: 1. Pasteurization and Sterilization (15 processes). 2. Removal of Radioactive Contaminants (8). 3. Buttermilk and Allied Products (10). 4. Modified Milk Products (21). 5. Dehydration of Skim Milk (28). 6. Dehydration of Whole Milk, Whey and Milk Blends (47). 7. Concentrated Milk (19). 8. Cream (22). 9. Butter (11). 313 pages. $35

EGGS, CHEESE AND YOGURT PROCESSING 1971
by G. Wilcox
Food Processing Review No. 17

One route to discovering the latest technology is via this Food Processing Review which serves to bring you timely, useful information. Brought to you in this one easy-to-use, comprehensive volume is commercial research and development done in the field from 1960 to 1970, gathered from the U.S. patent literature.

The Table of Contents shows the processes discussed in this book. The numbers in parentheses indicate the number of processes covered for each particular process, equipment or product.

Section I—Eggs: Whole Eggs (18); Egg Yolks (7); Egg Whites (27); Egg Products (10). Section II—Cheese and Yogurt: Cottage Cheese (26); Cheddar Type and Process Cheeses (45); Cream Cheese and Bakers' Cheese (8); Mozzarella, Provolone and Parmesan Cheeses (8); Miscellaneous Cheese Processes (17); Manufacture of Yogurt (5).

280 pages. $35

RICE AND BULGUR QUICK-COOKING PROCESSES 1970
by R. Daniels
Food Processing Review No. 16

This salient report in our Food Processing Review series summarizes with detailed process information the pertinent U.S. patent literature relating to quick-cooking processes for both rice and bulgur. The information provides needed know-how concerning processing of raw rice and wheat to obtain the more desirable refined forms.

63 specific processes covered. The numbers in ( ) indicate their distribution.

Rice Milling—Extraction—Polishing (13)
Quick-Cooking Rice (17)
Special Rice Processes (6)
Brown and Parboiled Rice (6)
Specialty Products with Rice (9)
Quick-Cooking Wheat Bulgur Products (12)

267 pages. $35

FRUIT JUICE TECHNOLOGY 1970
by M. Gutterson
Food Processing Review No. 15

This publication deals with the technology of the noncarbonated fruit juice industry from 1960 through 1970 as covered in the U.S. patent literature. Modern technology has studied and overcome many processing problems, resulting in a vast output of new and improved processing methods. It is oftentimes difficult to keep up with the latest technology. This book is designed to offer you such help.

In the abbreviated Table of Contents shown below, you can see the large amount of valuable material included. The numbers in ( ) indicate the number of processes discussed.

1. Manufacturing Techniques (23), 2. Concentration of Fruit Juices (37), 3. Stabilization Processes (19), 4. Dehydration (18), 5. Freeze Drying (24), 6. Flavors from Fruit Juices (10), 7. Miscellaneous Processes (9).
206 pages. $35

MEAT PRODUCT MANUFACTURE 1970
by Dr. E. Karmas
Food Processing Review No. 14

This Review concerns latest technology in preparing packaged meats in ready-to-cook and ready-to-eat forms.

The Table of Contents below indicates the many areas covered in this survey.

General Processing: Curing Methods and Ingredients, Increased Water Binding and Yield, Improved Curing Formulations, Integral Meats, Smoking, Thermal Processing and Sterilization, Miscellaneous Processing Methods.

Products: Bacon Production, Patty Type Products, Dehydrated Convenience and Snack Products, Modified and Novel Products.

273 pages. $35

MODERN BREAKFAST CEREAL PROCESSES 1970
by R. Daniels
Food Processing Review No. 13

Describes in detail production processes and equipment for the manufacture of modern breakfast cereals. These include both ready-to-eat and quick-cooking products.

Offers detailed practical information for the manufacture and production of these cereal products based on the U.S. patent literature. 61 processes included. Abbreviated Table of Contents follows.

Dough Cooking and Extrusion Processes
Treatment Prior To Puffing
Puffing Processes
Processes For Whole Cereal Grains
Cereal Shaping Processes
Sugarcoating Process
Fruit Incorporation and Nutritional Enrichment
Quick Cooking Cereal Products

217 pages. $35

STARCHES AND CORN SYRUPS 1970
by Dr. A. Lachmann

This report covers the field of starch production from many standpoints.

Wet milling is the primary method of starch production, therefore much of the material is concerned with this route to starch. Dry milling processes are also covered.

In addition, coverage of the current technological progress in hydrolyzing starches into dextrins, corn syrups and dextrose and starch fractionation into amylose is covered.

Contains 139 processes covering: The Manufacture of Starch, Treatment of Starch, Modified Starch, Pregelatinized Starch, Acid Hydrolysis of Starch to Sweeteners, Starch Hydrolyzing Enzymes, Enzymatic Starch Hydrolysates, Starch Hydrolysates Produced by Acid and Enzyme Treatments, Starch Fractionation. 275 pages. $35

FRESH MEAT PROCESSING 1970
by Dr. E. Karmas
Food Processing Review No. 12

This Food Processing Review, deals with 106 detailed processes covering essential developments in the fresh meat processing industry since 1960. The book provides a well-organized tour through the field; the processes included are well researched and presented as an easy-to-use guide to what is being done in this vital field today.

The material has been divided into two parts; processes for enhancing palatability, and preservation processes. The numbers in ( ) after each heading indicate the number of processes for each entry.

A. Palatability: Tenderness (33), Flavor and Tenderness (8), Flavoring (12), Color (13), Integral Texture (6). B. Preservation: Moisture Retention (9), Antimicrobial Treatment (10), Ionizing Radiation (7), Other Methods of Preservation (8). 236 pages. $35

SOLUBLE TEA PRODUCTION PROCESSES 1970
by Dr. N. Pintauro
Food Processing Review No. 11

This book describes production processes for producing soluble tea and offers a wealth of detailed practical information based primarily on the U.S. patent literature. Describes 73 specific processes in this field with substantial background information. The Table of Contents is listed below. The numbers in ( ) indicate the number of processes in that category.

Withering and Rolling (4)
Fermentation, Firing and Sorting (8)
Extraction (13)
Recovery of Aroma (10)
Tannin-Caffeine Precipitate (Cream) (15)
Filtration and Concentration (8)
Dehydration Process (6)
Agglomeration and Aromatization (9)

Illustrations. 183 pages. $35

BAKED GOODS PRODUCTION PROCESSES 1969
by M. Gutterson
Food Processing Review No. 9

This book describes 201 recent processes for the production of baked goods. Based on the patent literature, it offers an up-to-date comprehensive publication of manufacturing processes.

There is a substantial amount of information in this book relating to the use of various chemicals and related additives.

Contents: Bread, Yeast Leavened Products, Chemically Leavened Products, Leavening Agents, Air Leavened Products, Non-Leavened Products, Refrigerated Doughs, Emulsifiers and Dough Improvers, Miscellaneous, Indexes. Illustrations. 353 pages. $35

FOOD

DEHYDRATION PROCESSES FOR CONVENIENCE FOODS 1969
by R. Noyes
Food Processing Review No. 2

Describes 236 up-to-date dehydration processes for producing specific foods. Most detailed body of information ever published.

The detailed, descriptive process information in this book is based on 236 U.S. patents in the food dehydration field—issued between January 1960 and May 1968. This book serves a double purpose in that it supplies detailed technical information, and can be used as a guide to the U.S. patent literature on dehydration of foods. By indicating only information that is significant, and eliminating much of the legal jargon in the patents; this book then becomes an advanced commercially oriented review of food dehydration processes.

Dry Milk Products, Cheese and Yoghurt, Eggs, Fruit and Vegetable Juices, Fruits, Potatoes, Vegetables, Coffee, Tea, Miscellaneous. Many illustrations. 367 pages. $35

CHEMICAL

ISOCYANATES MANUFACTURE RECENT DEVELOPMENTS 1972
by M. W. Ranney
Chemical Process Review No. 63

Describes 127 improved methods based on 172 U.S. Patents since 1967

1. **Toluene Diisocyanates**
   Phosgenation
   Distillation
   Stabilization
2. **Aromatic Isocyanates**
   Polyisocyanates
   Nitro Conversions
   Sulfonyl Isocyanates
3. **Nonaromatics**
   Cycloaliphatics
   Unsaturateds
   Monoisocyanates
4. **Halogen Isocyanates**
5. **Polyureas and Silicones**
6. **General Processes**
   Syntheses
   Conversions
   Special Phosgenations
   Stabilizers
   Intermediates

258 pages. $36

ANTIFOAMING AND DEFOAMING AGENTS 1972
by T. G. Rubel
Chemical Process Review No. 60

Describes 105 patent-based processes.

1. **Foam Control in Aqueous and Nonaqueous Systems**
2. **Mechanical Means of Foam Control**
3. **Foam Inhibition in Lubricants, Fuels, Hydraulic Fluids, and Organic Liquids**
4. **Antifoamers for Wax Coatings**
5. **Foam Prevention in Pulp and Paper Production**
6. **Foam Reduction in Paints**
7. **Controlling Foam in Drilling Fluids**
8. **Phosphoric Acid Foam**
9. **Antifoamers for Gases**
10. **Breaking Detergent Foams**
11. **Foam Depressants for Dyes**
12. **Antifoaming of Proteins**
13. **Foam in Antifreezes**
14. **Defoaming Steam**
15. **Polymer Foam Suppression**
16. **Foam Prevention in Photoengraving Solutions**

251 pages. $36

SYNTHETIC LUBRICANTS 1972
by M. W. Ranney
Chemical Process Review No. 59

A review of 205 patent-based processes.

1. **Organic Esters**
   General Syntheses
   Inhibited Fluids
   Extreme Pressure Additives
   Greases
2. **Silicones**
   Fluid Syntheses
   Grease Formulations
   Solid Lubricants
3. **Polyglycols, Phosphates and Silicates**
4. **Polyphenyl Ethers**
   Additives
   Halogen-Containing Ethers
5. **Fluorocarbons**
   Grease Formulations
   Fluorinated Esters
   Thread Sealants
6. **Petroleum**
   Viscosity Index Improvers
   Dispersants
   Extreme Pressure Additives
   Grease Formulations
   Metalworking Formulas

245 pages. $36

ANTISTATIC AGENTS 1972
Technology and Applications
by K. Johnson

Antistatic agents described in 160 U.S. patents since 1965 are covered in this book with specific sections devoted to two large volume applications—Plastics and Textiles.

Antistatic agents are used in plastics such as phonograph records, bottles, and film wraps to reduce pickup of dust and dirt.

In the textile industry, charged fibers interfere with the spinning process, and the attraction of dust particles produces marks and soiling in weaving.

1. **Polyolefins (27)**
2. **Records and Films (17)**
3. **Fibers & Fabrics Treatment (29)**
4. **Further Fibers & Fabrics (27)**
5. **Techniques of Treatment (22)**
6. **Softener & Lubricant Formulations (20)**
7. **Other Applications and Syntheses (18)**

307 pages. $36

CHEMICAL

OPTICAL BRIGHTENERS 1972
Technology and Applications
by T. Rubel

Describes synthesis and use of fluorescent dyes that have the property of absorbing ultraviolet radiant energy and emitting energy in the visible range.

To the unaided human eye the fluorescence is not noticeable. Its presence, however, has the effect of brightening colors and making things whiter than white.

Based on 136 U.S. patents issued since Jan. 1965:

1. Optical Brighteners Added to Natural Materials (6 Processes)
2. To Synthetic Polymers (52)
3. To Natural & Synthetic Substances (42)
4. To Detergents (8)
5. To Fabric Softeners (5)
6. Brighteners for Papers and Photographic Materials (19)
7. Other Applications (4)
   Company Index
   Inventor Index
   U.S. Patent Number Index.

281 pages. $36

LIQUID FUELS FROM COAL 1972
by G. K. Goldman
Chemical Process Review No. 57

1. **Extractive Solvents (15)**
2. **Solvation (4)**
3. **Deashing (4)**
4. **Dual Solvent Systems**
5. **Multistage Systems (4)**
6. **Microwave and Ultrasonics (2)**
7. **Thermal Liquefaction**
8. **Separation of Suspensions**
9. **Noncatalytic Hydrogenation (4)**
10. **Catalytic Hydrogenation (13)**
11. **Hydroconversion Catalysts (13)**
12. **Ebullated Bed Processes (11)**
13. **Hydrocracking**
14. **Quadriphase Hydrogenation**
15. **Refining (2)**
16. **Jet Fuel Production**
17. **Underground Hydrogenation**
18. **Thermal Cracking (16)**
19. **Catalytic Depolymerization**
20. **Pyrolysis (5)**

101 Processes. 228 pages. $36

HYDROFLUORIC ACID MANUFACTURE 1972
by S. Weiss
Chemical Process Review No. 58

Reviews 166 U.S. patents pertaining to HF manufacture, including a chapter on fluosilicic acid.

The current expansion in HF capacity appears to stress the traditional fluorspar-sulfuric acid route in foreign (esp. Mexican) plants built or planned. Domestic producers, squeezed by rising costs and shrinking supplies of U.S. fluorspar, are becoming increasingly interested in HF recovery from waste products of phosphate fertilizer plants.

1. **HF from Fluorspar (38 processes)**
2. **HF from Waste Gases (26)**
3. **HF from Other Sources (11)**
4. **Refining by Distillation (25)**
5. **Purification in Solution (12)**
6. **Refining by Formation of Reversible Complexes (21)**
7. **Purification by Chemical Reactions (18)**
8. **Fluosilicic Acid Manufacture (15)**

254 pages. $36

SULFURIC ACID MANUFACTURE AND EFFLUENT CONTROL 1971
by M. Sittig
Chemical Process Review No. 55

102 processes of manufacture and 39 pollution control measures are outlined in this encyclopedic survey:

SO_2 from Sulfur (11 processes)
SO_2 from Waste Gases (11)
SO_2 from Hydrogen Sulfide (2)
SO_2 from Sulfide Ores (8)
SOx and H_2SO_4 from Sulfates (12)
H_2SO_4 from SO_2 HCl (2)
The Chamber Process (4)
Conversion of SO_2 to SO_3 (9)
SO_3 to H_2SO_4 and Oleum (5)
Integrated Contact Processes (14)
Unconventional Processes (2)
Concentration of H_2SO_4 (2)
Dilution of H_2SO_4 (1)
Purification of H_2SO_4 (2)
Recovery of Spent H_2SO_4 (17)
Removal and Recovery of SOx from Tail Gases (30)
Recovery of Acid Mists (9)
Future Trends

423 pages. $48

TRIMELLITIC ANHYDRIDE AND PYROMELLITIC DIANHYDRIDE 1971
by P. Stecher
Chemical Process Review No. 53

Trimellitic anhydride (TMA) being an anhydride and a carboxylic acid, can undergo many useful reactions; it can form esters and polyimides. Demand is rising sharply. Pyromellitic dianhydride (PMDA) yields polyimides and cured epoxy resins having very high temperature stability. This book gives 61 manufacturing processes and all the technology involved in making TMA and PMDA:

1. TM-Acid Synthesis (17)
2. TM-Acid Purification (3)
3. TMA Preparation (8)
4. TMA Purification (2)
5. TM-Double Anhydride (2)
6. PM-Acid Synthesis (6)
7. PMDA Preparation (5)
8. PMDA Purification (4)
9. Non-Hazardous Oxidations (1)
10. Derivatives of TMA & PMDA (13)

233 pages. $35

PHTHALOCYANINE TECHNOLOGY 1970
by Y. L. Meltzer
Chemical Process Review No. 42

Advances in phthalocyanine technology have been truly explosive during the past few years. New phthalocyanine products, processes and applications have poured forth from industrial, governmental and academic laboratories at a rapid pace. These advances in technology have made themselves felt in the market place and in government programs, and have contributed to corporate sales and profits. At the same time, however, competition has become more intense in the phthalocyanine field making it imperative to keep up with the latest technological advances.

Examines recent developments in phthalocyanine technology as reflected in U.S. patents and other literature. The first 23 chapters discuss up-to-date manufacturing processes for phthalocyanine pigments and dyes. Chapters 24 through 31 discuss unusual new applications for phthalocyanines. 390 pages. $35

CHEMICAL

SYNTHETIC PERFUMERY MATERIALS 1970
by M. Gutcho
Chemical Process Review No. 45

This Review shows you how to produce synthetic perfumery materials. It contains a valuable odor index.

The 152 U.S. patents included in this book are distributed among the 11 areas as shown below:

From Terpenic Materials (28)
Alcohols (11)
Esters (18)
Ethers (19)
Aldehydes (10)
Ketones (18)
Lactones, Pyrones, Substituted Phenols and Quinones (12)
Other Structures (7)
Naphthalene and Indene Derivatives (17)
Compounds with Scent of Ambergris or Irone (9)
Product Application (13)

273 pages. $35

ION EXCHANGE RESINS 1970
by C. Placek
Chemical Process Review No. 44

This report on ion exchange resins provides detailed information on 126 U.S. patents issued since 1960 concerning the composition and manufacture of ion exchange materials. This Review, by its organization, also provides a guide to these ion exchange resins by grouping them according to physical form, behavior characteristics, etc.

1. Anion Exchange Resins
2. Cation Exchange Resins
3. Resins For Removing Metals
4. Resins Having Mixed Properties
5. Specific Use Resins
6. Unconventional Materials
7. Process Emphasis
8. Properties of Ion Exchange
9. Ion Exchange Membranes
10. Emphasis on Shapes

329 pages. $35

RADIATION CHEMICAL PROCESSING 1969
by R. Whiting
Chemical Process Review No. 41

This book surveys the radiation processing field and is based on the U.S. patent literature since 1960. Over 250 separate processes are described.

Contents: Polyolefins, Other Polymers, Elastomers, Hydrocarbons, Organic Chemicals, Inorganic and Organic-Metallic Compounds, Other Processes. 377 pages. $35

PHOTOCHEMICAL PROCESSES 1969
by B. Albertson
Chemical Process Review No. 36

Describes 210 photochemical production processes in detail.

Introduction, Photohalogenation, Photonitrosation, Organic Photochemical Reactions, Inorganic Photochemical Reactions, Photopolymerization, Indexes. Illustrations. 185 pages. $35

CHLORINE AND CAUSTIC SODA MANUFACTURE RECENT DEVELOPMENTS 1969
by Dr. R. Powell
Chemical Process Review No. 33

Brine Electrolysis
Diaphragm and Mercury Cells
Recovery of Mercury
NaOH Production
Titanium Anodes
Sea Water Electrolysis
Cl_2 Production
Deacon Process Modifications
Numerous Illustrations,
48 processes, 265 pages. $35

AMINES, NITRILES AND ISOCYANATES PROCESSES AND PRODUCTS 1969
by M. Sittig
Chemical Process Review No. 31

Material covered includes: Manufacture of Amines, Manufacture of Mono-Nitriles, Acrylonitrile Derivatives, Isocyanate Manufacture, Future Trends. 62 illustrations. 201 pages. $35

CITRIC ACID PRODUCTION PROCESSES 1969
by R. Noyes
Chemical Process Review No. 37

Detailed descriptions of production processes for citric acid, based on the patent literature.

The Table of Contents is indicated below:

Processing, Iron Impurities Other Microorganisms, Recovery and Purification, Other Processes. Indexes. 157 pages. $24

NITRIC ACID TECHNOLOGY RECENT DEVELOPMENTS 1969
by Dr. R. Powell
Chemical Process Review No. 30

Ammonia Oxidation Process, Wisconsin Thermal Process, Nitrogen Fixation by Shock Waves, Nitrogen Fixation in a Nuclear Reactor, Absorption of Nitrogen Oxides in Water, Concentration of Dilute Nitric Acid Solutions, Direct Production of Concentrated Nitric Acid, Purification of Nitric Acid, Stabilizers for Nitric Acid. Numerous illustrations. 245 pages. $35

AROMATICS MANUFACTURE AND DERIVATIVES 1968
by M. Sittig
Chemical Process Review No. 17

Contents: Introduction, Production of Aromatics, Separation of Aromatics, Purification of Aromatics, Reactions giving Hydrocarbon Products, Other Reactions, Phenol Production, Styrene Manufacture and Derivatives, Future Trends. 73 illustrations. 232 pages. $35

INDUSTRIAL GASES MANUFACTURE AND APPLICATIONS 1967
by M. Sittig
Chemical Process Review No. 4

This book discusses conventional cryogenic air separation and purification techniques in considerable detail.

This book also discusses newer techniques such as adsorption using molecular sieves, and permeation using various membrane materials. 313 pages. 103 illustrations. $35

CHEMICAL

POLYMETHYLBENZENES 1969
by H. W. Earhart

Presents physical property data. Discusses the chemistry of the PMB's established chemical reactions, relative kinetic rate data, yields, etc. Known as well as suggested end-uses for numerous PMB's and derivatives are given, e.g. for benzene, toluene, xylene, mesitylene, pseudocumene, hemimellitene, durene, isodurene, prehnitene, penta-and hexamethylbenzene. 63 tables. 549 references. 158 pages. $20

ORGANIC CHEMICAL PROCESS ENCYCLOPEDIA
by M. Sittig

Second Edition 1969

A handy desk-top reference to organic chemicals and their industrial manufacturing processes.

Gives the key processing facts for instant reference to 711 industrial organic chemical processes—with 711 large flow diagrams. 712 pages—8½" x 11"—hard cover. $35

INORGANIC CHEMICAL AND METALLURGICAL PROCESS ENCYCLOPEDIA 1968
by M. Sittig

This book is organized in an unusual format. There is one inorganic chemical or metallurgical process on each page. At the top of the page an equipment drawing or flow diagram is shown, and underneath a description of the process is given. 883 pages—8½" x 11"—hard cover. $35

OTHERS

ELECTRON BEAM WELDING 1971
by Dr. R. Bakish

The material has been arranged for easy use in the seven broad areas shown, with the number of developments indicated. 1. Processes and Equipment (41), 2. Alternate Beam Generating System (8), 3. Beam Control (20), 4. Moveable Chambers (8), 5. Beam Protective Devices (6), 6. Viewing Devices (10), 7. End Products (13). 150 pages $35

RESISTOR MATERIALS 1971
by P. Conrad
Electronics Materials Review No. 12

A helpful guide to new resistor materials.

Metals and Other Elements and Alloys (48)
Metal-Metal Oxide Mixtures (9)
Inorganic Oxide Compositions (32)
Other Inorganic Compounds (21)
Organic Compositions (12)

217 pages. $35

CONTINUOUS CASTING OF STEEL 1970
33 Magazine

This volume of highly technical and economic data and information on continuous casting originally appeared as a series of articles over the years in 33 Magazine. Continuous Casting of Steel 1970 is as current as the World Continuous Casting Round/Up of August-September 1970. 341 pages. $20

BATTERY MATERIALS 1970
by P. Conrad
Electronics Materials Review No. 10

Describes 162 processes useful in the manufacture of modern batteries:

Aqueous Battery Systems (89)
"Dry" Battery Systems (29)
Inorganic Electrolyte Systems (6)
Organic Electrolyte Systems (6)
Solid Electrolyte Systems (18)
Molten Electrolyte Systems (11)
Radioactive Batteries (3)

171 pages. $35

COATINGS

POLYURETHANE COATINGS 1972
by K. Johnson

Reviews 157 U.S. patents issued mostly since 1960 related to paint vehicles, wet look, highly glossy fabric coatings, microporous products, etc.

1. COATING VEHICLES
   Isocyanates (9 processes)
   Carboxyl, Polyols (16)
   Modified Resins (13)
   Aqueous Systems (9)
   Catalysts & Crosslinking (5)
   Miscellaneous (17)
2. MICROPOROUS MATERIALS
   Solvent Processes (22)
   Pore-Forming Agents (9)
   Suede Substitutes (3)
   Others (3)
3. COATED FABRICS & PAPERS
   Water Repellency & Ability to Dryclean (8)
   Coated Fabrics (13)
   Papers (3)
4. MAGNETIC TAPES & OTHER SUBSTRATES
   Magnetic Tapes (6)
   Wire Coatings (3)
   Other Applications (18)

338 pages. **$36**

POWDER COATINGS AND FLUIDIZED BED TECHNIQUES 1971
by Dr. M. W. Ranney

Describes 166 processes. Due to the actuality of the subject all have been developed very recently. Numbers in ( ) indicate number of processes per chapter.

1. FLUIDIZED BED—SPRAY-POURING TECHNIQUES (27)
   Fluidized Bed Designs
   Spray-Powder, etc.
2. ELECTROSTATIC PROCESSES (8)
3. EPOXIES (19)
   Curing Agents
   Modified Epoxies
   Powdering Techniques
4. POLYOLEFINS (10)
   Primers & Surface Treatment
   Use of Copolymers
5. VINYLS (6)
6. OTHER RESINS (10)
7. PIPE COATINGS (28)
8. ELECTRICAL COMPONENTS (19)
9. OTHER APPLICATIONS (24)
10. INORGANIC & PARTICLE COATINGS (15)

249 pages. **$36**

PAPER COATINGS BASED ON POLYMERS 1971
by K. Johnson

178 Processes for coating paper stock elaborated during the last 10 years! Pigment binder and barrier coatings are discussed in detail e.g. acrylics give high gloss and good ink holdout, while silicone and solvent-based coatings allow considerable latitude in formulations.

1. POLYETHYLENE ETHYLENE COPOLYMERS—HOT MELTS (34 processes)
2. POLYVINYLIDENE CHLORIDE BARRIER COATINGS (10)
3. WATER-SOLUBLE COATINGS (30)
4. STYRENE-BUTADIENE ETHYLENE-PROPYLENE LATICES (27)
5. VINYL ACETATE LATICES (13)
6. ACRYLIC LATICES (23)
7. SILICONE AND SOLVENT-BASED COATINGS (24)
8. SPECIALTY COATINGS (17)
   Photographic Paper Coatings, Opaque Coatings, Metallized Coatings, Chemical Watermark Paper, Coatings for Erasable Paper, Mulch Sheets.

313 pages. **$36**

WIRE COATINGS 1971
by D. J. De Renzo

The good insulating properties of many thermoplastic and thermosetting polymers make them suitable for coating wires to be used as conductors in electrical apparatus. Other properties include abrasion resistance, impact strength, flexibility, solvent resistance, and high temperature stability.

The U.S. patent literature of the past ten years provides an excellent description of the many types of processes for organic wire coatings. Numbers in ( ) indicate the number of processes described in each chapter. 180 processes in all.

Acrylics (12), Epoxy Resins (9), Fluorinated Resins (4), Polycarbonates (5), Polyesters (31), Polyimides and Polyamides (26), Polyolefins (41), Polyspirones (5), Polyurethanes (8), PVA Resins (15), Silicones (7), Vinyl Chloride Polymers and Copolymers (13), Other Resins (4).

232 pages. **$35**

PAINT ADDITIVES 1970
by H. Preuss

This publication surveys the field of paint additives offered for sale by manufacturers in the United States. It gathers together for you in one useful volume a series of articles written by Mr. Preuss for METAL FINISHING from 1965 through 1970. It is designed to help lead the paint formulator through the maze of modern additives; placing at his fingertips needed information about their chemistry, properties, specifications, uses and applications.

Additives form an integral part of a coating. Some of the additives discussed in this book are: antiskinning, antifoaming, antifouling, antifreezing, dispersing, destaticizing, antiliverng, antisagging, curing, antisettling or suspension and moisture resistant agents; plasticizers; antioxidants; fire retardants; corrosion inhibitors; odorants and deodorants; and many others. 249 pages. **$18**

METAL COATING OF PLASTICS 1970
by Dr. F. Lowenheim

Describes 125 processes for applying metallic coatings to plastic articles on a production basis with reasonable reliability, and in such fashion that the metal is acceptably adherent to the substrate, and that the resulting products are useful for decorative or functional purposes.

254 pages. **$35**

ELECTRODEPOSITION AND RADIATION CURING OF COATINGS 1970
by Dr. M. W. Ranney

The advantages of electrodeposition: Pinhole-free coating, eliminating of fire hazards and air pollution problems, automated operation and fast through-put make this an attractive method. Radiation curing is quick, eliminates ovens and uses solvent-free vehicles. 96 Patent-based process on 170 pages. **$35**

POLYMERS AND ADHESIVES

WATER-SOLUBLE POLYMERS 1972
Technology and Applications
by Y. L. Meltzer
Chemical Process Review No. 64

Synthetic water-soluble polymers along with modified starches are making appreciable inroads into the traditional starch and and natural gum market. 139 Patent-based processes.

1. **Market Survey**
2. **Acrylamide Polymers**
3. **Acrylic and Methacrylic Polymers**
4. **Cellulose Ethers**
5. **Carboxymethyl Cellulose**
6. **Hydroxyethyl Cellulose**
7. **Hydroxypropyl Cellulose**
8. **Methyl Cellulose**
9. **Other Derivatives**
10. **Ethylene Oxide Polymers**
11. **Polyethylenimine**
12. **Polyvinyl Alcohol**
13. **Polyvinylpyrrolidone**
14. **Starch, Modified Starch and Derivatives.**

323 pages. **$36**

EPOXY RESIN HANDBOOK 1972
Blue Book No. 5

Contains technical data and specifications of commercial epoxy resins and their curing compounds, as manufactured or sold in the U.S. by 71 companies.

Epoxy resins, based on ethylene oxide (oxirane) or its homologs or derivatives, and nearly always used with hardeners or curing agents, are classified as thermosetting resins.

Consumption of uncured epoxy resins in the U.S. has increased from less than one million pounds in 1950 to about 140 million pounds in 1971.

All the information is taken directly from the manufacturers' hard-to-get data sheets and technical bulletins at no cost to, nor influence from, the manufacturers of the specified epoxy resins and compounds.

Like the other volumes in our Blue Book Series, this book is intended as a guide to U.S. firms with standard product lines. 426 pages. **$36**

PLASTICIZERS 1972
Blue Book No. 4

This book contains technical data and specifications of commercial plasticizers manufactured in the United States by 52 companies.

The U.S. market has plasticizer sales of about $250 million per year.

Like the other volumes in our Blue Book Series, this book is intended as a guide to U.S. manufacturers with standard product lines.

All the information is taken directly from the manufacturers' hard-to-get data sheets and technical bulletins at no cost to, nor influence from, the manufacturers of the specific plasticizers.

In each case enough information is presented to enable the researcher or chemical processor of plastics to judge from the data presented whether or not a given plasticizer type or grade will do the job for the intended application to the resin on hand. 282 pages. **$36**

POLYOLEFIN RESINS 1972
Blue Book No. 3

This informative data book on olefin polymers and their producers is the third volume in our comprehensive Blue Book Series. In keeping with our policy, this book attempts to provide a complete listing of commercially available, standardized resins offered by 22 U.S. producers.

The book describes the readily available higher-molecular-weight polymers of ethylene and propylene. Some elastomeric copolymers are also listed. Included are product specifications and applications.

All the information in this volume is taken directly from the manufacturers' hard-to-get data sheets and technical bulletins. In each case enough information is provided to enable the researcher to judge from the data presented whether or not a given polyolefin type or grade will do the job for the intended applications. 291 pages. **$36**

ADHESIVES 1972
Blue Book No. 2

The second volume in our comprehensive Blue Book Series. In keeping with our policy, this book attempts to provide a reasonably complete listing of commercially available, standardized products offered by American industry. The book describes the "on the line" adhesive products of 121 U.S. manufacturers arranged according to company names.

All the information was selected from the manufacturers' hard-to-get data sheets at no cost to, nor influence from, the manufacturers of the adhesives.

A primary purpose of the book is to present the significant, first line information about adhesives all in one place—saving you many hours of work trying to obtain specific information and facts. 407 pages. **$36**

POLYMER ADDITIVES 1972
Blue Book No. 1

Perhaps the most comprehensive listing of commercially available, protective additives ever offered to the plastics industry. Gives products of 86 U.S. manufacturers arranged according to company name. Each product is carefully indexed by chemical, generic, trivial, and trade name or registered trademark in the "one alphabet" index at the end of the book.

Abounds with antioxidants and stabilizers plus countless other protective aids. By listing the intended uses and physical properties, as well as the manufacturers and suppliers, this book intends to furnish a real service to the advancing polymer technology.

The data appearing in this book were selected by the publisher from manufacturers' literature at no cost to, nor influence from the manufacturers of the materials. 472 pages. **$36**

POLYMERS AND ADHESIVES

SYNTHETIC TURF AND SPORTING SURFACES 1972
by M. S. Casper

High labor costs have made natural turf surfaces almost prohibitively expensive. Increased interest in spectator sports make it desirable to provide sport surfaces which can be used despite weather conditions and seasonal variations. This review of 84 U.S. patents describes the present state of the art.

1. **Turf Substitutes (8)**
2. **Golf Installations (4)**
3. **Tennis Courts (2)**
4. **Race Tracks (2)**
5. **Resilient Surfaces (7)**
6. **Installation Aids (3)**
7. **Skiing Surfaces (29)**
8. **Golf Surfaces (22)**
9. **Skating Surfaces (4)**
10. **Tobogganing Slide**
11. **Aircraft Landing Pads**
12. **Lawn Elements**

260 pages. **$36**

RIGID FOAM LAMINATES 1972
by M. G. Halpern

This book concerns itself with the manufacture of rigid foam laminates, in which the foam cores are bonded to a great variety of outer coatings. 134 processes from U.S. patents.

1. Chemical Processes for the Preparation of Plastic Foams (7)
2. Nonplastic Expanded Materials (5)
3. Plastic Skin—Plastic Foam Laminates (15)
4. Polyurethane Foam Compositions (10)
5. Nonplastic Skin and Foamed Plastic Core Laminates (9)
6. Paper & Foam Compositions (6)
7. Containers & Packaging Materials (9)
8. Refrigerated Containers (9)
9. Acoustical Panels (5)
10. Roofing Applications (12)
11. Structural Panels (14)
12. Internal Reinforcements (12)
13. Adhesive Techniques (8)
14. Embossing & Decorating (4)
15. Miscellaneous (9)

271 pages. **$36**

SEALING AND POTTING COMPOUNDS 1972
by J. A. Szilard

Sealing and potting compounds are used to protect against ingress or egress of liquids or gases. In most cases the desired protection is against the pentration of moisture. Describes 166 sealant manufacturing processes based on polymer technology including silicones.

Products intended for:
General Use
Soil Treatment
Highways and Runways
Building Construction
Aircraft Construction
Pipe Joint Sealing
Automotive Use
Shoemaking
Swimming Pool & Aquarium Sealing
Shafts & Stuffing Boxes
Carton & Container Sealing
Electrical & Electronic Instruments

Also discusses special sealants for extremely high and low temperatures, lubricating sealants, encapsulants for printed and high frequency circuits, and many more. 288 pages. **$36**

FLEXIBLE FOAM LAMINATES 1971
by M. McDonald

Reviews the U.S. patent literature on the technology of flexible foam laminates from 1960 through early 1971. Altogether 101 processes in 6 chapters. In 1970 about 75 million pounds of polyurethane foam in the U.S. alone were bonded, mostly to fabrics, to form flexible laminates.

1. THERMAL METHODS (16 processes)
2. ADHESIVE METHODS (15)
3. FOAM-IN-PLACE METHODS (19)
4. FABRIC TO FOAM (17)
5. FLOOR COVERINGS (14)
6. MISCELLANEOUS (20)
   Packaging Materials, Foam on Cardboard, Sealing Strips, Sound Insulation, Polyester Foam Laminates, Polypropylene Foam to Metal, Deforming Foam Surfaces, Stretchable Foam Laminates.

265 pages. **$36**

POLYMERS IN LITHOGRAPHY 1971
by D. J. De Renzo

About 50% of all printing today is done by lithography, also named the planographic method. This has led to an extensive use of polymers for making the base plates and the sensitized plate coatings. Polymers are used also in etching solutions, lacquers, deletion fluids, inks, etc. 145 processes from the U.S. patent literature since 1965.

1. DIAZO TYPE PRESENSITIZERS (46 processes)
2. NONDIAZO PHOTOSENSITIVE LAYERS (17)
3. BASE PLATES AND COATINGS (32)
4. OTHER POLYMERS FOR PLATES (9)
5. LACQUERS, ETCHANTS, ETC. (18)
6. THERMOGRAPHIC PROCESSES (13)
7. ELECTROPHOTOGRAPHIC PROCESSES (9)
8. THE DRIOGRAPHIC PROCESS.

216 pages. **$36**

URETHANE FOAMS TECHNOLOGY AND APPLICATIONS 1971
by Y. L. Meltzer

Urethane foam production is growing at about twice the annual growth rate of the overall plastics industry. Rigid foams are used in refrigerators and freezers, and in sophisticated and efficient types of food processing and preserving equipment. By far the largest consumer is the building and construction industry; while flexible foams are leading the demand in the furniture, aviation and automotive industries. The book contains descriptions of 148 manufacturing processes of which 42 deal with application technology:

Raw Materials (39)
Special Additives (30)
Product Types (29)
Processing (8)
Applications (42)

448 pages. **$36**

REINFORCED COMPOSITES FROM POLYESTER RESINS 1972
by Dr. M. W. Ranney

1. POLYESTER INTERMEDIATES
   Hydroxy Intermediates (9)
   Carboxylic Acids-Anhydrides (14)
   Diels-Alder Adducts (8)
   Others (10)
2. ADDITIVES
   Catalysts (9)
   Accelerators-Promoters (12)
   Inhibitors (12)
   Thickeners (16)
   Color Controllers (10)
   Miscellaneous (3)
3. FLAME RETARDANTS
   Acid-Anhydrides (9)
   Hydroxy Intermediates (6)
   Other Reactants (16)
4. FORMULATIONS
   Acrylics (3)
   Nitrogen Compounds (4)
   Other Thermoplastics (6)
   Reactive Solvents (4)
   Low Profile Shrinkage (3)
5. PROCESSING
   Glass Fiber Treating (4)
   Other Processes (8)

166 Patent-based Processes on 324 pages. **$36**

PLASTIC PRINTING PLATES MANUFACTURE AND TECHNOLOGY 1971
by M. G. Halpern

Shows easy adaptation to traditional letterpress practice and automated equipment for plastic 3-dimensional relief plates.

1. PHOTOPOLYMERIZED PRINTING PLATES (2)
2. CELLULOSE POLYMERS (5)
3. OTHER POLYMERS (6)
4. ANCHOR LAYERS (6)
5. PHOTOINITIATORS (10)
6. MODIFICATIONS (22)
   Afterexposure Treatments
   Calendering Aids
   Speed & Contrast Aids
   Increasing Sensitivity
7. PHOTOCHEMICAL CROSSLINKING (9)
8. POLYAMIDE PLATES (12)
   Sensitizers
   Photomechanical Processes
9. PHOTOREPRODUCTION LAYERS (20)
10. INTAGLIO IMAGES AND ETCHABLE PLASTIC PLATES (7)
11. MOLDED PLATES (4)

294 pages. **$36**

VINYL AND ACRYLIC ADHESIVES INCLUDING PRESSURE SENSITIVES 1971
by K. Johnson

Includes 123 processes.

1. ACRYLICS (51)
   Pressure Sensitives
   Laminates
   Tire Cord Adhesives
   Anaerobics
2. POLYVINYL ACETATE (14)
   Hot Melts
   Wood Bonding
3. POLYVINYL ALCOHOL (10)
   Paper and Corrugated Board
   Water-Soluble Pressure Sensitives
   Cement Compositions
4. ETHYLENE COPOLYMERS (13)
   Hot Melts
   Atactic Polypropylene
5. OTHER VINYL POLYMERS (25)
   Iron-On Adhesives
   Fabric Bonding
   Pressure Sensitive Phenol-Aldehyde Resins
6. PRESSURE SENSITIVE TAPES (10)
   Release Coatings

287 pages. **$36**

EPOXY AND URETHANE ADHESIVES 1971
by Dr. M. W. Ranney

An ever increasing demand for epoxy adhesives makes them the leader of the industry. Urethane polymers and intermediates are also augmenting the adhesives market. Isocyanate monomers are applied to textile fibers, metals, and elastomers as primers for adhesion. Polymethylene polyphenyl isocyanates are used in formulations for bonding glass, metal, elastomers and wood to a variety of substrates. Polyurethanes with polyesters, polyamides, and elastomers meet many high performance requirements. 111 processes in 7 chapters:

Metal to Metal (25)
Fiber to Rubber (22)
Glass & Ceramics to Metal (18)
Polymer to Polymer (12)
Plastic to Metal (10)
Paper and Wood (13)
General Purpose Adhesives (11)

280 pages. **$36**

POLYIMIDE MANUFACTURE 1971
by Dr. M. W. Ranney
Chemical Process Review No. 54

The U.S. patent literature of the past ten years provides an excellent description of the many types of polyimides and their syntheses. This book is an attempt to collate and summarize those processes pertaining to the manufacture of polymers containing an imide grouping. Emphasis has been placed on practical, technically useful information. About 90 distinct processes of manufacture are described in eight chapters.

1. Polyamide Acids (20)
2. Polyimide-Esters from TMA (7)
3. Polyimide-Amides (11)
4. Modified Polyimides and Cross-Linking (11)
5. Specialty Intermediates (15)
6. Silicone-Fluorocarbon-Polysulfone Modifications (11)
7. Cellular Polyimides (5)
8. General Processing Techniques (9)

243 pages. **$35**

HOT-MELT ADHESIVES 1971
by M. McDonald

Reviews the U.S. patent literature from 1950 through early 1971. Hot-melts set fast enough to accommodate high-speed machinery in shoemaking, paper converting, bonding textiles (replacing hand and machine sewing), metal container sealing, etc. 63 processes:

1. **Hot-Melts for Shoemaking (14)**
2. **Bonding Paper & Paperboard (23)**
3. **Bonding Metals (7)**
4. **Bonding Plastics (6)**
5. **Bonding 2 or more Materials (7)**
6. **Textiles and Coated Substrates (6)**

238 pages. **$35**

POLYMERS AND ADHESIVES

FLAME RETARDANT POLYMERS 1970
by M. Ranney

Summarizes selected process technology for the use of fire retardant imparting additives and reactive intermediates for major polymeric plastic materials, with particular emphasis on recent technology in the areas of polyesters, polystyrene and polyurethane foam. There are 144 separate processes included, all based on the U.S. patent literature.

An abbreviated Table of Contents is listed. The numbers in parentheses indicate the number of processes included for each entry.

Polyethylene and Polypropylene (15)
Polystyrene (19)
Polyurethanes (50)
Polyesters (13)
Other Polymer Systems (26)
General Utility Additives (21)

263 pages. **$35**

POLYSULFIDE MANUFACTURE 1970
by C. Placek
Chemical Process Review No. 50

This report covers 73 processes dealing with polymers possessing the disulfide (-SS-) group. Basic Processes (14); Modified Polysulfide Polymers (13); Curing (12); Process Control (4); Physical Form (3); Single-Package Compositions (6). 141 pages. **$35**

POLYCARBONATES—RECENT DEVELOPMENTS 1970
by K. Johnson
Chemical Process Review No. 47

Part I—Aromatic Polycarbonates: Synthesis and Polymerization (17); Halogen-Containing Polycarbonates (9); Processing (20); Modified Polycarbonates (23); Applications (5). Part II—Aliphatic Polycarbonates: Cycloaliphatics (8); Linear Aliphatics (6). 298 pages. **$35**

FLUOROCARBON RESINS 1971
by Dr. M. W. Ranney
Chemical Process Review No. 51

1. Polytetrafluoroethylene (42). 2. Vinylidene Fluoride Elastomers (38). 3. Vinyl Fluoride. 4. Trifluorochloroethylene (15). 5. Fluorodienes (16). 6. Fluoroethers (13). 7. Fluorinated Nitroso Polymers (6). 8. Others (23). 9. General Processing Techniques (14). 226 pages. **$35**

ABS RESIN MANUFACTURE 1970
by C. Placek
Chemical Process Review No. 46

ABS (acrylonitrile-butadiene-styrene) resins make up one of the most rapidly growing segments of the polymer industry. **Straight ABS Materials; ABS Modified with Acrylic Derivatives; ABS from Alpha-Methylstyrene; Miscellaneous Modifiers; Modification of Properties; Process Variations.** 233 pages. **$35**

CELLULAR PLASTICS RECENT DEVELOPMENTS 1970
by K. Johnson

Describes 189 processes:
Polyolefins (15)
Polyvinyl Chloride (15)
Polystyrene (22)
Rubber (15)
Polyurethanes (71)
Polyesters and Epoxides (15)
Urea-Formaldehyde and Phenolic Resins (9)
Other Cellular Products (27)

280 pages. **$35**

ETHYLENE-PROPYLENE-DIENE RUBBERS 1970
by Dr. M. W. Ranney
Chemical Process Review No. 49

Summarizes patent literature relating to: Polyene Monomers—Polymer Synthesis; Catalysts and Activators; MW Regulators; Process Technology; Recovery Techniques; Modified Terpolymers; Adhesives; Miscellaneous Vulcanizates; Cross-Linking Agents. 272 pages. **$35**

COMPATIBILITY AND SOLUBILITY 1968
by I. Mellan

Normally, it requires laborious testing to determine compatibility of polymers, resins, elastomers, plasticizers, and solvents. Predictions made without testing or literature searching, are usually unreliable.

This book helps you evaluate proper materials by the use of 224 tables. 304 pages. **$20**

Noyes Data Corporation provides up-to-date industrial process publications for the food, chemical, textile, pollution control, and allied industries. Technology is changing very quickly, and our rapid publication techniques provide you with the latest technical know-how at the earliest possibility.

ENVIRONMENT AND POLLUTION

NONCATALYTIC AUTO EXHAUST REDUCTION 1972
by D. Post
Pollution Control Review No. 10

A review of 87 U.S. patents of recent origin. Unburned hydrocarbons and carbon monoxide are the principal components to be eliminated.

1. **Combustion**
   Initiation of Combustion
   Secondary Air Supply
   Mixing Devices
   Exhaust Gas Recycling
2. **Separation**
   Sorption
   Inertial Separation
   Electrostatic Separation
3. **Chemical Treatment**
   Removal of Lead
   Granular Reagents
   Gold-plated Fibers
4. **Other Methods**
   Ozone Injection
   Exhaust Pressure Control
   Fumes Collecting Tunnel
   Exhaust Monitoring

278 pages. **$36**

NITROGEN OXIDES EMISSION CONTROL 1972
by A. A. Lawrence
Pollution Control Review No. 9

Principal sources of air-polluting nitrogen oxides (NOx) are power plants and automobiles. How to control these emissions is shown in 79 patent-based processes of recent origin.

1. **Catalytic Conversion of Stack Gases**
2. **Adsorptive Techniques**
3. **Liquid Scrub Processes Purification of Waste Gases from Nitric Acid Processes**
4. **Combustion Techniques**
5. **Recovery and Utilization of NOx**
6. **Catalytic Conversion of Auto Exhaust NOx**
7. **Noncatalytic Means of NOx Reduction**
8. **Health Devices Removing NOx from Tobacco Smoke**

212 pages. **$36**

POLLUTION ANALYZING AND MONITORING INSTRUMENTS 1972

All together 157 companies are represented. About 350 instruments or other analytical equipment pieces are described.

The major listings include:
1. **Diagrams of the apparatus with a description of its components and accessory equipment.**
2. **Technical discussion of the analytical reactions involved.**
3. **Specifications.**
4. **Brief statement about the specific and all-around uses of the apparatus.**

The data appearing in this volume were selected by the publisher from each manufacturer's literature at no cost to, nor influence from the manufacturers of the equipment.

Supplies detailed technical data on the types of pollution measurements and analyses which can be made and lists the companies providing such instrumentation. 354 pages. **$36**

WASTE TREATMENT WITH POLYELECTROLYTES 1972
by S. Gutcho
Pollution Control Review No. 8

Polyelectrolytes are high-molecular-weight polymers. They are water soluble or water dispersible and may be anionic, cationic or nonionic. They are now being used in the treatment of sewage, the clarification of industrial waste water, and the purification of contaminated rivers and lakes. 144 Processes based on U.S. patents.

1. **Anionic Polyelectrolytes** (26 processes)
2. **Cationic Polyelectrolytes (52)**
3. **Nonionics (10)**
4. **Polyelectrolytes from Natural Sources (25)**
5. **Aspects and Methods (12)**
6. **Types & Syntheses (12)**
7. **Sewage Treatment (7)**

Attention is focused on phosphate removal, low ash sewage sludge formation, and on recycling of waste water for acceptable purity. 237 pages. **$36**

DETERGENTS AND POLLUTION 1972
Problems and Technological Solutions
by H. R. Jones
Pollution Control Review No. 7

Thanks to a complicated series of scientific disputes over alleged deleterious effects on environment and human health, there is probably more confusion about how to properly satisfy this market, than there is about any other class of chemical products.

188 Processes:
Manufacture of Linear Olefins
Purification Thereof
Manufacture of Linear Paraffins
Their Purification
Routes to Linear Alcohols
Their Purification
Alkylaromatic Hydrocarbons
Their Purification
Other Raw Materials
Sulfation and Sulfonation
Detergent Formulations
The "Builder" Problem
Removal from Sewage
Future Trends

268 pages. **$36**

ENVIRONMENTAL CONTROL IN THE INORGANIC CHEMICAL INDUSTRY 1972
by H. R. Jones
Pollution Control Review No. 6

Inorganic pollution control is still a field that must receive greater attention and action. Because of the nature of the industry obsolete units remain sufficiently profitable to continue in use.

Provides helpful directions for adequate pollution control. Commercial processes are shown, as well as detailed technology from the U.S. patent literature:
1. Wastewater Characteristics
2. Water Pollution Problems in the Manufacture of 33 Specific Chemicals
3. Actual Pollution Control Processes for Specific Wastewaters
4. Air Pollution Problems in the Manufacture of 15 Specific Chemicals
5. Actual Pollution Control Processes for Atmospheric Emissions
6. Awareness and Future Trends.

249 pages. **$36**

ENVIRONMENT AND POLLUTION

ENVIRONMENTAL CONTROL IN THE ORGANIC AND PETROCHEMICAL INDUSTRIES 1971
by H. R. Jones
Pollution Control Review No. 3

1. WASTE SOURCES
2. WATER USE
3. WATER POLLUTION
4. RECEIVING WATERS
5. REUSE OF WATER
6. PHYSIOLOGICAL EFFECTS
7. POLLUTION PARAMETERS
8. MONITORING
9. CHEMICAL CLASSIFICATION
10. SPECIFIC PRODUCTS
11. WASTE WATERS
12. INDUSTRIAL-MUNICIPAL TREATMENT PLANTS
13. INTERNAL IMPROVEMENTS
14. PHYSICAL TREATMENTS
15. CHEMICAL TREATMENTS
16. BIOLOGICAL TREATMENTS
17. OTHER DISPOSAL METHODS
18. WATER ECONOMICS
19. AIR POLLUTION
20. AIR POLLUTANTS REMOVAL
21. IMMEDIATE RECOMMENDATIONS
22. REFERENCES-INDEXES

257 pages. **$36**

HYDROGEN SULFIDE REMOVAL PROCESSES 1972
by P. G. Stecher
Pollution Control Review No. 5

Furnishes reliable and efficient methods of H_2S removal from gases, air, and liquids. 80 Patent-based processes.

1. **ABSORPTION FROM GASES WITH INORGANICS**
2. **WITH ORGANICS**
   Alkanolamines
   Other Amines
   Esters and Ethers
   Other Compounds
   Recycling Absorbents
3. **REMOVAL FROM GASES BY OTHER METHODS**
   Oxidation
   Adsorption
   Hydrotreating
   Use of Electrolysis
   H_2S Recycle in White Liquor Regeneration
   Ion Exchange Resins
   Molecular Sieves
   Adsorption plus Oxidation
4. **REMOVAL FROM LIQUIDS**
   From Fluid Hydrocarbons
   From Aqueous Solutions

288 pages. **$36**

SULFUR DIOXIDE REMOVAL FROM WASTE GASES 1971
by A. V. Slack
Pollution Control Review No. 4

Reviews the problems of smelter operators, power plants, refineries, sulfur acid plants, and Claus process sulfur plants.

1. THE PROBLEM
   Alternatives of Control
2.-3. THROWAWAY PROCESSES
   Dry Systems
   Wet Systems
4. ECONOMIC FACTORS IN RECOVERY
5. RECOVERY PROCESSES
6. ALKALIS AS ABSORBENTS
7. ALKALINE EARTH ABSORBENTS
8. METAL OXIDE SORPTION
9. ADSORPTION PROCESSES
10. CATALYTIC OXIDATION & REDUCTION
11. OTHER RECOVERY METHODS
   Organic Sorbents
   Fuel Gasification
   Gas Cleaning

200 pages. **$36**

MERCURY POLLUTION CONTROL 1971
by H. R. Jones
Pollution Control Review No. 1

In 1970 mercury pollution hit the headlines. Mercury ions were expected to react with other inorganic ions in the water, form precipitates, and sink harmlessly to the bottom. But when organic molecules from sewage or dead organisms are present, mercury reacts with these molecules to form toxic methyl mercury compounds which are excreted very slowly by fish and man. This book explains in detail what measures can be taken to prevent further pollution and to remove existing contamination. 13 Chapters: 1. The Fish Episode. 2. Production of Mercury. 3. Its Properties. 4. Uses of Mercury. 5. Its Toxicology. 6. Detection and Determination. 7. Air Pollution by Mercury. 8. Cleanup of Spilled Mercury. 9. Removal from Gases. 10. From Liquids. 11. From other Materials. 12. Legislation. 13. Trends and Problems. 251 pages. **$35**

HAZARDOUS CHEMICALS HANDLING AND DISPOSAL 1970
The Institute of Advanced Sanitation Research International

This publication is the record of a symposium of hazardous chemicals handling and disposal.
Hygiene Control—Handling Certain Hazardous Chemicals
Hazardous Chemicals Handling in "The Pharmaceutical Chemical Industry"
Effects of Hazardous Chemicals on Biochemical Oxygen Demand Tests of Stream Water Samples
Pesticide Handling in an Industrial Plant
Thermal Method for the Disposal of Hazardous Wastes
Land Disposal of Hazardous Chemicals
Hazardous Chemicals Disposal in a Large Chemical Complex
Research, Development and Application of New Biological Methods for Toxic Wastes Degradation and Disposal
Design and Tests of a Portable Cask for Explosive Chemicals

130 pages. **$15**

HAZARDOUS CHEMICALS HANDLING AND DISPOSAL 1971
The Institute of Advanced Sanitation Research International

This is the record of the Second Symposium.
Pesticide Container Decontamination
Fate and Effects of Pesticides
Separation of Organic and Inorganic Chemicals in a Waste Stream
Vector Problems in Waste Disposal
Plants Poisonous to Livestock
Land Application and Anaerobic Lagoon Disposal of Waste
Reverse Osmosis for Reclamation and Reuse of Chemical and Metal Waste Solutions
Thermal Methods for Destruction of Chemical Waste
Specialty Gases and Hazardous Wastes
Education of the Public
Hazardous Chemicals from Natural Sources
Maximum Allowable Concentrations and Water Quality
Biochemical Oxygen Demand and its Meaning

163 pages. **$20**

PAPER RECYCLING AND THE USE OF CHEMICALS 1971
by M. McDonald

Paper accounts for almost half of our solid waste (ca. 40 million tons per year accumulate in the U.S. alone). But paper can be reused. Treatment is mechanical and chemical, using many substances for removing printer's ink, plastic coatings, wax, adhesives, etc. The secondary fibers industry, on the threshold of an era of expanded growth, is consuming a steadily increasing amount of solvents, bleaches, and other chemicals discussed in this book. 68 processes.

Removing Ink (23)
Removing Coatings and Impregnants (19)
Dispersing Coatings (11)
Repulpable Adhesives and Coatings (5)
Miscellaneous (10)

304 pages. **$36**

CATALYTIC CONVERSION OF AUTOMOBILE EXHAUST 1971
by J. McDermott
Pollution Control Review No. 2

The need for controlling the exhaust gas emissions from gasoline, diesel, and jet engines is becoming increasingly urgent. The emission of carbon monoxide, nitrogen oxides, lead compounds, reactive olefins, and even of carcinogens such as benzopyrene, has become the concern of almost every legislature.

This book summarizes the U.S. patent literature relating to combustion catalysts, such as various metal oxides, platinum, and palladium. In general, the catalytic unit, mounted after the exhaust manifold, completes the oxidation of unburned hydrocarbons and converts the carbon monoxide to carbon dioxide. 94 processes and devices are described.

Catalytic Converter Design (31)
Catalyst Bed Design (17)
Catalysts (32)
Catalytic Units (14)

208 pages. **$36**

CORROSION RESISTANT MATERIALS HANDBOOK 1971
Second Edition
by I. Mellan

Corrosion, always an urgent and persistent problem, bothers and baffles us even more today, because of the quantity and complexity of chemicals in our polluted biosphere. This book will help you cut losses by enabling you to choose the proper commercially available corrosion resistant material. The index lists thousands of corrosive substances and refers you to specific recommendations in the 147 tables:

Synthetic Resins (90 tables)
Elastomers (17 tables)
Cements (11 tables)
Glass & Ceramics (4 tables)
Wood (3 tables)
Metals & Alloys (22 tables)

487 pages. **$25**

DESALINIZATION BY REVERSE OSMOSIS 1970
by J. McDermott

Summarizes 71 U.S. patents relating to reverse osmosis with emphasis on membrane technology. Chapters on Membrane Preparation; Equipment Design; Modified Reverse Osmosis Techniques; Vapor Permeation; Energy Sources; Solution Modifications; Submerged Units; Special Units; Detailed Engineering Outlays and Drawings. 209 pages. **$35**

DESALINIZATION BY FREEZE CONCENTRATION 1971
by J. McDermott

This volume is the second of a series of three, dealing with the major desalinization processes. The information contained will provide needed knowhow concerning the renewed interest in freeze concentration processes, such as vacuum freezing, surface freezing, hydrate processes and crystal washing. 207 pages. **$35**

DESALINIZATION BY DISTILLATION 1971 RECENT DEVELOPMENTS
by J. McDermott

The prospect of revitalizing arid areas, together with the problem of providing increased quantities of potable water has produced considerable research. In the past two years, over one hundred patents relating to distillation processes have been issued in the U.S. 194 pages. **$35**

SOUNDPROOF BUILDING MATERIALS 1970
by Dr. M. W. Ranney

There is increasing public awareness of the health menace of excess noise. The Walsh-Healy Act regulates noise in factories with government contracts. These factors, added to the current interest in other forms of pollution abatement, signify a rapidly growing market. Describes 100 patent-based processes. 217 pages. **$35**

OTHERS

MICROENCAPSULATION TECHNOLOGY 1969
by M. W. Ranney

Based on 81 U.S. patents describing encapsulating techniques for particles and droplets. **Phase Separation Methods; Interface Reactions—Polymerizations. Physical Methods (Multi-Orifice Centrifugal, etc.) and Applications (transfer sheets, etc.)** are the chapter headings. There is considerable potential for future application of microencapsulated products. 275 pages. **$35**

FIRE RETARDANT BUILDING PRODUCTS AND COATINGS 1970
by Dr. M. W. Ranney

The value of this report is indicated by the number of building materials here.

1. **Wood Impregnation**
2. **Fiberboard**
3. **Ceiling Tile and Panels**
4. **Asphaltic Products**
5. **Intumescent Coatings**
6. **General Formulations**
7. **Adhesives**

186 pages. **$35**

SUPERCONDUCTING MATERIALS 1970
by P. Conrad
Electronics Materials Review No. 11

Gives 99 detailed manufacturing and fabrication processes. Superconductive materials can improve long distance transmission of electric power, provide more compact memories for computers, improve magnets for physics research and thermonuclear power reactors.
135 pages. **$35**

PHOTOCONDUCTIVE MATERIALS 1970
by M. Sittig
Electronic Materials Review No. 8

Photoconductors have a number of important applications, such as television caméra tubes, solar cells, photoelectric cells, solid state light amplifiers, electrophotographic copying machines. 86 processes. 288 pages. **$35**

U. S. MARKETS

PHARMACEUTICAL AND COSMETIC FIRMS USA 1972
Second Edition

Lists about 750 leading U.S. companies with these categories:

Prescription Drugs
OTC Drugs
Biologicals
Vet. Pharmaceuticals
Hospital Supplies
Dental Supplies
Cosmetics
Private Formulas giving
Name & Address
Phone Numbers
Parent Company
Annual Sales
No. of Employees
Executives
Plant Locations
Divisions and Subsidiaries
Foreign Subsidiaries
Types of Products
Brand Names of OTC's and Cosmetics

2 Indexes 278 pages. **$24**

POLLUTION CONTROL COMPANIES U.S.A. 1972

Provides a marketing guide to the U.S. pollution control industry.

The first section is an alphabetical listing of ca. 1,500 companies or company units (divisions, subsidiaries, etc.) which manufacture or supply products useful in the areas of air, water, noise, and radiation pollution control and waste management.

The second section lists more than 500 companies and company units which provide such professional services as consulting, design engineering, and analyses of air or water pollutants.

The address and telephone number of each company is listed together with a brief description of the company's pollution control products or services, although these may constitute only a small portion of the company's business. 239 pages. **$24**

CHEMICAL GUIDE TO THE UNITED STATES 1971
Sixth Edition

Describes over 400 of the largest chemical firms in the U.S.: Those who actually carry out chemical syntheses in their plants. Companies and factual data about them are listed in alphabetic order, followed by an index which gives companies, subsidiaries and divisions, again by strict alphabetic arrangement.

Whenever available the following information is given in detail:

Name and Address with Zip Code
Ownership
Annual Sales
Number of Employees
Principal Executives and Titles
Plant Locations
Products
Subsidiaries and Affiliates
Internal Structure

Also gives information on closely held firms, joint ventures, and others that do not publish annual reports. 191 pages. **$20**

FOOD AND BEVERAGE PROCESSING INDUSTRIES 1971

The usefulness of this book derives from its organization. It is divided into 2 sections. The first section is an alphabetical listing of approximately 3,500 U.S. food firms giving the correct name, address and zip code. Also concise, pertinent information (where available) such as:

(a) Annual Sales
(b) Number of Employees
(c) Name of Chief Executive
(d) Product Types

The second section is arranged numerically according to zip code with the companies once again listed alphabetically within their proper zip code numbers, thereby providing you with an easy-to-use geographical index to the U.S. food industry.

It is a great help in marketing efforts by providing the means for forward geographical planning.

169 pages. **$20**

TEXTILES

ACRYLIC AND VINYL FIBERS 1972
by M. Sittig
Chemical Process Review No. 62

Acrylic fibers are the most wool-like of the noncellulosic fibers, while modified acrylic fibers have a silk-like feel.

1. **Acrylonitrile Manufacture**
2. **Acrylonitrile Polymers**
3. **Spinning Solutions**
4. **Spinning Fibers**
5. **Vinyl Chloride Manufacture**
6. **Polyvinyl Chloride**
7. **PVC Fiber Manufacture**
8. **Vinyl Acetate Manufacture**
9. **Polyvinyl Acetate**
10. **Polyvinyl Alcohol**
11. **PVA Fiber Production**
12. **Acrylic Ester Manufacture**
13. **Polyacrylic Esters**
14. **Vinylidene Chloride**
15. **Polyvinylidene Chloride**
16. **Fibers from It**
17. **Vinyl Fluoride Production**
18. **Polyvinyl Fluoride**
19. **Fibers from It**
20. **Future Trends**

Altogether 123 patent-based processes are described in detail. 331 pages. **$36**

POLYAMIDE FIBER MANUFACTURE 1972
by M. Sittig
Chemical Process Review No. 61

The polyamides or nylons are the oldest wholly synthetic fibers. This is a book about the actual production of nylon fibers based on no less than 223 U.S. patents.

1. **Introduction**
2. **Polyamide Compositions**
3. **Adipic Acid Manufacture**
4. **Higher Dibasic Acids**
5. **Alternate Acid Types**
6. **Adiponitrile Manufacture**
7. **Diamine Manufacture**
8. **Nylon Salt**
9. **Caprolactam—Conventional Routes**
10. **Caprolactam—Alternate Routes**
11. **Higher Lactam**
12. **Polylactam Processes**
13. **Polyamide Production**
14. **Fiber Production**
15. **Waste Recovery and Pollution Problems**
16. **Future Trends**

262 pages. **$36**

FLOCKED MATERIALS 1972
Technology and Applications
by E. L. Barden

Demand for flocked materials and articles covered with fibers or similar substances is increasing. Fibers and flakes are used to flock walls and ceilings, because the flocking is decorative, soundproofs, waterproofs, acts as a thermal insulator and is highly durable.

Plush toys, velvet greeting cards, good quality carpeting, even battery plates and synthetic turfs can be produced by flocking. 111 processes:

Electrostatic Methods (23)
Mechanical Flocking (4)
Types of Flock (3)
Flock Treatments (8)
Adhesives (13)
Decorative Effects (12)
Wearing Apparel (16)
Home Furnishings (11)
Automobiles (7)
Others (14)

294 pages. **$36**

POLYESTER FIBER MANUFACTURE 1971
by M. Sittig
Chemical Process Review No. 56

In 1970 polyester became the number one U.S. fiber, surpassing the consumption figure for nylon by another 100 million pounds. Predictions are that continuous processing will be common in 1975. A total of 116 processes is given.

1. INTRODUCTION
2. ECONOMICS
3. VARIOUS POLYESTER COMPOSITIONS (6)
4. RAW MATERIALS AND PURIFICATION (9)
5. DIMETHYL TEREPHTHALATE (8)
6. INTEGRATED POLYESTER PRODUCTION (5)
7. BIS (HYDROXYETHYL) TEREPHTHALATE (23)
8. PREPOLYMERS (8)
9. POLYCONDENSATION (28)
10. POLYMER AFTER-TREATMENT (2)
11. SCRAP RECOVERY (3)
12. FIBER PRODUCTION (24)
13. FUTURE TRENDS

214 pages. **$36**

SYNTHETIC PAPER FROM SYNTHETIC FIBERS 1971
by K. Johnson

This book provides a summary of the U.S. patent literature through 1970, relating to 74 processes for producing synthetic papers on conventional papermaking machinery:

1. **Cellulosics** (37 processes)
   - Cellulose Derivatives (15)
   - Cross-Linked Cellulose (4)
   - Blends with Other Fibers (10)
   - General (8)
2. **Polyamides** (8)
   - General (5)
   - Nylon + Cellulose (3)
3. **Polyacrylonitrile** (13)
   - Bonding (7)
   - General (4)
   - Blends (2)
4. **Synthetic Fibers** (16)
   - General Processing (10)
   - Specialty Papers (2)
   - Polytetrafluoroethylene (3)
   - Polyester (1)

236 pages. **$35**

NONWOVEN FABRIC TECHNOLOGY 1971
by M. McDonald

Nonwovens are structures produced by bonding or interlocking of fibers, accomplished by mechanical, chemical, thermal, or solvent means. Low cost is the primary advantage over woven or knitted products, resulting in a wide variety of disposable items from flush-away diapers to industrial uniforms. U.S. hospitals and other medical organizations bought over $100 million worth of disposable hospital gowns and other nonwovens in 1970 of what is thought to be a $800 million market. This book emphasizes those processes that turn out materials for making garments, draperies, upholstery, sheets, etc. 121 processes are described:

1. Resin Bonding (22 processes). 2. Spunbonding (23). 3. Needle Punching (28). 4. Fluid Pressure (13). 5. Heat Bonding (10). 6. Web Formation (7). 7. Miscellaneous (18). 240 pages. **$35**

TEXTILES

MULTICOMPONENT FIBERS 1971
by C. Placek

Discusses processes for producing multicomponent fibers, that is, those fibers that consist of two or more polymeric compounds spun together. The spinning of multicomponent fibers is one method used to produce bulked or crimped fibers. The difference in physical properties of the two filaments results in a bulked yarn.

This report covers 106 patents issued since 1960. The processes range over a wide spectrum of fiber technology—composition, properties, physical forms and spinning techniques. Numbers in ( ) indicate the number of processes described in each chapter.

Variations of the Same Polymer (24), Chemically Unrelated Components (10), Spontaneous Crimp (2), Permanent Crimp (12), Specific Properties of Fibers (8), Production of Sheath-Core Structures (10), Side-by-Side Components (7), Spinning Technology (33).

225 pages. **$35**

SPANDEX MANUFACTURE 1970
by M. McDonald
Chemical Process Review No. 48

This book covers methods of making spandex fibers, that is, the conversion of polyurethanes into fibers, as described in the U.S. patent literature. The U.S. patent literature has the most complete and comprehensive process information available, and as such, this publication will give you key processing information for this fast-growing fiber. The Table of Contents below gives the scope of this volume. The numbers in ( ) represent the number of processes included.

Wet Spinning Processes (13), Solvent Spinning Processes (19), Melt Spinning Processes (3), Chemical Composition and Raw Materials (19), Improving Resistance to Ultraviolet Light and Oxidation (9), Improving Dyeability of Spandex (9), Miscellaneous (8).

191 pages. **$35**

SOIL RESISTANT TEXTILES 1970
by Dr. M. W. Ranney
Textile Processing Review No. 5

Ideal soil release finishes must be capable of releasing stains readily and preventing redeposition of soil during laundering. Treatments should render manmade fibers and durable press reactants less attractive to oily stains and should be more easily wetted.

This report summarizes the developments in soil retardant and soil release finishes in both the carpet industry and in textile manufacture. It includes the newest technology associated with the use of acrylates and fluorochemical treatments. The numbers in ( ) following each treating agent indicate the number of processes covered for that particular compound.

Introduction: Metal Oxides and Salts For Carpet Treatment (15), Acrylic and Vinyl Polymers (10), Silicones (6), Fluorochemical Compounds (72), General Treatments (14). 216 pages. **$35**

WATERPROOFING TEXTILES 1970
by Dr. M. W. Ranney
Textile Processing Review No. 4

This Textile Processing Review summarizes the technology of water resistant treatments for textiles and fabrics as described in the U.S. patent literature since the early 1950's. 246 waterproofing processes are included—64 relate to use of fluorochemicals.

The numbers in ( ) after each entry in the Table of Contents, where the treatment processes are organized by the agent used, indicates the number of production processes for each agent.

Production Processes for Waterproofing Textiles using the following Agents: Metal Salts and Wax-Containing Formulations (44), Silicones and Alkyl Polysiloxanes (53), Organofunctional Silicones and Fluorosilanes (20), Acrylics (8), Nitrogen Containing Compounds (30), Fluorochemical Compounds (64), Elastomer, Vinyl, Polyolefin Vapor Permeable Fabrics (27), Miscellaneous Treatments (9). 353 pages. **$35**

CREASEPROOFING TEXTILES 1970
by Dr. M. W. Ranney
Textile Processing Review No. 2

Summarizes detailed process information relating to textile creaseproofing agents used to obtain wash and wear, or permanent press fabrics. Over 300,000 words, describes 343 processes in this field. Shows you chemical agents used, and processes by which they are applied.

Dimethylolethylene Urea and Related Compounds, Aldehyde-Urea Condensates, Uron Resins, Aminoplasts—Catalyst Performance, Melamine Derivatives, Triazones, Carbamates, Other Nitrogen-Containing Compounds, Phosphorus-Amino Compounds—Aziridines, Aminoplast-Thermoplastic Resin Compositions, General Processing Techniques and Formulations, Aldehydes, Acetals, Epoxies, Epihalohydrins, Sulfones, Sulfonium Salts, Cross-Linking Agents, Miscellaneous, Polymeric Coatings—Rubber, Vinyl, Silicones, Radiation Curing, Wool, Nylon, and Others. Indexes. 460 pages. **$35**

FLAME RETARDANT TEXTILES 1970
by Dr. M. W. Ranney
Textile Processing Review No. 3

Describes 177 commercial processes to produce flame retardant textiles and fabrics.

Most activity is based on chemical modification of cellulose through hydroxyl groups. Use of phosphoric acid, urea-phosphates, and other phosphorylating agents all confer flame retardant properties to cellulose. A significant portion of this book is devoted to the latest in application of phosphorus containing materials.

Numbers in ( ) indicate the number of processes described.

Ammonium Salts, Borates (12); Antimony, Titanium Metal Oxides (25); Amine-Phosphorus Products (21); Aziridines, APO, APS (21); Methylol-Phosphorus Polymers, THPC (27); Phosphonitrilic Chlorides (9); Triallyl Phosphates and Phosphonates (26); Silicones Isocyanates, Miscellaneous (10); Nylon, Acrylics (18). 373 pages. **$35**

OVERSEAS MARKETS

TEXTILE GUIDE TO EUROPE 1972
Second Edition

This improved edition describes about 2,600 of the leading yarn, fabric, knitwear and hosiery manufacturers in 18 countries of Western Europe.

Each chapter is preceded by a resume of the textile industry in the country concerned.

Where available the following information is given for each company:

Full Name
Complete Address
Telephone Numbers
Telex Numbers
Ownership
Sales
Number of Employees
Plant Locations
Principal Executives
Range of Products
Domestic Subsidiaries
Foreign Subsidiaries

In each section the salient features are pinpointed through the expedient of a statistical analysis.
290 pages. **$24**

ELECTRONICS INDUSTRY OF JAPAN 1972

Aimed at giving you a concise but detailed picture of the present state of this important industry. The first section gives details of 145 major companies.

Complete Name
Address
Founding Date
Capital & Assets
Sales & Profits
Executives
No. of Employees
Major Stockholders
Bankers
Agreements
Products

The second section gives more limited data on a further 714 companies. The third section gives statistics. The book is completed by an alphabetical list of Japanese brand names. Prepared in Japan. 147 pages. **$24**

FOOD GUIDE TO EUROPE 1972
Second Edition

Describes 1,600 of the leading food processing companies located in 18 countries of Western Europe. Where available and pertinent, the following information is given:

Name
Address
Telephone Number
Telex Number
Ownership
Sales
Number of Employees
Plant Locations
Principal Executives
Products
Domestic Subsidiaries
Foreign Subsidiaries

Each chapter includes a short resume of the food industry in the country concerned, giving the salient features of the industry and production and trade statistics. Prepared in our London office. 325 pages. **$24**

WORLD PHARMACEUTICAL FIRMS 1972

Lists alphabetically by country the names and full addresses of over 6,000 manufacturers of ethical drugs, pharmaceutical specialties and basic pharmaceutical materials in all the major producing areas of the world.

For easy reference the 52 countries are arranged in geographical areas:

North America
South America
Western Europe
Eastern Europe
Africa
Australasia
Asia
North Africa and Middle East

This book will enable companies dealing with the pharmaceutical sector to mail sales literature, help plan and evaluate area representation, assist in the preparation of market reports and provide useful references.

Prepared in our London office.
122 pages. **$24**

EUROPEAN PAINT MANUFACTURERS 1971

Developing technology in the field of polymer-based and solventless coatings has resulted in European acquisitions and mergers. Describes about 1,000 of the most important paint manufacturers in 17 countries of Western Europe. Entries are listed alphabetically by country and the following data are presented as fully as they were obtainable:

Name
Address
Telephone Numbers
Telex Numbers
Number of Employees
Principal Executives
Products
Plant Locations
Domestic Subsidiaries
Foreign Subsidiaries
Sales Volume

Prepared in our London Office.
155 pages. **$24**

EUROPEAN CHEMICAL DISTRIBUTORS 1971

A considerable proportion of the sales of the world's major chemical companies are made through distributors, particularly in foreign countries. The book is divided into a main section, arranged by European distributors, and an index of worldwide companies the 1,300 distributors are representing.

Includes:

Austria	Luxembourg
Belgium	Netherlands
Denmark	Norway
Finland	Portugal
France	Spain
Germany	Sweden
Greece	Switzerland
Iceland	Turkey
Ireland	United Kingdom
Italy	

The chemical companies represented by the distributors are cross-referenced in the index.
264 pages. **$24**

EUROPEAN PHARMACEUTICAL MARKET REPORT 1971
Second Edition

Gives statistical information from 1960 to the end of 1969 on the 5,000 million dollar drug industry of the EEC (European Economic Community) and of the EFTA (European Free Trade Association). Type of information included:

General and Historical
Growth Potentials
Production and Sales
Domestic and Foreign Trade
Distribution—Pricing
Market Structure
Structure of the Industry
Names and Addresses of Trade Associations and of over 1,000 Manufacturers
Research and Development
Government Legislation
Health Service Expenditures

This book, a greatly expanded version of the earlier successful study, will bring you up to date with the European pharmaceutical scene and will give you a coherent picture of this complex market. 158 pages. $36

TEXTILE INDUSTRY OF JAPAN 1971

Relates significant developments of nearly 900 companies:

Name and Address
Principal Officers
Employees
Capital and Sales
Total Assets
Bankers & Stockholders

Statistically analyzes the whole Japanese textile industry:

Production
Trade
Raw Materials
Consumption
Employment Pattern
Wages & Productivity
Important Trademarks

Increasing pressures from expanding developing countries and restricted imports by the U.S. are producing many changes evaluated here.

205 pages. $35

CHEMICAL GUIDE TO EUROPE 1971
Fifth Edition

Prepared in our London Office to give on-the-spot coverage. Describes ca. 1,100 companies in the 19 countries of Western Europe which together constitute a market almost as large as that of the U.S. Includes all major European Chemical Companies. Gives this information (where pertinent and available):

Name and Address
Telephone and Telex Numbers
Ownership
Plant Locations and Products
Internal Structure
Local Subsidiaries and Affiliates
Foreign Subsidiaries and Affiliates
Principal Executives
Annual Sales
Number of Employees

Also companies which are predominantly non-chemical, but have important chemical interests, are included. 288 pages. $20

EUROPEAN KNITWEAR AND HOSIERY MARKET REPORT 1970

17 countries included. Trend of activity established. Size structure of domestic market determined. Information on yarn consumption, knitwear and hosiery production, foreign trade, number of companies and employees engaged in manufacture. Comprehensive selection of statistical material. 166 pages. $35

X-RAY CONTRAST AGENTS 1971
by M. Gutcho

Describes 55 processes suitable for the preparation of new x-ray contrast agents.
1. Polyiodobenzoic Acid Derivatives (16). 2. Polyiodophthalic Acid Derivatives (5). 3. Iodophenyl Derivatives (12). 4. Radiopaque Formulations (9). 5. Miscellaneous Contrast Media (4). 6. Process Improvements (9). 7. Other Uses for Radiopaque Materials. 130 pages. $35

ARTIFICIAL KIDNEY SYSTEMS 1970
by M. Gutcho

Contains significant, detailed technical data, based on the patent literature relating to manufacturing, assembly, and operation of artificial kidney systems. Dialyzers in Artificial Kidney Systems (17); The Design of Dialyzer Parts (9); Dialysate Modifications (12); Lung and Kidney Machines (6); Dialyzers (6); Blood Purification Processes (2). 320 pages. $35

SUSTAINED RELEASE PHARMACEUTICALS 1969
by A. Williams

To ensure a long lasting, continuous and not too intensive action of a therapeutic agent in the human body, it is necessary to delay the absorption.

Aside from chemical alterations of the drug in question, the problem can be solved by the preparation of the so-called sustained release medicaments for which 89 processes are given here. 273 pages. $35

NUCLEOTIDES AND NUCLEOSIDES 1970
by S. Gutcho

Organic Synthesis of Nucleotides in General, and of Specific Nucleotides, Fermentation Procedures for Nucleotides in General and for Specific Nucleotides, Enzymatic Digestion of Nucleic Acids, Nucleotides Coenzymes, Cyclic Nucleotides, Dinucleoside Phosphates, Purification Techniques, General Procedures, Nucleotides as Flavor Enhancers. 200 pages. $35

KEY EUROPEAN INDUSTRIALS 1970

This important new directory describes 1,000 leading manufacturers in 19 countries of Western Europe. Prepared by our London office, it will provide the hard-pressed executive with a unique, on-the-spot guide to the activities of Europe's key industrial companies and groups. Find out who owns whom.

Full name and address
Telephone and telex numbers
Share capital, sales, profit
Number of employees
Principal executives
Range of products and activities
Domestic and foreign subsidiaries

Volume I — (EFTA) — Austria, Denmark, Finland, Iceland, Ireland, Norway, Portugal, Sweden, Switzerland, United Kingdom. 180 pages.
Volume II — (EEC) — Belgium, France, Germany, Greece, Italy, Luxembourg, Netherlands, Spain, Turkey. 182 pages.
2 Volumes—$35

PETROCHEMICAL INDUSTRY OF JAPAN 1970

The bulk of the book is a detailed guide to the 205 manufacturers in the industry. Companies are listed alphabetically, giving their address, capital, sales, number of employees, president and, where relevant, ownership. For every company, the existing and planned capacity of each product is listed, together with its plant location and expected completion date. 147 pages. $35

ELECTRONICS MANUFACTURERS OF WESTERN EUROPE AND U.S.A. 1970

A comprehensive guide to names and addresses of electronics manufacturers in 16 countries.

Enables you to mail sales literature and promotional material as effectively as possible; helps to plan and evaluate area representation more accurately; is a useful day-to-day work of reference. 131 pages. $19

EUROPEAN FOOD MARKET RESEARCH SOURCES 1970

Enables you to pinpoint publications most likely to be of assistance.

References are classified by Government statistics and reports, other statistics and reports, trade associations, food trade journals; other newspapers and periodicals, directories, advertising statistics, bank reviews. 111 pages. $19

ELECTRONICS GUIDE TO EUROPE 1969

Contains company profiles of over 600 leading electronics manufacturers in 14 countries. Arranged for easy comparison: Full name and address, ownership, principal executives, product range, domestic and overseas subsidiaries, plant location, latest sales figures, number of employees. 150 pages. $20

TETRACYCLINE MANUFACTURING PROCESSES 1969
(2 Volumes)

CTC, Oxytetracycline, TC, DMTC and DMCTC, 2N-Derivatives, Position 4 Derivatives, 6-Methylene Derivatives, 6-Deoxy Derivatives, Anhydrotetracyclines, 7-and/or 9-Derivatives, 11a-Halo Derivatives, 5a, 11a-Dehydrotetracyclines, 12a-Derivatives, Epimers, Mechanism Study intermediates. Indexes. 931 pages. (2 Volumes) $45

VITAMIN B₁₂ MANUFACTURE 1969
by R. Noyes
Chemical Process Review No. 40

Vitamin B₁₂ active substances are important therapeutic products for treatment of pernicious anemia. Also used for treatment of various other human ailments, and as a veterinary growth factor. This book offers various methods of producing vitamin B₁₂ active substances. 327 pages. $35

FIBRINOLYTIC ENZYME MANUFACTURING 1969
by T. Rubel
Chemical Process Review No. 38

Methods for production and purification of fibrinolytic agents, their precursors and activators. Emphasis is on urokinase and streptokinase which activate plasminogen to form plasmin.

Plasminogen and Fibrinolysin, Urokinase, Streptokinase and Streptodornase, Other Fibrinolytic Enzymes. Indexes. 139 pages. $24

VITAMIN E MANUFACTURE 1969
by T. Rubel
Chemical Process Review No. 39

This review relates the known methods for the preparation of tocopherols from natural products or by synthetic means, and conversion of non-alpha tocopherols.

Introduction; Tocopherols From Deodorizer Sludge, Conversions to Alpha Tocopherol, Synthesis of Tocopherols, Miscellaneous Related Processes, Indexes. 114 pages. $24

ANTIOBESITY DRUG MANUFACTURE 1970
by Dr. B. Idson

Describes 162 processes for drugs or compounds proposed or employed for the reduction of body weight. The majority of the preparative processes is for sympathomimetic drugs of the pressor amine or amphetamine type. These substances, sometimes in connection with carboxymethylcellulose, produce a diminution of the desire for food intake. 193 pages. $35

COSMETIC FILMS 1970
by M. Gutcho

The abbreviated Table of Contents indicates the four major films and the distribution of the patents within each area — the numbers in ( ).
Powders (21)
Creams and Lotions (12)
Nail Preparations (19)
Lipsticks (14)
143 pages. $20

AGRICULTURAL CHEMICALS MANUFACTURE 1971
by M. Sittig
Chemical Process Review No. 52

Agricultural chemicals, properly used, are essential for supplying the food requirements of the world's evergrowing population. Current attacks on the toxicity of today's pesticides notwithstanding, our agriculture saves about 5 dollars worth of produce for every dollar spent on the war against harmful pests. 172 manufacturing process and product descriptions are given:

1. Environmental Control in Manufacture
2. Manufacture of Intermediates (4)
3. Insecticides (80)
4. Herbicides (48)
5. Fungicides (25)
6. Nematocides (5)
7. Plant Growth Regulators (9)
8. Fertilizer Additives (1)
9. Future Trends
264 pages. $35

HORMONAL AND ATTRACTANT PESTICIDE TECHNOLOGY 1971
by Y. L. Meltzer

The need for non-chemical pesticides which are highly selective, nonpolluting and nontoxic, is urgent. Encouraging results are being obtained with insect hormones and hormonelike substances that interfere with the life cycle of noxious insects. 23 chapters based on the world's patent literature and technical articles: 1. Scope of the Problem. 2. Insect Hormones as Pesticides. 3. Insect Development and Hormones. 4. Juvenile Hormone. 5 Ecdysone. 6. Brain Hormone. Attractants for 7. Insects. 8. Bees. 9. Boll Weevils. 10. Cabbage Looper Moths. 11. Cockroaches. 12. Bombyx mori. 13. Flies. 14. Gypsy Moths. 15. Pink Bollworms. 16. Termites. 17. Yellow Jackets. 18. 10,12-Hexadecadiene Derivs. 19. Aliphatic Hydroxy Attractants. 20. Polyenols. 21. Review Articles. 22. Regulations. 23. Future Trends. 281 pages. $35

CONTROLLED RELEASE FERTILIZERS 1968
by Dr. R. Powell
Chemical Process Review No. 15

This book offers you complete technical data on numerous processes and products in this field. The two major approaches are (a) compounds of low solubility, and (b) coated granules.
Introduction; Compounds of Low Solubility, Coated Granules, Prevention of Nitrogen Losses, Rapid-Release Fertilizer. 279 pages. $35

FERTILIZER DEVELOPMENTS AND TRENDS 1968
by A. V. Slack

R&D Trends, Ammonia, Ammonium Nitrate, Sulfate, Urea, Slow Release Nitrogen, Other Nitrogen Fertilizers, Phosphoric Acid, Ammonium Phosphate, Nitric & Superphosphates, Thermal and Other Phosphate Processes, Potassium Fertilizers, Fluid Fertilizers, Bulk Blending, Minor Nutrients. 98 illustrations. 406 pages. $35

TOBACCO FLAVORING SUBSTANCES AND METHODS 1972
by S. Gutcho

1. Menthol Flavor
   Menthol Acetals
   Menthyl Ethers
   Menthyl Esters
2. Tobacco Flavor
   Alkyl Pyridines
   Substituted Pyrroles
   Piperitenone
   Turkish Tobacco Flavor
3. Other Flavors
   Mint
   Spicy
   Woody
   Camphor
4. Flavor Improvement
   Adsorbents
   Enhancers
5. Flavor Incorporation
   Microcapsules
   Polymers
   Inclusion Complexes
6. Synthesis of Flavors

Based on 86 U.S. patents 161 pages. $24

ELECTROLESS PLATING AND COATING OF METALS 1972
by J. McDermott

Electroless plating contributes its share toward reducing pollution. 201 patent-based processes mostly since 1965.

1. Copper
   Formulations
   Additives
   Circuit Boards
   Ferrous Substrates
2. Ni, Cr, Co
   Formulations
   Substrates
3. Al, Zn, Sn
   Vapor Techniques
   Powder Techniques
   Molten Baths
4. Precious Metals
5. Coating of Refractories
6. Coating for Ni Alloys
7. Flame Spray, etc.
   Mechanical Plating
   Other Coatings

303 pages. $36

DENTIFRICES 1970
by T. Jefopoulos

A guide to information available from U.S. Patent literature to therapeutic and cosmetic agents in dentifrices.
The numbers in ( ) indicate the number of processes covered.
Cleaning Agents (17)
Polishing Agents (14)
Prophylactic Compositions (27)
Fluorides (30)
Dentifrices for Dentists (6)
Other Dentifrices (30)
Improved Processing (2)
191 pages. $35

HAIR PREPARATIONS 1969
by A. Williams

Provides a detailed technological summary of recent developments based on 138 U.S. patents, since 1960, covering all aspects of hair preparations for the head, beard, eyelashes, and eyebrows.
Introduction, Dyeing, Bleaching, Waving, Setting, Shampoos-Rinses, Grooming-Tonics, Shaving Assistants, Other, Indexes. 208 pages. $35

AQUATIC HERBICIDES AND ALGAECIDES 1971
by J. H. Meyer

There is a growing need for control of aquatic weeds and algae. This vegetation menace threatens natural waterways and water supplies. The most effective control is by chemicals. This book gathers the latest technology for producing or using aquatic herbicides and algaecides, based on U.S. patent literature. Numbers in ( ) indicate the number of processes described, 108 in all.
Metal Compounds (22)
Chlorinated Hydrocarbons (7)
Other Halogenated Compounds (16)
The Halogens Themselves (4)
Sulfur-Containing Materials (12)
Quaternary Ammonium Compounds (8)
Amides and Imides (10)
Amines (6)
Acids and Their Derivatives (8)
Miscellaneous Organic Compounds (9)
Carriers (6)
177 pages. $35

AMMONIUM PHOSPHATES 1969
by Dr. M. W. Ranney
Chemical Process Review No. 35

This book describes recent processes for production of ammonium phosphates.
Introduction; Ammonium Orthophosphates, Diammonium Orthophosphates, Ammonium Polyphosphates, Metal Ammonium Phosphates, Ammonium Phosphate—Ammonium Nitrate Mixtures. Many illustrations. 278 pages. $35

NEW FERTILIZER MATERIALS 1968
by C. I. E. C.

Ureaform, Crotonylidene & Isobutylidene Diurea, Triple Superphosphate, Ammonium Phosphates, Nitrate of Potash, Potassium Phosphates and Metaphosphates, Magnesium, Sulfur Fertilizers, Oxamide, Urea Nitrate and Phosphate, Hydrides of Phosphorus, Red Phosphorus. Applications. 430 pages. $35

INDUSTRIAL SOLVENTS HANDBOOK 1970
by I. Mellan

A handbook with complete, up-to-date, pertinent data regarding industrial solvents.
821 tables contain pertinent data concerning physical properties of solvents and degrees of solubility of materials in these solvents. Numerous graphs included giving a great deal of data concerning various parameters. Also includes phase diagrams for multi-component products.
The vast amount of information contained in this book is shown in the abbreviated Table of Contents in the next column. The numbers in ( ) after each entry indicate the number of tables.
Hydrocarbon Solvents (14); Halogenated Hydrocarbons (30); Nitroparaffins (5); Organic Sulfur Compounds (5); Monohydric Alcohols (122); Polyhydric Alcohols (150); Phenols (6); Aldehydes (10); Ethers (53); Glycol Ethers (79); Ketones (44); Acids (18); Amines (124); Esters (61). 478 pages. $25

RELEASE AGENTS 1972
by M. McDonald

Deals with substances that prevent or reduce adhesion, useful in ploymer processing, plastics molding, glass molding, metal casting, pressure-sensitive tapes and labels, and in baking, frying and wrapping food. Reviews 110 patents issued since 1960.

1. Metal Casting
   Water-Based Agents
   Oil-Based Agents
2. Plastics and Rubber Processing
3. Concrete
4. Tablets
5. Glass
6. Fiber Materials
7. Food Products
8. Ice
9. Miscellaneous
   Powders
   Oven Coatings
   Plasma Spraying
   Polysiloxane Emulsions

253 pages. $36

INDUSTRIAL MEMBRANES 1972
by J. McDermott

1. REVERSE OSMOSIS MEMBRANES
   Cellulose Acetate (18)
   Other Cellulosics (5)
   Unit Fabrication (13)
2. OTHER DEIONIZING MEMBRANES
   Nylon Membranes (4)
   Other Polymers (8)
   Discrete Particles (5)
3. ION EXCHANGE MEMBRANES
   Preparations (22)
   Inorganic Membranes (4)
   Separation Processes (5)
4. FILTER MEDIA
   For Fluids (9)
   For Gases (3)
   Ultrafiltration (3)
   Other Applications (4)
5. SOLVENT SEPARATION
   Hydrocarbons (4)
   Water Removal (2)
6. MEDICAL & OTHER USES
   Medical & Biological (5)
   Other Uses (4)

118 Processes. 246 pages. $36